Fluid Flow
A FIRST COURSE IN
FLUID MECHANICS

THIRD EDITION

Fluid Flow

A FIRST COURSE IN FLUID MECHANICS

Rolf H. Sabersky *California Institute of Technology*

Allan J. Acosta *California Institute of Technology*

Edward G. Hauptmann *University of British Columbia*

Macmillan Publishing Company
NEW YORK

Collier Macmillan Publishers
LONDON

Macmillan Publishing Company
866 Third Avenue, New York, New York 10022

Collier Macmillan Canada, Inc.

Library of Congress Cataloging-in-Publication Data

Sabersky, Rolf H.
    Fluid flow.

    1. Fluid dynamics.   I. Acosta, Allan J.,
            II. Hauptmann, Edward G.   III. Title.
TA357.S2   1989   620.1′064   88-27221
ISBN 0-02-404960-3 (Hardcover Edition)
ISBN 0-02-946850-7 (International Edition)

Printing: 2 3 4 5 6 7 8   Year: 9 0 1 2 3 4 5 6 7 8

Contents

8

Flow over External Surfaces

9

Compressible Fluids—One-Dimensional Flow

10

Elements of Two-Dimensional Gas Dynamics

11

Flow in Open Channels

12

Turbomachines   416

13

Some Design Aspects of Turbomachines   457

Appendixes   487

Answers to Selected Problems   519

Index   529

Nomenclature

The following notation has been generally used. As is inevitable, perhaps, some symbols have more than one meaning. When this does occur, the different usages are separated in the text, and we hope thereby that no confusion results. Symbols that occur only infrequently and are explained in immediate context have not been included in this list.

A	area, constant	h	distance or height, enthalpy
a	velocity of sound, acceleration	h_f	head loss due to friction
		h_t	total head
B	Bernoulli constant, constant	i	$\sqrt{-1}$, incidence angle
b	breadth	$\mathbf{i, j, k}$	unit vectors in x-, y-, z-directions
C_D	drag coefficient		
C_L	lift coefficient	k	loss coefficient; also, cavitation coefficient, turbulent kinetic energy
C_p	pressure coefficient		
c	chord, constant		
c_b	blade coefficient	L	lift force
c_c	contraction coefficient	l	length
c_d	discharge coefficient	$\mathbf{M}$	momentum
c_f	shear stress coefficient	M	machine or shaft work per unit mass, Mach number (V/a)
c_p	specific heat at constant pressure		
c_v	velocity coefficient, also specific heat at constant volume	m	mass
		$\dot{m}$	mass rate of flow
D	drag force	N	number of vanes
d	diameter	N_s	optimum specific speed [Eq. (12.31)]
E	total energy of system		
e	energy per unit mass	$\mathbf{n}$	unit vector normal to surface (or streamline)
Eu	Euler number $(p_\infty - p_v / \frac{1}{2}\rho u_0^2)$	P	power; also, perimeter, dimensionless parameter
F	force, complex velocity potential, Blasius function [Eq. (8.4)]	p	pressure
		p_t	total pressure
Fr	Froude number $(V/\sqrt{gl})$	p_{th}	thermodynamic pressure (Sec. 2.7)
f	force per unit mass, friction factor	Q	volume flow rate (e.g., m^3/sec), heat
g	acceleration of gravity		
$\mathbf{H}$	angular momentum	q	volume flow rate per unit length; also, a general quantity, heat per unit mass
H	total head of a pump (turbine) (Chaps. 12, 13)		

R	specified radius; also, area divided by perimeter, universal gas constant, reaction ratio
Re	Reynolds number (Vl/ν)
r	radius
S	surface area (surface element dS); also, suction specific speed [Eq. 13.51)], entropy, variable
S_0	control surface; also, slope of channel bed
S_u	slope of channel bed for uniform flow
s	length along path
T	torque; also, temperature
t	time
U	force potential
U_0	specified velocity
u	velocity component in x-direction; also, peripheral velocity ($r\Omega$, Chaps. 12, 13), internal energy
u, v, w	velocity components in x-, y-, z-directions
u_τ	$\sqrt{\tau_0/\rho}$
V	velocity; also, average velocity in a pipe; also, absolute velocity (Chaps. 12, 13)
V_0	specified velocity
v	velocity component in y-direction
$\mathbf{v}$	vector velocity
v_a	axial velocity component
v_m	meridional velocity component
v_r, v_θ, v_z	velocity components in r-, θ-, z-directions
$\mathscr{V}$	volume
W	work
w	velocity component in z-direction, relative velocity (Chaps. 12, 13)
x, y, z	Cartesian coordinates

GREEK SYMBOLS

α	angle, also angle between absolute velocity and peripheral direction
α^*	angle of attack
β	angle; also, angle between relative velocity and peripheral direction
Γ	circulation
γ	ratio of specific heats; also, stagger angle (angle between chord line and normal to peripheral direction)
δ	boundary layer thickness
δ^*	deviation angle (Fig. 13-5)
ε_s	size of sand grain roughness
ε	size of surface roughness; also, drag-lift ratio, rate of dissipation of turbulent kinetic energy
η	surface elevation of a wave, Blasius variable [Eq. (8.4)]
η_p	pump efficiency
η_t	turbine efficiency
θ	angle; also, wetting angle; also, angle in polar coordinates
Θ	div $\mathbf{v}$
λ	second viscosity coefficient; also, wavelength
μ	absolute viscosity; also, Mach angle
ν	kinematic viscosity $= \mu/\rho$; also, Prandtl–Meyer function
ν_T	eddy viscosity [Eq. (7.11)]
ξ	variable; also, geometrical ratio [Eq. (12.29)]
ξ_x, ξ_y, ξ_z	rate of angular deformation (Sec. 2.8)
Π	power coefficient [Eq. (12.21)]
ρ	density
σ	surface tension; also, solidity
$\sigma_{xx}, \sigma_{yy}, \sigma_{zz}$	normal stress acting in x-, y-, z-directions, respectively
τ	shear stress; also, torque coefficient
τ_{ij}	shear stress acting on ith face of a cube in the jth direction
τ_0	shear stress at wall
φ	velocity potential
ϕ	angle; also, flow coefficient (v_m/u)
ψ	stream function
χ	torque coefficient
Ψ	head coefficient, gH/u^2
Ω	angular speed (rad/sec)
ω	vorticity (curl $\mathbf{v}$)

Fluid Flow

**A FIRST COURSE IN
FLUID MECHANICS**

CHAPTER 1

Introduction

1.1 Introductory Remarks

In a general way all forms of matter can be grouped into two classes: fluids and solids. The dictionary defines a fluid as "a substance which yields to any force, however small, tending to alter its shape." In somewhat more technical terms a fluid may be described as a substance which, when at rest, cannot sustain a shear force, that is, a force exerted tangentially to the surface on which it acts. Solids, on the other hand, can withstand such forces, and although they may remain at rest, displacements or strains of the material result. A shear force can be sustained in a fluid, however, when relative motions between the particles of the fluid take place. Under ordinary conditions this distinction between fluids and solids is readily apparent for such materials as water, air, or steel. However, it is difficult to determine whether some substances should be classified as solids or fluids. Asphalt, for example, behaves like a solid when forces are applied and taken off during short intervals of time, but it behaves like a fluid when these forces are exerted for a very long time. Even such seemingly rigid materials as the rocks of the earth's crust may behave as fluids under such conditions. In the following we shall concentrate only on substances clearly recognizable as fluids.

The concept of a fluid presented above includes both liquids and gases. A liquid differs from a gas in that it has a more-or-less definite volume, so that if a volume of liquid is put into a large container, an interface separating the liquid from its vapor, or another gas, is formed. A gas, on the other hand, normally expands to fill uniformly all the space available to it, and a surface is not formed. Both gases and liquids are composed of conglomerations of molecules or atoms, and in principle it is possible to study the properties and the behavior of a fluid from a molecular point of view. However, we shall not adopt this course, since many problems of great technical importance in science and engineering may be treated adequately and more simply without detailed knowledge of the molecular behavior of fluids. For our purposes a fluid is considered an infinitely divisible substance, instead of a collection of a vast quantity of molecules, and all properties such as the mass per unit volume are assumed to have a definite value at each point in space. Of course, there is a lower limit to the size of a volume element that can be so considered, for the notion of a definite property fails as we approach molecular dimensions. In most applications of fluid mechanics, the physical dimensions of the bodies about which a flow may be taking place are very large compared to the size of the molecules themselves and to the distance between molecular interactions. The properties of the flow can then be examined on a scale small compared to the object but still large compared to molecular sizes; hence even the smallest element that must be considered will contain a great many molecules. Even in this smallest region, therefore, the percentage change in the number of molecules contained will be small from one moment to another. An important consequence of this result is that the mass per unit volume, or density, as well as other properties, will approach a limit as smaller and smaller volumes are examined. Thus on the scale of the object to be studied the density and other fluid properties will vary smoothly from one point to another. The word "continuum" has been aptly used to describe this idea. The fluids treated herein are *continuous* in the foregoing sense.

At normal pressures a gas may be considered a continuum, but when the pressure is sufficiently low, the average distance between molecules may become large compared to the size of an object over which the flow takes place. Such a tenuous, or *rarefied*, gas is still a fluid, but it does not fulfill the requirements of a continuous fluid postulated above.

In this book we shall be concerned with the motion of a fluid continuum under the action of applied forces. Frequently, the term *fluid mechanics* or *fluid dynamics* is used to describe this science. The term *hydrodynamics* is usually reserved for the study of nearly incompressible fluids such as water—as the name implies. All aspects of fluid motion are, therefore, included in the subject of fluid mechanics, and it may be expected that this covers a wide range of important engineering and technological problems as well as scientific questions of a less applied nature. Only a few examples of each category need be mentioned to emphasize the point. For example, acoustics, theory of flight, meteorology, and lubrication are but a few of the many topics embraced by fluid mechanics; many major engineering problems—for instance, the design of harbors and breakwaters, propulsion and hydraulic machinery, and numerous chemical engineering processes as well—are almost wholly concerned with fluids.

All these subjects can be usefully treated with the concept of a fluid continuum. The motion and general behavior of the fluid is governed by the fundamental laws of mechanics and thermodynamics. No new principles other than these are required. The laws of mechanics used herein are Newton's laws of motion. Since these are usually stated for a particle of mass, the general approach in fluid mechanics is to single out a fluid "particle" and apply Newton's laws to this particle. The other laws observed in this book are the conservation of mass principle and the conservation of energy, or the first law of thermodynamics. These laws form the basis of fluid mechanics, just as they do for the mechanics of solids, and it is not surprising that there are many similarities between the two subjects. The principal difference arises from the basic characteristics of a fluid and a solid. The forces exerted on an element of a fluid are related to certain combinations of velocity gradients; the forces exerted on a solid are related to the deformations or strains. In each case the particular relationships are in essence empirical. For fluids one of the simplest sets of relationships is essentially an extension of Newton's law of shear; namely, that for an elementary parallel flow with a velocity gradient normal to the flow, the shear force per unit area is proportional to the velocity gradient. For solids the corresponding relationships may be considered a generalization of Hooke's law, which states that stress (or force per unit area) is proportional to strain.

Following the steps outlined above, the basic differential equations of fluid mechanics can be derived. The degree of complication of these equations depends very largely on the complexity of the empirical relationship between the forces on the element of fluid and the velocity gradients. As in all problems in the physical sciences it is never feasible to express such relationships with absolute certainty and with the inclusion of all possible variables. In fact, in order to treat any physical problem in a mathematical way it is necessary to imagine an abstract model of the object or the process to be treated. One such abstraction has already been introduced by proposing the continuum concept. For fluids it is often assumed in addition that the fluid particles cannot sustain any shear forces whatsoever. An idealized fluid of this kind is called a *perfect fluid*. Aside possibly from helium at temperatures near absolute zero, such fluids do not actually exist. Nevertheless, under certain conditions the behavior of an actual fluid closely approaches that of a perfect fluid, and therefore the results of the analyses of such a fluid may be of definite practical value. A similar but much less restrictive fluid model is one in which these forces are assumed to be linearly proportional to the velocity gradients; this type of fluid, which was mentioned earlier, is called *Newtonian*, in honor of Newton, who introduced the concept of such a fluid. The model of an *incompressible fluid* is based on the assumption that the fluid must be of constant density.

An idealized model is introduced and corresponding assumptions are made in order to simplify the governing equations and their solution. It is difficult to decide just when a simplifying assumption is permissible; deep insight into the problem in question as well as previous experience with similar problems are required. Whenever new assumptions are introduced or whenever the familiar ones are applied to essentially new problems, a careful verification by means of experiments has to be carried out. On the other hand, experiments may sometimes suggest simplifications that may be made in further analytical work. A close coordination of analytical with experimental work is therefore essential for advances in any science, and this is particularly true of fluid mechanics at the present time.

Let us conclude the introduction with a brief sketch of the historical development of fluid mechanics. The earliest significant contributions to the field were undoubtedly made by Archimedes, who lived in Syracuse between the years 285 and 212 B.C. Especially noteworthy was his analysis of the buoyancy of submerged bodies, which was applied successfully to the determination of the gold content of the crown of King Hiero I, in what was probably the first case of scientific criminology. Also in ancient times Roman engineers were concerned with hydraulic problems. They provided Rome with a complete water supply system but apparently had only a rather vague understanding of the relationship between flow and friction.

Although there was some experimentation with fluid flow during the Middle Ages, the next really significant event in the development of fluid mechanics did not occur until the seventeenth century when Newton stated his famous laws of motion. Sometime thereafter Leonhard Euler (1755) applied these laws to an element of fluid and derived the basic differential equations of motion that are still known by his name. During the same period Daniel Bernoulli (1738) discovered some of the basic energy relationships that apply to a liquid.

Euler limited himself to a perfect fluid that cannot sustain shear forces. At about this time Jean le Rond d'Alembert showed with equations essentially equivalent to Euler's that there is no drag force on a finite body—a situation so at variance with observed fact that he could only leave the explanation to future workers. This is the so-called *d'Alembert paradox*. Euler's equations, of course, will correspond to the flow of an actual fluid only when the flow is such that it conforms well to the assumptions. In many cases of practical importance these conditions are satisfied, but in others such assumptions are not at all justified. Solutions to Euler's equations, therefore, have to be used with great discretion. The d'Alembert paradox, for example, arose because the assumption of a perfect fluid was made for a type of flow for which this assumption is not justifiable.

In the period following publication of Euler's equations theoreticians made little attempt to determine whether the results of their analysis corresponded in any way to the actual flow of fluids. But in fairness it should be said that practicing engineers of the time made little attempt to determine just what the flow of real fluids is like. Because of the obvious lack of correspondence between theory and practical results hydraulic engineers did not look with favor upon an analytical approach. Indeed, for several decades practicing engineers restricted themselves to experimenting and testing. Theoreticians, on the other hand, were content to consider abstract analyses without, it would seem, a proper regard for the observed behavior of real fluids. Undoubtedly, the lack of communication between the two groups retarded the progress of hydraulics and fluid mechanics for many years.

The equations of motion were finally generalized by Navier (1827), and independently by Stokes (1845), to include shear forces (and therewith friction). The Navier–Stokes equations are restricted to Newtonian fluids. Many fluids, fortunately, satisfy this type of relationship very well. However, while they accurately represent the motion of many fluids, the equations are so complicated that exact solutions can be obtained only in a few cases.

Furthermore, there was at first a serious objection to the Navier–Stokes equations. The simplest expression for the pressure drop in a pipe obtained from the equations indicated that it should vary as the first power of the velocity. But experiments showed that while under some conditions the first power law held, under others the resistance varied more nearly as the square of the velocity. It remained for Osborne Reynolds in 1883 to establish by direct observation of the flow that two types of flow exist: one, at relatively low velocities, in which the particles slide smoothly along lines everywhere parallel to the wall and for which the first power law of resistance is observed to hold; and the second, at relatively higher velocities, in which the particles execute a sinuous and then finally nearly a random fluctuating motion about the mean velocity. The two types of motion are now called *laminar* and *turbulent*, respectively. Reynolds also pointed out that the existence of the two flow types depends not just on the velocity but rather on a parameter Vd/ν, where V is the average velocity in the pipe, d the diameter, and ν the so-called kinematic viscosity. The ratio Vd/ν is called the *Reynolds number* and is one of the basic parameters in fluid mechanics.

The phenomenon of turbulent flow is not limited to pipes; it is of fundamental importance for all fluid flows. Reynolds also established that the Navier–Stokes equations are applicable to turbulent flow provided that the random, nonsteady nature of the motion is taken into account. For such a complicated flow, however, he was not able to derive analytically the observed flow, a situation that still exists today. Much later it was determined from the equations as well as from experiments that steady laminar flow in a pipe can become unstable; that is, an initially small disturbance can commence to grow rapidly and ultimately lead to a turbulent flow.

The next significant advance came in 1904 when Prandtl recognized that many flow fields may be divided into two regions, one close to solid boundaries, the other involving the remainder of the flow. He then proposed that in the region close to the boundary, called the *boundary layer*, the effect of viscosity is most important, but that the main portion of the fluid can be treated as a perfect fluid. With this approach one then could treat problems which at the time would have been too complicated to solve by direct application of the Navier–Stokes equations to the complete flow field.

Prandtl's formulation of the boundary layer concept marked the beginning of a most fruitful period in fluid mechanics during which both experimental and analytical techniques were applied to further our understanding of fluid flow. Theodore von Kármán in the United States and Geoffrey I. Taylor in England were among the principal leaders throughout this phase.

More recently, the computer has had a very significant impact on the course of fluid mechanics. In many cases the Navier–Stokes equations can now be solved numerically throughout the flow field and progress is being made in obtaining solutions even for turbulent flows. The computational work often requires a major effort, but it is warranted in quite a number of instances. The fact that the computer allows the solution of very complex equations will lessen the need for simplifying assumptions, which will certainly have a major effect on the way in which fluid mechanics problems will be approached in the future.

The scope of fluid mechanics in engineering is extremely broad. It is essential, for example, for the design of pipelines, gas turbines, pumps, airplanes, oil drilling platforms, and water supply systems as well as in all kinds of applications involving heat transfer and combustion. It is truly one of the very fundamental engineering disciplines, and in this book we have attempted to provide a systematic and enjoyable introduction to the subject.

In the initial chapters the basic equations of fluid flow are derived. For the most part the derivations are sufficiently complete so that the reader may appreciate the fundamental assumptions involved and be aware of the limitations that must be imposed on the application of the results to various engineering problems.

Subsequently, several more specific areas of fluid mechanics and gas dynamics will be discussed. The various topics, of course, cannot be treated in great depth, but it is hoped that the reader will obtain a sound introduction to the subject matter as well as to the methods that are used to approach typical problems in the general field of continuum mechanics.

1.2 Definitions

In describing the motion of a fluid, it will be necessary to use several properties for the description of the physical state of the fluid. Brief definitions of some of these are given in this section with additional concepts introduced in the course of the development, as needed.

To describe these properties in a quantitative manner, a system of units has to be selected. Two sets of units are currently being used by engineers: the British (or engineering) system, commonly used in the English-speaking world since the Industrial Revolution; and the Système International (French for International System), or simply SI, which many nations are now adopting. The SI units are commonly associated with the older "metric" system, but care must be taken not to confuse the two, particularly with regard to the measurement of force. In this book we emphasize SI units throughout because of their wide use in current classroom instruction; because of the continued use of British units in engineering practice, some examples and problems in this system are also included. A more complete description of the SI and British unit systems is given in Appendix 1.

As the first property let us consider *pressure*. Pressure is defined as force per unit area exerted in a direction normal to that area. In SI units the pressure is expressed in newtons per square meter (N/m^2). This unit has been given the special name of pascal (Pa), which means that 1 N/m^2 is the same as 1 Pa. In British units the pressure may be expressed as pounds force per square foot (lbf/ft^2); it is often given in terms of pounds force per square inch, or "psi" ($lbf/in.^2$). In addition, the abbreviations "psig" and "psia" are also in common use: the former refers to "gage"

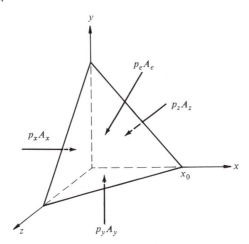

FIG. 1.1 The pressure forces on a prism of fluid at rest.

pressure, that is, the pressure being measured minus the pressure of the surrounding atmosphere (this is the pressure measured by a simple pressure gage); the latter, psia, designates the absolute pressure, that is, the actual force per unit area without adjustment. In the SI system no special recognition is made of the surrounding atmosphere; all pressures are absolute.

In a fluid at rest pressure is independent of direction. To verify this statement, let us consider a very small body of fluid in the form of a prism, as indicated in Fig. 1.1. The prism is small enough so that the pressure over each face may be assumed constant.

In addition to the pressure forces, forces such as gravity may act on the mass of fluid inside the prism. These forces, called *body forces*, are proportional to the total mass of the fluid in the prism. The components of this force per unit mass will be designated by the symbols f_x, f_y, and f_z. No other forces will have to be considered here; in particular, shear forces are absent because a fluid at rest cannot sustain such forces. Since the element of fluid is at rest, the summation of all forces in any direction must be equal to zero. Selecting the x-direction, for example, we obtain for the summation of forces:

$$\sum F_x = p_x A_x - p_e A_e \cos \alpha + f_x \rho \, \frac{A_x x_0}{3} = 0,$$

where

p_x = average pressure on the surface lying in the y-z plane,

A_x = area in the y-z plane,

p_e = average pressure on the inclined surface,

A_e = area of the inclined surface,

α = angle between the normals to the surfaces A_x and A_e,

f_x = x-component of a body force (force per unit mass),

x_0 = intercept of prism faces on the x-axis.

The quantity $A_x x_0 / 3$ in the equation above represents the volume of the prism; $f_x \rho (A_x x_0 / 3)$, therefore, is the component of the body force in the x-direction. Recognizing now that from the geometry of the prism

$$A_x = A_e \cos \alpha$$

and dividing through by A_x, we find that the equation reduces to

$$p_x - p_e + f_x \rho \frac{x_0}{3} = 0. \tag{1.1}$$

We now consider successively smaller prisms, and as the prism shrinks, x_0 will tend toward zero. In the limiting case, when the prism shrinks to a point, x_0 goes to zero and Eq. (1.1) reduces to

$$p_x - p_e = 0.$$

Since the arguments presented for the forces in the x-direction may be repeated for the other directions as well as for other orientations of the prism, it is concluded that *for a fluid at rest the pressure is equal in all directions*. It may be noted here that for special cases in which shear forces are negligible, the same conclusion may be reached even when the fluid is in motion.

Let us next consider the *density*. Density is defined as the mass per unit volume, and it is designated by the symbol ρ. In the SI system it is given in kg/m^3. The corresponding unit in the English system is slugs/ft^3, where the slug is that mass which is accelerated at the rate of 1 ft/sec^2 when acted upon by a force of 1 lbf.

The density (as well as most other properties) of a substance is a function of the temperature and pressure. The density of a gas is strongly dependent on these latter properties, but the density of liquids and solids is only slightly affected by them. Frequently, for example, the density of a liquid or solid may be assumed constant. We then speak of an *incompressible* fluid. Figures 1.A and 2.A in Appendix 2 show the effect of pressure and temperature on the density of water.

Closely related to the concept of density is that of *specific weight*. The specific weight of a fluid is the force with which a gravitational body such as the earth attracts a unit volume of the fluid. In SI the specific weight is expressed as N/m^3, while the corresponding units in the British system are lbf/ft^3. The specific weight can be readily obtained from the density by multiplying by the gravitational constant g. Thus the specific weight is equal to the product ρg. Since g is a quantity that varies from one part of the earth to another and with altitude as well, specific weight is not as useful as density. For this reason the use of density, the more basic property, is preferred.

Let us next consider the *viscosity*. The viscosity of a fluid is associated with the resistance to the sliding motion of one fluid layer over another. An example illustrating the action of viscosity is afforded by a thin film of fluid between two long parallel plates in relative motion. The film within the small clearance between the shaft and housing of a concentric journal bearing as shown in Fig. 1.2 will serve the same purpose and may be more easily imagined. In both cases the flow is essentially unidirectional (the effect of curvature may be neglected in the journal bearing) and the velocity parallel to the wall, say u, can only depend on the distance from the wall. For many ordinary fluids in such a one-dimensional flow the shear stress at any point in the fluid may be expressed by means of the relation

$$\tau = \mu \frac{du}{dy}, \tag{1.2}$$

where du/dy is the velocity gradient at the point being considered, τ is the shear stress, and μ, the constant of proportionality, is called the *absolute viscosity*. If the

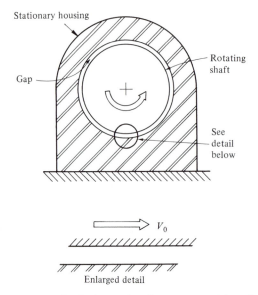

FIG. 1.2 Flow in a narrow annulus between two large concentric cylinders. This flow approximates that between two parallel horizontal plates.

fluid is compressible and if significant changes of volume take place, another viscous stress can, incidentally, arise that is proportional to the volumetric change. The pertinent coefficient for this type of stress, called the *second*, or *bulk*, *viscosity*, is discussed in Chapter 2.

If μ is independent of du/dy, the fluid is said to be Newtonian. For water, air, alcohol, and many other fluids, the viscosity is practically independent of the velocity gradient, but there are fluids in which the viscosity does change significantly with the magnitude of the velocity gradient. In such cases the concept of a viscosity becomes much less useful. Even for a Newtonian fluid, however, the viscosity may still change with pressure and temperature. The effect of pressure is usually small, but the effect of temperature may be very significant. The effects of temperature and pressure on the viscosity of water and other fluids are given in Figs. 3.A and 4.A and Table 1.A of Appendix 2. From its definition, it can be seen that absolute viscosity has the dimensions of (force/area) × (time). In the SI system, the units of absolute viscosity are therefore (N/m²) × (s), or alternatively, Pa · s; in the British system the corresponding units of absolute viscosity are (lbf/ft²) × (sec).

In analyzing fluid flows, the ratio μ/ρ occurs frequently and is given the symbol v. This ratio has units that include only length and time and is therefore called the *kinematic viscosity*. In the SI system the kinematic viscosity v has units of m²/s, while the corresponding British system units for v are ft²/sec.

A further comment regarding the behavior of real fluids must be made here to appreciate the peculiarities of these flows: Experiments have shown that for any fluid that can be considered continuous the velocity of the fluid touching a solid surface is the same as that of the surface. In other words, the fluid "sticks" to a solid surface and does not "slip" relative to it. This *no-slip condition* was subject to some considerable controversy by nineteenth-century theoreticians for it was—and perhaps is now—difficult to see, for example, why fluids that do not "wet" adjacent solid surfaces should obey the no-slip condition. Nevertheless, we shall take it as an experimental fact that ordinary liquids and gases that can be treated as continua obey this condition.

To familiarize the reader further with the effect of viscosity, the flow between flat plates mentioned earlier will be discussed in more detail in Example 1.1.

EXAMPLE 1.1 _____

The space between two very long parallel plates separated by a distance h is filled with fluid of constant viscosity μ. The upper plate moves steadily at a velocity V_0 relative to the lower one and the pressure is everywhere constant. The velocity distribution between the plates and the shear stress distribution in the fluid are to be found. These problems are solved by selecting at one instant an element of the fluid of some arbitrary length dx and height y above the stationary plate and considering the forces on this *free body*.

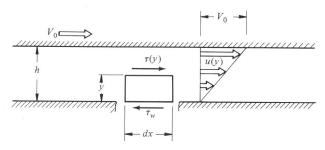

Example 1.1

Solution: The shear stress exerted by the fluid above the element is τ while the restraining shear stress exerted by the wall on the element is τ_w in the opposite direction. The element is not accelerating nor is there any net force due to pressure, since it is presumed to be constant. Thus from statics we must have

$$\tau = \tau_w = \text{const.}$$

But

$$\tau = \mu \frac{du}{dy} = \text{const.}$$

Hence

$$u = \frac{\tau}{\mu} y + \text{const.}$$

The no-slip condition requires that $u = 0$ at $y = 0$, and $u = V_0$ at $y = h$. Hence the integration constant is zero and

$$\tau = \frac{\mu V_0}{h}.$$

The velocity consequently is given by

$$u = V_0 \frac{y}{h}.$$

This kind of flow between two plates is often called *Couette flow*.

PROBLEMS

1.1 (a) Rework Example 1.1 for the case of a non-Newtonian fluid in which μ is equal to $\mu_0(du/dy)^n$, where μ_0 is a constant and n is a constant exponent.

(b) Find an expression for the shear stress τ as a function of (V_0/h), which can be shown to be the rate of angular strain of a fluid element (Section 2.7).

(c) Carefully sketch a graph of the shear stress versus the rate of angular strain V_0/h for the three cases $n > 0$, $n = 0$, and $n < 0$ (fluids with $n > 0$ are called *dilatant* while those with $n < 0$ are termed *pseudoplastic*).

1.2 A type of mud used as a lubricant while drilling oil wells may behave as an *ideal plastic* fluid: It remains solid under applied shear stress until a critical value τ_c is reached; at higher shear stress it flows like a Newtonian fluid with constant viscosity μ.

(a) If the gap described in Example 1.1 is filled with such a fluid and a shear stress $\tau = \tau_c$ applied to the upper plate, a small layer of thickness h_0 next to the bottom is assumed to become fluid while the portion above moves at a speed V. Sketch the velocity profile in the gap for this case. How does the flowing height h_0 change with increasing speed V of the upper plate?

(b) At what speed V_0 does the entire gap become fluid? What is the shear stress at a speed $V_1 > V_0$?

As the last in this group of properties, let us consider *surface tension*. Under conditions where gravity or other extraneous forces are negligible, a particle of unconfined liquid assumes a perfectly spherical shape because of the attracting forces between the liquid molecules. In the interior of the fluid a molecule is surrounded by many others, and on the average the attracting force is uniform in all directions. At the surface, however, there is no outward attraction to balance the inward pull, since there are only a few molecules outside. Thus the surface molecules are subject to an inward attraction, and molecules near the surface are also subjected to an inward attraction, though of smaller magnitude. It is easy to see then that it takes a certain amount of work to bring a molecule from the center of a spherical droplet to a point near the surface. When the droplet surface is increased, say by the introduction of more fluid, molecules will have to be moved from the interior of the sphere to the surface to account for the larger area there. From the foregoing reasoning, a definite amount of work will have to be done. It turns out that this work is proportional to the increase in surface area. Thus the creation of free surfaces involves the expenditure of energy. This energy, which is expressed in terms of work per unit area and has the units of force per unit length, is called *surface tension* and will be denoted by the symbol σ. The units for surface tension in the two unit systems are N/m and lbf/ft, respectively.

The name *tension* gives rise to the misconception that a physical tension actually exists in a liquid free surface. Although problems involving free surfaces may be mathematically treated as if there were a membrane with a uniform tension of σ, it is well to realize that this is only a question of convenience and that no such "skin" containing a tension actually exists. For purposes of computation, however, the free surface may be considered to act as a membrane with a uniform tensile force of σ pulling parallel to the surface.

A surface tension will always exist whenever there is a density discontinuity, as, for example, between water and air or water and oil. The magnitude of the surface

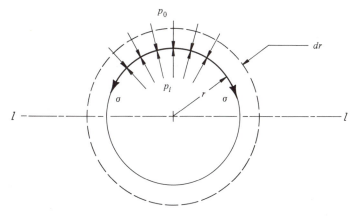

FIG. 1.3 Forces due to pressure and surface tension acting on a spherical droplet.

tension will depend upon the nature of both substances—liquid and liquid or liquid and gas—and in general it is a function of temperature and pressure. The surface tension of several interfaces is shown in Table 2.A of Appendix 2. The effect of pressure is slight except near the critical point, where surface tension vanishes.

An important aspect of surface tension is that it creates a pressure jump across the interface whenever the surface is curved. This leads to the phenomena of capillary waves and capillary rise among others. It is convenient to calculate the magnitude of the pressure jump with the concept of a surface tension. For example, let us consider the spherical droplet in Fig. 1.3. The pressure difference is given by $p_i - p_o$, where the subscript i refers to the interior of the droplet and o to the outside. Suppose that the radius of the droplet is increased from r to $r + dr$ by introduction of more fluid. The work required and therefore the increase in energy is evidently the increase in surface area times the surface tension, or

$$\sigma \, dA = \sigma 8\pi r \, dr,$$

where A represents the surface area $4\pi r^2$.

This work will also be equal to the pressure difference $p_i - p_o$ times the corresponding area times the distance dr over which this force acts; that is, the work is $4\pi r^2 (p_i - p_o) \, dr$. Upon equating these two expressions, we find that the pressure difference is

$$p_i - p_o = \frac{2\sigma}{r} \tag{1.3}$$

for a spherical surface.

It is not too hard to derive an expression equivalent to Eq. (1.3) for a general surface. The result is

$$p_i - p_o = \sigma\left(\frac{1}{r_1} + \frac{1}{r_2}\right), \tag{1.4}$$

where r_1 and r_2 are the local principal radii of curvature of the surface. The sum of these reciprocal radii is called the mean curvature, and it may be shown that this sum does not depend on the orientation of any particular coordinate system used to express it. For a sphere the two radii are equal, and we have Eq. (1.3). For a circular cylinder one of them is infinite, and we obtain

$$p_i - p_o = \frac{\sigma}{r}.$$

The two radii of curvature may be of opposite sign; that is, they may be on opposite sides of the surface. It is convenient to remember the pressure difference as a positive quantity. The pressure p_i must then be taken as being on that side of the surface having the smaller radius of curvature.

When a drop of liquid such as a raindrop is placed on a solid surface, it is found that the interface between the liquid and the air (or other medium surrounding the droplet) appears inclined at some angle to the surface as shown in Fig. 1.4. This angle is known as the *wetting angle* or *contact angle* and is indicated by θ in the diagram. The contact angle depends basically on the relative attraction of the molecules of the three media involved, and it is therefore a function of the characteristics of all three substances. Right at the point of intersection a small layer of liquid, perhaps a few molecules thick, is usually adsorbed to the solid surface; strictly speaking, the contact angle is indeterminate at that point. This region is only a few molecules in extent, so the concept of contact angle is still useful for most engineering purposes. However, it should be expected that the angle is sensitive to slight contaminants—such as detergents—and the cleanliness of the solid surface. Also, the angle seems to depend on whether the liquid is receding or advancing over the solid. The contact angle for any particular situation then will show rather large variations. For water, air, and clean glass it is zero degrees. Thus a water droplet will spread out smoothly on a piece of clean glass. But on dusty or slightly greasy glass it will not. For water, air, and paraffin wax, the contact angle is 105° and for mercury, air, and glass it is about 129°.

These phenomena result in the rise of liquid observed when a small tube is dipped into the liquid. When the rise is much larger than the radius of the tube, the interface separating the liquid from adjacent fluid (called the *meniscus*) may be assumed nearly spherical and the rise may be approximated by the formula

$$h = \frac{2\sigma}{gr(\rho_1 - \rho_2)} \cos \theta, \tag{1.5}$$

where h is the rise, g the acceleration of gravity, ρ_1 the density of liquid in the tube, and ρ_2 the density of the surrounding medium. Evidently, when the contact angle exceeds 90°, a depression rather than a rise is obtained.

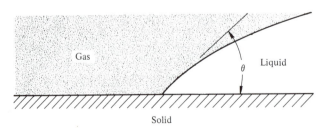

FIG. 1.4 Region of contact between a liquid, a solid, and a gaseous substance. The angle of contact is indicated.

PROBLEMS

1.3 For a pressure of 1 atm, find the kinematic viscosity of water at 10°C and at 65°C. (See Appendix 2.)

1.4 To what temperature does a light crude oil have to be heated to have the same kinematic viscosity as water at 20°C; as a SAE 10 oil also at 100°C? (See Appendix 2.)

1.5 The fluid in Example 1.1 is heated in such a way that the viscosity decreases linearly from the value μ at the lower surface to $\mu/2$ at the upper surface. Calculate the ratio of shear stress in this case to that corresponding to constant viscosity.

1.6 A shaft 1 in. in diameter and 1 ft long is concentrically located in a journal of the same length and 1.010 in. in diameter. Fluid corresponding to SAE 30 oil at 100°F (Appendix 2) fills the gap between the shaft and journal. Compute the horsepower required to turn the shaft at 8000 rpm.

1.7 Derive Eq. (1.5).

1.8 Show that the radius of curvature of the surface R in the case described by Eq. (1.5) is given by $R = r/\cos\theta$.

1.9 A clean glass capillary with a diameter of 2.50 mm is immersed in water. What will be the capillary rise above the main water level if the water temperature is 20°C?

1.10 The height h of the water in the glass tube shown is 50 mm. The diameter of the tube is 2.5 mm. What would the height in the tube be if the surface tension of water were zero? To what height would the water rise if the tube were 1.0 mm in diameter? For each tube what would the percentage error be in the height measurement if the capillary effect were not taken into account? Let the water temperature be 20°C.

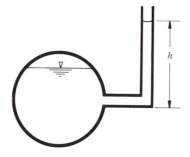

Problem 1.10

1.11 (a) A capillary tube with an internal diameter of 1 mm is placed vertically in a bucket of water. How high will the level in the capillary tube rise above the level in the bucket if the contact angle at the inner wall of the tube is 5° and the surface tension is 0.7 N/m?

(b) Consider a smaller capillary with the same contact angle and surface tension. If the water will vaporize below a pressure of 1.7 kPa, what is the maximum capillary height that can be reached, and what size of capillary is needed?

1.12 Assuming that the interface of a spherical droplet acts like a membrane of uniform tension σ, construct a free-body diagram of the upper half of the droplet above line l–l shown in Fig. 1.3 and derive Eq. (1.3).

1.13 Small gas bubbles are observed to form in a liquid when the pressure is reduced, as, for example, when a bottle of soda water is opened. Assuming that the liquid is water at 0°C, what is the pressure difference between the inside and outside of the bubble if the diameter is 0.1 mm?

1.14 A soap bubble hangs from a horizontal circular ring of radius 3 cm. The mass of the soapy water comprising the bubble is 0.0014 kg. Assume that the bubble is spherical and *neglect* any contact angle effects at the junction of the ring and the bubble.

(a) Find the angle A between a tangent to the bubble at the ring and the vertical (see illustration) if the surface tension of the soapy water is 0.05 N/m (and $g = 9.81$ m/s²).

(b) Find the radius R of the soap bubble.

(c) Find the thickness t of the bubble if the density of the fluid is 1000 kg/m³.

(d) Could you devise a way to measure surface tension of a liquid with this arrangement? What would you measure?

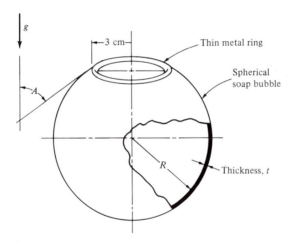

Problem 1.14

1.15 A plane wall is immersed in a large body of water. Taking surface tension into account, determine the shape of the water surface. Assume the surface deflections to be sufficiently small to allow the curvature to be approximated by

$$\frac{1}{r} = \frac{d^2 y}{dx^2}.$$

The contact angle at the wall, θ, is given. Also determine the maximum height h.

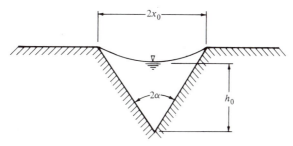

Problem 1.15

1.16 A liquid of density ρ (much larger than that of air) is trapped in a long narrow groove on a surface as shown. The groove has flat sides inclined at an angle 2α apart. Show that the liquid depth at the center is given by

$$h_0 = \frac{\cot(\theta + \alpha)}{\sqrt{\rho g/\sigma}\ \sinh(\sqrt{\rho g/\sigma}\ x_0)}.$$

Hint: Study the development suggested in Problem 1.15.

Problem 1.16

REFERENCES

1. H. Rouse and S. Ince, *History of Hydraulics*, Iowa Institute of Hydraulic Research, University of Iowa, Iowa City, Iowa, 1957.
2. N. K. Adam, *The Physics and Chemistry of Surfaces*, 3rd ed., Oxford University Press, London, 1941 (especially Chaps. 1 and 9).

CHAPTER 2

The Basic Equations

2.1 Introductory Concepts

In this chapter two basic laws of physics are applied to the flow of fluids. These are the principle of conservation of mass and Newton's law of motion. We assume while developing these laws that the fluid is continuous in the sense in which this word was defined in the introduction. Problems in which the mean free path is an appreciable fraction of a significant physical dimension of the flow field are, therefore, excluded. In all the work to follow we assume that for the purpose of taking derivatives and performing other similar mathematical operations, the fluid can be considered to be subdivided into small elements without limit and still remain a continuum. The equations that result have been shown to hold with great accuracy for a wide variety of fluids whenever the foregoing condition is observed.

Before we proceed into the derivation of the equations, we must decide on the way in which the fluid flow is to be described. Every particle of fluid in a flow has an instantaneous value of velocity and density as well as of other properties. As the particle moves about in the fluid, its velocity as well as its density, and so on, will change from point to point and from time to time. Rather than follow the detailed history of each individual mass particle as it moves about, we find it more convenient to describe the flow by giving the velocity components, pressure, density, and so on, of the flow at every point of space as a function of time. Thus we emphasize the velocities, density, and so on, at fixed points in space, rather than the history of these quantities for each individual mass particle. The latter procedure is called the *Lagrangian* description and the one we shall follow is called the *Eulerian* description.

In this description we specify any desired location in the flow and then state the magnitude of the variables that may be of interest. These locations are described in a coordinate system identifying the position in a three-dimensional space. The most common such system is the Cartesian coordinate system of the three orthogonal straight axes, x, y, and z. We shall also use cylindrical coordinates r, θ, and z. We then expect the variables to depend on these coordinates and perhaps on time as well. The variables generally will be the pressure p, density ρ, and the components of fluid velocity. In the Cartesian system these will be labeled u, v, w for the x, y, z axes, respectively, with a similar notation for the other systems. Thus we will expect that, for example, the pressure p will be a function of x, y, z and of time. This relationship will be indicated by the notation $p(x, y, z; t)$.

Sometimes it is convenient to think of a vector quantity such as the position vector $\mathbf{r}$ having components x, y, and z in the Cartesian system and a velocity vector $\mathbf{V}$. Then we expect $\mathbf{V}(\mathbf{r}, \mathbf{t})$, and so on.

2.2 The Continuity Equation

The conservation of mass principle is basic to mechanics; it merely states that mass cannot be created (or destroyed). (We neglect here all effects of special relativity and Einstein's famous mass–energy equation.) Thus, for example, a rigid impermeable tank of fluid may be expected to contain the same amount of matter for all possible motions. But if the tank is punctured and fluid escapes, the amount of mass or matter inside the tank clearly diminishes as some of the original mass flows out. The conservation principle then requires that the decrease of matter (or its rate of decrease) in the tank is equal to the outflow. In this example, the conservation principle was applied to the entire mass involved, and this kind of application of the principle is often quite useful. We shall also be interested in applying the principle to an infinitesimal volume within the flow, where variations in density and velocity take place. The principle of conservation of mass when applied to such an infinitesimal region will lead to a differential equation which is usually called the "continuity" equation.

To derive this continuity equation, we now apply the principle of conservation of mass to a small volume of space through which a flow takes place. This volume is an imaginary volume *fixed* in position and offering no resistance of any kind to the flow. It may be imagined, for example, as a thin wire cage. For convenience we shall adopt a Cartesian coordinate system (x, y, z), and in the interests of simplicity we shall treat only a two-dimensional flow as shown in Fig. 2.1 in which there is no component of flow along the z-axis. Sections normal to the z-axis, therefore, have an identical flow pattern, so that it is sufficient to consider a unit width in the z-direction. The fluid velocity in the x-direction will be designated by u, and that in the y-direction by v, while the density will be indicated by ρ. Both the velocity components and density are functions of position and time.

The principle of conservation of mass requires that the net outflow of mass from the volume be equal to the decrease of mass within the volume. This is readily calculated with reference to Fig. 2.1. The flow of mass per unit time and area through a surface is the product of the velocity normal to the surface and the density. Thus the x-component of the mass flux per unit area at the center of the volume is ρu. This flux, however, changes from point to point as indicated in

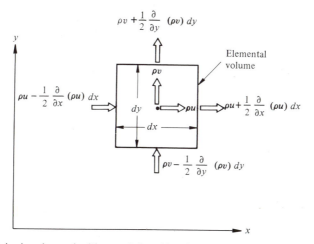

FIG. 2.1   Sketch showing velocities and densities for mass flow balance through a fixed volume element in two dimensions.

Fig. 2.1. The net outflow of mass per unit time, therefore, is

$$\left\{\rho u + \frac{1}{2}\frac{\partial}{\partial x}(\rho u)\,dx\right\}dy + \left\{\rho v + \frac{1}{2}\frac{\partial}{\partial y}(\rho v)\,dy\right\}dx - \left\{\rho u - \frac{1}{2}\frac{\partial}{\partial x}(\rho u)\,dx\right\}dy$$

$$- \left\{\rho v - \frac{1}{2}\frac{\partial}{\partial y}(\rho v)\right\}dx,$$

and this must equal the rate of mass decrease within the element

$$-\frac{\partial \rho}{\partial t}\,dx\,dy.$$

Upon simplification this becomes

$$\frac{\partial \rho}{\partial t} + \frac{\partial}{\partial x}(\rho u) + \frac{\partial}{\partial y}(\rho v) = 0,$$

and were we to have included the z-direction

$$\frac{\partial \rho}{\partial t} + \frac{\partial}{\partial x}(\rho u) + \frac{\partial}{\partial y}(\rho v) + \frac{\partial}{\partial z}(\rho w) = 0 \tag{2.1}$$

would have been obtained, w being the velocity in the z-direction.

Equation (2.1) is called the *continuity equation*. If the density is a constant (and the fluid therefore incompressible), the continuity equation becomes

$$\frac{\partial u}{\partial x} + \frac{\partial v}{\partial y} + \frac{\partial w}{\partial z} = 0. \tag{2.1a}$$

This equation is valid whether the velocity is time dependent or not. Incidentally, a flow for which all velocities and properties at a given location are independent of time is called *steady*. It follows that for a steady flow all partial derivatives in respect to time are equal to zero ($\partial/\partial t = 0$).

Of course, if the details of the flow are not to be treated, the conservation principle itself can be applied directly to the region of interest. For example, if the flow does not change with time, the mass flowing past any one section of a duct or conduit is the same as that past any other section. Thus with reference to Fig. 2.2,

$$\dot{m} = \int_{A_2} \rho_2 u_2\,dA_2 = \int_{A_1} \rho_1 u_1\,dA_1,$$

where u is the component of velocity perpendicular to the area A.

These remarks are easily generalized for any arbitrarily shaped volume with the aid of vector methods. Let **v** be the vector velocity and **n** the outward-pointing unit vector normal to the surface of an arbitrary volume (Fig. 2.3). Then, if dS is an element of the surface S enclosing the volume, the mass outflow per unit time through the surface of the volume is

$$\int_S \rho\mathbf{v}\cdot\mathbf{n}\,dS,$$

and this must be equal to the rate of decrease of mass contained within the fixed volume

$$-\int_\mathscr{V} \frac{\partial \rho}{\partial t}\,d\mathscr{V},$$

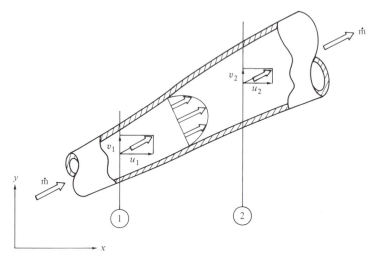

FIG. 2.2   Mass flow through two fixed reference cross sections in a duct.

where $\mathscr{V}$ designates the volume. The surface integral is transformed into a volume integral by Gauss's theorem† and upon transposing we have

$$0 = \int_{\mathscr{V}} \left[\frac{\partial \rho}{\partial t} + \nabla \cdot (\rho \mathbf{v}) \right] d\mathscr{V}.$$

This relation must hold, of course, for all arbitrary volumes, a condition that is satisfied only if

$$\frac{\partial \rho}{\partial t} + \nabla \cdot (\rho \mathbf{v}) = 0. \tag{2.?}$$

In a Cartesian coordinate system, $\mathbf{v} = \mathbf{i}u + \mathbf{j}v + \mathbf{k}w$, where u, v, and w are veloci components in the x-, y-, and z-directions and $\mathbf{i}$, $\mathbf{j}$, $\mathbf{k}$ are unit vectors also in the y-, and z-directions, respectively. For this coordinate system the divergence te $\nabla \cdot (\rho \mathbf{v})$ may be expanded to read

$$\frac{\partial}{\partial x}(\rho u) + \frac{\partial}{\partial y}(\rho v) + \frac{\partial}{\partial z}(\rho w),$$

so that the expression above is precisely the continuity equation [Eq. (2.1)].

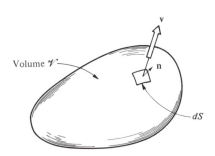

FIG. 2.3   Flow through an arbitrary fixed volume.

† For a summary of vector relations, see Appendix 3.

PROBLEMS

2.1 Derive the continuity equation in cylindrical polar coordinates from Eq. (2.1) by the coordinate transformations $z = z$, $x = r \cos \theta$, $y = r \sin \theta$.

$$Ans.: \quad \frac{\partial \rho}{\partial t} + \frac{1}{r} \frac{\partial \rho v_r}{\partial r} + \frac{1}{r} \frac{\partial \rho v_\theta}{\partial \theta} + \frac{\partial \rho v_z}{\partial z} = 0.$$

2.2 Derive the continuity equation in spherical polar coordinates (r, ϕ, θ) by considering the outflow of mass from a small volume bounded by the surfaces corresponding to θ, $\theta + d\theta$; ϕ, $\phi + d\phi$; r, $r + dr$. Follow the approach used in deriving Eq. (2.1) in Cartesian coordinates.

2.3 A thin "ocean" of liquid (considered incompressible) covers a sphere of radius R to a depth h ($h \ll R$) which depends on the angle from the polar axis θ, longitude ϕ, and time t. Assume that the velocity components v_θ, v_ϕ do not vary with the depth. By considering the flow through a small volume element $R^2 h \sin \theta \, d\theta \, d\phi$, obtain the form for the continuity equation for such a flow.

2.4 A vertical, cylindrical tank closed at the bottom is partially filled with an incompressible liquid. A cylindrical rod of diameter d_i (less than tank diameter, d_o) is lowered into the liquid at a velocity V. Determine the average velocity of the fluid escaping between the rod and the tank walls (a) relative to the bottom of the tank, and (b) relative to the advancing rod.

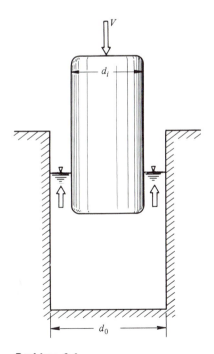

Problem 2.4

2.5 Determine if the following flows of an incompressible fluid satisfy the continuity equation.

(a) $u = \left[\dfrac{1}{x^2 + y^2} - \dfrac{2x^2}{(x^2 + y^2)^2} \right] V_0 r_0^2,$

$v = \dfrac{-2xy}{(x^2 + y^2)^2} V_0 r_0^2.$

V_0 is a reference velocity and r_0 is a reference length. Both are constants.

(b) $u = \dfrac{-2xyz}{(x^2 + y^2)^2} V_0 r_0,$

$v = \dfrac{(x^2 - y^2)z}{(x^2 + y^2)^2} V_0 r_0,$

$w = \dfrac{y}{x^2 + y^2} V_0 r_0.$

2.6 For the flow of an incompressible fluid the velocity component in the x-direction is

$$u = ax^2 + by,$$

and the velocity component in the z-direction is zero. Find the velocity component v in the y-direction. To evaluate arbitrary functions that might appear in the integration, assume that $v = 0$ at $y = 0$.

2.7 An incompressible fluid flows through the circular pipe shown in the figure at the rate of Q m³/s.
(a) If it is assumed that the velocities at stations 1, 2, and 3 are uniform, what are the velocities, given that the diameters of the pipe at the three sections are A, B, and C, respectively?
(b) Compute numerical values for the case that

$$Q = 0.4 \text{ m}^3/\text{s},$$

$$A = 0.4 \text{ m},$$

$$B = 0.2 \text{ m},$$

$$C = 0.65 \text{ m}.$$

(c) Find the velocities for the case that $Q = 0.4$ m³/s at station 1, but the density changes in such a way that $\rho_2 = 0.6\rho_1$, $\rho_3 = 1.2\rho_1$. The flow is steady in all cases.

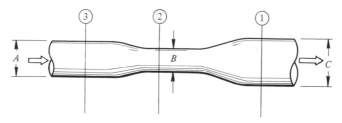

Problem 2.7

2.8   The incompressible continuity equation in polar coordinates is

$$\frac{1}{r}\frac{\partial(rv_r)}{\partial r} + \frac{1}{r}\frac{\partial v_\theta}{\partial \theta} = 0.$$

(a) On the basis of this equation alone, what is the most general type of flow possible if $v_\theta = 0$? Sketch the flow.

(b) Similarly, what is the most general type of flow if $v_r = 0$? Sketch the flow.

2.9   A compressible fluid is caused to flow through a tube of constant diameter in such a way that the velocity along the axis is given by

$$u = \frac{u_1 + u_2}{2} + \frac{u_2 - u_1}{2}\tanh x,$$

where u_1 and u_2 are the velocities when x is minus or plus infinity, respectively. The density does not change with time at any point. The density at $x = -\infty$ is $\rho = \rho_1$. Obtain an equation for the distribution of density along the tube.

2.10   Water flows at a steady rate into a tank supported on scales as shown. The pipe is straight and filled with fluid at all times and its end is always submerged. The pipe is supported externally. It is desired to calibrate a flow meter by taking measurements of elapsed time and weight gain on the scales. If the reading so obtained is $\dot{w}$ lbf/sec, is this also the rate of weight flow through the meter? If not, determine the correct value, using

$$\text{diameter tank} = a,$$

$$\text{outside diameter, pipe} = b,$$

$$\text{inside diameter, pipe} = c,$$

$$\text{density of water} = \rho.$$

All velocities are negligibly small, and pressures can be computed from static considerations.

Hint: Find the force on the scales by considering the pressure on the bottom of the tank.

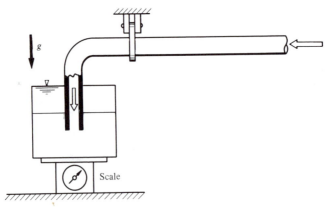

Problem 2.10

2.3 Description of Fluid Motion: Streamlines, Stream Function, and Trajectories

Fluid flow is revealed to an observer in ordinary experience as a series of lines or patterns made visible by smoke trails, or disturbances on the surface of flowing water. This streaming or flowing motion seems a rather natural way to explain what is perceived by the eye, and everyone has an intuitive feeling for the term "streamline" as a word in everyday use. The idea expressed by this term, however, refers to only one of several ways of describing fluid motion; another useful notion is based on the trajectory or "pathline" that a particular particle of fluid might be expected to follow as the flow progresses. One can imagine the particle to be marked by dye or some other substance to make these paths visible. A still different notion defines a "streakline"; that is, the streak made visible by some flow tracer material being injected at a particular location in a flow. Perhaps the most common example of a streakline is that of the plume leaving the smokestack of a power plant or ship. These physical perceptions of flow are very graphic but not yet sufficiently precise for our purposes. For this we need to define these concepts more carefully. A *streamline* is a curve everywhere parallel to the direction of flow, that is, the local velocity vector. A *pathline* is the trajectory of a particular particle of fluid. A *streakline*, however, is the trace or locus of the particles passing through a fixed location. In one very important situation all of these curves are the same; this is the case when the velocities at a fixed point do not change with time (i.e., for a steady flow). When velocities change, all of these curves are different and their calculation becomes quite a challenge given the Eulerian velocity components.

The flow descriptions above give one an idea of the geometry of the flow patterns, but do not give pressures or accelerations, which are dynamic quantities. For this reason we speak of the "kinematics" of the flow, which also includes the notion of rotation, shearing, and dilation (or expansion) of a fluid element. These concepts are defined and discussed in Section 2.7. In the present section we continue with the discussion of streamlines and the stream function. We use the Cartesian notation of the preceding section for the velocity components; imagine that we have a curve which is everywhere parallel to a two-dimensional velocity vector with components u, and v. Then the idea of parallelism of the elements dx and dy of this curve to the velocity vector is simply expressed by

$$\frac{dy}{dx} = \frac{v}{u}. \tag{2.3}$$

A similar but not as graphically transparent result for three dimensions is

$$\frac{dx}{u} = \frac{dy}{v} = \frac{dz}{w}. \tag{2.4}$$

It is important to note that the boundaries of stationary solid surfaces are always streamlines, since the fluid cannot cross solid surfaces.

For the special case of two-dimensional motion, the concept of streamlines can be related to the continuity equation. By limiting ourselves to incompressible flow, the continuity equation may be written

$$\frac{\partial u}{\partial x} + \frac{\partial v}{\partial y} = 0. \tag{2.5}$$

This equation can be satisfied automatically by introducing a new function defined by the equations

$$u = \frac{\partial \psi}{\partial y},$$

$$v = -\frac{\partial \psi}{\partial x}, \tag{2.6}$$

where ψ is a function of x, y, and possibly time and is called the *stream function*. Equation (2.6) contains no more information than the original differential equation, Eq. (2.5). However, the velocity components u and v determined by Eq. (2.6) automatically satisfy the continuity relationship. The total differential of ψ (using the chain rule of partial differentiation) is

$$d\psi = \frac{\partial \psi}{\partial x} \, dx + \frac{\partial \psi}{\partial y} \, dy$$

or with Eq. (2.6)

$$d\psi = -v \, dx + u \, dy.$$

If $d\psi$ is set equal to zero, we obtain the previously derived equation of a streamline. That is, lines of constant ψ represent streamlines.

Although the stream function has been introduced as a mathematical convenience, it has a more important physical interpretation. Let us consider a two-dimensional, incompressible flow in a channel. The conservation of mass principle states that the amount of flow per unit depth between the fixed walls of a channel is the same for all cross sections, or

$$Q = \int_A u \, dA,$$

where A is the cross-sectional area of the duct measured perpendicular to u and Q is the flow rate. In two dimensions if lines of constant x are taken as the cross sections and y_1 and y_2 are the coordinates of the channel top and bottom, then the integration can be carried out at constant x, giving for the flow rate per unit width

$$q = \int_{y_1}^{y_2} u \, dy,$$

and of course this is the same for all stations. The difference in stream function between the same two points y_1 and y_2 at a constant value of x can be obtained by integrating the first of Eqs. (2.6) with respect to y only. This is

$$\psi_2 - \psi_1 = \int_{y_1}^{y_2} u \, dy,$$

which is precisely the same as the flow rate per unit width just obtained.

We may conclude from our results that lines of constant stream function are everywhere parallel to the flow and that the numerical difference in ψ between two streamlines is equal to the flow rate per unit width passing between the streamlines.

EXAMPLE 2.1

Consider the velocity field

$$u = U,$$

$$v = V \cos \frac{2\pi}{\lambda} (x - Ut),$$

where U and V are constants and λ is a length. This flow will be used to derive a stream function and illustrate the differences between streamlines, pathlines, and streaklines.

(a) Stream Function

Equation (2.6) states that the velocity components u and v are derived from the stream function ψ. Integrating the equation for u with respect to y, and the equation for v with respect to x, we obtain

$$\psi = \int u \, dy + f(x, t),$$

$$\psi = - \int v \, dx + g(y, t),$$

where f and g are arbitrary functions of x and y, respectively. Substituting the expressions for the given velocity components, we obtain

$$\psi = Uy + f(x, t)$$

$$= -V \frac{\lambda}{2\pi} \sin \left[\frac{2\pi}{\lambda} (x - Ut) \right] + g(y, t).$$

By examination, the expression for the stream function must be

$$\psi = Uy - V \frac{\lambda}{2\pi} \sin \left[\frac{2\pi}{\lambda} (x - Ut) \right].$$

(b) Streamlines

A streamline is a line on which ψ is a constant, say ψ_0. Solving for y from the expression for the stream function above yields

$$y = \frac{\psi_0}{U} + \frac{V\lambda}{2\pi U} \sin \left[\frac{2\pi}{\lambda} (x - Ut) \right].$$

Thus at one instant of time, streamlines are lines that vary sinusoidally with x as shown in part (a) of the figure. Note that the argument $(x - Ut)$ remains constant if x increases as $Ut(x = Ut)$; this implies that a *stationary* observer will see the streamline pattern shown in part (a) of the figure move to the right at a speed U.

(c) Pathlines

To illustrate the concept of a pathline, we choose to follow a particular particle located at the origin at time t_0 (the same procedure could be used for any other particle starting from a different position or at a different time). For our particle, the x-component of velocity is $u = U$, so that after a short interval of time dt_1, the particle moves in the x-direction to a new position,

$$dx_1 = U \, dt_1.$$

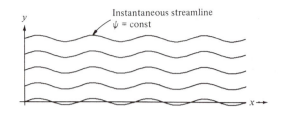

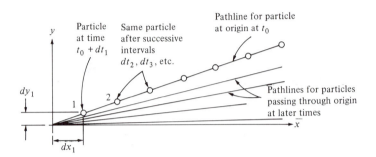

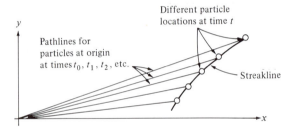

Example 2.1 (a) Instantaneous streamlines at the time $t = 0$. The instantaneous velocity is tangent to a streamline. (b) Generation of a pathline for a particle passing through the origin. The particle moves in a straight line; the slope depends on the time the particle is at the origin. (c) Formation of a streakline by marking all the different particles passing through the origin at successive times.

The vertical component of velocity as the particle starts from the origin is

$$v = V \cos \frac{2\pi}{\lambda} (-Ut_0),$$

so that after the same time interval dt_1, the new y-coordinate is

$$dy_1 = V \left[\cos \frac{2\pi}{\lambda} (-Ut_0) \right] dt_1.$$

The particle is now located at point 1 on part (b) of the figure. If a second time interval dt_2 passes, the additional movement in the x-direction is $dx_2 = U \, dt_2$. The vertical movement during the same interval is $dy_2 = v \, dt_2$, where care must be taken to evaluate the velocity v at the current particle location. Thus

$$dy_2 = V \left\{ \cos \frac{2\pi}{\lambda} \left[dx_1 - U(t_0 + dt_1) \right] \right\} dt_2,$$

or

$$dy_2 = V\left\{\cos\frac{2\pi}{\lambda}(Ut_0)\right\}dt_2;$$

that is, the particle moves at constant speed $V\cos(2\pi/\lambda)(Ut_0)$ in the y-direction.

This procedure can be repeated for many time steps up to a time $t = t_0 + \sum(dt_1 + dt_2 + \cdots)$, so that the x-position of the particle will be $x = \sum(dx_1 + dx_2 + \cdots) = U(t - t_0)$. The vertical position y becomes $y = \sum(dy_1 + dy_2 + \cdots) = V[\cos(2\pi/\lambda)(Ut_0)](t - t_0)$. Referring once again to part (b) of the figure, we see that the particle path is a straight line passing through the origin; the slope of such a pathline depends on when the particle is at the origin (the time t_0).

(d) Streaklines

We stated previously that a streakline is the collection or locus of *different* particles passing through a fixed location, which once again we choose as the origin for purposes of this example. Part (c) of the figure shows the concept of a streakline and how it differs from a pathline. Consider a starting time $t_0 = 0$, and assume that a time t has elapsed. The particle passing through the origin at $t_0 = 0$ will have moved along its pathline (the ray with slope arctan V/U) to a point as shown. A particle leaving at a later time t_1 moves on a different pathline with slightly lesser slope; its total distance traveled is also less, since the elapsed time is reduced to $t - t_1$. Similarly, particles passing through the origin at still later times t_2, t_3, and so on, can be located on their pathlines at time t, and the collection of these particle locations becomes the streakline shown. Keeping track of these particles and plotting their locations is easily accomplished using a simple spreadsheet program and a personal computer. The result of such a computation for this case is shown in Fig. 2.4.

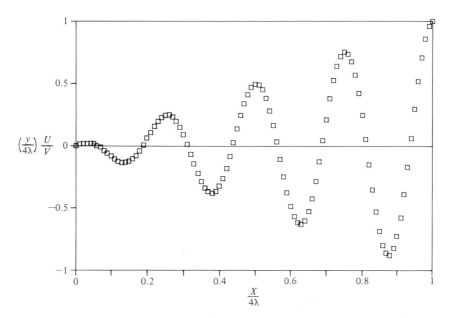

FIG. 2.4 Streaklines for Example 2.1 shown at a time $t = 4\lambda/U$. The particles have all been "marked" at the origin at regular time intervals; at the instant shown, 100 particles have passed through the origin.

PROBLEMS

2.11 Sketch neatly the streamlines for the functions $\psi = x^2 + y^2$ and $\psi = x^2 - y^2$, showing the direction of flow in each case.

2.12 Verify that in polar coordinates the relations corresponding to Eq. (2.6) are $v_r = \partial\psi/r\,\partial\theta$, $v_\theta = -\partial\psi/\partial r$.

2.13 Verify that ψ is a constant along a streamline using polar coordinates in your argument.

2.14 (a) For three-dimensional axisymmetric ($\partial/\partial\theta = 0$) incompressible flow, show that a suitable definition of the stream function in cylindrical coordinates is

$$v_r = -\frac{1}{r}\frac{\partial\psi}{\partial z}, \qquad v_z = \frac{1}{r}\frac{\partial\psi}{\partial r} \qquad \text{(see Problem 2.1).}$$

(b) Sketch the streamlines for the functions $\psi = ar^2$ and $\psi = br^2 z$.

2.15 Show that in a *compressible* steady flow the following equations suitably define the stream function:

$$\rho u = \rho_0\,\frac{\partial\psi}{\partial y},$$

$$\rho v = -\rho_0\,\frac{\partial\psi}{\partial x},$$

(where ρ_0 is a reference density) and repeat the discussion following Eq. (2.6).

2.16 Show that the following flow satisfies the continuity equation, find the stream function and sketch neatly a few of the streamlines for $r > r_0$ and $r < r_0$.
Note: $r = r_0$ is a streamline.

$$v_r = V_0\left(1 - \frac{r_0^2}{r^2}\right)\cos\theta,$$

$$v_\theta = -V_0\left(1 + \frac{r_0^2}{r^2}\right)\sin\theta.$$

V_0 is a reference velocity.

2.17 The x-velocity component of a two-dimensional, steady, incompressible flow is given by

$$u = e^{-x}\cosh y + 1.$$

(a) Find the y-component of velocity, v, assuming that $v = 0$ on $y = 0$.
(b) Calculate the stream function from the velocity components and sketch the streamlines $\psi = 0$ and $\psi = 1$.

2.18 (a) Find the equation for the pathline of a particle that is at a position (x_0, y_0) at a time t_0 for the flow described in Example 2.1.
(b) For the flow in Example 2.1, show that the equation of the streakline for particles passing through the origin can be written as

$$y = \frac{V}{U}\,x\left[\cos 8\pi\left(1 - \frac{x}{4\lambda}\right)\right].$$

2.19 The stream function for an *unsteady*, two-dimensional incompressible flow is

$$\psi = Axyt,$$

where A is a constant.
(a) Sketch the streamlines for this flow at an instant of time t.
(b) Derive the expressions for the instantaneous x and y velocity components, $u(x, y, t)$ and $v(x, y, t)$.
(c) Show that the pathline of a particle whose position is (x_0, y_0) at a time $t = 0$ is given by the expressions

$$x = x_0 e^{At^2/2},$$
$$y = y_0 e^{-At^2/2}.$$

2.4 Equations of Motion Neglecting Viscosity

The dynamic behavior of fluid motion is governed by a set of equations called the *equations of motion*. For the present we shall develop them neglecting viscosity, so that the resulting equations apply only to a perfect fluid. These equations are obtained by applying Newton's laws in a manner convenient for the treatment of fluid flows. This will be done by singling out an infinitesimal mass of fluid and applying the appropriate equations of mechanics. In distinction to the approach followed in the derivation of the continuity equation, we shall consider here a definite amount of mass and shall follow its motion. For notational convenience a two-dimensional Cartesian coordinate system will be used as before.

From Newton's law of motion we may write

$$F_x = ma_x, \tag{2.7}$$

where F_x is the force in the positive x-direction acting on a particle of mass m and a_x is the acceleration in the x-direction. When the equations of motion of a particle are written as in Eq. (2.7), the reference frame coordinate system must not be accelerating or rotating. Such a reference frame is said to be *inertial*. Accelerating, or noninertial, reference frames may be used, but additional terms depending on the acceleration of the frame must be included. The "particle" in the present case is a portion of the flow and therefore moves about with the fluid. The particle is shown in Fig. 2.5 in two positions at two slightly different instants of time. As before, the velocity in the x-direction is u and that in the y-direction is v. The acceleration a_x in the x-direction is

$$a_x = \frac{du}{dt}.$$

However, u depends on x, y, and time. The total change in u between the two positions shown in Fig. 2.5 is (from the rules of partial differentiation)

$$du = \frac{\partial u}{\partial t} dt + \frac{\partial u}{\partial x} dx + \frac{\partial u}{\partial y} dy,$$

neglecting terms of order higher than the first. The infinitesimal changes in position dx, dy of the particle are given, however, by the trajectory equations $dx = u\, dt$, $dy = v\, dt$, since the particle moves with the flow of which it is a part. (We are, in fact, considering the particle to be a point mass moving with the fluid, and dx and

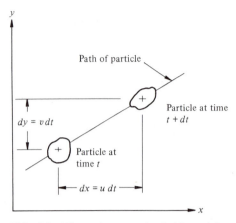

FIG. 2.5 The trajectory of a fluid particle.

dy are changes in the coordinates of its center of gravity.) The acceleration, therefore, becomes

$$a_x = \frac{du}{dt} = \frac{\partial u}{\partial t} + u \frac{\partial u}{\partial x} + v \frac{\partial u}{\partial y} \tag{2.8}$$

for the x-direction. In deriving Eq. (2.8), we determined the change in u as the particle moved with the fluid. The type of derivative obtained in Eq. (2.8) is therefore a special one. For this reason the group of differential operators on the right of Eq. (2.8) is given the special symbol

$$\frac{D}{Dt} = \frac{\partial}{\partial t} + u \frac{\partial}{\partial x} + v \frac{\partial}{\partial y}. \tag{2.9}$$

Thus

$$a_x = \frac{Du}{Dt}.$$

The special derivative given by Eq. (2.9) is sometimes called the *derivative following the fluid*. It is easy to see from it that the rate of change of any quantity, let us say q whether it is a vector or a scalar quantity, is (in three dimensions)

$$\frac{Dq}{Dt} = \frac{\partial q}{\partial t} + u \frac{\partial q}{\partial x} + v \frac{\partial q}{\partial y} + w \frac{\partial q}{\partial z}.$$

Thus, for example,

$$a_y = \frac{Dv}{Dt}, \quad a_z = \frac{Dw}{Dt}$$

or

$$\mathbf{a} = \frac{D\mathbf{v}}{Dt}. \tag{2.8a}$$

The continuity equation, Eq. (2.2), can also be expressed in terms of this derivative as

$$\frac{1}{\rho} \frac{D\rho}{Dt} + \nabla \cdot \mathbf{v} = 0.$$

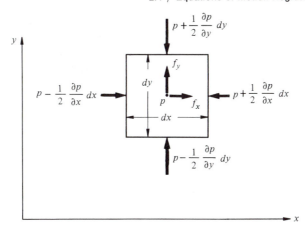

FIG. 2.6   Forces on a fluid element in the absence of friction.

In Eq. (2.8a) we have an expression for the acceleration of a particle of fluid as it moves about. This acceleration is the result of forces that act on the particle as given by Newton's laws of motion. The principal forces with which we are here concerned are those due to pressures acting on surfaces and to forces acting directly on the mass of the particle. The latter ones are often called *body* forces. Their magnitude is proportional to the mass of the element, as was pointed out in Chapter 1. Their components per unit mass will be designated by f_x, f_y, and f_z. The best-known force of this type is that due to gravity. In determining the pressure forces we will designate the pressure at the center of the element as p. The pressure at the right-hand face will then be $p + \frac{1}{2}(\partial p/\partial x)\, dx$, and the pressure at the other faces may be similarly determined. The complete force system (shown in two dimensions for simplicity) is illustrated in Fig. 2.6, where dx and dy now refer to the dimensions of the fluid particle.

We can now evaluate all the terms of Eq. (2.7). For the x-direction we have

$$ma_x = \rho \; dx \; dy \; \frac{Du}{Dt} = F_x = -\frac{\partial p}{\partial x} \; dx \; dy + \rho f_x \; dx \; dy.$$

Upon division by the mass of the element ($\rho \; dx \; dy$) the equation of motion in the x-direction becomes

$$\frac{Du}{Dt} = -\frac{1}{\rho}\frac{\partial p}{\partial x} + f_x \tag{2.10a}$$

and in the y-direction

$$\frac{Dv}{Dt} = -\frac{1}{\rho}\frac{\partial p}{\partial y} + f_y. \tag{2.10b}$$

The flow component in the third dimension z can easily be added to these results by including the term $w\partial/\partial z$ in Eq. (2.9) and for the equation of motion in the z-direction we have evidently

$$\frac{Dw}{Dt} = -\frac{1}{\rho}\frac{\partial p}{\partial z} + f_z. \tag{2.10c}$$

Equations (2.10a) through (2.10c) describe the motion of a perfect fluid in an inertial reference frame. No assumption was made about density, so that these results apply to a compressible fluid as well as to an incompressible one. These important equations were derived by Euler in 1755 and are called, therefore, the *Eulerian* equations. In a subsequent chapter we deal with some properties of these equations.

Before proceeding further, it is desirable to obtain the Eulerian equations in vector form. Although this can be done by adding the Cartesian components of Eqs. (2.10), it is more instructive to rederive them by vector methods.

Let us consider the mass within an arbitrary volume $\mathscr{V}$. From Newton's laws, the force acting on a given mass is equal to the product of this mass times its acceleration. For the mass contained in the given volume the terms representing this product become

$$\int_{\mathscr{V}} \rho \frac{D\mathbf{v}}{Dt} \, d\mathscr{V}.$$

This integral now must equal the total force arising from both the body force vector $\mathbf{f}$ and the pressure acting on the surfaces of the volume. It then follows that

$$\int_{\mathscr{V}} \rho \frac{D\mathbf{v}}{Dt} \, d\mathscr{V} = \int_{\mathscr{V}} \rho\mathbf{f} \, d\mathscr{V} - \int_{S} p\mathbf{n} \, dS,$$

where $\mathbf{n}$ is the unit vector normal to the surface element dS. The surface integral is then again transformed by a form of Gauss's theorem (Appendix 3) to a volume integral. Upon collecting all terms within the integral, the equation of motion may be written as

$$\int_{\mathscr{V}} \left(\rho \frac{D\mathbf{v}}{Dt} - \rho\mathbf{f} + \text{grad } p \right) d\mathscr{V} = 0.$$

Since the volume element is arbitrary we require the integrand to vanish, or

$$\frac{D\mathbf{v}}{Dt} = -\frac{1}{\rho} \text{ grad } p + \mathbf{f}. \tag{2.11}$$

Equations (2.10a) through (2.10c) are recovered by writing Eq. (2.11) in component form. The vector Eq. (2.11) can be put into several different forms by making use of several well-known properties of vectors. First let us make use of the conventional "del" operator to write Eq. (2.11) as

$$\frac{D\mathbf{v}}{Dt} = \frac{\partial \mathbf{v}}{\partial t} + (\mathbf{v} \cdot \nabla)\mathbf{v} = -\frac{1}{\rho} \nabla p + \mathbf{f}. \tag{2.12}$$

An alternative expression of Eq. (2.12) of great usefulness follows from the vector identity derived from Eq. (A-3.12e) of Appendix 3,

$$\tfrac{1}{2}\nabla(\mathbf{v} \cdot \mathbf{v}) = \tfrac{1}{2}\nabla V^2 = (\mathbf{v} \cdot \nabla)\mathbf{v} + \mathbf{v} \times \text{curl } \mathbf{v}.$$

Substituting this into Eq. (2.12), we have

$$\frac{\partial \mathbf{v}}{\partial t} + \frac{1}{2} \nabla V^2 + \frac{1}{\rho} \nabla p = \mathbf{f} + \mathbf{v} \times \text{curl } \mathbf{v}, \tag{2.13}$$

a relation we use in Chapter 3.

PROBLEMS

2.20 In a given two-dimensional, nonsteady flow field, the following velocity components are measured at a given instant of time.

x	y	u	v
0	0	20	10
1	0	22	15
0	1	14	5

Velocity measurements made at the point $x = y = 0$ at two instants of time are

t	u	v
0	20	10
$\frac{1}{2}$	30	10

u, v are in meters per second, t is in seconds, and x, y are in meters. Find the instantaneous acceleration in the x- and y-directions at $x = y = 0$. Assume all changes with time or distance to be linear.

2.21 Obtain the equations of motion in plane, polar coordinates:

$$\frac{\partial v_r}{\partial t} + v_r \frac{\partial v_r}{\partial r} + \frac{v_\theta}{r} \frac{\partial v_r}{\partial \theta} - \frac{v_\theta^2}{r} = -\frac{1}{\rho} \frac{\partial p}{\partial r} + f_r,$$

$$\frac{\partial v_\theta}{\partial t} + v_r \frac{\partial v_\theta}{\partial r} + \frac{v_\theta}{r} \frac{\partial v_\theta}{\partial \theta} + \frac{v_\theta v_r}{r} = -\frac{1}{\rho r} \frac{\partial p}{\partial \theta} + f_\theta,$$

by considering the forces on a small element bounded by the lines corresponding to r, $r + dr$, and θ, $\theta + d\theta$.

2.22 By writing out the components in a Cartesian x-, y-, and z-coordinate system, show that Eq. (2.12) reduces to Eqs. (2.10).

2.23 Consider the stream function $\psi = x^2 - y^2$ for a fluid of constant density (see also Problem 2.11).
   (a) Calculate the total acceleration vector **a** and show that it is proportional to the radius vector.
   (b) Use the result in part (a) to integrate the equation of motion assuming no friction and no body forces and find the pressure as a function of radius.

2.24 Assume that a perfect and incompressible fluid is flowing horizontally over a large surface. The flow is steady. Let the x-direction be in the direction of flow and let the y-direction be parallel to the action of the gravitational forces. Gravitation is the only body force acting. Find the relation between p and y.

2.25 Consider a perfect and incompressible fluid flowing in circular path about a center. The flow is purely two-dimensional and steady. There are no body forces acting.
   (a) Find from Euler's equation the differential relation between the pressure, the tangential velocity v_θ, and the radial distance.
   (b) Find this same relation by considering the equilibrium of a small element of fluid in this special case.

2.26 Using the answer to Problem 2.25, find the relation between pressure and radial distance for the cases that (a) $v_\theta = \Omega r$; (b) $v_\theta = c/r$. Both c and Ω are constants, and $v_r = 0$ for both flows.

2.27 The velocity components of an *unsteady* flow of an incompressible, inviscid fluid of density ρ in a straight, horizontal pipe parallel with the x-axis are

$$u = U(t), \qquad v = w = 0,$$

where $U(t)$ is a function of time t only. The acceleration of the fluid in the x-direction for this case is

$$a_x = \frac{dU}{dt}.$$

Neglecting all body forces, find the difference in pressure, dp, between two points along the pipe separated by a length L in the x-direction.

2.28 A cylindrical container is rotated at constant angular velocity ω. The contents of the cylinder, a compressible fluid, rotate with the cylinder so that the velocity at any point is given by $v_\theta = \omega r$. The density of the fluid is related to the pressure by the equation

$$\rho = Ap,$$

where A is a constant. Find the pressure distribution as a function of radius r, assuming that the pressure at the center, p_c, is given.

2.29 Consider a spherical planetary body of radius R which consists of a fluid of uniform density ρ. The variation of pressure p within such a sphere is given by

$$\frac{dp}{dr} = -\rho g,$$

where r is the radius of a point in the interior. If the acceleration due to gravity g varies linearly with the radial position r and has a value g_0 at the surface, find an expression for pressure in the interior in terms of ρ, g_0, r, and R.

2.5 Hydrostatics

(a) General Relations

A most important application of fluid mechanics is to a fluid that has no motion at all; that is, the fluid is stationary and the fluid acceleration is zero. Then we may say that the flow is static, and the name *hydrostatics* is natural, since these considerations were first applied to water. The density, however, may also vary, as in the atmosphere or ocean. Hydrostatics is a time-honored subject; it is vital for the analysis of floating bodies such as ships, of the measurement of pressure, and of the forces on dams. In the subsections that follow, we consider the distribution of pressure under gravity, forces on surfaces subject to hydrostatic pressure, and submerged bodies and floating bodies. In all cases our starting point is the Euler equation, Eq. (2.11), with zero velocity:

$$\frac{1}{\rho} \nabla p = \mathbf{f}. \tag{2.14}$$

In Cartesian coordinates this equation is

$$\frac{1}{\rho}\frac{\partial p}{\partial x} = f_x, \qquad \frac{1}{\rho}\frac{\partial p}{\partial y} = f_y, \qquad \text{and} \qquad \frac{1}{\rho}\frac{\partial p}{\partial z} = f_z.$$

From Eq. (2.14) one sees immediately that the derivatives of the pressure are proportional to the components of the force field, or, in vector language, the pressure gradient is proportional to the force field vector. Lines of constant pressure, or *isobars*, are therefore everywhere perpendicular to the force field.

As the most common example of the force fields we experience, let us consider the force of gravity. Taking y positively upward from the surface of the earth, we have

$$f_y = -g,$$
$$f_x = 0,$$

and from Eqs. (2.14),

$$\frac{\partial p}{\partial y} = -\rho g, \qquad (2.15)$$

where g is the local gravitational constant.

(b) Distribution of Pressure

Equation (2.15) is the basic fluid static relationship. To proceed further, we need some knowledge of the density as a function of height and pressure. The important and historical case is that of constant density, for which we have the result

$$p = -\rho g y + \text{const.} \qquad (2.16)$$

The pressure is then said to be distributed *hydrostatically*, decreasing linearly with increasing height. From Eq. (2.16) stem all the facts about buoyant forces on submerged and floating bodies and manometric measuring instruments. Several examples of these are afforded in the problems that follow and the examples below.

A second case of great importance is one in which the density throughout the fluid may have different values, yet does not depend on the pressure. Salt water with varying salinity would be one example of such a fluid. Let us consider such a fluid in a gravitational field. The relation shown in Eq. (2.15) applies, and differentiating this equation with respect to x yields

$$\frac{\partial^2 p}{\partial x \, \partial y} = -g \frac{\partial \rho}{\partial x}.$$

As f_x is zero, the x-component of Eq. (2.14) shows that

$$\frac{\partial p}{\partial x} = 0,$$

and similarly

$$\frac{\partial^2 p}{\partial x \, \partial y} = -g \frac{\partial \rho}{\partial x} = 0.$$

It has to be concluded, therefore, that ρ is only a function of y. Similarly, p also is only a function of y, and lines of constant pressure and constant density are, therefore, parallel to each other and parallel to the x-axis. When the fluid consists of distinct components such as oil and water, the two will separate out into two layers, and the fluid is then called *stratified*. If there is a continuous density change as in salt water, one may still think of infinitesimal horizontal layers of fluid, together forming a stratified fluid.

(c) Manometric Measurement of Pressure

Equation (2.15) provides the means to measure pressure by the elevation of liquid levels in a tube in a known gravitational acceleration; devices based on this principle are called *manometers*.

EXAMPLE 2.2 _____

The pressure difference $p_1 - p_2$ is to be determined for the manometer shown below, with the densities ρ and elevations h of the fluids in the connecting tubes given.

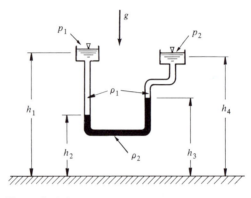

Example 2.2

Solution: Using the relation $p = \rho g h$, where h is the distance in the direction of gravity, we have

$$p_1 + \rho_1 g(h_1 - h_2) - \rho_2 g(h_3 - h_2) - \rho_1 g(h_4 - h_3) = p_2$$

or

$$p_2 = p_1 - (h_3 - h_2)(\rho_2 - \rho_1)g + \rho_1 g(h_1 - h_4).$$

Frequently, $h_1 = h_4$, and then this device is called a *U-tube* manometer. Suppose further that the heavier fluid is mercury at 20°C and the lighter fluid is water at 20°C with $h_3 = 2$ m, $h_1 = 1$ m, and $h_1 = h_4$. Then from Appendix 2,

$$p_2 - p_1 = (9.81)(2 - 1)(13,586 - 1000) = 123.3 \text{ kPa}.$$

(d) Pressure Distribution Under Acceleration

The case of a constant acceleration or acceleration as a known function of position is readily incorporated into the results already established for hydrostatics. To show this, the acceleration term merely has to be transposed to the right-hand side of Eqs. (2.11).

The equation corresponding to Eq. (2.14) becomes

$$\frac{1}{\rho} \nabla p = \mathbf{f} - \frac{D\mathbf{v}}{Dt}, \tag{2.14a}$$

where $D\mathbf{v}/Dt$ is a vector that may be designated by $\mathbf{a}$. In Cartesian coordinates, Eq. (2.14a) becomes

$$\frac{1}{\rho} \frac{\partial p}{\partial x} = f_x - a_x,$$

$$\frac{1}{\rho} \frac{\partial p}{\partial y} = f_y - a_y,$$

and

$$\frac{1}{\rho} \frac{\partial p}{\partial z} = f_z - a_z.$$

Lines of constant pressure are then perpendicular to the combined force field, consisting of the components of $f_x - a_x$, $f_y - a_y$, and $f_z - a_z$, or in vector form $\nabla p = \rho(\mathbf{f} - \mathbf{a})$, where $\mathbf{f}$ and $\mathbf{a}$ are the force field and acceleration vectors, respectively.

EXAMPLE 2.3

A liquid of density ρ_1 having a free surface or surface of constant pressure floats on a heavier one of density ρ_2 in the presence of gravity. The upper fluid is made to rotate at angular velocity Ω_1 and the lower one at angular velocity Ω_2. The heights y_1, y_2 of the two surfaces relative to their respective heights at the center of rotation are to be found.

Solution: The upper surface being a surface of constant pressure must be perpendicular to the force field $-\mathbf{j}g + \mathbf{i}_r r\Omega_1^2$; thus

$$\frac{dy_1}{dr} = \frac{r\Omega_1^2}{g}$$

and

$$y_1 = \frac{r^2 \Omega_1^2}{2g},$$

which is the equation of a parabola.

The interface between the two fluids is not, however, a surface of constant pressure. Nevertheless, in the lower fluid we have, from Eq. (2.14a),

$$\nabla p_2 = \rho_2(-\mathbf{j}g + \mathbf{i}_r r\Omega_2^2)$$

and in the upper fluid

$$\nabla p_1 = \rho_1(-\mathbf{j}g + \mathbf{i}_r r\Omega_1^2),$$

so that at the interface

$$\nabla(p_2 - p_1) = -\mathbf{j}(\rho_2 - \rho_1)g + \mathbf{i}_r r(\rho_2 \Omega_2^2 - \rho_1 \Omega_1^2).$$

Furthermore, along the interface, the pressures must also be equal, so that

$$p_2 - p_1 = 0.$$

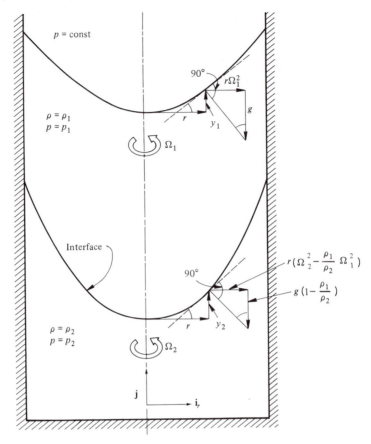

Example 2.3

It may then be reasoned that the slope of the interface, dy_2/dr_2, must be perpendicular to the gradient $\nabla(p_2 - p_1)$. If this were not so, there would be a component of this gradient along the surface which would be incompatible with the requirement that the pressure difference $(p_2 - p_1)$ is equal to zero all along the surface. We may then write for the slope of the interface

$$\frac{dy_2}{dr_2} = \frac{r(\Omega_2^2 - (\rho_1/\rho_2)\Omega_1^2)}{g(1 - \rho_1/\rho_2)}.$$

After integration this result may be written as

$$y_2 = y_1 \frac{\Omega_2^2/\Omega_1^2 - \rho_1/\rho_2}{1 - \rho_1/\rho_2},$$

which shows that the interface is also a parabola but with a different scale than that of the free surface.

PROBLEMS

2.30 Sometimes it is necessary to deal with "stratified" fluids that are essentially incompressible but have a density varying smoothly from one point to another.

This may occur when fluids of varying density are mixed or when the temperature or concentration of a dissolved substance varies with position. Often in the motion of such a fluid it may be assumed that the density of a particle of fluid does not change as it moves about. Show then that the continuity equation for such a fluid is still $\nabla \cdot \mathbf{v} = 0$, even though the flow may not be steady.

2.31 For a constant-temperature atmosphere (which incidentally does not exactly correspond to the normal atmospheric conditions) the specific weight of the atmosphere may be expressed as a function of the pressure approximately as follows:

$$\rho g = \frac{p}{8443} \text{ N/m}^3,$$

where p is the pressure in N/m^2. Find the pressure of this atmosphere as a function of elevation. Take the pressure at zero elevation to be 100 kPa.

2.32. A vertical mine shaft extends 3000 m down into the earth and contains static air, whose temperature increases linearly with depth from 20°C at the surface to 45°C at the bottom. Find the air pressure at the bottom of the shaft. (The pressure is 100 kPa at the surface; the air is *not* incompressible, but may be considered to be a perfect gas; see Appendix 4.)

2.33 A tank car 3 m long and 1.5 m in diameter and completely filled with water ($\rho = 1000 \text{ kg/m}^3$) is accelerated at 3 m/s^2 to the right. If the pressure at the top of the tank, midway between the ends, is atmospheric, determine the location and magnitude of the maximum and minimum pressures. Also determine what the slope of the water surface within the tank would be if it were only partially filled, assuming now that the tank is open to the atmosphere and the tank walls are sufficiently high to allow the water to assume a free surface without spilling.

2.34 A 0.6-m-long capillary tube closed at one end with the other open to atmosphere is completely filled with water and rotated about the closed end at n revolutions per second. Assuming that the water does not run out of the tube and that no voids are formed inside the liquid, compute the value of n required to reduce the pressure at the closed end to zero.

2.35 An ordinary U-tube manometer is mounted on a vehicle in order to fashion a simple accelerometer. The plane formed by the two legs of the U-tube is

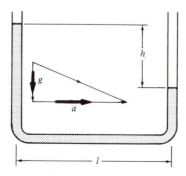

Problem 2.35

aligned with the direction of motion as shown. If the vehicle accelerates in the x-direction at a constant acceleration a, what is the relationship between a, h, L, and gravity g?

2.36 The pressure p_s and density ρ_s of the atmosphere on the surface of the planet Venus are 9.2 MPa and 63 kg/m³. Up to an altitude of about 40 km the atmosphere behaves adiabatically; that is, the pressure and density are related by the expression $p = C\rho^\gamma$, where γ is the ratio of specific heats ($\gamma = 1.2$ approximately for Venus) and C is a constant. Assume that the acceleration due to gravity has a constant value of 8.7 m/s² on Venus.
(a) Find an expression for the pressure p as a function of the altitude y and the constants p_s, ρ_s, γ, and g.
(b) Evaluate the pressure at an altitude of 30 km.

2.37 A U-tube mercury manometer shows a deflection of 0.2 m. Except for the mercury, the fluid in the manometer is filled with water. Assuming that the density of water is 1000 kg/m³ and that of mercury is 13,600 kg/m³, find the pressure difference between the two points at which the manometer connections are made.

2.38 A mercury manometer is connected to a large reservoir of water as shown. The difference in elevation between the free surface of the reservoir and the mean position of the two mercury levels in the manometer is y; the difference in elevation between the two mercury levels is $2x$. Find the ratio x/y if the density of mercury is 13.6 times that of water.

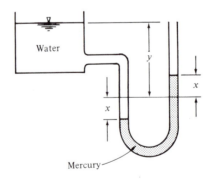

Problem 2.38

2.39 A vertical cylinder containing a liquid of density ρ_L has air of density ρ_A injected into its base so that a uniform, bubbly mixture is formed. Any unit volume of the mixture contains a volume α of air and a volume $1 - \alpha$ of liquid (the quantity α is called the *void fraction*).
(a) What is the effective density ρ of the mixture in terms of α, ρ_L, and ρ_A?
(b) The legs of an inverted air–liquid U-tube manometer are connected to the cylinder at two points a height H apart as shown. The levels in the manometer legs register a difference h. Assume that the air bubbles cannot enter the manometer tubes, which are therefore filled with pure liquid. Find a relationship between the void fraction α and the lengths h and H. (Since $\rho_A \ll \rho_L$, you can assume that $\rho_A = 0$ for this part of the problem.)

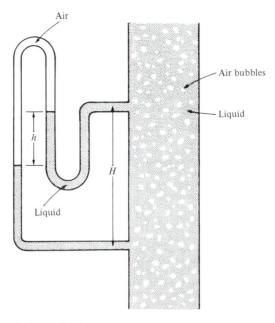

Problem 2.39

2.40 A differential manometer is used to measure the pressure rise across a pump. The heavy liquid in the manometer is mercury with a density approximately 13.6 times that of water and all other lines are filled with water ($\rho = 1000$ kg/m^3). The observed manometer deflection is 762 mm and the manometer leads are connected to the pump as shown (see illustration). What is the pressure rise $p_2 - p_1$ in kPa?

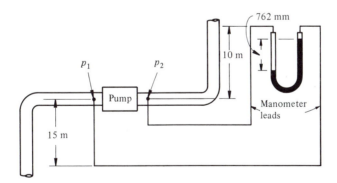

Problem 2.40

2.41 Matter is attracted to the center of the earth with a force proportional to the radial distance from the center. Using the known value of g at the surface ($g = 9.81$ m/s^2), where the radius is 6440 km, compute the pressure at the earth's center, assuming the material behaves like a liquid, and that the mean density is 5600 kg/m^3. (A tube of constant diameter rather than a spherical segment may be considered for convenience.) Obtain first a formula in symbols before numerical values are inserted.

(e) Forces on Structures and Submerged Bodies

The hydrostatic equation also permits calculation of the force exerted on surfaces submerged in the liquid arising from the distribution of hydrostatic pressure. The pressure force is always perpendicular to the bounding surface; the orientation of the surface must therefore be known to determine the direction of the force. The hydrostatic pressure varies with depth, and the normal to the surface (which gives the direction) may change with position. The force components must then be derived by integration as shown in Example 2.4.

EXAMPLE 2.4 _____

The end of a reservoir has the shape of a quarter circle of radius R_0. It is hinged at the bottom and restrained by a horizontal strap at the top. The reservoir is filled with fluid of specific weight ρg. Determine the strap force T per unit width of reservoir.

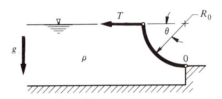

Example 2.4

Solution: The pressure at an arbitrary point on the circle is $p = \rho g R_0 \sin \theta$ and the moment of the elemental force $p R_0 \, d\theta$, which is normal to the cylindrical surface about point O is $p R_0^2 \cos \theta \, d\theta$. Equating this to the restraining moment $T R_0$ about O, we have

$$T = \frac{1}{R_0} \int_0^{\pi/2} \rho g R_0^3 \sin \theta \cos \theta \, d\theta = \frac{1}{2} \rho g R_0^2.$$

Note that the vertical reaction at point O is just the weight of the fluid displaced by the cylinder but pointing downward and that the horizontal reaction there is zero.

_____

It is instructive to indicate a general result for an arbitrary surface element of area dA whose unit normal $\mathbf{n}$ points out of the surface; the element of force is then just

$$dF_x = -(\mathbf{i} \cdot \mathbf{n} \, dA)p = p \, dA_x,$$

where dA_x is the projection of area dA onto a plane perpendicular to the x-axis. The force F_x is the integral of this expression and is seen to be the hydrostatic force on the x-projection of area A, as Example 2.4 shows. Similar results follow for the other axes.

A particularly useful result follows for a submerged body. We take the y-axis to be vertical as in Eq. (2.15); the upper and lower surface elements of the body surface each have the projection dA_y on a plane perpendicular to the y-axis. The upper and lower surface elements dA_y have a net upward, or buoyant, force of

$$dF_y = \rho \, dA_y(h_l - h_u)g,$$

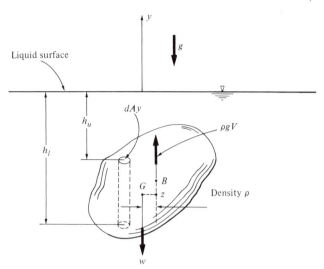

FIG. 2.7   Forces on a submerged body.

where h_l is the lower depth and h_u is the upper depth of the surface element, as shown in Fig. 2.7.

The integral of this result may be seen to be the specific weight of the fluid times the volume of the body. More simply, the buoyant force is just the weight of the fluid volume displaced by the body; this is known as *Archimedes' principle*. In addition to this important principle, we have from mechanics the result that the surface pressures giving rise to the buoyancy act at the *centroid* or center of mass of the displaced fluid; this is simply called the *center of buoyancy*. The body itself may have a mass M and the center of this mass is located at point G as sketched in Fig. 2.7. The center of buoyancy is located at point B. The positions of G and B will not be the same if the density of the body is not uniform, as assumed in the present example. The buoyant force $\rho g V$ acts upward through B and the weight of the body W acts downward at G. The body weight W may just equal the buoyant force, in which case we say that the body is neutrally buoyant. Even in this situation the body is not in a state of equilibrium because it tends to turn under the moment created by the two parallel forces W and $\rho g V$. The perpendicular distance between these forces is L. This distance is called the *righting arm*. At equilibrium L must be zero and for stability G should be as low as possible relative to B because then a slight rotation will cause a restoring moment.

(f) Floating Bodies

For a floating body, the weight of the body will, of course, exactly balance that of the buoyant force. At equilibrium the center of gravity and the center of buoyancy will lie on a vertical line. Floating bodies can, however, tip or rotate; the shape of the displaced volume can therefore change, unlike that of the submerged body, and the position of the center of buoyancy may move. If the weight is constant, the volume of the displaced fluid must remain constant, but because the center of buoyancy moves, a moment may be required to rotate the body. This situation is shown in Fig. 2.8 and illustrated in the following example.

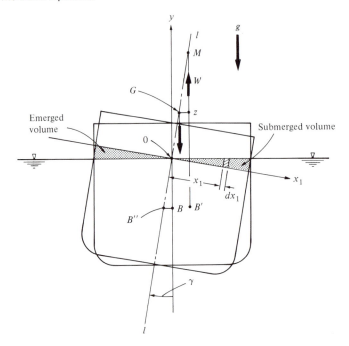

FIG. 2.8 Tilted floating body; the angle of inclination γ is very small compared to unity.

EXAMPLE 2.5

The two-dimensional rectangular shape shown is submerged into a liquid of density 1000 kg/m³. The centroid of a right triangle is known to be one-third the altitude of the base. We can then readily find that the distance between A and $B1$, the intercept on a horizontal line extended from the interface with the buoyant force is

$$\tfrac{4}{3} \cos \gamma + \tfrac{1}{3} \sin \gamma = 1.3416 \text{ m,}$$

where γ is the angle of tilt between the rectangle and the horizontal and

$$\gamma = \arctan \tfrac{1}{2}.$$

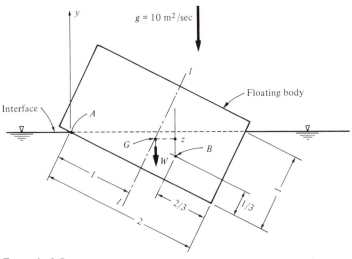

Example 2.5

The buoyant force is the volume (for a unit depth) times the specific weight $1000 \times 10 = 10,000$ N/m^3. This force would create a moment around point A of $13,416$ N · m.

Let the floating body now be nonuniform, but having a total weight of $10,000$ N. For simplicity, also assume that the center of gravity lies along the line $l-l$ and that it is designated as G in the example. Then we can see that if G lies to the left of the buoyant force through B, or its vertical intercept $B1$, there will be a moment tending to rotate the body into a horizontal position; if it lies to the right of $B1$, there will be a moment tending to increase the angle of tilt γ. In the first case the equilibrium is said to be stable and in the second, unstable. The important quantity, as in all floating bodies, is the "righting arm." A floating body will have a stable inclination provided that the righting arm produces a moment that tends to reduce the angle of inclination.

An interesting situation is that of a symmetric floating body with no inclination (i.e., with its centerline vertical and the center of mass on the centerline). Thus G and B lie along the same vertical line and the body is in equilibrium. Now let us *imagine* the body is tilted through a small angle from the vertical, γ. We wish to determine if the floating body remains stable after such an infinitesimal tilt. The initial position of the center of buoyancy B is shown in Figure 2.8; after a tilt of angle $\gamma \ll 1$, the center of buoyancy becomes B'. In this tilting, some of the volume V emerges on the left of the centerline $l-l$, and an equal part is submerged on the right if γ is sufficiently small (otherwise, the intersection of the plane of the water surface with the body, called the waterplane, will have to shift up or down to maintain a constant buoyant force). The position of the new center of buoyancy, B', can be found by taking moments of the volume V in its tilted configuration about the axis through the point 0 normal to the paper. The purpose of the calculation of B' is to determine the relative location of G, the center of mass, and the new center of buoyancy B' to determine if the body is floating stably. It is somewhat easier to calculate the new position B' by considering the change in the center of buoyancy from its original position on the centerline of the body (shown here as point B'') to the new point B'. We now take moments of the volume of V about the axis through 0, with $B''B'$ representing the distance between B' and B'':

$$(B''B')V = \int x_1^2 \gamma \, dx_1 s(x_1),$$

where $s(x_1)$ is the depth into the paper of the waterplane and the integration is carried out over the complete width, which includes both the submerged and emerged volumes. The distance $B''B'$ has a very simple geometric interpretation, namely

$$B''B' = \gamma \frac{I}{V},$$

where I is the moment of inertia of the area of the waterplane about the axis through 0. The line of action of the buoyant force through B' intersects the centerline $l-l$ at point M. Remembering that $\gamma \ll 1$, the distance $B''M$ is equal to $B''B'/\gamma$, so that

$$B''M = \frac{I}{V}.$$

The distance between the center of mass G and the point M is clearly crucial to the righting moment arm L; as long as G is below M, a restoring righting moment is present. Thus the original horizontal floating position is stable as long as

$$\frac{I}{V} - BG > 0,$$

where BG denotes the distance between B and G (again, remember that the angle $\gamma \ll 1$, so that $B''M = BM$). The point M has a separate and interesting geometric significance; it is termed the *initial metacenter*, and the distance GM is the *initial metacentric height*. The position of the metacenter will change as the angular displacement increases and a curve may be defined for the metacenter as a function of this angle. The relations above are useful in determining initial stability as long as the small-angle approximation is valid (about 6°). Beyond that, the righting moment must be calculated from first principles as in Example 2.5 to determine stability.

PROBLEMS

2.42 A hemispherical shell of diameter d rests on a glass plate and is filled with a liquid of specific weight ρg through a small tube at the top. Find the least weight of the shell that will prevent it from being lifted by the pressure of the enclosed liquid. (There is a short way to solve this problem.)

2.43 A cylindrical can with one end open is observed to be floating on a liquid of density ρ with the open end down. The can is of weight W and is supported by air that is trapped in the can. The can floats out of the fluid a distance h. If the air is assumed to follow Boyle's law, that is, pressure times volume is a constant, determine the force F necessary just to submerge the can. The internal cross-sectional area of the can is A.

Note: The thickness of the wall is assumed to be zero and the hydrostatic pressure due to the atmosphere may be neglected.

Hint: The distances x_1 and x_2 may be used as auxiliary quantities, but they should not appear in the final answer.

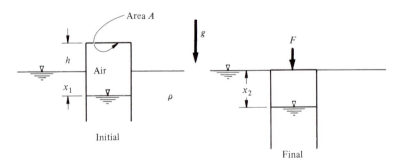

Problem 2.43

2.44 A cylindrical canister with an open bottom is pushed into a large body of water and held in such a position that the top surface is exactly at the water level. The force necessary to do this is F_0. The cylinder contains a gas, the pressure of which varies in such a way that $pV = $ constant, where V is the gas volume. The canister is now submerged further by applying a force F. Let the distance from the main water level to the top of the canister be y, and the distance from the top of the canister to the water level inside the can be x. Initially, $x = x_0$ and $y = 0$; the cross-sectional area of the can is A.

(a) Determine F in terms of F_0, x_0, y, and A.

(b) Indicate (by means of a graph) the variation of F versus y.

2.45 A plate of weight 1 kN per unit width is suspended at one end by a hinge at the water level of the reservoir shown below. The bottom end is free to move as in the following sketch. Calculate the angle of repose, θ, of the plate, using 1000 kg/m^3 as the density of the water. ($g = 9.81$ m/s^2.)

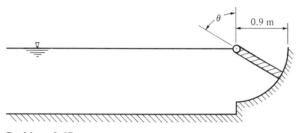

Problem 2.45

2.46 A body of water is held in place by a hinged gate as shown. The water level is below the elevation h_0 of the hinge. The gate has a weight w per unit length (normal to the page) and is of uniform thickness. The gate is held closed solely due to its own weight, and will open when water pressure acts to overcome the effect of the weight. What is the value of w at which the gate opens if the water level reaches a height αh_0, where α is between 0 and 1.0?

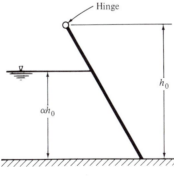

Problem 2.46

2.47 Show that the horizontal hydrostatic force on a surface submerged beneath a liquid is equal to the product of the specific weight of the liquid, the area of the surface projected on a vertical plane, and the distance from the centroid of this area to the liquid surface.

2.48 The L-shaped gate shown in the figure can rotate about the hinge. As the water level rises, the gate will open when the level reaches a critical height h_c. If the length of the lower horizontal arm is 1 m, find the critical height h_c. (Neglect the weight of the gate itself.)

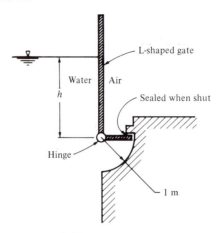

Problem 2.48

2.49 A wall with given dimensions retains a body of water as shown. The weight of the wall is W per unit breadth (normal to the page).

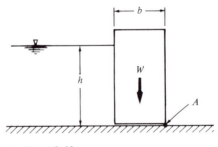

Problem 2.49

(a) Determine the net moment about point A and solve for the minimum weight W which would prevent overturning of the wall. Assume that no water can penetrate underneath the wall.

(b) Find the minimum weight W if it is assumed that water can penetrate freely under the wall up to the point A, but there is an effective seal at A.

2.50 A rectangular block of concrete is used as a dam for a reservoir of water. The block has a height a, a breadth b, and the water depth is $3a/4$. The density of concrete can be taken as $\frac{5}{2}$ times that of water. Find the ratio a/b above which the block will be overturned. (Assume that the water has seeped under the dam, so that there is a thin film of water under the dam, but there is no leakage past point A, as shown in the illustration for Problem 2.49.)

2.51 A rigid dam is composed of material of density ρ_d. The dam height is h. What is the minimum breadth b of the dam necessary to prevent it from tipping about the point O when the water of density ρ reaches the top of the dam? Assume that the maximum hydrostatic pressure acts over the bottom of the dam (i.e., there is seepage beneath the dam).

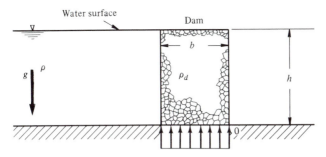

Problem 2.51

2.52 Prove Archimedes' principle ("A submerged body loses in weight an amount equal to the liquid it displaces") for a fluid of nonconstant density by imagining that the interior of the region occupied by the body is filled with the same fluid as on the outside at the same level.

2.53 A watertight bulkhead 6 m high forms a temporary dam for some construction work. The top 3 m behind the bulkhead consist of seawater with a density of 1020 kg/m³, but the bottom 3 m, being a mixture of mud and water, can be considered a fluid of density 1500 kg/m³. Calculate the total horizontal load per unit width and the location of the center of pressure measured from the bottom.

2.54 A special balloon (proposed for exploring the atmosphere of Mars) has the shape shown in the sketch. The main section is a cylinder of diameter D and length l, mated with matching hemispheres on top and bottom (the length l is *much* larger than D). A tiny hole at the bottom allows the pressure on the inside and outside of the balloon to be equal at that point.
  (a) If the balloon is filled with a light gas of density ρ, while the external atmosphere has a density ρ_0, what is the lift L that the balloon can develop?
  (b) Suppose that the upper hemisphere is made of thin material which can sustain a tension σ per unit length. Find the maximum value of lift $L(\text{max})$ for a fixed value of D, and the corresponding balloon length $l(\text{max})$.

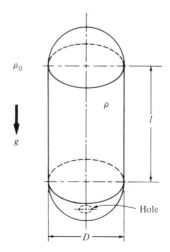

Problem 2.54

2.55 A very lightweight plastic plate of length l_0 is hinged as shown. The water level is at a distance h above the hinge. The plate comes to rest at an angle θ. The plate may be considered weightless for our purpose. Its cross-sectional area is constant and equal to A. A concentrated weight (e.g., a metal rod) is placed at the tip (or along the edge of the plate). The water density is ρ. Derive an expression for the weight W in terms of h, θ, A, ρ, g, and l_0.

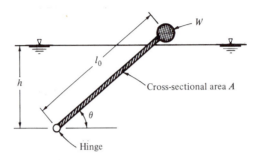

Problem 2.55

2.56 A rectangular barge with outside dimensions of 20 ft width, 60 ft length, and 10 ft height weighs 350,000 lb. It floats in salt water ($\gamma = 64$ lb/ft³) and the center of gravity of the loaded barge is 4.5 ft from the top. Due to a gust of wind the barge lists 10° and the center of buoyancy moves 1.28 ft in the horizontal direction. Locate both the center of buoyancy when floating on an even keel and the metacentric height for the 10° list.

2.57 A thin-walled rectangular channel section (open at the ends) floats upside down in a pool of water. As indicated in the figure, the vertical sides are submerged to a depth h and the width of the channel section is B. If the center of mass of the channel section happens to lie on the waterline, find the particular value of the ratio B/h above which this floating configuration is stable.

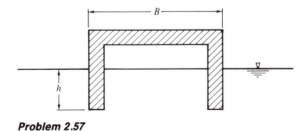

Problem 2.57

2.6 Equations of Motion for a Viscous Fluid

In a previous section the equations of motion for a nonviscous fluid were derived. In a viscous fluid the surface forces acting on an element of fluid are considerably more complicated. These are of two types: there is a normal force or normal stress similar to the pressure but it may not be the same in all directions, and there are shear forces or stresses whose direction is parallel to the surface on which they act.

The total force per unit area is a vector and it is shown in Fig. 2.9 acting on an element of surface area dS that has an outward-pointing unit normal vector **n**. The

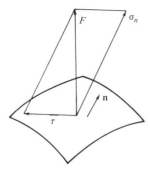

FIG. 2.9 Stress vectors acting on element of surface area.

symbols σ_n and τ represent the normal stress and shear stress vectors, respectively. The stress system acting on an element of the surface is conveniently visualized by showing in detail the stress components on a small cube (Fig. 2.10). We use the notation common in elasticity and denote the normal stresses by σ and the shear stresses by τ. Two subscripts are attached to each of the stress symbols: the first indicates the direction of the normal of the surface on which the stress acts and the second indicates the direction in which the stress acts. For example, τ_{yx} denotes a stress acting in the x-direction on a surface whose normal points in the y-direction. This is therefore a shear stress. Similarly, σ_{xx} denotes a normal stress. The stresses and shear forces on each surface are reported in terms of a right-handed coordinate system in which the outwardly directed surface normal indicates the positive direction.

In general, the various stresses change from point to point and give rise to net surface forces on a small element of fluid that tend to accelerate it. Such an element is shown in two dimensions in Fig. 2.11. In order to obtain a common reference point, the stresses at the center of this element are set equal to σ_{xx}, τ_{xy}, σ_{yy}, and so on. The forces acting on each surface are then obtained taking into account the variation of the stresses with distance. The net force in the x-direction for a unit depth in the z-direction evidently is

$$\frac{\partial}{\partial x}(\sigma_{xx}) \, dy \, dx + \frac{\partial}{\partial y}(\tau_{yx}) \, dx \, dy,$$

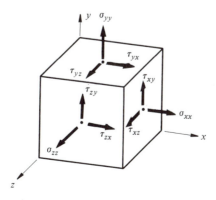

FIG. 2.10 Normal stresses and shear stresses acting on an element of fluid.

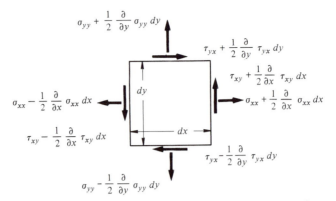

FIG. 2.11 Stresses on the surfaces of an element of fluid. All stresses may be functions of position. At the center of the element the stresses are taken to be σ_{xx}, σ_{yy}, τ_{xy}, and τ_{yx}.

and if three dimensions were taken into account, we would obtain

$$\frac{\partial}{\partial x} (\sigma_{xx}) \, dx \, dy \, dz + \frac{\partial}{\partial y} (\tau_{yx}) \, dx \, dy \, dz + \frac{\partial}{\partial z} (\tau_{zx}) \, dz \, dx \, dy.$$

These surface forces now replace the role of the pressure in the development leading to Eqs. (2.10). Upon dividing by the volume element we have the equation of motion in the x-direction:

$$\rho \frac{Du}{Dt} = \rho f_x + \frac{\partial}{\partial x} \sigma_{xx} + \frac{\partial}{\partial y} \tau_{yx} + \frac{\partial}{\partial z} \tau_{zx}. \tag{2.17a}$$

Similarly, we find

$$\rho \frac{Dv}{Dt} = \rho f_y + \frac{\partial}{\partial x} \tau_{xy} + \frac{\partial}{\partial y} \sigma_{yy} + \frac{\partial}{\partial z} \tau_{zy} \tag{2.17b}$$

for the y-direction and

$$\rho \frac{Dw}{Dt} = \rho f_z + \frac{\partial}{\partial x} \tau_{xz} + \frac{\partial}{\partial y} \tau_{yz} + \frac{\partial}{\partial z} \sigma_{zz} \tag{2.17c}$$

for the z-direction.

The next and most vital step involves the relationship between the stress system and the motion of the fluid. It has been found experimentally that to a high degree of accuracy the stresses in many fluids are related linearly to the derivatives of the velocities. Furthermore, most fluids have no preferred direction in space—they are said to be *isotropic*. In addition, the stresses do not explicitly depend on the position coordinate and the velocity of the fluid.

If the conditions above are assumed to apply exactly and if we also assume that

$$\tau_{xy} = \tau_{yx}, \quad \tau_{yz} = \tau_{zy}, \quad \tau_{zx} = \tau_{xz} \tag{2.18}$$

(an assumption discussed in detail in Section 2.7), then a unique form of the relationship between the stresses and the velocity gradients can be developed. Rather than present this rather tedious mathematical development, we shall simply state the result first and then subsequently examine the characteristics and physical implications of the given equations.

The relations between stresses and velocity derivatives are

$$\sigma_{xx} = \left(\lambda - \frac{2}{3}\mu\right)\Theta + 2\mu\frac{\partial u}{\partial x} + A,$$

$$\sigma_{yy} = \left(\lambda - \frac{2}{3}\mu\right)\Theta + 2\mu\frac{\partial v}{\partial y} + A,$$

$$\sigma_{zz} = \left(\lambda - \frac{2}{3}\mu\right)\Theta + 2\mu\frac{\partial w}{\partial z} + A, \tag{2.19}$$

$$\tau_{xy} = \tau_{yx} = \mu\left(\frac{\partial u}{\partial y} + \frac{\partial v}{\partial x}\right),$$

$$\tau_{xz} = \tau_{zx} = \mu\left(\frac{\partial u}{\partial z} + \frac{\partial w}{\partial x}\right),$$

$$\tau_{yz} = \tau_{zy} = \mu\left(\frac{\partial v}{\partial z} + \frac{\partial w}{\partial y}\right), \qquad \text{where } \Theta \equiv \frac{\partial u}{\partial x} + \frac{\partial v}{\partial y} + \frac{\partial w}{\partial z}.$$

These equations contain the three parameters μ, λ, and A. According to the assumption of a linear relationship between stresses and velocity gradients, all three parameters must be independent of the velocity gradients, but they may still depend on such fluid properties as the temperature and the density. As a consequence μ, λ, and A may also depend on time and position.

Let us examine these parameters now somewhat further in order to obtain a better physical understanding of their meaning. We shall first discuss the quantity A and for this purpose consider the particularly simple situation in which the fluid is at rest or in uniform motion. All the shear stresses are then equal to zero and the three normal stresses are all equal to A; that is,

$$\sigma_{xx} = \sigma_{yy} = \sigma_{zz} = A.$$

For this very simple flow then, A is just the negative of the fluid pressure, which in this instance is equal for all directions, as was shown before. For the flow under consideration the fluid is in an equilibrium condition of the kind that may be treated in thermodynamics, and the pressure $(-A)$ may be identified with the *thermodynamic pressure*. This pressure, which we shall designate by p_{th}, is the one related to the density and temperature by the thermodynamic relation usually called the equation of state. (For a so-called perfect gas, for example, this equation takes the form $p = \rho RT$, where R is a constant; see also Appendix 4.) Let us proceed now and consider a very general flow of a fluid that follows the stress relations of Eq. (2.19). Velocity gradients will now exist. Nevertheless, A is independent of the velocity gradients, and we must conclude, therefore, that the quantity A still retains its previous meaning and that it is, therefore, identical to the negative value of the thermodynamic pressure; that is,

$$A \equiv -p_{\text{th}}.$$

Often in fluid mechanics it is more convenient to introduce as another concept the *average* pressure, which is defined as

$$\bar{p} = -\tfrac{1}{3}(\sigma_{xx} + \sigma_{yy} + \sigma_{zz}).$$

Substituting from Eq. (2.19), it is seen that this average is equal to

$$\bar{p} = -\lambda\Theta + p_{\text{th}}.$$

This equation shows first that the average pressure is equal to the so-called thermodynamic pressure for an incompressible fluid since then Θ is zero. Next, substituting the relation for $\bar{p}$ into the equations for the stresses [Eq. (2.19)], one obtains

$$\sigma_{xx} = -\bar{p} - \frac{2}{3}\mu\Theta + 2\mu\frac{\partial u}{\partial x},$$

$$\sigma_{yy} = -\bar{p} - \frac{2}{3}\mu\Theta + 2\mu\frac{\partial v}{\partial y},$$

$$\sigma_{zz} = -\bar{p} - \frac{2}{3}\mu\Theta + 2\mu\frac{\partial w}{\partial z},$$

$$\tau_{xy} = \tau_{yx} = \mu\left(\frac{\partial u}{\partial y} + \frac{\partial v}{\partial x}\right), \quad (2.20)$$

$$\tau_{xz} = \tau_{zx} = \mu\left(\frac{\partial u}{\partial z} + \frac{\partial w}{\partial x}\right),$$

$$\tau_{yz} = \tau_{zy} = \mu\left(\frac{\partial v}{\partial z} + \frac{\partial w}{\partial y}\right).$$

An inspection of Eqs. (2.20) shows that by introducing the average pressure it has been possible to eliminate both A and λ from the relationships of Eq. (2.19). It is in this form that the expressions for the normal stresses are most frequently used. For convenience the bar over the pressure term is often omitted, and we shall also follow this practice in the subsequent text. One should not forget, however, that the pressure as used in Eqs. (2.20) represents an average and that this average pressure is in general not equal to the pressure as defined in thermodynamics.

The distinction between the thermodynamic pressure and the average pressure is not always carefully made in the literature. This may be explained by the fact that for most fluids under most circumstances the values of p_{th} and $\bar{p}$ are not greatly different, and it may often be quite permissible to substitute one for the other. However, the assumption must always be reexamined for each problem, and special care is advisable when dealing with new types of problems, for example, those that may arise when large velocity and temperature gradients occur. The flow within a shock wave is an example of such a problem.

To become further acquainted with the stress–deformation relationships, it is of interest to consider the very simple flow in which only the velocity component u exists, and ρ is constant. For this simple flow Eqs. (2.20) shows that all shear stresses are zero with the exception of τ_{yx} and τ_{xy}, which are equal to

$$\tau_{xy} = \tau_{yx} = \mu\frac{\partial u}{\partial y}.$$

The expression above is now exactly the one previously called Newton's law of shear, and μ is seen to be the viscosity as defined previously by this relation.

The general stress–strain relationships of Eqs. (2.20) are seen to reduce to Newton's law of shear for this simple flow. Conversely, one may look at the set of relations of Eqs. (2.20) as an extension of Newton's law of shear. Similarly, a "Newtonian" fluid should be redefined more broadly as one that follows the relations given in Eqs. (2.20).

The viscosity μ is more precisely called the first viscosity coefficient, since the stress–deformation relations when given in the form of Eqs. (2.19) (i.e., before the introduction of the concept of the average pressure) also contain the parameter λ. As can be seen from the structure of the equations, λ has the same dimensions as μ, and it is, therefore, appropriately called the second viscosity coefficient. In the development and application of fluid mechanics up to the present time, the second viscosity coefficient plays much less of a role than the first. For the incompressible fluids the term in λ disappears completely and for compressible fluids it is of significance mainly in a few specialized problems such as the analysis of the shock wave structure, which was mentioned previously. Few direct measurements of λ are available. With the use of kinetic theory it can be shown, however, that for a perfect monatomic gas λ is equal to zero.

For our purpose we shall now use the expressions for the stresses as given in Eqs. (2.20) in which the concept of the average pressure has been introduced. By substituting these expressions into the equations of motion as given by Eqs. (2.17), one obtains

$$
\rho \frac{Du}{Dt} = -\frac{\partial p}{\partial x} - \frac{2}{3}\frac{\partial}{\partial x}(\mu\Theta) + 2\frac{\partial}{\partial x}\left(\mu\frac{\partial u}{\partial x}\right) + \frac{\partial}{\partial y}\left\{\mu\left(\frac{\partial u}{\partial y} + \frac{\partial v}{\partial x}\right)\right\}
$$
$$
+ \frac{\partial}{\partial z}\left\{\mu\left(\frac{\partial u}{\partial z} + \frac{\partial w}{\partial x}\right)\right\} + \rho f_x,
$$

$$
\rho \frac{Dv}{Dt} = -\frac{\partial p}{\partial y} - \frac{2}{3}\frac{\partial}{\partial y}(\mu\Theta) + \frac{\partial}{\partial x}\left\{\mu\left(\frac{\partial u}{\partial y} + \frac{\partial v}{\partial x}\right)\right\} + 2\frac{\partial}{\partial y}\left(\mu\frac{\partial v}{\partial y}\right)
$$
$$
+ \frac{\partial}{\partial z}\left\{\mu\left(\frac{\partial v}{\partial z} + \frac{\partial w}{\partial y}\right)\right\} + \rho f_y, \tag{2.21}
$$

$$
\rho \frac{Dw}{Dt} = -\frac{\partial p}{\partial z} - \frac{2}{3}\frac{\partial}{\partial z}(\mu\Theta) + \frac{\partial}{\partial x}\left\{\mu\left(\frac{\partial u}{\partial z} + \frac{\partial w}{\partial x}\right)\right\} + \frac{\partial}{\partial y}\left\{\mu\left(\frac{\partial v}{\partial z} + \frac{\partial w}{\partial y}\right)\right\}
$$
$$
+ 2\frac{\partial}{\partial z}\left(\mu\frac{\partial w}{\partial z}\right) + \rho f_z,
$$

in which the bar over the pressure term has been dropped.

The three equations above are the famous Navier–Stokes equations. These equations play a central role in modern fluid mechanics, and nearly all analytical work involving a viscous fluid is based on them.

As can be seen by reviewing the derivation, the Navier–Stokes equations are founded on Newton's law of motion and on Newton's law of viscous friction in its extended form. The equations so far have not been restricted to either constant density or constant viscosity. They are valid for viscous compressible flow with varying viscosity.

If the viscosity can be assumed constant, the Navier–Stokes equations may be simplified and rearranged to give

$$
\frac{Du}{Dt} = -\frac{1}{\rho}\frac{\partial p}{\partial x} + \nu\left(\frac{\partial^2 u}{\partial x^2} + \frac{\partial^2 u}{\partial y^2} + \frac{\partial^2 u}{\partial z^2}\right) + \frac{1}{3}\nu\frac{\partial\Theta}{\partial x} + f_x,
$$

$$
\frac{Dv}{Dt} = -\frac{1}{\rho}\frac{\partial p}{\partial y} + \nu\left(\frac{\partial^2 v}{\partial x^2} + \frac{\partial^2 v}{\partial y^2} + \frac{\partial^2 v}{\partial z^2}\right) + \frac{1}{3}\nu\frac{\partial\Theta}{\partial y} + f_y, \tag{2.22}
$$

$$
\frac{Dw}{Dt} = -\frac{1}{\rho}\frac{\partial p}{\partial z} + \nu\left(\frac{\partial^2 w}{\partial x^2} + \frac{\partial^2 w}{\partial y^2} + \frac{\partial^2 w}{\partial z^2}\right) + \frac{1}{3}\nu\frac{\partial\Theta}{\partial z} + f_z.
$$

These are conveniently summarized in vector notation as

$$\frac{D\mathbf{v}}{Dt} = -\frac{1}{\rho}\nabla p + \nu\nabla^2\mathbf{v} + \frac{1}{3}\nu\nabla\Theta + \mathbf{f}. \tag{2.23}$$

If the density is constant, Θ will, of course, be zero, and the third term on the right-hand side will disappear. Equation (2.23) is also sometimes written in a different form, which can be obtained by substituting for $\nabla^2\mathbf{v}$ by means of the vector identity $\nabla^2\mathbf{v} = \nabla\Theta - \nabla \times (\nabla \times \mathbf{v})$.

By performing the tedious but straightforward mathematical steps of coordinate transformation, the Navier–Stokes equation as well as the stress–deformation relationships [Eqs. (2.20)] may be expressed in terms of different coordinate systems. In a cylindrical coordinate system with the coordinates r, z, and θ, the Navier–Stokes equation takes the form

r-direction

$$\frac{\partial v_r}{\partial t} + v_r\frac{\partial v_r}{\partial r} + \frac{v_\theta}{r}\frac{\partial v_r}{\partial \theta} + v_z\frac{\partial v_r}{\partial z} - \frac{v_\theta^2}{r}$$

$$= -\frac{1}{\rho}\frac{\partial p}{\partial r} + \frac{1}{3}\nu\frac{\partial\Theta}{\partial r} + \nu\left\{\frac{1}{r}\frac{\partial}{\partial r}\left(r\frac{\partial v_r}{\partial r}\right) + \frac{1}{r^2}\frac{\partial^2 v_r}{\partial\theta^2} + \frac{\partial^2 v_r}{\partial z^2} - \frac{v_r}{r^2} - \frac{2}{r^2}\frac{\partial v_\theta}{\partial\theta}\right\} + f_r,$$

θ-direction

$$\frac{\partial v_\theta}{\partial t} + v_r\frac{\partial v_\theta}{\partial r} + \frac{v_\theta}{r}\frac{\partial v_\theta}{\partial \theta} + v_z\frac{\partial v_\theta}{\partial z} + \frac{v_r v_\theta}{r}$$

$$= -\frac{1}{\rho r}\frac{\partial p}{\partial \theta} + \frac{1}{3}\nu\frac{1}{r}\frac{\partial\Theta}{\partial \theta} + \nu\left\{\frac{1}{r}\frac{\partial}{\partial r}\left(r\frac{\partial v_\theta}{\partial r}\right) + \frac{1}{r^2}\frac{\partial^2 v_\theta}{\partial\theta^2} + \frac{\partial^2 v_\theta}{\partial z^2} - \frac{v_\theta}{r^2} + \frac{2}{r^2}\frac{\partial v_r}{\partial\theta}\right\} + f_\theta,$$

$$\tag{2.24}$$

z-direction

$$\frac{\partial v_z}{\partial t} + v_r\frac{\partial v_z}{\partial r} + \frac{v_\theta}{r}\frac{\partial v_z}{\partial \theta} + v_z\frac{\partial v_z}{\partial z}$$

$$= -\frac{1}{\rho}\frac{\partial p}{\partial z} + \frac{1}{3}\nu\frac{\partial\Theta}{\partial z} + \nu\left\{\frac{1}{r}\frac{\partial}{\partial r}\left(r\frac{\partial v_z}{\partial r}\right) + \frac{1}{r^2}\frac{\partial^2 v_z}{\partial\theta^2} + \frac{\partial^2 v_z}{\partial z^2}\right\} + f_z,$$

and the stresses are given by

$$\sigma_{rr} = -p - \frac{2}{3}\mu\Theta + 2\mu\frac{\partial v_r}{\partial r},$$

$$\sigma_{\theta\theta} = -p - \frac{2}{3}\mu\Theta + 2\mu\left\{\frac{1}{r}\frac{\partial v_\theta}{\partial\theta} + \frac{v_r}{r}\right\},$$

$$\sigma_{zz} = -p - \frac{2}{3}\mu\Theta + 2\mu\frac{\partial v_z}{\partial z},$$

$$\tag{2.25}$$

$$\tau_{r\theta} = \mu\left\{r\frac{\partial}{\partial r}\left(\frac{v_\theta}{r}\right) + \frac{1}{r}\frac{\partial v_r}{\partial\theta}\right\},$$

$$\tau_{rz} = \mu\left\{\frac{\partial v_z}{\partial r} + \frac{\partial v_r}{\partial z}\right\},$$

$$\tau_{\theta z} = \mu\left\{\frac{\partial v_\theta}{\partial z} + \frac{1}{r}\frac{\partial v_z}{\partial\theta}\right\}.$$

2.7 Discussion of the Newtonian Law of Friction

The extended form of Newton's law of shear was given in Eq. (2.19) without a derivation. Now we discuss some of the features of this set of equations to indicate how the original assumptions have been incorporated and we attempt to give some physical interpretation to the grouping of the terms in the equations.

First, let us consider the assumptions involving the equality of the shear stresses $\tau_{xy} = \tau_{yx}$, and so on. For this purpose consider an element of fluid as shown in Fig. 2.12. Let us concentrate our attention on the shear stresses that contribute to a torque about an axis through the center of the element and parallel to the z-axis. The torque produced by these forces about the center of gravity of the element can be directly computed and is seen to be

$$T = \tau_{xy} \, dx \, dz \, dy - \tau_{yx} \, dy \, dz \, dx$$

or

$$T = (\tau_{xy} - \tau_{yx}) \, dx \, dy \, dz.$$

Applying Newton's law, the torque may be equated to the product of angular acceleration, $\ddot{\alpha}$, and moment of inertia, both to be taken in respect to the previously specified axis of rotation. We then obtain

$$(\tau_{xy} - \tau_{yx}) \, dx \, dy \, dz = \frac{\rho}{12} \, dx \, dy \, dz(dx^2 + dy^2)\ddot{\alpha}_z$$

or

$$(\tau_{xy} - \tau_{yx}) = \frac{\rho}{12} (dx^2 + dy^2)\ddot{\alpha}_z.$$

Recalling now that dx and dy are infinitesimal quantities, it must be concluded from the equation above that the angular acceleration of every infinitesimal element would tend toward infinity as dx and dy approach zero unless $\tau_{xy} = \tau_{yx}$. It is therefore concluded that τ_{xy} must equal τ_{yx}, and this fact is reflected in the expressions for the shear stresses. The argument for the shear stresses in the other two directions is, of course, identical.

Next, let us examine the type of relationship that might exist between the velocity derivatives and the various stresses. In general, a fluid element is acted upon by surface shears as well as by normal stresses. By a simple geometrical reorientation of the coordinate axes, it is always possible, however to determine one coordinate

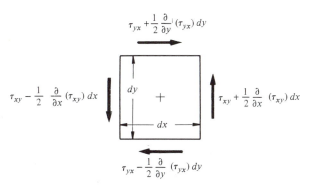

FIG. 2-12 Shear forces contributing to the torque on a fluid element.

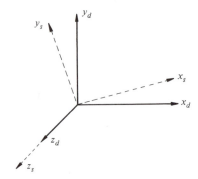

FIG. 2.13 Hypothetical position of the principal axes of stress in respect to the principal axes of deformation.

system in which all shear stresses, vanish, leaving only the normal stresses. This set of axes is called the *principal axes.*

In a completely analogous way a second set of principal axes may be found in respect to which the element possesses only velocity gradients in the direction of the axes (i.e., $\partial u/\partial x$, $\partial v/\partial y$, and $\partial w/\partial z$). We now turn to the assumption by which it was postulated that the fluid was isotropic. For such a fluid it may be stated that the two sets of principal axes must be the same. This can be illustrated by an example. Consider a fluid element undergoing deformation in such a way that the principal axes of deformation are oriented as shown in Fig. 2.13. We shall designate these axes as x_d, y_d, and z_d. Let us further assume for simplicity that the deformation along the y_d- and z_d-axes is equal to zero. Let us postulate now for a moment that the resulting principal stress axes (x_s, y_s, and z_s) are oriented as indicated by the dashed lines in Fig. 2.13. One then must conclude that the deformation in the x-direction causes a stress pattern which would be unsymmetrical about the x_d-axis. This, however, could only be explained by the fluid having directional properties, a conclusion which would violate the concept of isotropy. On the basis of arguments of this type, it can be shown that for an isotropic fluid the principal axes of deformation must coincide with the principal axes of stress.

Examining an element in terms of its principal axes simplifies the description of the state of the element greatly. However, in no way does it restrict the generality of our considerations, as principal axes can always be found. Using now the further assumptions that the relations between stress and deformation are linear and that the stresses are independent of the explicit values of the coordinates x, y, z, we can deduce that the stresses in the three principal directions must be of the form

$$\sigma_{xx} = c_1 \frac{\partial u}{\partial x} + c_2 \frac{\partial v}{\partial y} + c_3 \frac{\partial w}{\partial z} + A,$$

$$\sigma_{yy} = c_1 \frac{\partial v}{\partial y} + c_2 \frac{\partial u}{\partial x} + c_3 \frac{\partial w}{\partial z} + A,$$

$$\sigma_{zz} = c_1 \frac{\partial w}{\partial z} + c_2 \frac{\partial v}{\partial y} + c_3 \frac{\partial u}{\partial x} + A.$$

The same constants appear in each equation, again because the fluid is isotropic, and it must consequently be possible to designate arbitrarily any of the three axes as the x-axis without influencing the results. The quantity A is unchanged by any transformation of axis and $(-A)$ represents the thermodynamic pressure as dis-

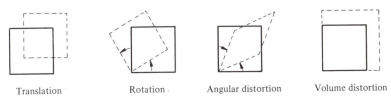

Translation   Rotation   Angular distortion   Volume distortion

FIG. 2.14   Basic components of the motion of a fluid element.

cussed before. As to the other constants one more simplification is possible, again on the basis of isotropy. Examining σ_{xx} first, it can be said that deformations in the y-direction and in the z-direction must influence σ_{xx} in the same way. Both the y- and z-directions are normal to the x-axis, and isotropy requires that one not be affected differently from the other. The consequence of this argument then is that c_2 and c_3 must be equal. Aside from the thermodynamic pressure, this then leaves two arbitrary constants in the relationship between the stresses and deformations. This is exactly the number of constants that appear in Newton's extended law of shear as given in Eq. (2.19). The appearance of just two constants cannot, therefore, be considered purely accidental, and similarly the equality of the shear stresses τ_{xy} and τ_{yx}, and so on, must be considered an essential characteristic of the fluid motion.

Finally, let us examine the physical meaning of the groupings of differentials which appear in Eq. (2.19). For this purpose we have to study briefly the kinematics of the flow field. Let us consider the possible motions of an element of the fluid, for example, the cube shown in Fig. 2.10. As the fluid particles comprising the original cube move about in the flow, it is clear that the element can be translated, rotated, and distorted. These possibilities are shown in Fig. 2.14. A mere translation of a fluid flow cannot result in a stress, since by a change of coordinate system the element can be brought to rest. Similarly, a uniform rotation or solid body rotation of a fluid cannot result in a stress since, with respect to the rotating system, there is no relative motion. Thus it appears that the stresses owe their existence to the distortion or deformation of the element. It is convenient to consider two separate aspects of this deformation. The first of these is the angular distortion of an element of constant volume as in an incompressible fluid. This will be designated as a *shear strain*; this strain results in a change in the angle between two adjacent sides of an element. The other aspect is the change in volume. Because of the importance of the various components of the fluid motion, we shall calculate the rotation, shear strain, and dilation of the two-dimensional volume element shown in Fig. 2.15. We need to examine only the motions of two adjacent sides of the element of initial length dx and dy over a time increment dt.

Let us consider first the angular deformation of the two sides which results in the angular displacements α and β. This deformation may be subdivided into two parts. The first consists of an angular motion of both sides through the same angle $\frac{1}{2}(\alpha - \beta)$ and in the same direction (counterclockwise in Fig. 2.14). This motion of the two sides corresponds to the motion that would have been performed by the two sides of a solid body rotating through an angle $d\theta = (\alpha - \beta)/2$. This angle is therefore called the angle of rotation of the fluid element. The second part of the angular motion consists of an angular displacement of $d\varepsilon = \frac{1}{2}(\beta + \alpha)$, in which the two sides move in opposite directions (in Fig. 2.15 the side dy in a clockwise direction and the side dx in a counterclockwise direction). This motion corresponds to a uniform angular distortion, and $\frac{1}{2}(\beta + \alpha)$ may therefore be called the *angle of distortion*. The rotation through an angle $\frac{1}{2}(\alpha - \beta)$ followed by the distortion in the

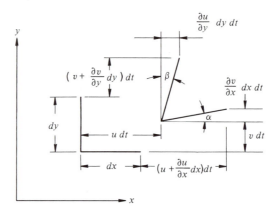

FIG. 2.15 Change in length and position of the sides of a two-dimensional fluid element in motion.

amount $\frac{1}{2}(\beta + \alpha)$ leave the two sides in an angular position α and β, respectively, as indicated in Fig. 2.15.

We shall next assume α and β to be small compared to unity and we shall express these angles in terms of the quantities of Fig. 2.15. It follows that

$$\alpha = \frac{\partial v}{\partial x}\, dx\, dt\, \frac{1}{dx}$$

and

$$\beta = \frac{\partial u}{\partial y}\, dy\, dt\, \frac{1}{dy}.$$

The average rate of rotation in the positive or counterclockwise sense defined as

$$\Omega_z = \frac{1}{2}\frac{\alpha - \beta}{dt}$$

may then be found to be equal to

$$\Omega_z = \frac{1}{2}\left(\frac{\partial v}{\partial x} - \frac{\partial u}{\partial y}\right). \tag{2.26a}$$

The rotation Ω_z takes place about an axis parallel to the z-axis. This has been indicated by the subscript z. In a similar fashion we can show that

$$\Omega_x = \frac{1}{2}\left(\frac{\partial w}{\partial y} - \frac{\partial v}{\partial z}\right) \tag{2.26b}$$

and

$$\Omega_y = \frac{1}{2}\left(\frac{\partial u}{\partial z} - \frac{\partial w}{\partial x}\right). \tag{2.26c}$$

The terms Ω_x, Ω_y, Ω_z are the components of a vector that is recognized to be one-half of the curl of the velocity vector, that is,

$$\mathbf{\Omega} = \mathbf{i}\Omega_x + \mathbf{j}\Omega_y + \mathbf{k}\Omega_z = \tfrac{1}{2}\nabla \times \mathbf{v}.$$

The vector $2\mathbf{\Omega} = \boldsymbol{\omega} = \nabla \times \mathbf{v}$ is called the *vorticity*.

We now return to the angular distortion. The total angular distortion in the xy-plane associated with the element in Fig. 2.15 was shown to be $\frac{1}{2}(\alpha + \beta)$. Following steps similar to the above, the rate of angular distortion in the xy-plane is found to be

$$\xi_z = \frac{1}{2}\left(\frac{\partial v}{\partial x} + \frac{\partial u}{\partial y}\right).$$

Analogously, the angular distortion in the yz-plane is

$$\xi_y = \frac{1}{2}\left(\frac{\partial w}{\partial x} + \frac{\partial u}{\partial z}\right)$$

and in the yz-plane it is given by

$$\xi_x = \frac{1}{2}\left(\frac{\partial v}{\partial z} + \frac{\partial w}{\partial y}\right).$$

The remaining deformations are associated with linear and volumetric strains resulting from an increase in the volume of the element. For example, let us consider one-dimensional motion in the x-direction. As a result of a variation in velocity in the direction of motion an element of initial length l is stretched in the time increment dt to the new length $l + dl$, where $dl = l(\partial u/dx)\,dt$. The rate of strain in this direction is therefore

$$\frac{1}{l}\frac{dl}{dt} = \frac{\partial u}{\partial x}.$$

Similarly, the linear strain rates in the direction of the y- and z-axes are $\partial v/\partial y$, $\partial w/\partial z$, respectively. Finally, it is easy to show that the rate of increase in volume per unit volume of the element is

$$\nabla \cdot \mathbf{v} = \frac{\partial u}{\partial x} + \frac{\partial v}{\partial y} + \frac{\partial w}{\partial z} = \Theta.$$

The quantity Θ, the divergence of the velocity vector, is frequently called the dilation. Apart from a translation, the state of the motion is completely characterized by the rotation vector and the various deformation components above. It is plausible, therefore, that the stresses should be expressible in terms of the dilation, the components of the angular distortion, and the components of the rates of strain. This is precisely the form in which Eq. (2.19) has been written.

2.8 Boundary Conditions

Having presented the differential equations for the motion of a fluid, let us say a word about the necessary boundary conditions. By first considering a solid surface, it is clear that the velocity component normal to the surface must be equal to that of the surface itself since otherwise there would be flow through the surface. The conditions that should be imposed on the tangential component of the velocity at a surface are not as evident and have, in fact, been the subject of some controversy in earlier years. On the basis of many experiments with liquids and gases at ordinary pressures and temperatures, the conclusion has been drawn that the fluid adheres to the surface. The tangential component as well as the normal component of the fluid at a surface, therefore, are equal to the corresponding components of the surface itself. In other words, there is no relative motion or "slip" between the fluid and the solid. If the solid is at rest, the fluid, of course, will also have zero motion.

There is an exception to the "no-slip" condition that arises whenever the mean free path of the molecules of the fluid becomes appreciable when compared to an important physical dimension of the body over which or through which the flow takes place. In this event the difference in tangential velocities between the fluid and solid is not zero but rather is proportional to the surface shear stress. These considerations are only important in problems involving the flow of highly rarified gases, as might occur, for example, at extreme altitudes.

At a free surface or interface between two immiscible fluids, the shear stress must be continuous. If an appreciable surface tension results from the presence of the interface, however, the normal component of the stress vector is discontinuous by an amount

$$\Delta p = \sigma\left(\frac{1}{r_1} + \frac{1}{r_2}\right),$$

as discussed in Chapter 1.

We shall now consider a few elementary examples of real fluid flows.

EXAMPLE 2.6

Poiseuille Flow in a Channel. We consider the steady (Section 2.2) flow of a viscous, incompressible fluid in an infinitely long, two-dimensional stationary channel of breadth h with no body forces present. The flow is everywhere parallel to the x-axis and the y-axis is placed at the bottom of the channel. The velocity profile and shear stress distribution are to be determined.

Example 2.6

Solution: With $v = w = 0$ the equations of motion [Eqs. (2.22)] become

(a) $$0 = -\frac{1}{\rho}\frac{\partial p}{\partial x} + v\frac{\partial^2 u}{\partial y^2},$$

(b) $$0 = -\frac{1}{\rho}\frac{\partial p}{\partial y} = -\frac{1}{\rho}\frac{\partial p}{\partial z},$$

in the x-, y-, and z-directions, respectively, and the continuity equation [Eq. (2.1)] is

(c) $$\frac{\partial u}{\partial x} = 0.$$

In addition, we have the boundary conditions $u = 0$ at $y = 0, h$. From Eq. (c) it follows that u can be a function of y only. Furthermore, Eq. (b) shows that p cannot depend on y or z, but only on x. Finally, from (a) it is seen that

(d) $$\frac{dp}{dx} = \mu\frac{d^2 u}{dy^2}.$$

The right-hand side of this equation can only depend on y and the left side only on x. These conflicting requirements can only be satisfied if both sides are equal to the same constant. Thus the pressure gradient must be a constant for this flow. Equation (d) is now easily integrated, and we obtain

$$u = \frac{1}{2\mu} \frac{dp}{dx} y^2 + Ay + B,$$

where A and B are constants. When these are evaluated with the use of the boundary conditions, we obtain

$$u = -\frac{1}{2\mu} \frac{dp}{dx} (hy - y^2).$$

Thus the velocity profile is *parabolic* with the maximum velocity at the center of the channel. The shear stress is

$$\tau_{yx} = \mu \frac{du}{dy} = -\frac{1}{2} \frac{dp}{dx} (h - 2y),$$

so that on the upper surface the stress $\tau_0 = (h/2)\, dp/dx$ acts on the fluid. Similarly, on the lower surface the stress on the fluid is $\tau_0 = -(h/2)\, dp/dx$ (recall the convention of positive shear stresses). Of course, these stresses could have been determined directly from statics since there is no acceleration of the fluid.

The existence of the motion in this case depends on the pressure gradient dp/dx. The volume flow rate per unit depth of channel is

$$q = \int_0^h u\, dy = -\frac{h^3}{12\mu} \frac{dp}{dx},$$

and we see that a discharge in the positive x-direction requires a negative pressure gradient or a pressure that decreases in the direction of flow. The pressure drop in a channel of length l can be expressed in terms of the average velocity $q/h = V$ as follows:

$$\Delta p = \frac{12\mu l}{h^2} V = \frac{24v}{Vh} \left(\frac{l}{h}\right) \left(\rho \frac{V^2}{2}\right).$$

The first term on the right-hand side may be called the *friction factor f*. For the special case treated, f is inversely proportional to Vh/v, a parameter which was said earlier to appear frequently in the treatment of real fluid flows, and which was called the Reynolds number. The second term shows that the pressure decrease is proportional to the length expressed in terms of channel heights, and the last term on the right-hand side, $\frac{1}{2}\rho V^2$, which has the units of pressure, is called the *dynamic pressure*.

It may be pointed out that in solving Example 2.6 it was assumed that the flow was steady. This assumption is not justified if the Reynolds number is above a certain critical value. In that case a type of flow known as *turbulent* flow will take place; turbulent flow is discussed in Chapter 7.

EXAMPLE 2.7

Impulsive Motion of an Infinite Flat Plate. We consider now a somewhat more complicated problem. A stationary plate of infinite length is immersed in a still fluid of viscosity μ and constant density ρ. At the time $t = 0$, the plate is made to move at the constant velocity U_0 in its own plane and the subsequent motion of the surrounding fluid $u(y, t)$ is to be determined. The quantities in parentheses indicate the variable upon which u depends. Since the plate is of infinite length, no quantities change with x.

Solution: The boundary conditions are

$$u(0, t) = U_0 \text{ for } t > 0 \text{ and } v = 0 \text{ at } y = 0 \text{ (i.e., the no-slip condition).}$$

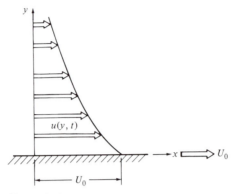

Example 2.7

The initial conditions are

$$u(y, 0) = 0 \qquad \text{for } t = 0.$$

From the continuity equation we have

$$\frac{\partial u}{\partial x} + \frac{\partial v}{\partial y} = 0.$$

Since $\partial u/\partial x = 0$, it follows that $\partial v/\partial y = 0$; therefore, v is also zero everywhere as it is zero at the boundary. We shall take the pressure to be constant throughout the fluid. The equations of motion (2.22) now become

(a)
$$\frac{\partial u}{\partial t} = v \frac{\partial^2 u}{\partial y^2}.$$

This partial differential equation is also the equation that governs the transient conduction of heat in a solid and the diffusion of one gas into another, and it is commonly known as the *diffusion equation*. Many formal methods are available for the solution of this equation. We notice, however, that there is no typical physical length in the problem formulated above, so that it might be suspected that u will be a function of some combination of y and t. That is, we shall look for a new independent variable, z, say, so that u becomes a function of z rather than y and t separately.

If we should be able to succeed in finding such a variable, we shall also be in a position to convert the partial differential equation to a total differential equation, which can then be solved by standard methods. A variable that accomplishes this modification for the equation above is

(b)
$$z = \frac{y}{\sqrt{2vt}},$$

where the factor $2v$ has been included only to simplify some of the coefficients that are going to occur. After performing the required differentiations and substitutions, Eq. (a) becomes

(c)
$$-z\frac{du}{dz} = \frac{d^2u}{dz^2},$$

which is readily integrated to

(d)
$$u(y, t) = c_1 \int_0^{y/\sqrt{2vt}} e^{-z^2/2} \, dz + c_2,$$

where c_1 and c_2 are constants. The boundary condition $u(0, t) = U_0$ requires $c_2 = U_0$, whereas the initial condition $u(y, 0) = 0$ requires that

$$0 = c_1 \int_0^\infty e^{-z^2/2} \, dz + U_0.$$

The infinite integral has the value $\sqrt{\pi/2}$, hence $c_1 = -U_0\sqrt{2/\pi}$, so that finally, we have

$$u(y, t) = U_0 \left\{ 1 - \sqrt{\frac{2}{\pi}} \int_0^{y/\sqrt{2vt}} e^{-z^2/2} \, dz \right\}$$

as the sought-for solution. By introducing a new variable, say ξ, where $\xi = z/2$, the solution may also be written as

$$u = U_0 \left\{ 1 - \frac{2}{\sqrt{\pi}} \int_0^{y/\sqrt{4vt}} e^{-\xi^2} \, d\xi \right\}.$$

In this form the quantity in braces may be identified with the so-called *complementary error function*, which is a tabulated function.† From tables for this function it is seen that u is reduced to 1 percent of U_0 when $y/\sqrt{vt} = 4$ or when $y = 4\sqrt{vt}$.

This interesting result shows that the effect of viscosity diffuses from the solid boundary into the fluid and that the distance at which a given effect occurs varies with the square root of vt. The elapsed time t may be expressed in terms of the distance x through which the plate travels (i.e., $t = x/U_0$), so that we can also write

$$y \sim x\sqrt{\frac{v}{U_0 x}}.$$

The diffusion of viscous effects from solid surfaces into the flow is a recurring feature of real fluid flows.

In closing this chapter, we note that we have discussed relationships between six quantities consisting of the three velocity components, pressure, density, and viscosity. So far we have developed four equations relating these variables, namely, the three equations of motion and the continuity equation. These equations are sufficient to describe the flow if the density and viscosity are constant, for then there are only four unknowns. If the density is not constant but happens to be a known function of the pressure alone, the same statement can still be made. In general, however, the density and viscosity are functions of pressure and temperature as

† For numerical values, see H. B. Dwight, *Table of Integrals and Other Mathematical Data*, Macmillan, New York, 1961.

described by the equation of state for the particular substance. A fifth relation connecting the temperature to the other variables then becomes necessary as well as two additional relations describing the dependence of density and viscosity on temperature and pressure. The additional physical law that has to be added in this case is the first law of thermodynamics, which relates the mechanical work performed by an element of the fluid to the change in internal or thermal energy of the substance and the amount of heat transferred to the element. The resulting differential equation, which is usually called the *energy equation*, is essential in discussing problems involving heat transfer or high-speed flows in which density and viscosity changes have to be taken into account. For a brief discussion of the thermodynamic aspects of fluid mechanics, see Chapter 9 and Appendix 4.

It should be noted that since we can consider the motion of a fluid as completely described through the equations outlined above, it is possible in principle to compute the flow directly by numerical methods. The essential idea in numerical computation of a flow is to consider once again an infinitesimal element of fluid and its relationship to other adjacent fluid elements. If the flow and properties of all the surrounding elements are specified, the fundamental laws already outlined can be used to find the flow and properties of the element itself. In most practical problems, conditions are specified only on a distant surface or boundary, so that the flow must be divided into many such small elements comprising the entire fluid. The equations are then solved for *all* the elements at the same time. This usually requires a large number of elements for sufficient accuracy, and therefore a correspondingly larger number of simultaneous computations. This was a formidable task for the early workers in fluid mechanics mentioned in Chapter 1.

The continuing development of methods for performing numerical approximations to differential equations, together with the advances in computing capability, make it possible now to solve Navier–Stokes equations for certain types of simple flows on a personal computer (Reference 1). The reader should be aware that despite the apparent ease with which such results are obtained, a great deal of understanding of the flow is still necessary *beforehand* to ensure that the proper numerical methods and approximations have been used, and that sufficient accuracy and precision in the computations has been maintained. In particular, methods used with success in one specific problem may not be as suitable to another, as yet unsolved, problem. Care should always be taken when using results of a flow computed by a method unknown to the user.

It seems clear that exciting achievements in the computation of fluid flows will continue in the future. The full exposition of numerical methods in current use would require detailed study beyond the intent of our present effort. The interested reader should consult the references at the end of this chapter for more information.

PROBLEMS

2.58 Liquid of density ρ and viscosity μ flows down a wide plate inclined at the angle θ to the horizontal under the influence of gravity. The depth of the liquid normal to the plate is h. The flow is steady and everywhere parallel to the plate. The viscosity of the air in contact with the upper surface of the liquid may be neglected. Determine the velocity profile parallel to the plate, the shear stress at the wall, and the average velocity.

2.59 A uniformly thick bed of mud rests on a long incline, as shown. The mud behaves as an ideal plastic fluid (see Problem 1.2), for which $\tau = \tau_c + \mu(du/dy)$.

(a) Using the equation of motion, show that for either the static $(V = 0)$ or steady uniform flow cases $(\partial/\partial x = 0)$, the shear stress is

$$\tau = (\rho g \sin \theta)(H - y).$$

(b) Find the critical slope θ_{cr} for flow to begin.
(c) If the slope exceeds θ_{cr}, a uniform plug of depth h will move at the speed V. Show that

$$V = \frac{\rho g \sin \theta}{2\mu} H^2 \left(1 - \frac{\sin \theta_{cr}}{\sin \theta}\right)^2; \qquad h = H \frac{\sin \theta_{cr}}{\sin \theta}.$$

(State all assumptions and boundary conditions.)
(d) Find the total mud flow rate Q.

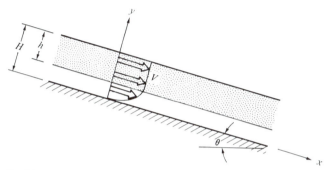

Problem 2.59

2.60 (a) A viscous fluid of viscosity μ is made to flow in concentric circles about the origin with the velocity distribution $v_\theta(r)$. A typical element is shown in the accompanying sketch. Using this element, determine the shear stress τ_{yx}, which is the same as $\tau_{\theta r}$ in polar coordinates when y is set equal to zero.
(b) Determine the same for the purely radial flow $v_r(\theta)$ and also calculate σ_{xx} or σ_{rr} when y is set equal to zero.

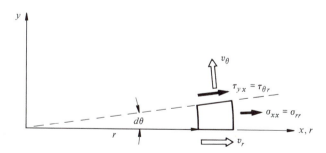

Problem 2.60

2.61 By making use of the vector form of the Navier–Stokes equations and the properties of the vector operators of this equation, derive Eq. (2.24) in cylindrical polar coordinates (r, θ, z) with corresponding velocity components v_r, v_θ, v_z.

2.62 Consider the steady laminar incompressible flow between two parallel plates as shown below. The upper plate moves at velocity U_0 to the right and the lower plate is stationary. The pressure gradient is zero. The lower half of the region between the plates (i.e., $0 \le y \le h/2$) is filled with fluid with density ρ_1 and viscosity μ_1, and the upper half ($h/2 \le y \le h$) is filled with fluid of density ρ_2 and viscosity μ_2.

(a) State the condition that the shear stress must satisfy for $0 < y < h$.

(b) State the conditions that must be satisfied by the fluid velocity at the walls and at the interface of the two fluids.

(c) Obtain the velocity profile in each of the two regions and sketch the result for $\mu_1 > \mu_2$.

(d) Calculate the shear stress at the lower wall.

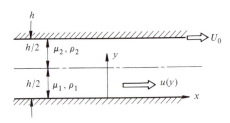

Problem 2.62

2.63 A cylinder of radius a rotates with angular velocity Ω concentrically inside a larger stationary one of radius b. Fluid of viscosity μ and constant density ρ fills the gap between the cylinders. Assume that the flow is steady and the motion is purely circular so that no quantities change with angular position. Obtain the velocity distribution between the cylinders and the shear stress on the two cylinders in two ways: (a) by setting up the equations of equilibrium for an annulus consisting of the inner cylinder and an outer one of arbitrary radius r and applying the results of Problem 2.60, and (b) by solution of the Navier–Stokes equations in cylindrical coordinates [Eqs. (2.24)].

2.64 A flow is described by the equation $v_\theta = Ar + B/r$.

(a) Calculate the average angular rate of rotation of the fluid elements as a function of the radius r.

(b) Calculate the rate of shearing deformation. A and B are constants.

(c) If the fluid has viscosity coefficients λ, μ, what would the stresses be at $r = 1$?

2.65 A viscous flow of viscosity μ is given by the stream function

$$\psi = -Axy.$$

(a) Sketch the streamlines *accurately* near the point $x = y = 0$. Take A to be positive.

(b) Assuming that there is no body force **f**, integrate the Navier–Stokes equations for this special case and obtain a relationship between the pressure, A, x, and y.

(c) If the average pressure at the point $x = y = 0$ is p_0, what is the normal stress σ_{xx} at the same point?

(d) Assume that the line $x = 0$ is a *solid* surface. Can the stream function above represent correctly the flow near $x = y = 0$? Explain very briefly.

2.66 Consider again the planar flow of an incompressible fluid given by

$$\psi = Axyt.$$

Assume that the flow is inviscid and that there are no body forces. Find the pressure, p, within the flow as a function of A, ρ (fluid density), x, y, and t by integrating the equations of motion. (The result contains an arbitrary constant which could be evaluated by assuming that the pressure at any one point in the flow—for example, at the origin—is known.)

2.67 An incompressible liquid is contained in the annulus between an inner rotating cylinder of radius r_i and an outer stationary one. The outer radius r_o is much larger than r_i (r_o may essentially be taken to be infinite). The inner cylinder rotates at a constant angular speed ω, so that the surface velocity is $V_{\theta i} = \omega r_i$. The velocity at $r = r_o$ may be taken to be zero. The flow is steady, the components v_z and v_r are assumed to be zero, and all derivatives with respect to θ are also zero. The flow is laminar and the viscosity is constant.
(a) Determine the velocity v_θ as a function of the radius r.
(b) What is the torque (surface force times radius) required to turn the cylinder?

2.68 Consider the two equations of motion for the planar flow of an incompressible, frictionless fluid under the action of conservative body forces. By eliminating the pressure from these two equations and using the equation of continuity ($\partial u/\partial x + \partial v/\partial y = 0$), show that

$$\frac{\partial \omega}{\partial t} + u\frac{\partial \omega}{\partial x} + v\frac{\partial \omega}{\partial y} = 0,$$

where ω is the magnitude of the vorticity ($\omega = \partial v/\partial x - \partial u/\partial y$). What can you conclude about the vorticity of a particular element of fluid in such a flow?

2.69 Beginning with the Navier–Stokes equations in vector form [Eq. (2.23)], show that the results of Problem 2.68 can be extended to three dimensions for an incompressible fluid (with f a conservative body force) as

$$\frac{D\Omega}{Dt} = \nu\nabla^2\Omega.$$

Hint: Perform the $(\nabla\times)$ operation to Eq. (2.23), and note that it is interchangeable with the time derivative d/dt.

2.70 Consider the flow as in Example 2.7 again; however, now the plate speed is given by

$$U(t) = U'e^{Kt},$$

where U' and K are constants. By assuming that the velocity $u(y, t)$ can be expressed as a product of two functions $f(y)$ and $g(t)$, find the shear stress that must be applied to the plate to achieve this motion. State your answer in terms of K, U', ρ, μ, and t.

2.71 Bubbles are seen to form in soda pop when the pressure is released on the bottle. The bubbles expand with time, and it is often assumed that their growth is purely *spherical* (i.e., the motion is *radial*). The radial velocity of the surface of the bubble is $V(t)$ and the radius of the bubble is R. R is a function of time and (by definition) $V(t) = dR/dt$.

(a) Assuming the surrounding liquid to be incompressible, determine the radial velocity v_r at a radius $r > R$.

(b) Assuming that there is a surface tension σ of the interface and a viscosity μ of the liquid (the viscosity of the gas inside may be neglected), calculate the pressure p_b *inside* the bubble in terms of the pressure p in the liquid just outside the bubble, the radius R, the viscosity μ, and V.

Hint: Consider the forces acting across a small section of the bubble wall.

REFERENCES

1. A. A. Busnaina, *General Flow, Fluid Flow Simulation Software*, Kern International, Pembroke, Mass., 1986.
2. D. A. Anderson, J. C. Tannehill, and R. H. Pletcher, *Computational Fluid Mechanics and Heat Transfer*, McGraw-Hill, New York, 1984.
3. A. K. Gupta and D. G. Lilley, *Flowfield Modeling and Diagnostics*, Abacus Press, Tunbridge Wells, Kent, England, 1985.

3

The Bernoulli Equation

3.1 Introduction

The equations of motion were developed in Chapter 2; as is readily apparent, these equations are very complicated. A somewhat similar situation occurs in the motion of mass particles studied in mechanics. In this case under certain circumstances the equation of motion of a particle can be integrated to result in a statement about the kinetic energy and the potential energy of the particle. The resulting equation is a limited form of the energy principle stemming from the more general first law of thermodynamics. A completely equivalent situation occurs in the flow of fluids. In the sections to follow, the fluid equations of motion are integrated in space. At first these equations are integrated only along a particular line in space. Later, as we shall see, fewer restrictions are placed on this integration, provided that the flow has certain properties. In all cases an "energy" equation results, which is most useful in physically understanding the flow and its application to engineering. Alternatively, we could use the first law directly to obtain these relations for one-dimensional flows; this approach is most useful for compressible fluid flow, and for that reason we defer a full discussion of the energy equation to Section 9.4. Most of the material of the present chapter is therefore restricted to incompressible fluids.

3.2 A Simple Form of the Bernoulli Equation

Even without the complication of viscosity the equations of motion are very complex and are not capable of solution except in special cases. The reason for this difficulty is that they are nonlinear in the terms $u\, \partial u/\partial x$, and so on. However, under certain restrictions the equations of motion can be integrated once. As a first example let us consider the steady two-dimensional flow of an incompressible, inviscid fluid in the absence of body forces. The Eulerian equations [Eqs. (2.10)] for this case are

$$u\frac{\partial u}{\partial x} + v\frac{\partial u}{\partial y} = -\frac{1}{\rho}\frac{\partial p}{\partial x}$$

$$u\frac{\partial v}{\partial x} + v\frac{\partial v}{\partial y} = -\frac{1}{\rho}\frac{\partial p}{\partial y}.$$

This set of equations is readily integrated along a streamline. To this end we multiply the first equation of dx and the second by dy, where dx and dy represent the projections on the x and y axes, respectively, of an element of the streamline. By virtue of the definition of a streamline [Eq. (2.4)] the elements dx and dy are related to the velocity components by the equation

$$\frac{dy}{dx} = \frac{v}{u}.$$

If v in the first equation above and u in the second are eliminated by means of this relation and if the two equations are added, we obtain

$$\frac{1}{2}\frac{\partial}{\partial x}(u^2 + v^2)\,dx + \frac{1}{2}\frac{\partial}{\partial y}(u^2 + v^2)\,dy = -\frac{1}{\rho}\left(\frac{\partial p}{\partial x}\,dx + \frac{\partial p}{\partial y}\,dy\right).$$

Both the left- and right-hand sides are seen to be perfect differentials. Furthermore, the magnitude of the velocity vector is

$$V^2 = u^2 + v^2,$$

so that we can write

$$d\left(\frac{V^2}{2} + \frac{p}{\rho}\right) = 0, \tag{3.1}$$

since ρ is constant. Equation (3.1) can now be integrated to obtain

$$\frac{V^2}{2} + \frac{p}{\rho} = \text{const} \tag{3.2a}$$

or

$$\frac{V_1^2}{2} + \frac{p_1}{\rho} = \frac{V_2^2}{2} + \frac{p_2}{\rho}, \tag{3.2b}$$

where the subscripts 1 and 2 refer to two points *on the same streamline*. Equation (3.2) is also valid in three dimensions, where $V^2 = u^2 + v^2 + w^2$.

This result establishes a direct connection between the pressure and velocity, provided that the latter is known and shows that the velocity is greatest where the pressure is least. Equation (3.2b) is the simplest form of the Bernoulli equation, named in honor of Daniel Bernoulli (1700–1782). The integrals to be developed in the following sections of this chapter may be regarded as extended forms of the Bernoulli equation.

EXAMPLE 3.1

Consider the flow of a liquid through the tube shown below. Let us determine the pressure difference between points 1 and 2 as a function of the flow rate Q and density ρ. The velocity will be assumed constant over the cross sections, and the pressure will also be assumed uniform over each of these two areas A_1 and A_2. It is also assumed that there is no viscosity.

Solution: Since the density is constant, the continuity equation requires that

$$V_1 A_1 = V_2 A_2 = Q,$$

where Q is the volume flow rate.

The Bernoulli equation [Eq. (3.2b)] is given by

$$\frac{V_1^2}{2} + \frac{p_1}{\rho} = \frac{V_2^2}{2} + \frac{p_2}{\rho}.$$

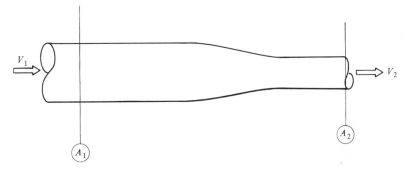

Example 3.1

Eliminating V_1 and solving for the pressure difference, we obtain

$$p_1 - p_2 = \rho \frac{V_2^2}{2}\left[1 - \left(\frac{A_2}{A_1}\right)^2\right] = \frac{\rho}{2}\frac{Q^2}{A_2^2}\left[1 - \left(\frac{A_2}{A_1}\right)^2\right].$$

Thus the pressure difference is a function of the flow rate, the density, and the areas. A tube of the type considered in this example is frequently used as a device to measure flow rates and it is called a *Venturi tube*. However, for a real fluid that has viscosity the flow rate calculated for a given pressure difference must be multiplied by a correction coefficient. Fortunately, this coefficient for many technical applications differs only slightly from unity.

3.3 Conservative Body Forces

We shall next include in the Eulerian equations a special body force having the property that the force can be expressed as the gradient of a scalar function. Thus

$$f_x = \frac{\partial U}{\partial x},$$

$$f_y = \frac{\partial U}{\partial y},$$

(3.3)

where U is a function of x and y. Such a force is said to be a *conservative force* and U is called the *force potential*. Since U is a function of x and y, we have

$$dU = \frac{\partial U}{\partial x}\,dx + \frac{\partial U}{\partial y}\,dy = f_x\,dx + f_y\,dy.$$

When such a force is included in the equations of motion, we have in place of Eq. (3.1),

$$d\left(\frac{V^2}{2} + \frac{p}{\rho} - U\right) = 0,$$

(3.4a)

or

$$\frac{V_1^2}{2} + \frac{p_1}{\rho} - U_1 = \frac{V_2^2}{2} + \frac{p_2}{\rho} - U_2.$$

(3.4b)

This expression is again limited to points along the same streamline in a steady, inviscid, incompressible flow.

3.4 Barotropic Fluids

A barotropic fluid is one for which the density is a single-valued function of the pressure, that is, $\rho = \rho(p)$. Although, generally, the density is a function of both pressure and temperature, there are important cases in which the foregoing simple relationship applies. From thermodynamics it may be shown, for example, that in the absence of friction and heat transfer, the density of a pure substance becomes a function of pressure alone. For a barotropic fluid, instead of Eq. (3.1) we may now only write

$$d\left(\frac{V^2}{2}\right) + \frac{dp}{\rho} = 0.$$

But since $\rho = \rho(p)$, dp/ρ may be regarded as merely another differential function of p (or ρ). Let us call it di, that is,

$$di = \frac{dp}{\rho}$$

or

$$i = \int \frac{dp}{\rho} + \text{const.}$$

Hence instead of Eq. (3.1), one obtains

$$d\left(\frac{V^2}{2} + i\right) = 0,$$

and upon integrating,

$$\frac{V_2^2}{2} + i_2 = \frac{V_1^2}{2} + i_1, \tag{3.5}$$

where the subscripts 1 and 2 again refer to two points along the same streamline.

In the special case of the isentropic flow of a perfect gas, for example, the relation between p and ρ takes the form $p/\rho^\gamma = \text{const}$ and i is equal to the *enthalpy* of the gas (see Appendix 4).

3.5 Nonsteady Flow

So far we have omitted the nonsteady term, but this term may also be included in Eq. (3.4). We must then add to the equation leading to (3.1) the terms

$$\frac{\partial u}{\partial t}\, dx$$

for the x-direction and

$$\frac{\partial v}{\partial t}\, dy$$

for the y-direction. But these are the two components of

$$\frac{\partial V}{\partial t}\, ds,$$

where V is the magnitude of the complete velocity vector and ds is an element of the streamline. Thus in place of Eq. (3.4a), we obtain

$$d\left(\int_0^s \frac{\partial V}{\partial t}\, ds + \frac{V^2}{2} + \frac{p}{\rho} - U\right) = 0, \qquad (3.6a)$$

in which the integration of the nonsteady term is carried out along a given stream-line at a given instant of time starting from an arbitrary reference point. Equation (3.6a) shows that the quantity in parentheses inside the differential must be a con-stant along the streamline. This constant is sometimes called the Bernoulli "constant" and is given the symbol B. The word "constant" has been set in quota-tion marks because the integration has been carried out along a special path at a given instant of time, so that B could still be a function of time and may be written as $B(t)$ to emphasize this point.

Integrating Eq. (3.6a) between two points of the streamline leads to the expres-sion

$$\int_1^2 \frac{\partial V}{\partial t}\, ds + \frac{V_2^2}{2} + \frac{p_2}{\rho} - U_2 = \frac{V_1^2}{2} + \frac{p_1}{\rho} - U_1. \qquad (3.6b)$$

This expression is now applicable to two points on a given streamline in a non-steady flow of an incompressible fluid in the presence of conservative forces.

The force potential most usually considered is that due to gravity. Let us take the y-axis to be positive when pointing upward and normal to the surface of the earth. The force per unit mass due to gravity is therefore directed downward and is of magnitude g. Thus

$$f_x = 0,$$

$$f_y = -g,$$

and hence

$$U = -gy.$$

Substituting in Eq. (3.6), we obtain the equation that takes into account the effect of the gravitational potential

$$\int_1^2 \frac{\partial V}{\partial t}\, ds + \frac{V_2^2}{2} + \frac{p_2}{\rho} + gy_2 = \frac{V_1^2}{2} + \frac{p_1}{\rho} + gy_1. \qquad (3.7)$$

It has to be pointed out here that the integral of $\partial V/\partial t$ is not always easy to evalu-ate. It can, however, be accomplished in several special cases, and it is mainly for this reason that Eqs. (3.6b) and (3.7) are useful.

EXAMPLE 3.2

Let us consider the efflux of liquid of density ρ through an orifice of area A_0 at the bottom of a large tank of cross-sectional area A_t. The depth of liquid at the given instant is y. The pressure exerted on the liquid in the tank is maintained at p_t and the pressure outside the orifice is the atmospheric pressure p_a. The action of gravity is to be considered, but frictional effects may be neglected. The velocity of efflux V_j is to be determined for these conditions.

Solution: A few possible streamlines have been indicated in the accompanying figure. Let us pick one of these and write the Bernoulli equation between a point on

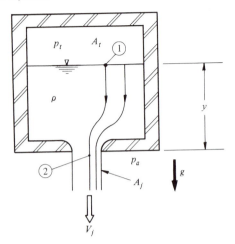

Example 3.2

the water surface (1) and one in the jet (2), both of which points are imagined to lie on the same streamline. Let us assume that the nonsteady effects are negligible. Thus the appropriate Bernoulli equation from Eq. (3.7) is

(a)
$$\frac{V_j^2}{2} + \frac{p_a}{\rho} = \frac{V_1^2}{2} + \frac{p_t}{\rho} + gy,$$

with the reference datum for elevation being taken at the plane of the orifice. We now notice that an identical equation would have been obtained for any other streamline, so that we may conclude that the velocity V_j is uniform throughout the jet, provided that V_1 and y are constant over the upper surface. From the continuity equation we may then write

$$V_1 A_t = V_j A_j,$$

and by eliminating V_1 from the Bernoulli equation, we obtain the expression

(b)
$$V_j^2 = 2\left[gy + \frac{1}{\rho}(p_t - p_a) \right] \frac{1}{1 - (A_j/A_t)^2}.$$

Let us now examine the validity of our assumption regarding the steadiness of the flow. As the flow continues, y decreases with time, since the tank is being emptied so that V_j is an implicit function of time. We must now verify whether or not it was permissible to neglect the nonsteady terms. Let us assume that the nonsteady effect of the flow within the region of the orifice is negligible. The integral of Eq. (3.7) may then be approximated by

(c)
$$\int_1^2 \frac{\partial V}{\partial t}\, ds \approx \frac{dV_1}{dt} y = y \frac{dV_j}{dt} \frac{A_0}{A_t},$$

where the velocity at any cross section in the tank has been assumed uniform and the total length of each streamline has been approximated by y. For the expression of V_j as given in Eq. (b) to be accurate, this nonsteady term must be small compared to the other terms in the Bernoulli relation. To simplify our comparison, let us take $p_t = p_a$. Then we must require the acceleration term given by (c) to be much smaller than the gravity term gy. That is,

(d)
$$\left| \frac{A_0}{A_t} y \frac{dV_j}{dt} \right| \ll gy \quad \text{or} \quad \left| \frac{A_0}{A_t} \frac{dV_j}{dt} \right| \ll g.$$

Provided that the nonsteady, or acceleration, term is small, we can approximate V_j by its steady value

$$V_j^2 = 2gy \frac{1}{1 - A_0^2/A_t^2},$$

so that

$$\frac{dV_j}{dt} = \frac{dy}{dt} \sqrt{\frac{g}{2y}} \frac{1}{\sqrt{1 - A_0^2/A_t^2}}.$$

But $dy/dt = -V_1 = -V_j(A_0/At)$. Thus

$$\frac{dV_j}{dt} = - \frac{A_0/A_t}{1 - A_0^2/A_t^2} g,$$

and from (d) the nonsteady terms can be neglected if

(e)
$$\frac{(A_0/A_t)^2}{1 - (A_0/A_t)^2} \ll 1.$$

If the pressure difference due to acceleration is in error by 2 percent, the velocity will be in error by about 1 percent. To remain within these limits of error, it is seen from Eq. (e) that it is necessary to satisfy the inequality

$$\frac{(A_0/A_t)^2}{1 - (A_0/A_t)^2} < 0.02$$

or

$$\frac{A_0}{A_t} < \frac{1}{7}.$$

Thus the nonsteady terms in a problem of this type can be safely neglected when $A_0/A_t < 0.1$. However, if a substantial length of pipe is attached to the orifice, this may no longer be true (see Problem 3.2).

EXAMPLE 3.3 _____

We now consider a problem in which the nonsteady terms cannot be neglected. Such an example is afforded by an open U-tube, as shown in the figure. Fluid of density ρ fills both sides of the tube, but the initial height of the fluid on the left-hand side is l_1 and that on the right-hand side is l_2. The cross-sectional area of both tubes is the same. The connection between the tubes is assumed to be of negligible length. We now want to determine the subsequent motion of the fluid in the tubes, assuming no friction.

Solution: We write the Bernoulli equation in the form given in Eq. (3.7) between two points on a given streamline. As the first point (1) we select one on the surface of the left tube and as the second we take a corresponding point (2) on the surface of

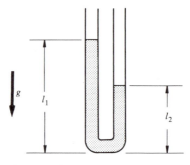

Example 3.3

the right-hand side. Thus

$$\int_1^2 \frac{\partial V}{\partial t} \, ds + \frac{V_2^2}{2} + gl_2 = \frac{V_1^2}{2} + gl_1.$$

Again we assume the velocity to be essentially uniform over each cross section, so that we would have obtained the same equation for any of the streamlines that we might have selected. We may then set

$$|V_1| = |V_2| = V$$

and evaluate the Bernoulli equation, obtaining simply

$$(l_1 + l_2) \frac{dV}{dt} = g(l_1 - l_2),$$

since dV/dt is not a function of the distance s.

Now let $l_1 + l_2 = l$, where l is the total constant length of fluid in the tube and realizing that $V = dl_2/dt$, the equation above can be written as

$$\frac{d^2 l_2}{dt^2} + \frac{2g l_2}{l} = g.$$

This differential equation is identical to the one for a pendulum oscillating with a frequency $f = (1/2\pi)\sqrt{2g/l}$. We expect the fluid in the U-tube, therefore, to carry out an equivalent oscillatory motion at this frequency.

PROBLEMS

The properties of the various fluids used are to be evaluated at 20°C and 1 atmosphere pressure unless they are otherwise specified.

3.1   The velocity of a water jet issuing from a fire hose is to be 15 m/s. If the nozzle is located 9 m above the pump, what does the pressure have to be at the pump discharge? Assume no friction (usually not a good assumption) and a ratio of 4 : 1 for the hose to nozzle area.

3.2   A pipe of uniform diameter and of length l is connected to a large vessel in which the pressure is maintained at a constant value p_1. A valve at the end of the pipe prevents flow. The length l is very much greater than the pipe diameter and the vessel is sufficiently large, so that the velocity anywhere within the vessel may be neglected. The atmospheric pressure is p_a, and an incompressible, frictionless fluid of density ρ completely fills the pipe and vessel. Elevation changes are negligible. The valve is suddenly opened and flow commences.

(a) Determine the flow velocity at the end of the pipe as a function of time.

(b) If $l = 150$ ft, $p_1 = 30$ psia, $p_a = 14.7$ psia, and the fluid is water, how much time is required to achieve 90 percent of the ultimate velocity?

Hint: Note that the velocity in the pipe is a function of time only.

3.3 A water tank has an orifice at the bottom of the tank. The cross-sectional area of the jet at the point where it leaves the orifice (at $x = 0$) is A_0. The height of the water in the tank is h, and this height is kept constant by continuously refilling the tank. The tank cross section is much larger than the orifice area. Neglecting friction, surface tension, and so on, find the cross-sectional area A of the jet as a function of x.

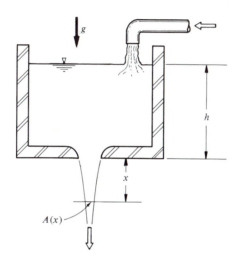

Problem 3.3

3.4 A suction device is arranged as shown. Find the flow rate through the main pipe at the instant at which the suction will begin.

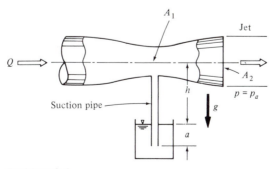

Problem 3.4

$$A_2 = 4A_1, \qquad h = 3 \text{ ft},$$
$$A_1 = 0.5 \text{ in.}^2, \qquad a = 2 \text{ ft}, \qquad g = 32.2 \text{ ft/sec}^2$$

The fluid is incompressible and friction may be neglected. The external atmospheric pressure is uniform throughout. Give the answer in ft^3/sec.

3.5 A blunt symmetrical body is moving at a constant velocity V in the positive x-direction in an infinite frictionless incompressible fluid of density ρ, otherwise at rest. The fluid at the nose of the body must also move at velocity V in the same direction. What is the difference in pressure between the point on the nose and a point infinitely far upstream where the velocity is zero?

Note: This flow is nonsteady. The problem may be solved by application of Eq. (3.7) or may be reduced to a steady flow by coordinate translation. It is instructive to solve it by both methods.

3.6 The long pipe leading from the reservoir shown in the sketch is the frustrum of a cone whose vertex lies at a distance x_1 into the reservoir from the entrance to the pipe. The reservoir is sufficiently large so that the height y to the free surface may be assumed constant with time and so that any point in the pipe may be considered to be at the same elevation. The system is initially filled with a frictionless incompressible fluid up to the end of the pipe, which is capped. The exit area of the pipe is A_2. Find an expression for the time required after removing the cap for the efflux velocity V_2 to reach ηV_0, where V_0 is the steady-state value of V_2 and $\eta < 1$. Express t as a function of $x_1, x_2, y,$ and η.

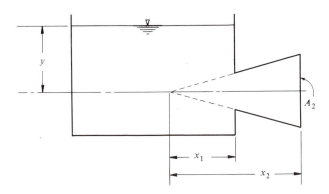

Problem 3.6

3.7 A frictionless fluid of density ρ_1 enters a chamber where it is heated, so that the density decreases to ρ_2. The light fluid then escapes through a chimney that has a height y. Except for the heating process, treat the fluid as incompressible. What is the velocity V in the stack? The velocity entering the heating chamber is negligible, and the cold fluid, which also has a density ρ_1, completely surrounds the chimney.

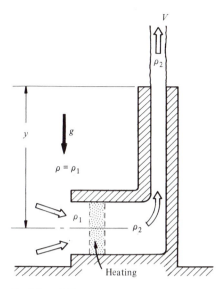

Problem 3.7

3.8 Two very long parallel plates of length $2L$ are separated a distance b. The upper plate moves *downward* at a steady rate V. A nonviscous and incompressible fluid of density ρ fills the gap between the plates. Fluid is squeezed out between the plates, and since the flow is symmetrical, the velocity parallel to the plate at the center is zero. Assume that $b \ll L$ and that the velocity u parallel to the plate is substantially constant across the gap. Treat the flow as being one-dimensional and parallel to the x-axis.

(a) Show that the velocity at any point x from the center is approximately $u = Vx/b$.

(b) Noting that b changes with time and assuming that the pressure outside the plates is zero, obtain an expression for the pressure at any point x along the plate. Neglect gravity.

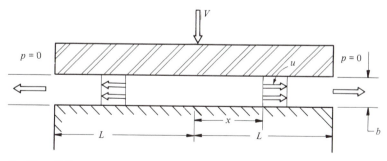

Problem 3.8

3.9 A certain two-dimensional incompressible, steady flow is given by the stream function

$$\psi = A \ln r + B\theta,$$

where r and θ are the usual polar coordinates and A and B are positive constants.

(a) Find the components of fluid velocity and show that the continuity equation is satisfied.

(b) Sketch a sufficient number of streamlines so that the flow pattern becomes clearly evident.

(c) Find the radial and tangential components of fluid acceleration.

(d) Find the distribution of pressure as a function of r and θ by *two* methods (i.e., starting with the differential equation of motion directly and using the Bernoulli equation).

3.10 Consider the steady flow of a frictionless fluid in a gravitational field for which the body force is inversely proportional to the square of the distance from a fixed central point. Assume that the density can be expressed as a function of pressure in the form $\rho = \rho_0(p/p_0)^a$, where a is a constant different from unity. Derive the Bernoulli equation in this system for two points on a streamline.

3.11 A closed tube is shaped as shown. It is filled with two fluids, of densities ρ_a and ρ_b, respectively. The length of tube occupied by fluid a is l_a, and that occupied by fluid b is l_b. The tube diameter is constant throughout. The fluid is then displaced by some means, so that the interface on the right-hand side is raised a distance x above the equilibrium position. It is then released. Write a differential equation for x to describe the subsequent motion.

Hint: Call the interfaces points 1 and 2; then let the pressure at the points be p_1 and p_2 and write Bernoulli equations between these points following a streamline in each medium. The pressures p_1 and p_2 are, however, not given and should not appear in the final answer. Neglect friction.

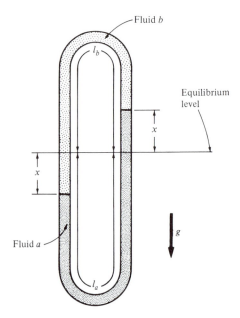

Problem 3.11

3.12 If a small amount of explosive is detonated in a large volume of liquid, a spherical bubble is created that rapidly grows with time. The flow may be considered incompressible, nonviscous, and spherically symmetric. The radius of the bubble $R(t)$ is a function of the internal pressure p_1 and the pressure p_∞ a great distance away, where the velocity can be considered to be zero. We want to get a differential equation relating $R(t)$, p_1, p_∞, and the density ρ.

(a) Determine the radial velocity V at any radius r in terms of R and the velocity at the bubble surface $\dot{R}$.

(b) With the nonsteady Bernoulli equation, obtain the desired equation.

Note: Do not try to solve this equation.

3.13 Water flows vertically downward through a Venturi meter as shown. The dimensions of the Venturi and the levels in the mercury manometers are as shown in the diagram. Estimate the flow rate through the Venturi in ft³/sec. (Take careful account of the water in the manometer tubes; their inlets are 2 ft apart.)

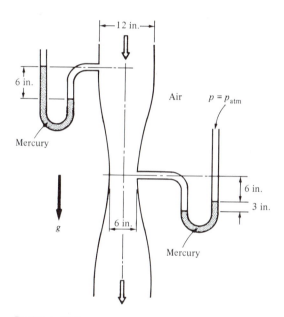

Problem 3.13

3.14 Incompressible fluid ($\rho = 1000$ kg/m³) fills a constant-diameter pipe, one end of which is fitted with a piston. The piston undergoes the displacement $x = x_0 \sin \omega t$. By application of the unsteady Bernoulli equation, obtain a formula for the pressure difference $(p_1 - p_2)$, where p_1 is at the piston face and p_2 is at a distance L down the pipe. Express your answer in terms of p, L, x_0, ω, and so on.

3.15 A circular plate is forced down at a steady velocity V_0 against a flat surface. Frictionless fluid of density ρ fills the gap h. *Assume* that $h \ll R_0$, the plate radius, and that the radial velocity $V_R(R, t)$ is *constant* across the gap.

(a) From continuity considerations, obtain a formula for $V_R(R, t)$ in terms of R, V_0, and h.

(b) Noting that $h = h(t)$, evaluate $\partial V_R / \partial t$.

(c) Substitute into the Bernoulli equation and calculate the pressure distribution, assuming that $p(R = R_0, t) = 0$.

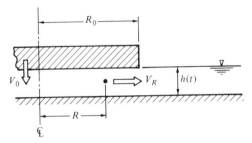

Problem 3.15

3.16 A sphere of diameter d is placed at the centerline inside a circular pipe of diameter D that contains flowing water. The flow may be taken as incompressible (density ρ) and steady. All frictional effects can be neglected. A U-tube manometer with both legs filled with water has a heavier fluid in the U-tube portion (density ρS). The manometer fluid levels are a height h apart. If the one manometer tap is on the pipe wall directly opposite the sphere center while the other is far upstream, find the volumetric flow rate in the pipe Q m^3/s, assuming that the fluid velocity V at both manometer tap locations is uniform across the unobstructed part of the pipe and parallel to the pipe axis.

3.17 Consider the flow of an incompressible liquid, $\rho = 1000$ kg/m^3, in the branched pipe shown. Assume steady flow and no gravity or friction. Calculate the following differences.

(a) $p_c - p_b$.
(b) $p_b - p_a$.

Assume that the Bernoulli constant B is constant everywhere.

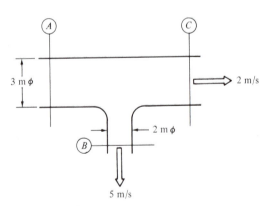

Problem 3.17

3.18 Water flows in a horizontal, frictionless channel under the action of gravity ($g = 9.81$ m/s^2) through an "undershot weir," as shown. The upstream and downstream depths are 0.2 and 0.1 m, respectively. Assume that the Bernoulli constant B is constant everywhere and calculate the upstream and downstream velocities.

Note: V_1 and V_2 are constant throughout depths h_1 and h_2.

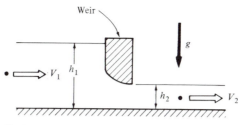

Problem 3.18

3.6 The Bernoulli Equation in Irrotational Flow

In the preceding sections the equations of motion were integrated along streamlines to obtain various forms of the Bernoulli equation excluding, so far, any effects of friction or work addition. From the results it was shown that certain groups of terms—for example,

$$\int_0^s \frac{\partial V}{\partial t}\, ds + \frac{V^2}{2} + \frac{p}{\rho} - U$$

in Eq. (3.7)—were at at any one instant constant along a streamline. This instantaneous constant was called the Bernoulli constant B, and it was stated that B could be a function of time. In general, however, the function B—even at a fixed instant of time—must be expected to differ from streamline to streamline. Nevertheless, we have also seen in several of the foregoing examples and problems that there are many cases in which the constant B is the same for all streamlines of the flow. We shall now examine in more generality the characteristics of B for a frictionless, incompressible fluid. The equation pertinent to investigating this point is again the Eulerian equation of motion, Eq. (2.13):

$$\frac{\partial \mathbf{v}}{\partial t} + \frac{1}{2}\nabla V^2 + \frac{1}{\rho}\nabla p = \mathbf{f} + \mathbf{v} \times \boldsymbol{\omega}, \tag{2.13a}$$

in which the definition

$$\boldsymbol{\omega} = \nabla \times \mathbf{v} = \text{curl } \mathbf{v}$$

has been used. Let us now integrate the above equation along an arbitrary path that is *not necessarily* a streamline. The increment of such a path is represented by the vector element $d\mathbf{r}$. To obtain this integral, the dot product of $d\mathbf{r}$ and each term of Eq. (2.13a) is formed. The nonsteady term can also be included conveniently if we note that

$$(d\mathbf{r} \cdot \nabla)\int \frac{\partial \mathbf{v}}{\partial t} \cdot d\mathbf{r} \equiv \frac{\partial \mathbf{v}}{\partial t} \cdot d\mathbf{r}.$$

The result of this operation is

$$(d\mathbf{r} \cdot \nabla)\left[\int \frac{\partial \mathbf{v}}{\partial t} \cdot d\mathbf{r} + \frac{V^2}{2} + \frac{p}{\rho}\right] = \mathbf{f} \cdot d\mathbf{r} + (\mathbf{v} \times \boldsymbol{\omega}) \cdot d\mathbf{r}. \tag{3.8a}$$

We shall further restrict ourselves to conservative forces, as defined by Eq. (3.3), which in vector notation can be written as

$$\mathbf{f} = \nabla U.$$

Introducing this relationship into Eq. (3.8a) above, we have

$$(d\mathbf{r} \cdot \nabla)\left[\int \frac{\partial \mathbf{v}}{\partial t} \cdot d\mathbf{r} + \frac{V^2}{2} + \frac{p}{\rho} - U\right] = (\mathbf{v} \times \mathbf{\omega}) \cdot d\mathbf{r}. \tag{3.8b}$$

Let us examine this equation now first for the special case that *dr does* as before represent an element of the streamline. In that event the dot product on the right-hand side of Eq. (3.8b) is equal to zero, since *dr* is parallel to **v** and (**v** × **ω**) is a vector perpendicular to **v**. It then follows that at any given instant the quantity in brackets is constant along a streamline. This constant (which still could be a function of time) is the one designated by the symbol $B(t)$ in connection with Eq. (3.6a). Thus we so far simply have recovered the result previously contained in Eqs. (3.6a) and (3.6b).

Next, however, we shall regard *dr* as an arbitrary element of distance that is not along a streamline. In that case the right-hand side is still equal to zero if either $\mathbf{\omega} = \nabla \times \mathbf{v} = 0$ or if **ω** is parallel to **v**. In either of those two special cases we can conclude that the function inside the brackets of Eq. (3.8a)—which we still shall call the Bernoulli function $B(t)$—is the same for all points of the flow field and not only along points of a given streamline. This conclusion follows from the fact that *dr* now represents an arbitrary element of path, and the path of integration for the left-hand side of Eq. (3.8a) may therefore be selected so as to connect any two desired points of the field. As a consequence, whenever **ω** = 0, or **v** is parallel to **ω**, the Bernoulli equation for an incompressible fluid takes the form

$$\int \frac{\partial \mathbf{v}}{\partial t} \cdot d\mathbf{r} + \frac{V^2}{2} + \frac{p}{\rho} - U = B(t), \tag{3.8c}$$

which at any given instant is valid all throughout the fluid. Following equivalent steps it can be shown that for a barotropic fluid $[\rho = \rho(p)]$ under the same conditions

$$\int \frac{\partial \mathbf{v}}{\partial t} \cdot d\mathbf{r} + \frac{V^2}{2} + \int \frac{dp}{\rho} - U = B(t). \tag{3.8d}$$

There is also a third case when (**v** × **ω**) · *dr* is zero; namely, when *dr* lies in the surface perpendicular to **v** × **ω**. Therefore, **v** and **ω** at each point in a fluid flow define the tangent plane of this surface, which is appropriately called a *Bernoulli surface*. Thus the Bernoulli function $B(t)$ is the same constant everywhere on a Bernoulli surface at a given instant of time. Of course, $B(t)$ may be different on an adjacent Bernoulli surface at the same instant of time. The Bernoulli surface is everywhere defined by the directions of the two vectors **v** and **ω**. Lines parallel to **v** are, of course, streamlines. We may term lines everywhere parallel to **ω** "vortex" lines by analogy to streamlines. We can now imagine a *Bernoulli surface* to consist of a mesh—rather like a fisherman's net—of vortex lines and streamlines.

Of the special cases mentioned, the one corresponding to **ω** = 0 is of particular interest. As was shown in Section 2.7, **ω** may be identified with the rotation of the individual fluid particles, and there is an absence of such rotational motion, therefore, when **ω** = 0. Such flows are appropriately called *irrotational*. Irrotational flows form an important class of flows, because many actual flows or at least portions of them may be considered irrotational, and under these conditions certain simplifica-

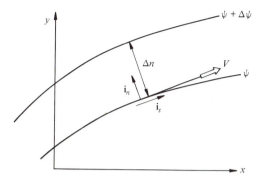

FIG. 3.1 Sketch used in deriving the relationship between vorticity and the variation of Bernoulli constant.

tions in the analytical approach become feasible. Chapter 6 is devoted largely to a discussion of irrotational flows.

When the flow is two-dimensional† and steady, a simple but interesting connection between the variation of B in the flow and the vorticity may be pointed out. In a two-dimensional x-y coordinate system with unit vectors $\mathbf{i}$, $\mathbf{j}$ in the x and y directions, respectively, the vorticity vector $\boldsymbol{\omega}$ has only one component,

$$\boldsymbol{\omega} = \mathbf{k}\omega_z = \mathbf{k}\left(\frac{\partial v}{\partial x} - \frac{\partial u}{\partial y}\right).$$

Then from Eq. (2.13a), recalling that the flow is steady,

$$\nabla B = \mathbf{v} \times \boldsymbol{\omega} = -\mathbf{k} \times \mathbf{i}_s V\omega_z,$$

where $\mathbf{i}_s$ is a unit vector pointing in the direction of the flow and parallel to the streamline ψ, as shown in Fig. 3.1. Also, $\mathbf{k}$ is a unit vector pointing out of the xy-plane in Fig. 3.1, so that $\mathbf{k} \times \mathbf{i}_s = \mathbf{i}_n$ is a unit vector normal to the streamline. Thus the gradient of B is normal to the streamline and we can write

$$\nabla B = \mathbf{i}_n \frac{\partial B}{\partial n} = -\mathbf{i}_n V\omega_z.$$

But the magnitude of the velocity V is

$$V = \lim_{\Delta n \to 0} \frac{\Delta \psi}{\Delta n} = \frac{\partial \psi}{\partial n}.$$

Hence, for two-dimensional flows,

$$\frac{\partial B}{\partial \psi} = -\omega_z. \tag{3.9}$$

Equation (3.9) shows again that if $\omega_z = 0$, B is constant and independent of ψ. We have previously shown that the value of B may change from streamline to streamline [i.e., $B = B(\psi)$]. Thus from Eq. (3.9) ω_z is a function of ψ only, so that ω_z is also constant along streamlines. It should be remembered that this result is limited to two-dimensional, incompressible (or barotropic) flows of nonviscous fluids subjected only to conservative body forces.

---

† The remainder of this section may be omitted without loss of continuity.

PROBLEMS

3.19 The inlet of a two-dimensional duct is of width b and contains a frictionless fluid of constant density ρ. The velocity in one half of the duct is V_1 and that in the other half is $V_1/2$. The two streams do not mix. The outlet of the duct is of width $b/2$. The flow is parallel at the duct entrance and exit so that the pressure is uniform across these two sections. By writing the continuity equation and Bernoulli equation, determine the pressure change $p_1 - p_2$ across the duct contraction and the individual dimensions b_1, b_2 of the two fluid streams at the exit section.

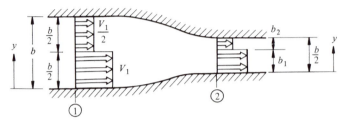

Problem 3.19

3.20 Verify by calculation that Eq. (3.9) is satisfied for the flow $v = 0$, $u = ky$, k being a constant.

3.21 Consider the flow in Problem 3.19 except that the inlet velocity is $u = V_1 y/b$. With the use of Eqs. (3.9) and (2.26a) show that the outlet velocity distribution is $3V_1/4 + V_1 y/b$, where y is measured as shown.

3.22 For an axisymmetric flow (with no swirling or tangential velocity) show that the analog of Eq. (3.9) is

$$\frac{\partial B}{\partial \psi} = -\frac{\omega_\theta}{r},$$

where ω_θ is the vorticity component in the tangential direction and the stream function is defined by the equation $v_z = 1/r \, \partial \psi/\partial r$, and so on. From this result show that the velocity profile becomes "flatter" when an axisymmetric flow is passed through a contracting nozzle.

3.23 A "stratified" flow takes place in the channel shown. There is *no* gravity or friction present and the flow is steady. The dense stream is increased to a speed of 2 m/s by a contraction of the duct. The pressure across the duct is constant. Calculate the velocity of the lighter fluid.

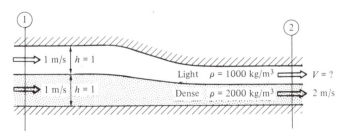

Problem 3.23

3.24 Consider the annular flow between two long, concentric cylinders described in Problem 2.63. The inner cylinder of radius a rotates with angular velocity Ω inside a larger, stationary cylinder of radius b. A Newtonian fluid fills the gap; the tangential velocity profile is

$$v_\theta = \frac{a^2 \Omega}{a^2 - b^2} \left(r - \frac{b^2}{r} \right).$$

(a) Find an expression for the change in Bernoulli function dB/dr.

(b) Integrate the result to get an expression for the pressure at various radial locations.

(c) Check the result of part (b) by direct integration of the Navier–Stokes equations for this flow.

3.7 Extended Bernoulli Equation

Let us consider a more general case where body forces are present that are not conservative and, in addition, where shear and normal stresses arising from friction are also present. As in the preceding derivations, the equations of motion are to be integrated along a streamline. The nonconservative forces will be indicated symbolically by $\mathbf{f}_{nc}$ and the viscous contributions, that is, the viscous part of the Navier–Stokes equations, by $\mathbf{f}_f$. We then add to the right-hand side of Eq. (3.8) or (3.1) the terms

$$\mathbf{f}_{nc} \cdot d\mathbf{s} + \mathbf{f}_f \cdot d\mathbf{s}.$$

If these are included in the Bernoulli equation (3.4), we obtain

$$\int_1^2 \mathbf{f}_{nc} \cdot d\mathbf{s} + \int_1^2 \mathbf{f}_f \cdot d\mathbf{s} + \frac{V_1^2}{2} + \frac{p_1}{\rho} + gy_1 = \frac{V_2^2}{2} + \frac{p_2}{\rho} + gy_2,$$

where $d\mathbf{s}$ is an element of the streamline.

The first integral may be interpreted as the useful work done on the fluid by the force $\mathbf{f}_{nc}$ in traversing from point 1 to point 2. Unlike a conservative force, it must be known at every point along the path. Such a force can be caused by moving solid surfaces in the fluid, such as the vanes of a pump or turbine or by electromagnetic forces. In the latter case the effect is that of a body force. In the former the action is more indirect, because a vane surface comes in direct contact with only a portion of the fluid. The force $\mathbf{f}_{nc}$ should then be regarded rather as the effective force per unit mass brought about by the additional pressure and shear stress due to the action of the vanes. This work done by the forces $\mathbf{f}_{nc}$ may be positive or negative, that is, it may be done on or by the fluid. The first case would correspond to a pump being located somewhere between stations 1 and 2, and the second would correspond to a turbine. In either event the integral refers to the mechanical performance of work and we shall merely indicate this by writing

$$\int_1^2 \mathbf{f}_{nc} \cdot d\mathbf{s} = \pm W_s.$$

The symbol W_s refers to the work done per unit mass, and the subscript s is used to indicate that it is usually transmitted by a rotating shaft. The plus sign refers to a pump and the negative sign to a turbine. The second integral arises from the friction in the fluid and may be interpreted as the work done by the fluid in overcoming viscous resistance, and it is not therefore available for performing useful work. In analogy to the mechanical work W_s we shall call the viscous term W_f, and it will always have the sense of a "loss," that is, $W_f = -\int_1^2 \mathbf{f}_f \cdot d\mathbf{s}$. Although W_s or W_f can only rarely be calculated, the separate identification of these terms is useful, since these quantities can be determined experimentally.

We shall now adopt a convention that is useful in solving engineering problems involving the Bernoulli equation. This is that the flow conditions downstream or after the flow process are always designated by 2 and those upstream or before by 1. Then we have the "extended" Bernoulli equation

$$\frac{V_2^2}{2} + \frac{p_2}{\rho} + gy_2 = \frac{V_1^2}{2} + \frac{p_1}{\rho} + gy_1 + W_s - W_f. \tag{3.10}$$

Each of the terms in this equation represents work or energy per unit mass. This is the form frequently encountered in thermodynamics. In hydraulic engineering practice, however, it is customary to express Eq. (3.10) in terms of work per unit *weight*,

$$\frac{V_2^2}{2g} + \frac{p_2}{\rho g} + y_2 = \frac{V_1^2}{2g} + \frac{p_1}{\rho g} + y_1 + M - h_f. \tag{3.11}$$

Each of these terms represent energy per unit weight and has the dimensions of length. M and h_f are symbols for the pump work and friction loss, respectively, in these units. An example of the friction loss h_f is given for a simple case in Example 2.3.

Each of the terms in Eq. (3.11) has come to have a name in engineering practice. The vertical coordinate y is called the *elevation head*. The quantity $p/\rho g$ is called the *static pressure head* and $V^2/2g$ is called the *velocity head*. The sum of the terms in parentheses in Eq. (3.11) is often called the *total head* and is given the special symbol h_t. We then may write Eq. (3.11) simply as

$$h_{t2} = h_{t1} + M - h_f. \tag{3.12}$$

This equation may be expressed in words as:

□ *The total head after the flow process is equal to the total head before the process plus the mechanical work done minus the loss due to friction.*

Aeronautical engineers frequently multiply Eq. (3.10) by the density to obtain Bernoulli's equation in terms of work per unit volume. In this form each term has the dimensions of a pressure; for that reason the term $\rho V^2/2$ is spoken of as the dynamic pressure and $(p + \rho V^2/2 + \rho gy)$ is called the *total pressure* in analogy to the total head. The total pressure is also known as the *stagnation pressure*, since it is the pressure that would result if the stream were brought to rest (hence stagnated) with no losses.

Each of the terms in Eq. (3.10) represents energy per unit mass. When multiplied by the mass rate of flow, $\dot{m}$, these terms then in S.I. units give *power* in watts. Similarly, the terms in Eq. (3.12) are energy per unit weight and, when multiplied by the weight rate of flow ρgQ, have the same units of power. The function of a pump is to increase the total head of the flow, Δh_t. Clearly, there will be a power associated with this increase in the amount

$$P = \rho g \, \Delta h_t Q = \Delta p_t Q, \tag{3.13}$$

where Q is the volumetric flow rate in cubic meters per second, Δp_t the total pressure increase in N/m^2, and Δh_t the total head increase in meters. P is called the *fluid* power. A completely equivalent expression exists when the energy is expressed per unit mass as in Eq. (3.10).

The fluid power is positive for a pump as defined above and negative for a turbine. The actual power, P_{act}, required for a pump is always larger than the fluid power above because of friction, internal leakage, and a variety of other losses. The ratio P/P_{act} is called the pump "efficiency." A turbine, however, extracts power from the flow, so a smaller output is to be expected. Thus the turbine efficiency is defined as the ratio P_{act}/P.

Finally, it should be emphasized again that the concepts of head, friction loss, and so forth, developed in this section apply only to incompressible fluids and cannot be extended without modification to compressible fluids.

3.8 Engineering Applications of the Bernoulli Equation

Let us now consider the application of the Bernoulli equation to several typical problems of engineering interest. Among these are problems that involve the flow in pipe or pipe circuits, including various fittings, Venturi meters, and so on. In these examples it will often be possible to assume that the flow is "one-dimensional," that is, there is no variation in velocity normal to the axis of the conduit and the velocity is equal to the average velocity over the cross section. Now, as we have seen, the Bernoulli equation holds in general only along an individual streamline. With the assumptions above, however, there will be no difference between the flows along the various streamlines within the conduit. The Bernoulli equation written for any one streamline, therefore, will automatically apply to the entire flow in the conduit. For many problems of the type mentioned, the approach above leads to results of acceptable accuracy. It is inherent in the assumptions, however, that no details of the flow pattern—such as the velocity distribution normal to the axis of the duct—can be obtained in this manner.

(a) Flow in Pipe Circuits

Let us first consider the flow through piping circuits and illustrate the suggested method of solution by two examples.

EXAMPLE 3.4 _____

Consider the pump shown, transporting water between the two reservoirs. From measurements it is known that the friction losses between stations 1–2 and stations 3–4 are both $h_{f1-2} = h_{f3-4} = 5$ m. The elevations are $y_1 = 20$ m and $y_5 = 40$ m, and the discharge rate is $Q_p = 0.5$ m³/sec. The cross-sectional area of the pipe at point 4 is 0.04 m², and the pump efficiency is 75 percent. From this information the head across the pump and the power necessary to drive the pump are to be determined.

Solution: To determine the head across the pump (M), it is only necessary to write the extended Bernoulli equation between points 5 and 1. We then have

$$h_{t(5)} = h_{t(1)} + M - h_{f1-2} - h_{f3-4} - h_{f4-5}.$$

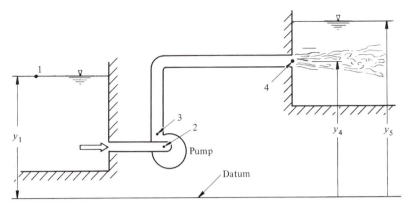

Example 3.4   Pump operating between two reservoirs.

But at points 1 and 5 (the upper water surface), the pressure is the same and the velocities are zero. Thus $h_{t5} - h_{t1} = y_5 - y_1$; hence

$$M = y_5 - y_1 + h_{f1-2} + h_{f3-4} + h_{f4-5}.$$

All quantities are known except h_{f4-5}. The discharge from the pipe at point (4) is a parallel jet and eventually the velocity in the jet is dissipated. The loss term h_{f4-5} can be found by writing Bernoulli's equation between points 4 and 5, that is,

$$h_{t5} = y_5 = h_{t4} - h_{f4-5} = \frac{V_4^2}{2g} + \frac{p_4}{\rho g} + y_4 - h_{f4-5}.$$

However, if the fluid in the reservoir surrounding the jet is stationary, then the pressure there is hydrostatically distributed; that is,

$$\frac{p_4}{\rho g} + y_4 = y_5,$$

hence

$$h_{f4-5} = \frac{V_4^2}{2g}$$

and

$$M = y_5 - y_1 + h_{f1-2} + h_{f3-4} + \frac{V_4^2}{2g}.$$

Thus

$$M = 40 - 20 + 5 + 5 + \frac{(0.5/0.04)^2}{2 \times 9.81} = 38.0 \text{ m.}$$

The hydraulic (or fluid) power is found by multiplying the head by the weight rate of flow. It follows that

$$P = (38.0 \text{ m})(1000 \text{ kg/m}^3)(0.5 \text{ m}^3/\text{s}) = 19.0 \text{ kW.}$$

The pump efficiency is the ratio of the hydraulic power produced to the required brake power, and, therefore,

$$P_{\text{act}} = \frac{19.0}{0.75} = 25.3 \text{ kW.}$$

EXAMPLE 3.5

A pipeline of constant diameter d and length l is attached to a water reservoir, as shown. The elevation of the water reservoir is y_1 and the outlet of the pipe is at y_2. The flow takes place in a normal gravitational field. There occurs a frictional loss in total head of $f V^2/2g$ for a length of pipe equal to one diameter, where f is assumed to be a constant. The possible loss at the pipe entrance is to be neglected. Determine the outlet velocity if $f = 0.02$, $l/d = 1000$, and $y_1 - y_2 = 100$ m. Also find the distribution of static pressure and of total head.

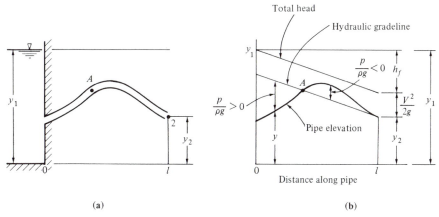

Example 3.5   (a) Pipeline discharging water from a reservoir into the atmosphere; (b) head and pressure as a function of the distance along the pipe.

Solution: We have by applying Eq. (3.11) or (3.12) between points 1 and 2,

$$\frac{V^2}{2g} + \frac{p_a}{\rho g} + y_2 = h_{t2} = h_{t1} - h_f = y_1 + \frac{p_a}{\rho g} - \frac{l}{d} f \frac{V^2}{2g}.$$

Thus

$$V^2 = 2g \frac{y_1 - y_2}{1 + f(l/d)},$$

and the discharge velocity is consequently

$$V = \sqrt{\frac{2(9.81)(100)}{1 + 20}} = 9.67 \text{ m/s}.$$

To determine the pressure at any point along the pipe, we again write the Bernoulli equation, this time between station 1 and the arbitrary station x, from which we obtain

$$h_{tx} = h_{t1} - \frac{x}{d} f \frac{V^2}{2g},$$

where x is the length of pipe between points 1 and x. The equation shows that the *total head* decreases linearly with the distance along the pipe. Similarly, we may show that the group of terms

$$\frac{p}{\rho g} + y = h_t - \frac{x}{d} f \frac{V^2}{2g} - \frac{V^2}{2g}$$

is also a linear function of the distance along the pipe. This group $(p/\rho g + y)$ is sometimes called the *piezometric head*, and the graph of the piezometric head along the pipe is called the *hydraulic grade line*. From the foregoing equation it is seen that the hydraulic grade line is independent of the actual elevation of the pipe. However, if the elevation of the pipe at any section exceeds that of the hydraulic grade line, the static pressure will fall below the atmospheric pressure. The pressure at point *A* in the illustration, for example, is at atmospheric pressure. Pressures below that of the surrounding atmosphere are usually avoided in large pipelines because of the risk of collapsing the pipe.

(b) Orifices and Pitot Tubes

The flow rate and pressure drop through simple openings in solid surfaces or walls or obstructions in the flow through pipes must often be considered; these are generally called "orifices." Often, the pressure drop that occurs may itself be used as a measure of the flow rate. In Example 3.1 it was seen that the pressure difference in a Venturi tube could be used to infer flow rates. Similarly, the flow around bodies produces a pressure distribution. This distribution may be of interest in itself, or as is the case for the Pitot tube, a certain pressure difference may be used as a measure of the flow speed approaching the body. In this section we discuss these simple but very useful flows from an engineering point of view.

The exact flow through or about these objects may actually be quite complicated. For this reason, it has become customary in these cases to make simplifying assumptions—such as the absence of friction, one-dimensional flow, and so on—and to compute idealized answers by applying the Bernoulli equation. The idealized quantities are then corrected by coefficients that are either determined experimentally or from more exact theories. These coefficients are of two types. The first accounts for the effects of viscosity in changing the velocity determined from the Bernoulli equation without friction. The second takes into account the geometric distortion of the fluid flow not considered in one-dimensional hydraulic theory.

As an example of the definition and use of these coefficients, let us consider the flow through a sharp-edged orifice from a reservoir under a head *h* (Fig. 3.2). If friction were absent, the viscosity in the jet would become uniform at some distance from the orifice, and from Bernoulli's equation this velocity would be equal to

$$V_i = \sqrt{2gh}.$$

The subscript *i* indicates that this is an ideal velocity occurring only in the absence of friction. This velocity would of course not be attained exactly at the orifice where the curvature of the streamlines is still important but only at some distance down-

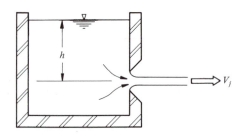

FIG. 3.2 Flow from a reservoir through a sharp-edged orifice.

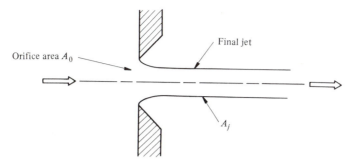

FIG. 3.3 Appearance of an ideal jet issuing from a sharp-edged orifice.

stream where the streamlines are parallel. The appearance of the ideal jet would be as shown in Fig. 3.3.

The discharge from this ideal jet is equal to the ideal velocity multiplied by the area of the jet. As seen from Fig. 3.3, the area of the jet is not equal to the area of the orifice, but they are of course related. Let $A_{jet} = c_c A_0$, so that the discharge is equal to

$$Q_{ideal} = c_c A_0 \sqrt{2gh}, \tag{3.14}$$

where c_c is called a *contraction coefficient* and A_0 is the orifice area. The appearance of the actual jet differs somewhat from that of the ideal one above, since the jet will be slowed down due to friction with the surrounding medium and consequently will have a tendency to spread out. Due to the initial curvature of the streamlines near the orifice and the spreading effect, the jet usually has a minimum size a few diameters downstream. This minimum section is called the *vena contracta*. At the vena contracta the streamlines are parallel, and the velocity profile will approach most closely that of the ideal jet. Because of friction, however, the velocity of the actual jet will differ from that of the ideal one and the actual velocity may be written as

$$V_{actual} = c_v V_{ideal} = c_v \sqrt{2gh}, \tag{3.15}$$

where c_v is an experimental coefficient called the velocity coefficient. The discharge of the actual jet will then be

$$Q_{actual} = c_v c_c A_0 \sqrt{2gh}, \tag{3.16}$$

where $(c_c A_0)$ is the area of the vena contracta. Because of the action of viscosity c_c may be slightly different from the value for the ideal jet. The discharge from the jet is also sometimes written as

$$Q_{actual} = A_0 c_d \sqrt{2gh}, \tag{3.17}$$

in which c_d is called the *discharge coefficient*. By comparison with Eq. (3.16) it is seen that $c_d = c_v c_c$ in this case. Approximate values for the coefficients of a sharp-edged orifice are

$$c_v = 0.98, \qquad c_c = 0.61, \qquad c_d = 0.60.$$

A sketch of the actual flow through a sharp-edged orifice is shown in Fig. 3.4. Additional information for other types of orifices is given in Fig. 3.5.

It is of importance to emphasize that a velocity coefficient less than unity implies the existence of friction and therefore a loss of total head. When the head loss is to be determined, it is more convenient to introduce the concept of a head loss coeffi-

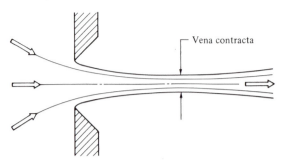

FIG. 3.4 Appearance of an actual jet issuing from a sharp-edged orifice.

cient rather than to work with a velocity coefficient. For example, for the orifice of Fig. 3.2 we can express the head loss as

$$h_f = k \frac{V^2}{2g},$$

where k is called the head loss coefficient. The coefficients k and c_v are, of course, related and we shall now show that

$$k = \frac{1}{c_v^2} - 1. \tag{3.18}$$

First let us recall that for any orifice the *ideal* velocity may be expressed as

$$V_i = \sqrt{2g\left(h_t - \frac{p}{\rho g}\right)},$$

where h_t is the total head just upstream of the orifice and p is the static pressure at the orifice. By definition

$$V = c_v \sqrt{2g\left(h_t - \frac{p}{\rho g}\right)}.$$

We can also write the extended Bernoulli equation as

$$h_t - k \frac{V^2}{2g} = \frac{V^2}{2g} + \frac{p}{\rho g}.$$

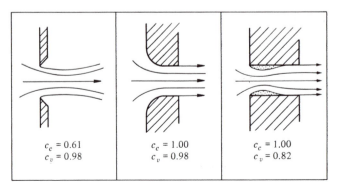

FIG. 3.5 Flow through various types of orifices. Typical values of the velocity and contractional coefficients are noted for each type.

Introducing the definition of c_v above, we have

$$\frac{p}{\rho g} = h_t - \frac{1}{c_v^2}\frac{V^2}{2g}$$

and combining the two expressions

$$k = \frac{1}{c_v^2} - 1,$$

which was to be shown. The above definition of velocity coefficient avoids confusion that may arise when applying the velocity coefficient to piping systems.

EXAMPLE 3.6 _____

An orifice of area A_0 and velocity coefficient $c_v = 0.8$ is installed in a pipe of area $A_p = 2A_0$. The pipe is attached to a dam as shown and the water level in the dam is 100 m higher than the outlet from the reservoir. The contraction coefficient of the orifice is unity. There are no other losses in the pipe or between the end of the orifice and the pipe. The velocity leaving the pipe V_p is to be determined.

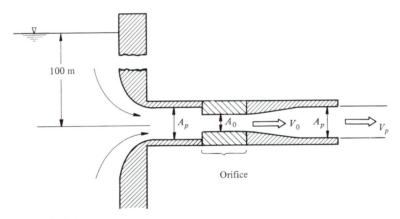

Example 3.6

Solution: We shall solve this first by using the velocity coefficient directly. From the Bernoulli equation the velocity V_0 downstream of the orifice is

$$V_0 = c_v\sqrt{2g\left(h_t - \frac{p}{\rho g}\right)},$$

where p is the static gage pressure at the end of the orifice.

To obtain the pressure at the end of the orifice, we write Bernoulli's equation again between the orifice outlet and the end of the pipe:

$$\frac{p}{\rho g} + \frac{V_0^2}{2g} = \frac{V_p^2}{2g},$$

or since $A_0 V_0 = A_p V_p$,

$$\frac{p}{\rho g} = -\frac{V_p^2}{2g}\left[\left(\frac{A_p}{A_0}\right)^2 - 1\right],$$

and finally,

$$V_0 = \frac{A_p}{A_0} V_p = c_v \sqrt{2g\left[h_t + \frac{V_p^2}{2g}\left(\frac{A_p^2}{A_0^2} - 1\right)\right]}.$$

Solving for V_p, we have

$$V_p^2 = \frac{2gh_t}{1 + A_p^2/A_0^2(1/c_v^2 - 1)},$$

or

$$V_p = \sqrt{\frac{(2)(9.81)(100)}{1 + 4(1/0.64 - 1)}}$$

$$= 24.6 \text{ m/s}.$$

The other approach consists of writing the Bernoulli equation, including the loss term between the pipe outlet and the reservoir surface, that is,

$$h_t - k\frac{V_0^2}{2g} = \frac{V_p^2}{2g}.$$

But

$$k = \frac{1}{c_v^2} - 1$$

and

$$V_0 = \frac{A_p}{A_0} V_p,$$

or

$$V_p^2 = \frac{2gh_t}{1 + A_p^2/A_0^2(1/c_v^2 - 1)},$$

as before.

Consider now the Venturi meter of Example 3.1. The ideal rate of discharge was calculated to be

$$Q_{ideal} = \frac{A_2\sqrt{(2/\rho)(p_1 - p_2)}}{\sqrt{1 - A_2^2/A_1^2}}.$$

If we define a discharge coefficient again as

$$c_d = \frac{Q_{actual}}{Q_{ideal}},$$

we obtain

$$Q_{actual} = \frac{c_d A_2\sqrt{(2/\rho)(p_1 - p_2)}}{\sqrt{1 - A_2^2/A_1^2}}.$$

Note, however, that in this case c_d is not the product of a velocity coefficient times a contraction coefficient. The discharge coefficients for Venturi meters are determined

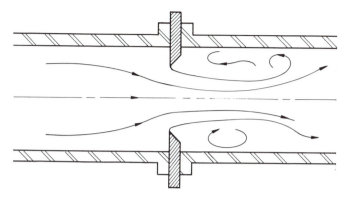

FIG. 3.6   Measuring orifice in a pipeline.

experimentally. Typical values of c_d for a well-designed meter operated with water range from 0.96 to 0.995. The Venturi meter lends itself to very accurate measurements and causes only a relatively small total pressure loss.

It is also possible to use an orifice in a pipeline for the purpose of measuring discharge (see Fig. 3.6). The main flow will be similar to that in a Venturi meter, and the equation for the discharge is, of course, the same as for the Venturi meter. In general orifices are simpler to install. Their coefficients are known fairly accurately; however, orifice plates cause a substantially higher head loss than Venturi meters. Discharge coefficients for both Venturi meters and this type of orifice are discussed in more detail in Chapter 5, where some numerical values will also be given.

Let us now consider a small, bent, open tube that is inserted into the steady flow in a pipe. The open end points in the upstream direction as shown in Fig. 3.7. This device is known as a *Pitot tube*—after Henri Pitot (1695–1771)—or *impact probe*. Immediately upstream of the tube the pressure at the wall measured with gage A is p_0 and the pressure measured by gage B attached to the tube is p_1. The problem is to find the velocity V. The pressure at the upstream end of the tube evidently must be the stagnation pressure, since the velocity there must be zero. The process of stopping the flow at the nose of such a tube is done with negligible friction, provided that the Reynolds number based on the tube diameter (Vd/v) exceeds 100. Therefore, the stagnation pressure in the bent tube is also equal to the upstream total pressure.

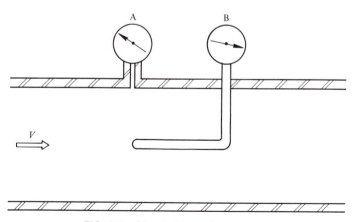

FIG. 3.7   Simple Pitot tube in a pipe.

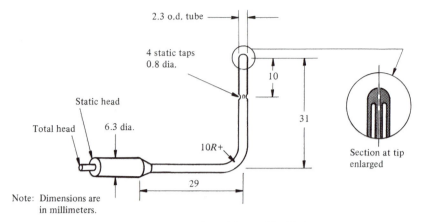

FIG. 3.8 Typical Prandtl-type Pitot tube.

The pressure read by gage B is then equal to the total pressure of the stream minus that due to the elevation of the gage. The pressure read by gage A is the static or free stream pressure of the upstream flow minus that due to the elevation of the gage. If both gages A and B are at the same elevation, then the difference $p_1 - p_0$ is equal to the dynamic pressure $\frac{1}{2}\rho V^2$.

It is generally convenient to locate the measuring point for the free stream pressure p_0 on the tube itself. This has been done by experimentally determining the point (about four diameters downstream of the tip) on the bent tube, where the surface pressure is nearly equal to p_0. The two pressures, static and dynamic, can be

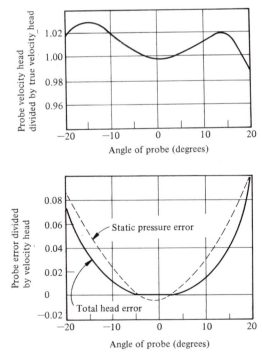

FIG. 3.9 Effect of angular position on the readings of a typical Prandtl-type Pitot tube. (From S. J. Markowski and F. M. Moffatt, "Instrumentation for Development of Aircraft Components," *Soc. Auto. Eng. Trans.*, vol. 2, 1948.)

measured with an instrument called a *Prandtl tube*, in which the static pressure is measured as above and a second tube concentric with the first is arranged to measure the total pressure. By connecting the two tubes to the high- and low-pressure sides of a differential manometer, the velocity head may be obtained directly. A typical construction of such a flow measuring instrument is shown in Fig. 3.8. The instrument gives accurate readings only of course when it points in the direction of the flow. Fortunately, however, small deviations from this direction do not cause large errors. Experimental data giving Pitot tube readings as a function of the angular position of the tube are shown in Fig. 3.9.

PROBLEMS

When not otherwise specified, the properties of the fluids are to be evaluated at 20°C and 1 atmosphere pressure.

3.25 A certain rocket motor that delivers a thrust of 50,000 lbf consumes 180 lbm of nitric acid per second. This acid is to be injected into the chamber through six sharp-edge orifices in parallel, which are all supplied by the same large manifold. The pressure drop through the orifices is to be 100 psi. The specific weight of the acid is 94 lbf/ft³, and it is assumed to be incompressible. Find (a) the diameter of each orifice, (b) the velocity of the acid jets, (c) the diameter of each orifice if the entrance were rounded.

3.26 The pump shown in the accompanying figure pumps 0.03 m³ of water per second from the reservoir through the pipe and nozzle assembly. The flow is frictionless except for the nozzle that has a velocity coefficient of 0.8. The contraction coefficient of the nozzle is 1.0. The elevations of the various stations indicated in the figure are

$$y_1 = 3 \text{ m},$$

$$y_2 = 1.5 \text{ m},$$

$$y_4 = y_5 = y_3 = 2.4 \text{ m}.$$

The cross-sectional areas of the pipe are

$$A_2 = 0.01 \text{ m}^2,$$

$$A_4 = A_3 = 0.01 \text{ m}^2,$$

$$A_5 = 0.003 \text{ m}^2.$$

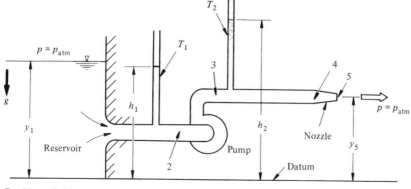

Problem 3.26

The subscripts refer to the corresponding station numbers. Find (a) the pump head in meters; (b) the power required if the pump were 100 percent efficient; (c) the heights h_1, h_2 to which the liquid in columns T_1, T_2 rises.

3.27 A short length of pipe is connected to a reservoir of height h. A nozzle of exit area A_j is connected to the pipe, the pipe cross-sectional area being three times that of the nozzle exit. The velocity coefficient of the nozzle is c_v. The head loss in the pipe (including any entrance losses) is given by $h_f = k(V^2/2g)$, where k is a given loss coefficient and V is the velocity in the pipe.
(a) Obtain an algebraic expression for the final jet velocity V_j.
(b) Find the numerical value for the velocity V_j, if $h = 15$ m, $c_v = 0.9$, and $k = 0.5$.

3.28 The water pressure in a large water main is maintained at a gage pressure p_0, independent of the flow rate. The velocity in the main can be neglected. A relatively small line of cross-sectional area A_p branches off from the water main. This line terminates in a nozzle of area A_n, which has a velocity coefficient c_v, and a contraction coefficient of unity. The loss coefficient for the entire pipeline between the main and the nozzle (including entrance effects at the main) is k, based on the velocity in the pipe.
(a) Find the jet velocity.
(b) Find the power in the jet associated with the kinetic energy.
(c) Find the area A_n for which this power is a maximum.

3.29 A piping arrangement (see the figure) delivers a jet of water to a Pelton turbine. Water enters the pipe from the reservoir at point A and flows to the 45° bend at point B. The loss coefficient between A and B (including entrance effects) is $k_{AB} = 7$ based on the velocity in the pipe. The flow then proceeds to the nozzle at C. The loss coefficient k_{B-C} is 3, again based on the velocity in the pipe. The pipe is of constant diameter $d = 1$ m. The jet diameter is 0.2 m and the velocity coefficient of the nozzle is 0.97.
(a) Compute the jet velocity V_j.
(b) Compute the maximum obtainable power from the jet.
(c) Find the pressure at B.

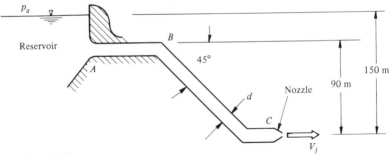

Problem 3.29

3.30 A turbine is fed from a water reservoir whose level is 173 m above the level of the discharge pipe of the turbine. The discharge pipe discharges into the atmosphere and the velocity at this point is 3 m/s. The friction loss in the entire pipeline (except for the losses in the turbine) is 15 m when the flow is 10 m³/s. If the power output under these conditions is 14 MW, find the efficiency of the turbine.

3.31 Oil is flowing in a pipeline at a rate of 20 ft³/sec. The friction drop in the pipeline is 50 ft per mile. Available for the pumping stations are 500-hp pumps with an efficiency of 80 percent. At what intervals should the pumping stations be located? Assume that the density of the oil is the same as that of water at 68°F.

3.32 Consider the flow from a large tank through a short tube as shown. The entrance to the short tube can be regarded as a sharp-edged orifice. Find the pressure at the vena contracta in terms of the pressure in the tank. If the fluid is water and the vapor pressure is 3.5 kPa, find the tank pressure at which the pressure in the vena contracta becomes equal to the vapor pressure. When the pressure reaches this value, the water boils and the vapor formed will induce the jet to leave the walls, making it spring clear, as shown in the figure. The atmospheric pressure is 100 kPa.

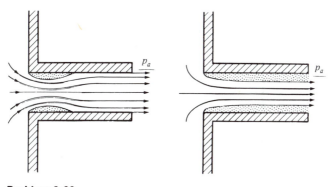

Problem 3.32

3.33 (a) Water flows from the tank through the converging–diverging nozzle, as shown. The discharge coefficient of the orifice as a whole is $c_d = 0.80$ and the loss coefficient up to section A is $k = 0.01$ based on the velocity in the narrow throat. The ratio of the discharge area to the throat area is 5. Find the pressure at section A in terms of the water level h and the atmospheric pressure p_a. Assume the discharge contraction coefficient c_c to be unity and assume the height h to be much larger than the nozzle exit diameter d.

(b) If the pressure at A becomes equal to the vapor pressure of the water, the water in this section will start to vaporize. The pressure at A will then remain unchanged with further increase in h. This condition is known as

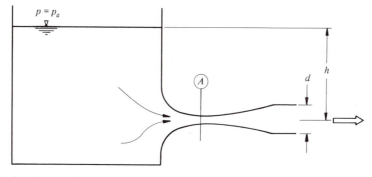

Problem 3.33

cavitation. Find the flow rate at which cavitation will begin if the atmospheric pressure $p_a = 14.7$ psia, the vapor pressure $p_v = 0.5$ psia, and the discharge area is 1 in.^{2}.

3.34 Consider the piping circuit shown in the accompanying diagram. The head imparted by the pump is M, the tank level is at a height h, the cross-sectional areas of the various pipe sections are A_{12}, A_{23}, and A_{24}, respectively, and the corresponding loss coefficients are k_{12}, k_{23}, and k_{24}. The notation k_{12} designates the loss coefficient between stations 1 and 2 and is based on the velocity in this section. The other two coefficients are defined analogously. All pipes are assumed to lie in a horizontal plane and the height h is so much larger than the diameters of the pipes that elevation differences within the pipe may be neglected.

For this piping circuit set up equations so that the flow rates from the two pipes may be compared in terms of the given data. (Do not solve the equations.)

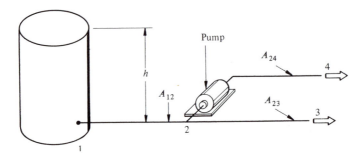

Problem 3.34

3.35 A cylindrical tank with a cross-sectional area A_1 is provided with a sharp-edged orifice of area A_2. The coefficient of discharge of the orifice is c_d. The water level is initially at a height y_0 above the orifice. Find the time for the water in the tank to decrease to a height y_1 above the orifice. The area $A_1 \gg A_2$, so that the velocity of the water surface in the tank may be neglected. Neglect also changes in the kinetic energy of any fluid within the tank, including that in the neighborhood of the orifice.

3.36 Calculate the discharge from a rectangular orifice of width b and height a if the height of the water level above the top of the orifice is h, where h is of the same order of magnitude as a. Assume a discharge coefficient c_d for all sections of the orifice. The velocity of the approaching fluid may be neglected.

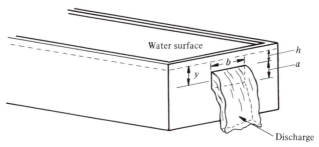

Problem 3.36

3.37 A fire nozzle is to be used at an elevation of 9 m above the level of a reservoir. The velocity of the water jet is to be 15 m/s. The ratio of hose to nozzle area is 4. The loss coefficient between point A and the inlet to the pump is 5, and the loss coefficient between the discharge side of the pump and the *entrance* to the nozzle is 6. The velocity coefficient of the nozzle is $c_v = 0.90$, and the contraction coefficient is 1.0. The area of the inlet pipe is the same as that of the hose.
(a) Find the net head to be supplied by the pump.
(b) If the pump efficiency is 70 percent, find the power required to drive the pump for a discharge of 14 kg/s.

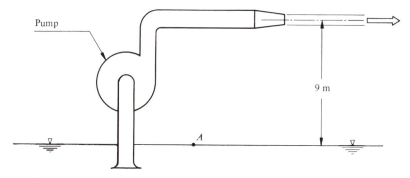

Problem 3.37

3.38 The tailrace (discharge point) of a turbine installation is a distance h below the water level of a dam, as shown. The head lost to friction in the piping, and so on, is $k(V^2/2g)$, where V is the water velocity at the tailrace. The cross-sectional area of the pipe A is the same everywhere and the water density ρ is constant.
(a) What is the head absorbed by the turbine?
(b) What is the power developed by the turbine assuming that the turbine has an efficiency of 90 percent?
(c) What is the condition for maximum power output from the turbine assuming constant h, a constant k, and a constant turbine efficiency?
(d) If the *overall* efficiency of the installation is defined as the ratio of the head abstracted from the flow by the turbine to the elevation head h, what is the efficiency for maximum power output?

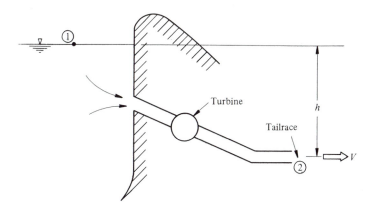

Problem 3.38

3.39 **(a)** A light sports car of mass M is on a hydraulic rack. The cross-sectional area of the cylinder is A, and the car is at height y_0 above the ground. The fluid in the cylinder is oil of a density ρ. The car is to be lowered by opening an orifice located at ground level. The area of the orifice is A_0 and the discharge coefficient is c_d. Find the time it takes for the car to reach the ground after the orifice is opened. Neglect the effect of the nonsteady term in the Bernoulli equation and neglect all losses except those which are taken into account by the discharge coefficient.

(b) What orifice area A_0 is required if the car is to reach the ground 15 s after the valve has been opened if the numerical values of the above-mentioned quantities are

$$M = 500 \text{ kg}, \qquad A = 0.1 \text{ m}^2, \qquad h_0 = 2 \text{ m},$$

$$c_d = 0.60, \qquad \rho = 800 \text{ kg/m}^3.$$

3.40 A steady force F is exerted on a piston (see the figure). The area of the piston is A_p and the jet area is A_j. Assume that the flow to the right of the piston is steady and friction losses in the portion upstream of the nozzle may be neglected. The fluid density is ρ and the atmospheric pressure is p_a. If the velocity coefficient of the nozzle is c_v, what is the jet velocity?

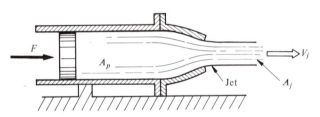

Problem 3.40

3.41 **(a)** Under what conditions does the following formula hold?

$$\frac{p_1}{\rho g} + \frac{V_1^2}{2g} + y_1 = \frac{p_2}{\rho g} + \frac{V_2^2}{2g} + y_2.$$

(b) What additional assumption is being made if this equation is applied to the flow through pipes, as was done in many of the foregoing problems?

3.42 A pump takes water from a reservoir (see the figure), and discharges it at an elevation 30 m higher. The friction losses in the piping are given by $k(V^2/2g)$, where V is the velocity in the pipe and k is the loss coefficient, which is considered constant. The contraction coefficient at the pipe discharge $c_c = 1.0$. The

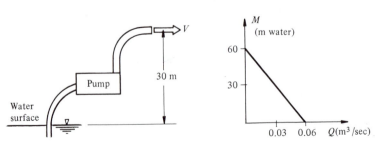

Problem 3.42

relation between the total head rise across the pump (M) and the flow rate through the pump (Q) is given by the accompanying chart. If the liquid is water, $k = 20$, and the pipe diameter is 0.1 m,

(a) What is the flow rate through the pump?

(b) What is the power required to drive the pump if $\eta_{pump} = 70$ percent?

3.43 A wind tunnel consists of a contracting nozzle, a test section in which experiments are conducted, a system of piping to recirculate the flow, and a fan to drive the air (see the figure). The throat of the tunnel is 10 ft dia., and the upstream diameter is 20 ft. Assume that the nozzle velocity coefficient is 0.98, and that 0.2 of the velocity head in the throat is lost to friction in the remainder of the circuit.

(a) What is the pressure drop Δp in psi across the nozzle if the velocity in the throat is 235 ft/sec? The specific weight of air is 0.075 lbf/ft³, and air may be considered incompressible at these velocities.

(b) If the fan efficiency is 85 percent, what horsepower is required to drive the fan of this tunnel?

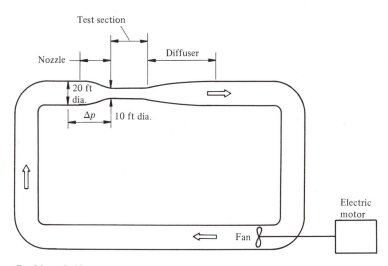

Problem 3.43

3.44 Suppose in Problem 3.42 that the total length of piping is 30 m, that the friction loss can still be given by the relation $k(V^2/2g)$, and that initially the pipe is full of water but has zero velocity. The pump is then started and brought quickly to speed. It may be assumed that the pump characteristic is at all times given by the diagram of Problem 3.42.

(a) Write the nonsteady Bernoulli equation, including the pump head M and head loss between a point on the water surface and the discharge to obtain a differential equation for the velocity of flow in the pipe.

(b) Determine the initial acceleration of the water mass.

(c) Based on the value of the initial acceleration of the fluid, how far above the surface of the reservoir can the pump be located without the pressure difference being more than 80 kPa below atmospheric pressure at that point? Assume that the length of pipe leading from the water reservoir to the inlet of the pump is equal to the elevation of the pump above the water surface.

3.45 In order to traverse poorly surfaced terrain a truck is designed to float on an air cushion. Assume the "air cushion" to be contained within a canvas-like skirt that is fastened around the periphery of the truck. Air, supplied by a compressor to the cushion, escapes through the clearance between the end of the skirt and the ground. The skirt is stretched in a rectangular shape 3×9 m. The mass of the loaded truck is 9000 kg and the average ground clearance is 2.5 cm. The volume of the cushion is sufficiently large so that the velocities in the central portion of the cushion are very low. The air may be treated as incompressible ($\rho = 1.2$ kg/m^3), but this assumption should be checked.
(a) Determine the airflow rate needed to maintain the cushion.
(b) Find the power given by the compressor to the cushion air.

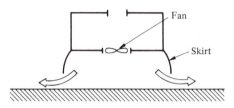

Problem 3.45

3.46 A water tunnel consists of a closed loop of conduit. Of this, a length l_1 has a cross section A_1 and a shorter length l_2 has a cross section A_2. The latter corresponds to the test section. The loss coefficient in the first portion is equal to k_1 [$h_{f1} = k_1(V_1^2/2g)$], and in the test section it is k_2 [$h_{f2} = k_2(V_2^2/2g)$]. The velocity in the *test section* at normal operating conditions is equal to V_0. Normally, the water is circulated by a pump. Assume now that the pump is suddenly shut off. The pump impeller, however, continues to turn, because it is now driven by the water and, in fact, acts as a turbine. In this mode of operation, it absorbs a head M.
(a) Find an expression for the time required for the water in the tunnel to come to rest. For convenience define a suitable equivalent length such as

$$l = l_2 + \frac{A_2}{A_1} l_1$$

and similarly, an equivalent loss coefficient

$$k = k_2 + k_1 \left(\frac{A_2}{A_1}\right)^2.$$

(b) Simplify this expression for the case that

$$\frac{kV_0^2}{gM} \gg 1$$

and evaluate the time numerically for

$$l_1 = 36 \text{ m}, \qquad k_1 = 16.0, \qquad A_1 = 0.2 \text{ m}^2,$$
$$l_2 = 3 \text{ m}, \qquad k_2 = 6.0, \qquad A_2 = 0.05 \text{ m}^2,$$
$$V_0 = 30 \text{ m/s}, \qquad M = 0.6 \text{ m}.$$

3.47 A pump has the following operating characteristic when operated at a constant speed and for a given fluid:

$$\Delta p = A - BQ,$$

where Δp is the pressure increase across the pump and Q is the discharge rate. The velocity head at the intake and discharge from the pump is included in the term Δp. The pump discharges through a hose with a nozzle attached to the end. The loss coefficient of the hose is k (based on the velocity in the hose) and the cross-sectional area of the hose is A_1. The jet cross section is A_2 and the nozzle coefficient is c_v. The pump takes water from an open reservoir at sea level and the nozzle discharges into the atmosphere, also at sea level.
(a) Find the discharge rate from the nozzle.
(b) By means of a sketch, indicate by two curves the pressure loss in the circuit downstream of the pump and the pressure increase in the pump, both as a function of flow rate. Mark the operating point that corresponds to the solution in part (a).

3.48 A nozzle with an exit diameter of 2.5×10^{-2} m is to be used at an elevation of 40 m. The velocity of the jet is to be such that it would reach a point 20 m above this elevation (neglect air friction). The pump delivering the water is at zero elevation, and this pump receives water from a hydrant at 2×10^5 N/m². The losses at the entrance to the hose are

$$h_{f_1} = 0.50 \frac{V^2}{2g},$$

where V is the average velocity in the hose. The losses in the hose are

$$h_{f_2} = 0.025 \frac{V^2}{2g} \frac{l}{d},$$

where V is again the average velocity in the hose and where l/d is assumed to be 2500. The nozzle velocity coefficient $c_v = 0.97$. Furthermore, the cross section of the hose is 10 times that of the nozzle and the contraction coefficient of the nozzle is $1.0 (A_j/A_n = c_c = 1.0)$.
(a) Determine the water flow rate.
(b) Determine the power that the pump has to deliver to the water.

3.49 A river flows at a velocity V_1 that is assumed to be constant over the cross section. The river then discharges into the atmosphere through a submerged orifice that has a rectangular cross section. The width of the orifice is b and its height is h. The top edge of the orifice is located at a distance c below the water surface at point 1. The contraction coefficient of the orifice is c_c and friction losses may be neglected. The pressure throughout the jet as it issues from the orifice is atmospheric. All pressures are to be measured in respect to the atmospheric pressure (i.e., set $p_a = 0$) and elevations are to be measured in respect to the *lower edge* of the submerged orifice. Designate any arbitrary distance above this edge by y.
(a) What is the total head $(V_1^2/2g + p_1/\rho g + y_1)$ at any point in the flow at section 1?
(b) Write the Bernoulli equation for a streamline that starts at section 1 and leads to a point at an elevation y of the stream issuing from the orifice.

(c) From part (b) develop a relation for the discharge through an element of the orifice ($b\,dy$). Next obtain an equation for the discharge through the orifice.

(d) From part (c) derive the discharge through a weir by letting $c = 0$ and assuming that $V_1^2/2g \ll h$.

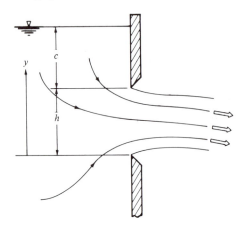

Problem 3.49

3.50 A pipe (or hose) is attached to a large water reservoir. The head loss in the pipe is equal to $k(V_p^2/2g)$, where V_p is the velocity in the *pipe* and k is a given constant. A nozzle is attached to the pipe and water issues from that nozzle into the atmosphere. The top of the tank is also exposed to the atmosphere. The water level in the tank is maintained at a height h above the nozzle. The cross-sectional *area* of the pipe is A_p, and the ratio of the *area* of the jet to that of the pipe is r ($A_j/A_p = r$). Neglect nozzle losses and other losses.

(a) Determine an expression for the jet velocity in terms of h, k, r, and g.

(b) The energy in the jet per unit weight is equal to its velocity head. With this in mind, develop an expression for the *power* of the jet. (This power could be converted, for example, by letting the jet drive a turbine.)

(c) The area of the pipe A_p is given, but that of the jet can be changed by adjusting the nozzle (or providing new nozzles). The ratio r is, therefore, variable. Find r in terms of the loss coefficient k so as to maximize the power in the jet. (As a general comment: k will usually be large enough so that $r < 1$, and the nozzle is of normal shape.)

(d) Write an expression for the maximum power in terms of h, A_p, k, ρ, and g.

3.51 Water is supplied from a single reservoir to two different levels of a house. The lower level is a distance h below the reservoir surface and the upper level is a distance b above the lower level (see the illustration). The head loss from the reservoir to the branch point B is given by $k_0 V_0^2/2g$, where V_0 is the velocity in the section of pipe between the reservoir and the branch point. The cross-sectional area of that portion of pipe is A_0. The head loss from the branch point B to the lower outlet is $k_1 V_1^2/2g$. The velocity V_1 and the cross-sectional area A_1 apply to the section of pipe from B to the lower outlet. Finally, the head loss from B to the *upper* outlet is $k_2 V_2^2/2g$. The area A_2 and the velocity V_2 apply to this last section of pipe. The water issues with a velocity V_1 from the lower outlet, and with V_2 from the upper outlet. All the loss coefficients and elevations are given.

(a) It is now specified that the discharge at the two levels is to be the same $(V_1 A_1 = V_2 A_2 = Q_1 = Q_2 = Q)$. Q is given and the area A_2 is also given. Find the necessary area A_1 to meet this requirement.

Hint: Bernoulli's equation holds along "streamlines." It is advised to use two "typical" streamlines each starting from the reservoir and ending at the respective outlets.

(b) The inhabitant of the upper story now wants to "help" his lower neighbors to save water. For that purpose the upper inhabitant installs a pump on his level as shown. The pump is designed to add a head M_p to each unit weight of water flowing through it. How large does M_P have to be so that *nothing* will flow out of the lower outlet? (Give your answer only in terms of h, b, and one or more of the k's.)

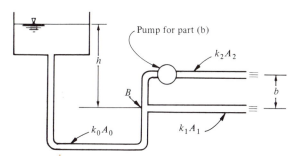

Problem 3.51

3.52 An incompressible fluid is flowing from a large tank through a long pipe. The outflow velocity is monitored by a Pitot tube. The pressure differential of the Pitot tube Δp (total pressure minus static pressure) is fed into a control system that instantaneously creates a pressure p_t in the tank such that p_t is equal to the atmospheric pressure plus α times the Pitot tube differential $(p_t = p_a + \alpha \, \Delta p)$.

(a) Develop a differential equation for the velocity V as a function of time.

(b) Solve the differential equation for V if the initial velocity is V_i.

(c) Sketch carefully V versus t for three cases: $\alpha = 1$; $\alpha > 1$; and $\alpha < 1$.

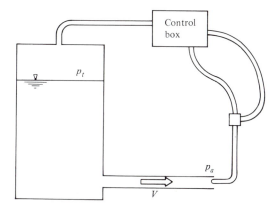

Problem 3.52

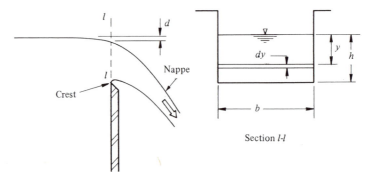

FIG. 3.10   Front and side views of a rectangular weir in a channel.

(c) The Weir

Last we discuss a problem of flow measurement in an open channel. The flow in the channel is restricted by inserting a plate containing a cutout, as shown in Fig. 3.10. Water issues from the cutout, and the total flow may be estimated from a knowledge of the depth of the stream over the crest of the cutout. Such a device is known as a *weir*. The exact flow is complicated, but it will again be possible to arrive at a simple idealized answer, which may then be corrected by means of empirical coefficients.

To calculate the discharge over the weir, we shall assume that the pressure along the line *l–l* is equal to that of the surrounding atmosphere. For this assumption to be valid, it is necessary that the pressure beneath the nappe (the overflow over the weir) also be equal to the atmospheric pressure and that the curvature of the streamlines is negligible. Then if the drawdown d (the drop of the water surface below that in the approach channel) is neglected, the velocity at any depth y is

$$V = \sqrt{2gy},$$

neglecting, in addition, friction and the velocity of the oncoming flow. We shall now furthermore assume that the discharge coefficient, c_d [Eq. (3.17)], is the same for all cross sections, so that the total discharge becomes

$$Q = c_d \int V \, dA = c_d b \int_0^h \sqrt{2gy} \, dy = \tfrac{2}{3} c_d b (2g)^{1/2} (h)^{3/2}.$$

The coefficient c_d again must be determined experimentally. It depends on the ratio h/b, although the value of $c_d = 0.62$ will be satisfactory for most applications. A correction for the velocity of approach can also be made. Coefficients for weirs of various types are available in Reference 1, for example.

PROBLEM

3.53 Calculate the discharge from a triangular orifice when the water level forms the top of the orifice. Assume again that c_d is the same for all sections. Such a device is called a triangular weir. It is also used to measure the discharge of channels by measuring the height h. For a triangular weir with an angle of $\alpha = 90°$ a discharge coefficient of $c_d = 0.58$ will yield good values for water, provided that $h > 0.1$ m.

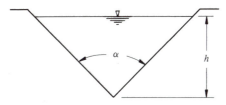

Problem 3.53

In one way or another the preceding examples have all dealt with the determination of a velocity from the measurement of a pressure difference and have therefore been based on the Bernoulli equation. They are more than simple examples of the use of this equation. In fact, they represent various useful methods of measuring velocity and hence rate of flow in incompressible fluid systems. It should be mentioned, however, that numerous other methods of velocity measurement are also available, and we shall mention a few. Velocities can be determined by timing the drift of particles or of chemical solutions suspected in a stream; modest flow rates can be measured by the rate at which containers of known volume are filled. Freewheeling propellers suspended in a stream turn at rates nearly proportional to the fluid velocity and when calibrated are used to determine velocities in rivers, streams, and ducts of air-conditioning systems.

Frequently, other physical phenomena that depend upon the velocity are used as velocity indicators. For example, one, which is of widespread use for measuring velocities of gases such as air, works on the principle that the rate of transferring heat to a flowing medium is dependent on the velocity. Another, which provides a signal directly proportional to the fluid velocity, is based on a result of electromagnetic theory that the EMF produced in a moving medium is directly proportional to the velocity of the medium and the magnetic field. More recently, optical methods have been applied with great success for direct measurement of the velocity components. A brief review of some of these methods and others may be found in References 3–6.

†3.9 Limitation of the Incompressibility Assumptions

The preceding examples and problems have all treated incompressible fluids or fluids of constant density. This is a reasonable approximation for liquids, especially when the flow is steady, since they are nearly incompressible. But frequently, the assumption of constant density is also made for the flow of gases, which are obviously very compressible. Problem 3.43 affords a typical situation in which the incompressibility approximation is often made for the flow of a gas. But even liquids are compressible to some degree, and before one can use the incompressibility assumption either for liquids or gases with certainty, one must have some criterion to ensure that the error is small. An estimate of this error can be made fairly easily for a barotropic fluid, that is, for a fluid whose density is only a function of pressure. We shall outline the steps for such an estimate in the following discussion. In the

† This section may be omitted without loss of continuity.

simplest case of steady flow with no body forces, we can write the Bernoulli equation as

$$\frac{V^2}{2} + \int \frac{dp}{\rho} = \text{const}, \tag{3.5a}$$

as in Eq. (3.5). Since $\rho = \rho(p)$, we can expand the reciprocal of the density in a McLaurin series

$$\frac{1}{\rho} = \frac{1}{\rho_1} + \frac{d}{dp}\left(\frac{1}{\rho}\right)_1 (p - p_1) + \cdots.$$

Since we are dealing with fluids that may be considered almost incompressible, the series can be terminated after the second term. The subscript 1 indicates that the quantities are evaluated at a reference point where the pressure is p_1 and density ρ_1. The derivative of the density in respect to the pressure can be written as

$$\frac{d}{dp}\left(\frac{1}{\rho}\right) = -\frac{1}{\rho^2}\frac{d\rho}{dp} = \frac{-1}{\rho^2(dp/d\rho)} = -\frac{1}{\rho^2 a^2},$$

where the symbol a^2 stands for the quantity $dp/d\rho$. With the above expansion for the density, Eq. (3.5a) can readily be integrated for small density changes, yielding the result

$$\frac{V^2}{2} + \frac{p}{\rho_1} - \frac{1}{\rho_1^2 a_1^2}\left(\frac{p^2}{2} - p_1 p\right) = \text{const.} \tag{3.5b}$$

To make this equation more meaningful, let us consider a definite problem: Suppose that a stream of velocity V_1, static pressure p_1, and density ρ_1 is brought to rest or stagnated. The velocity V_2 is zero, and the pressure there is measured to be p_2. We want to calculate V_1 knowing only p_1, p_2, and a_1. From Eq. (3.5b) we find after a little rearrangement that

$$\frac{V_1^2}{2} = \frac{p_2 - p_1}{\rho_1}\left(1 - \frac{p_2 - p_1}{2\rho_1 a_1^2}\right).$$

Of course, if the fluid were truly incompressible, a_1^2 would be infinite and then we would have merely $V_1 = \sqrt{(2/\rho_1)(p_2 - p_1)}$, as from the usual incompressible Bernoulli equation. Let us anticipate now that the true velocity will not differ greatly from this incompressible value. We may then surely approximate the correction term in the above equation by replacing the pressure difference $p_2 - p_1$ by its incompressible value $\rho V_1^2/2$. Thus, approximately,

$$V_1^2 = \frac{2}{\rho_1}(p_2 - p_1)\left(1 - \frac{V_1^2}{4a_1^2}\right).$$

We now see that the incompressible assumption is permissible when calculating velocities from pressure differences when the term $V_1^2/4a_1^2$ is small compared to unity. The term a_1^2 is an important property of the fluid. It may be shown that it is always positive. When multiplied by the density, it becomes equal to the *bulk compressibility* modulus,[†] which gives the pressure change for the fractional change in density. Furthermore, and of particular interest for the present discussion, a_1 is equal to the velocity with which sound waves are propagated in a compressible medium (as shown in Chapter 9).

---

† Values of this property may be found in Appendix 2.

The discussion above was limited to barotropic fluids. Nevertheless, for the purpose of the error analysis above, fluids may be considered barotropic for many engineering applications, including, for example, the flow of air over and around an airplane, the flow of gas through a nozzle, and the flow of water in conduits or over obstacles.

From the foregoing discussion it may be seen that compressibility effects will be small when the ratio of the fluid velocity to the velocity of sound is small compared to unity. For example, if V_1/a_1 is equal to 0.3, the foregoing equation indicates that the error in velocity incurred by using the incompressible assumption is about 1 percent. For air at room temperature this limitation corresponds to a fluid velocity of 100 m/s; in water for which the velocity of sound is about 1500 m/s, velocities as high as 450 m/s are allowable without causing significant errors.

Although a measure of the compressibility effect has been established for steady flows, we have not developed any such criterion for nonsteady motions, such as the acceleration of water masses in pipes. This problem is somewhat more difficult to treat than the one involving steady flow. It turns out, however, that if the time required for the nonsteady incompressible phenomenon to take place is much longer than the time required for a sound wave to traverse the portion of the flow in which substantial changes in velocity take place, then the incompressible Bernoulli equation may be used.

REFERENCES

1. H. Rouse, ed., *Engineering Hydraulics*, Wiley, New York, 1950, Chap. 3.
2. R. L. Daugherty and A. C. Ingersoll, *Fluid Mechanics*, 5th ed., McGraw-Hill, New York, 1954, Chap. 5.
3. V. Streeter, ed., *Handbook of Fluid Mechanics*, McGraw-Hill, New York, 1961, Chap. 14.
4. American Society of Mechanical Engineers, *Fluid Meters*, ASME, New York, 1959.
5. N. P. Cheremisoff, ed., *Encyclopedia of Fluid Mechanics*, Vol. I, Sec. III, Gulf, Houston, 1986.
6. F. A. Durst, A. Melling, and J. H. Whitlaw, *Principles and Practice of Laser-Doppler Anemometry*, 2nd ed., Academic Press, New York, 1981.

CHAPTER 4

Momentum Theorems

4.1 Introduction

Newton's law for a given particle or system of particles of fixed mass can be expressed in the form $F_x = d(M_x)/dt$, where F_x is the force on the particle in the x-direction and M_x is the momentum of the particle also in the x-direction. Thus, in words, this equation reads: The force exerted on the particle in a given direction is equal to its rate of change of momentum in that direction. Since the law is equally as applicable to a single particle as to a system of particles, it can be applied to the flow of fluids as well.

As an introductory example consider the deflection of an incompressible, steady fluid jet by a stationary channel. Let the jet speed be constant and equal to V. The density is ρ, the jet cross-sectional area is A, and θ is the angle through which the jet is deflected. The flow is pictured in Fig. 4.1.

To solve this problem, it must again be remembered that Newton's law applies to a particle (or a set of particles) of a fixed amount of mass. Accordingly, we must select a portion of the fluid jet with a fixed amount of mass and determine its rate of change of momentum. For this purpose let the dashed lines in the jet of Fig. 4.1 represent fixed surfaces through which flow takes place. They are chosen to coincide with the original position of the fluid mass, while the dotted lines represent its position at a time dt later. The momentum of the given fluid mass at the two instants can be determined in terms of the momenta in regions I, II, II, which will be designated by $\mathbf{M_I}$, $\mathbf{M_{II}}$, and $\mathbf{M_{III}}$. The initial momentum of the entire fixed fluid mass will be denoted by $\mathbf{M_1}$ and the momentum of the same fluid mass after the elapse of the time increment dt will be denoted by $\mathbf{M_2}$. The letters x and y will be

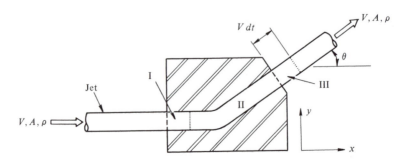

FIG. 4.1 A jet being deflected by a channel—illustrating the momentum principle.

affixed as second subscripts to indicate momentum components. It then follows that

$$\mathbf{M_2} = \mathbf{M_{II}} + \mathbf{M_{III}},$$

and the momentum of the same mass when it coincides with the fixed dashed lines is

$$\mathbf{M_1} = \mathbf{M_I} + \mathbf{M_{II}}.$$

The volume of both regions I and III is $VA\,dt$, since the velocity is constant, so that

$$M_{I_x} = \rho AVV\,dt$$

and

$$M_{III_x} = \rho AVV\,dt\,\cos\theta.$$

The change in momentum is then equal to

$$dM_x = M_{2_x} - M_{1_x} = \rho AV^2(\cos\theta - 1)\,dt,$$

since the term M_{II_x} cancels out. The force exerted on the fluid in the x-direction is

$$F_x = \frac{dM_x}{dt} = \rho AV^2(\cos\theta - 1).$$

In the same way the force in the y-direction is found to be

$$F_y = \frac{dM_y}{dt} = \rho AV^2\,\sin\theta.$$

The forces F_x and F_y are the forces that must be exerted on the fluid in order to deflect the jet through the angle θ. These forces are the components of a force vector $\mathbf{F}$ that is in the direction of the resultant momentum change of the fluid. The reaction $\mathbf{R}$ experienced by the channel is, of course, equal in magnitude and opposite in direction to the force on the fluid. The force system is shown in Fig. 4.2, and for simplicity it is assumed that the pressure at the ends of the jet and that surrounding the channel is zero. From Fig. 4.2b it is seen that the force necessary to hold the channel in place is simply $\mathbf{F}$.

Although all problems involving the use of the momentum theorem for fluid flow can be solved in the foregoing manner, this procedure is not too direct. For such problems it is convenient to express Newton's law in a form suitable for a fixed volume rather than a fixed amount of mass. In the problem just worked, for example, it would be simpler to consider only the flow contained within the fixed volume of the dashed lines of Fig. 4.1. The present results leads one, in fact, to anticipate the form of the desired law: the term ρAV^2 may be interpreted as

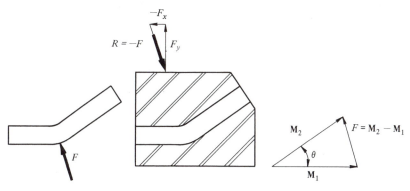

(a) Force on fluid in jet   (b) Force system on channel   (c) Momentum diagram

FIG. 4.2 Force and momentum diagrams for the jet of Fig. 4.1.

momentum flow or flux (mass × velocity/time), so that the force on the fluid may be said to be equal to the net outflow of momentum through the fixed surface shown. The volume determined by these fixed surfaces we shall call a *control volume*.

The results of the foregoing paragraphs can be conveniently summarized by the use of vectors. Momentum is a vector quantity; consequently, momentum flux is also a vector, having the direction of the velocity vector and the magnitude of $\rho A V^2$. If we let $\dot{\mathbf{M}}$ denote momentum flux vector ($\rho A V \mathbf{v} = \rho Q \mathbf{v}$), then the force on the fluid in the above example is the vector difference between the leaving momentum flux vector $\dot{\mathbf{M}}_2$ and the initial momentum flux vector $\dot{\mathbf{M}}_1$. The magnitudes of both vectors are the same, and they are inclined at an angle θ to each other, as shown in Fig. 4.2c. In this way the force on the fluid or reaction on the channel can be determined graphically.

4.2 General Derivation of the Momentum Theorem

Because of the importance of the momentum theorem in mechanics, it is desirable to derive it in more general terms. We begin by considering the momentum change of an arbitrary quantity of mass originally contained within the volume $\mathscr{V}_0$ shown in Fig. 4.3. This volume $\mathscr{V}_0$ coincides with the volume containing the mass under consideration at the initial instant. The volume $\mathscr{V}_0$ constitutes the *control volume* for this more general case. It should again be regarded as a volume fixed in space through which flow may take place unrestrictedly. The surface enclosing the control volume will be called the *control surface* and it will be designated by S_0. The theorem to be developed will be expressed in reference to the control volume and the control surface.

The volume containing the mass is shown at two instants of time. The mass elements comprising the original volume move with the fluid, and at succeeding increments in time the volume will be distorted and even change size if the fluid is compressible. The force acting on the mass contained within the volume by Newton's law is equal to the rate of change of the momentum of the mass. The momentum at any time evidently is

$$\mathbf{M} = \int_{\mathscr{V}'} \rho \mathbf{v} \, d\mathscr{V},$$

where the symbol $\mathscr{V}'$ indicates that the integral is to be evaluated in the region of space containing the set of mass particles selected for analysis. We have therefore

$$\mathbf{F} = \left(\frac{\Delta \mathbf{M}}{\Delta t} \right)_{\lim \Delta t \to 0} = \frac{d\mathbf{M}}{dt} = \frac{D}{Dt} \int_{\mathscr{V}'} \rho \mathbf{v} \, d\mathscr{V}. \tag{4.1}$$

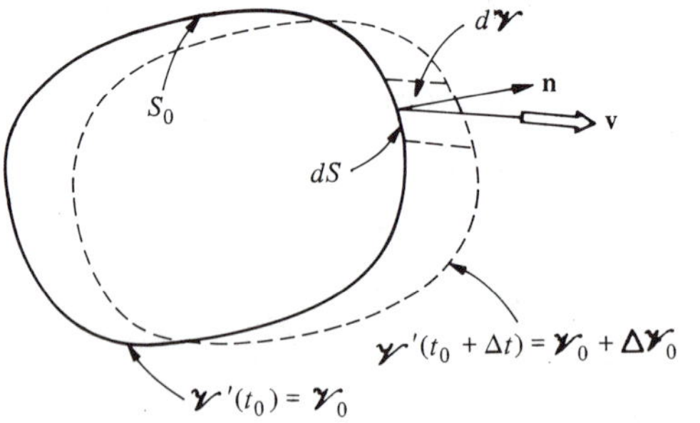

FIG. 4.3 Arbitrary body of fluid as used in the derivation of the momentum theorem. The fixed volume $\mathscr{V}_0$ represents the control volume.

This integral will now be transformed into other integrals more convenient for calculation. To do this, we write

$$\mathbf{F} = \lim_{\Delta t \to 0} \frac{\int_{\mathscr{V}_0 + \Delta \mathscr{V}} (\rho \mathbf{v})_{t_0 + \Delta t} \, d\mathscr{V} - \int_{\mathscr{V}_0} (\rho \mathbf{v})_{t_0} \, d\mathscr{V}}{\Delta t}, \qquad (4.2)$$

where $\mathscr{V}_0$ is the volume originally occupied by the given fluid mass and $\mathscr{V}_0 + \Delta \mathscr{V}$ is the volume occupied by the same mass at the end of the time interval Δt. Equation (4.2) represents the definition of the time derivative. We see that contributions to the integral arise from the rate of change of the product $\rho \mathbf{v}$ with time and also from the fact that volume occupied by the fixed mass is changing with time. Now if the time increment is small, we may write $\rho \mathbf{v}$ in the series

$$(\rho \mathbf{v})_{t_0 + \Delta t} = (\rho \mathbf{v})_{t_0} + \frac{\partial}{\partial t} (\rho \mathbf{v})_{t_0} \, \Delta t$$

and terminate the series with the second term. Higher-order terms will drop out when Δt is made to approach zero. The volume at time $t_0 + \Delta t$ is written

$$\mathscr{V}'_{t_0 + \Delta t} = \mathscr{V}_0 + \Delta \mathscr{V},$$

where $\mathscr{V}_0$ is the volume containing the given mass at time t_0. $\mathscr{V}_0$ is *not* a function of time and is a well-defined but arbitrary region in space. Equation (4.2) now becomes

$$\mathbf{F} = \int_{\mathscr{V}_0} \frac{\partial (\rho \mathbf{v})}{\partial t} \, d\mathscr{V} + \lim_{\Delta t \to 0} \int_{\Delta \mathscr{V}} (\rho \mathbf{v})_{t_0} \frac{d\mathscr{V}}{\Delta t}, \qquad (4.3)$$

and the integral over the volume increment $\Delta \mathscr{V}$ must be determined. The increment $\Delta \mathscr{V}$ consists of the volume region shown between the solid and dotted portions of Fig. 4.3. The volume element of $d\mathscr{V}$ is best expressed as an element of surface area dS of $\mathscr{V}_0$ and the perpendicular distance between the slightly separated surfaces of Fig. 4.3 (i.e., $d\mathscr{V} = dS \, dn$). But

$$dn = \mathbf{v} \cdot \mathbf{n} \, \Delta t,$$

where $\mathbf{n}$ is an outward-pointing unit vector normal to dS and $\mathbf{v}$, of course, is the velocity vector. The last term of Eq. (4.3) becomes, therefore,

$$\lim_{\Delta t \to 0} \int_{\Delta \mathscr{V}} (\rho \mathbf{v})_{t_0} \frac{\mathbf{v} \cdot \mathbf{n} \, \Delta t \, dS}{\Delta t} = \int_{S_0} \rho \mathbf{v} \mathbf{v} \cdot \mathbf{n} \, dS,$$

in which S_0 is the surface of the fixed volume $\mathscr{V}_0$. Thus we obtain the relation

$$\mathbf{F} = \frac{D}{Dt} \int_{\mathscr{V}'} \rho \mathbf{v} \, d\mathscr{V} = \int_{\mathscr{V}_0} \frac{\partial}{\partial t} (\rho \mathbf{v}) \, d\mathscr{V} + \int_{S_0} \rho \mathbf{v} \mathbf{v} \cdot \mathbf{n} \, dS, \qquad (4.4a)$$

where $\mathscr{V}_0$ and S_0 are the volume and surface of the fixed control volume. This relation is known as the *momentum* theorem, or *impulse* theorem. The limits of integration of the first term are fixed. Therefore, the differentiation can be performed on the integral rather than the integrand, so that we can also write

$$\mathbf{F} = \frac{\partial \mathbf{M}}{\partial t} + \int_{S_0} \rho \mathbf{v} \mathbf{v} \cdot \mathbf{n} \, dS, \qquad (4.4b)$$

where

$$\mathbf{M} = \int_{\mathscr{V}_0} \rho \mathbf{v} \, d\mathscr{V}. \tag{4.4c}$$

It should be emphasized that there is a careful distinction to be made between Eqs. (4.1) and (4.4). The first represents the rate of change of momentum of a fixed mass, although the volume containing the mass changes with time. This volume $\mathscr{V}'$ may be said to be a *material* volume, since it encloses a given amount of matter. The second, Eq. (4.4), expresses the momentum change in terms of two integrals in respect to a volume *fixed* in space. The first term of (4.4b) is the time rate of change of the momentum contained within a fixed volume and the second term is the integral of the quantity $\rho \mathbf{v} \mathbf{v} \cdot \mathbf{n}$ over the surface of the fixed volume. The product of $\rho \mathbf{v} \cdot \mathbf{n}$ is the mass flow per unit time and unit area perpendicular to the surface of the volume. Thus $\rho \mathbf{v} \mathbf{v} \cdot \mathbf{n}$ is the rate of momentum passing through a unit area of the surface in the direction of $\mathbf{v}$. The surface integral of Eq. (4.4) is therefore the net momentum outflow or flux through the surface of the volume. This formulation is particularly convenient for certain types of problems in which the velocity vector is known on the surface of a region of interest, but detailed knowledge of the flow field throughout the region is not available. The momentum theorem tells us that the force on the surfaces of, or within the volume of an arbitrary region of space is given by Eq. (4.4a). This important law may be summarized in words as:

□ *The time rate of change of momentum contained in a fixed volume plus the net flow rate of momentum through the surfaces of this volume are equal to the sum of all the forces acting on the volume.* (4.5)

The vector law expressed by Eq. (4.4a) or the statement (4.5) may also be expressed in terms of its directional components. For example, in Cartesian coordinates the force in the x-direction on the volume contained within S_0 may be written out in the form

$$F_x = \frac{\partial}{\partial t} \int_{\mathscr{V}_0} \rho u \, dx \, dy \, dz + \int_{S_0} \rho u [u \, dy \, dz + v \, dx \, dz + w \, dx \, dy].$$

In evaluating the second integral in this form of the equation, a distinction has to be made between inflow and outflow as they make a negative and positive contribution, respectively. The proper sign may be obtained from the vector form of the momentum theorem [Eq. (4.4a)].

The momentum theorem, like the energy equation, is one of the important principles of fluid mechanics. As pointed out in the derivation, it is not restricted to incompressible flow, and all types of forces, including those caused by friction, are permitted. Since the form of Newton's law is the same in any uniformly translating (but not accelerating) reference system, the same holds true for the momentum theorem.

In conclusion, let us emphasize that in the treatment of problems by means of the momentum theorem, the first step has to be the selection of the control volume and the corresponding control surface. Both the control volume and the control surface are imaginary concepts and flow can pass through them without being obstructed in any way. The selection of the control volume is essentially arbitrary. To illustrate, in the example in Section 4.1, a suitable control surface might include the dashed lines and the surfaces of the channel between these lines. However, other control surfaces could also be used, and with some experience the control surface best suited for any particular problem may be found.

EXAMPLE 4.1 Application to the Stationary Rocket Motor

As an example of the use of the momentum theorem, the problem of the stationary rocket motor will be studied. Consider the rocket motor shown in the sketch below. The exit pressure at the nozzle is p_e, and the entire motor is surrounded by atmospheric pressure p_a. The velocity of the entering propellant as well as any forces due to the pressure and stresses in the feed pipes may be neglected. We are interested in the net thrust or reaction on the motor itself rather than that on the fluid. For this purpose it is convenient to take a control surface S_0 surrounding the rocket, as shown in Fig. 4.4.

Solution: If the flow is steady, the net flux of momentum out of the surface S_0 is found to be $\rho_e V_e^2 A_e$. The rocket is held in place by the force R. The other forces on the control surface or volume are the pressure forces $(-p_e A_e)$ at the exhaust and the corresponding unbalanced atmospheric force at the left surface, $p_a A_e$. Thus the momentum theorem gives

$$F = \rho V_e^2 A_e = R - A_e(p_e - p_a)$$

or

$$R = \rho_e V_e^2 A_e + A_e(p_e - p_a).$$

The thrust of the rocket motor pushes against the supports necessary to keep it stationary and it is equal and opposite to the restraining force R. In some cases the term $p_e - p_a$ is zero, but in general it does not vanish and sometimes it is quite important. The momentum term of the equation above can be written as $\dot{m}V_e$, where $\dot{m}$ is the mass rate of flow. In this form the concept of momentum flow is readily apparent.

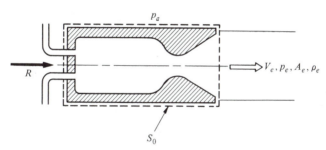

FIG. 4.4 Control volume suitable for computing the thrust of a rocket motor. The surface of the control volume is designated by S_0.

EXAMPLE 4.2 Hydraulic Catapult

A hydraulic catapult has been proposed as a method of launching airplanes. The catapult consists of a high-velocity jet of water that impinges on a bucket attached to the airplane. The bucket is assumed to reverse completely the flow direction without altering its velocity. The jet velocity is V_j and the airplane velocity is V. The area of the jet is A, the fluid density is ρ, and the airplane mass is M. The thrust on the bucket and the acceleration of the airplane are to be determined.

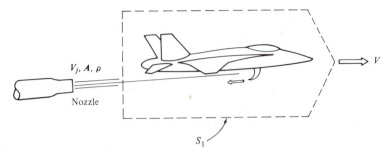

FIG. 4.5a   Control volume suitable for computing the force on a hydraulically launched airplane.

Solution: We neglect the effect of gravity and consider first a control surface S_1 surrounding the entire airplane and bucket as shown in Fig. 4.5a. There is no external force on this surface. We shall assume that S_1 is moving uniformly at velocity V, the instantaneous velocity of the airplane. The bucket and the water influenced by the action of the bucket are subject to the acceleration produced by the flow and their motion is therefore not steady. Nevertheless, we shall assume that the steady Bernoulli relation applies to the flow relative to the bucket, and in so doing it is necessary to keep in mind the type of restrictions mentioned in Example 3.2. Relative to S_1 we then have the velocity vectors $(V_j - V)$ and $-(V_j - V)$ entering and leaving S_1. We shall also choose the control volume so that the length of the jet between S_1 and the bucket is small, so that the fluid velocity leaving the bucket may be assumed equal to that leaving the surface of S_1, as shown. It is also supposed that the mass of liquid in contact with and accelerating with the bucket is small compared to M, so that we may neglect the change with time of the momentum of the jet within the control volume. Then application of Eq. (4.5) gives

$$0 = M\frac{dV}{dt} - 2\rho A_j(V_j - V)^2$$

or

$$\frac{dV}{dt} = \frac{2\rho A_j}{M}(V_j - V)^2$$

for the acceleration of the airplane.

Now let us take a different control surface S_2, surrounding the jet as before and the bucket as well but not the aircraft as shown in Fig. 4.5b. From Eq. (4.5) we have

$$F = -2\rho A_j(V_j - V)^2,$$

where F is the force on the bucket. By drawing a free-body diagram of the airplane, one will see that the *thrust* exerted on the airplane is equal and opposite to F, so

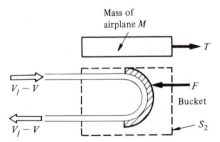

FIG. 4.5b   Detail of jet acting on the reversing bucket of the hydraulic launcher.

that

$$T = -F = 2\rho A_j(V_j - V)^2,$$

which, of course, is the force causing the acceleration computed above.

PROBLEMS

When not otherwise specified the properties of the fluids are to be evaluated at 20°C and 1 atmosphere pressure.

4.1 A large tank is filled with water up to a height h above a small orifice. The area of the orifice is A, the coefficient of contraction is c_c, and the velocity coefficient is c_v. Find the force on the tank due to the jet.

4.2 A stationary rocket is discharging hot gases at a temperature of 3300°C at a velocity of 2100 m/s. The flow rate of the gases is 1000 kg/s. Find the thrust of the rocket assuming that the nozzle exit pressure p_e equals the atmospheric pressure p_a.

4.3 A turbojet engine draws in 40 lbm of air per second, at a velocity of 400 ft/sec, and it discharges the air at 2000 ft/sec. Both velocities are measured relative to the engine. What is the thrust delivered by this engine?

4.4 A tank has an orifice as shown. With such an orifice, there is practically no flow along the tank walls, so that the pressure along the walls is uniform around the circumference, with the exception of the place where the orifice is located. Assuming the fluid to be frictionless and incompressible, find the contraction coefficient for this orifice. This type of orifice is called a *Borda mouthpiece*.

  Hint: Equate the hydrostatic force on the tank to the thrust as determined from the momentum theorem.

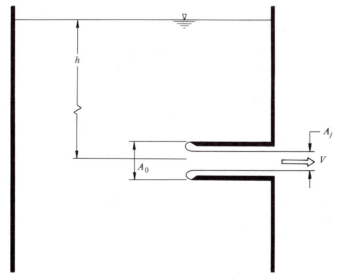

Problem 4.4

4.5 Water discharges through a well-rounded, frictionless orifice under a head h_1. (The water level in the tank may be considered constant.) The jet impinges on a large flat plate that covers the orifice of a second tank in which the water stands at a height h_2 above the orifice. The area of both orifices is the same. If h_2 is given, find h_1 such that the force of the jet is just sufficient to balance the force on the plate (see the accompanying figure).

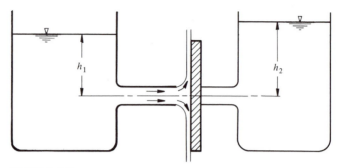

Problem 4.5

4.6 **(a)** Water is issuing from a pressurized tank, as shown. The rate of outflow increases with time, because the valve is being opened. All friction constants are known. Does the Bernoulli equation as stated in Eq. (3.6b) apply between a point on the water surface of the tank and a point in the jet? Explain why or why not.

(b) Does the momentum theorem as stated in Eq. (4.5) apply, using the control surface indicated?

(c) Would the momentum theorem apply if the fluid were air instead of water and the density would change along the pipe?

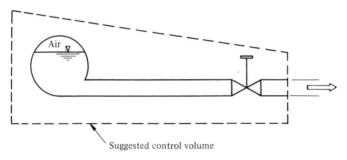

Suggested control volume

Problem 4.6

4.7 A turbojet engine in a wind tunnel receives air at a velocity of 120 m/s and a density of $\rho = 1.0$ kg/m³. The velocity is uniform and the cross-sectional area of the approaching jet is 0.1 m². The velocity of the exhaust jet, however, is not uniform but is given by the equation

$$v = 2V_0\left(1 - \frac{r^2}{r_0^2}\right),$$

where r_0 is the radius of the jet cross section. The density of this jet is $\rho = 0.5$ kg/m³ and $V_0 = 550$ m/s.

(a) Show that V_0 is the average velocity of the exhaust jet.

(b) Find the thrust on the turbojet.

(c) What would the thrust be if the exhaust jet had been of uniform velocity V_0?

4.8 An incompressible fluid flows steadily through a two-dimensional infinite row of fixed vanes, a few of which are shown. The vane spacing is a. The velocities and pressures are constant along stations 1 and 2, and are given by V_1, V_2, p_1, and p_2. The directions of the two velocities are β_1 and β_2, respectively. Find the reactions R_x and R_y necessary to keep one vane in place. (Use the continuity equation as one of the relations between variables.)

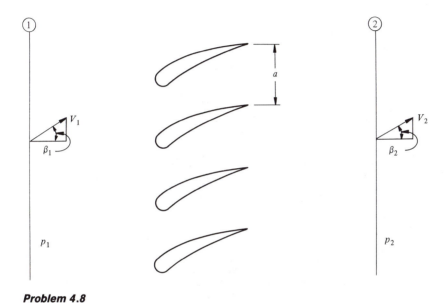

Problem 4.8

4.9 An incompressible fluid flows from a tank through a long pipe of cross-sectional area A and length l. The inflow to the tank is so regulated that the outflow velocity can be expressed by the relation

$$V = V_0 - at$$

at least for a certain interval during the outflow process. Any variations of velocity over the pipe cross section can be neglected.

(a) Find the horizontal component of the force required to hold the tank in place. Assume that all velocities in the tank proper are negligible compared to the velocities in the pipe. The discharge coefficient of the open end of the pipe is $c_d = 1.0$.

(b) What is the numerical value of the force if

$$a = 1.5 \text{ m/s}^2, \qquad l = 15 \text{ m,}$$

$$A = 0.03 \text{ m}^2, \qquad V = 6 \text{ m/s (instantaneous velocity),}$$

$$\rho = 1000 \text{ kg/m}^3.$$

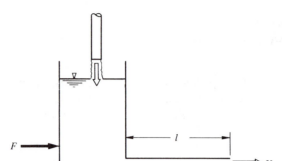

Problem 4.9

4.10 A frequently used hydraulic brake consists of a movable ram that displaces water from a slightly larger cylinder as shown in the accompanying sketch. The area of the cylinder is A_c and the cross-sectional area of the ram is A_r. The ram velocity V does not change with time. Assume that the gap between the cylinder and the ram is much smaller than the displacement of the ram x.

(a) Determine the pressure at the end of the cylinder (where the velocity is assumed to be zero) and the jet velocity V_j.

(b) Find the force F on the ram in terms of A_r, A_c, and V. Assume that the cylinder is initially full of water and that gravity effects are negligible.

Hint: Use a coordinate system in which the ram is stationary.

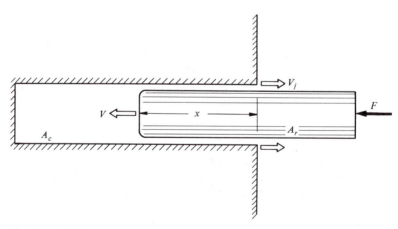

Problem 4.10

4.11 Fluid of uniform velocity U_0 and constant pressure flows over a flat plate. Due to the action of viscosity, the fluid adjacent to the place is slowed down and at the end of the plate the velocity component parallel to the plate is distributed as $u = U_0 f(y/y_0)$, where y_0 is the distance from the plate for which $u = U_0$. The pressure may be assumed constant. Show that the drag force on the plate per unit width is

$$D = \int_0^{y_0} \rho(U_0 - u)u \, dy.$$

Note: The streamlines are not all straight.

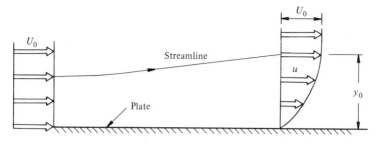

Problem 4.11

4.12 Water issues from a nozzle that is connected to a 90° elbow. A flexible hose connects the elbow to the rest of the piping. The elbow and nozzle are held in place by two forces as shown. The flexible connection cannot take any restraining force. The following data are given:

$$\text{jet area} = A_j = 0.25 \text{ in.}^2,$$

$$\text{cross-sectional area of elbow} = 1.00 \text{ in.}^2,$$

$$\text{flow rate} = Q = 0.05 \text{ ft}^3/\text{sec},$$

$$\text{nozzle coefficient} = c_v = 0.80,$$

$$\text{specific weight} = \rho g = 62.4 \text{ lbf/ft}^3.$$

(a) Find the gage pressure at the entrance to the elbow.
(b) Find the forces F_x and F_y.
Neglect all losses except the friction in the nozzle and assume that the flow does not change with time.

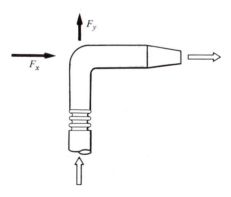

Problem 4.12

4.13 A jet of water is shot vertically upward as shown. The nozzle is connected to a reservoir of height h. The jet impinges on a flat plate of weight W. The area of the jet issuing from the nozzle is A and the velocity coefficient of the nozzle c_v.
(a) Compute the jet velocity at the exit from the nozzle. Also compute the discharge rate.

(b) Compute the velocity at a height h_1. Assume that there are no losses in the jet between the exit from the nozzle and the elevation h_1.

(c) Compute the weight W of the plate that can be held in place at the elevation h_1. (Assume lateral stability.)

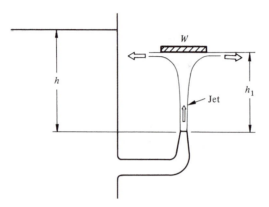

Problem 4.13

4.14 A reducing nozzle is attached to a fire hose. The inlet diameter of the nozzle is d_1, and the discharge diameter is d_2. The pressure at the inlet is p, and the atmospheric pressure is p_a. Assume that the contraction coefficient $c_c = 1.00$ and that the loss coefficient based on the jet velocity is equal to k.

(a) Find analytical expressions for the jet velocity as well as the force that must be exerted on the nozzle to keep it stationary. State clearly the direction of this force and do not neglect the action of the atmospheric pressure.

(b) Obtain numerical answers for the two quantities in part (a) if

$$d_1 = 8 \text{ cm}, \qquad d_2 = 2 \text{ cm},$$

$$p_1 = 1 \text{ MPa}, \qquad p_a = 100 \text{ kPa},$$

$$\rho = 1000 \text{ kg/m}^3, \qquad k = 0.05.$$

4.15 A motorboat is to be jet-propelled and the proposed arrangement is as indicated. It is known that the drag D of the boat for a speed of $V_b = 11$ m/s is $D = 900$ N. The diameter d_1 of all the piping is 15 cm throughout, and the diameter of the nozzle exit d_5 is 7.5 cm. The loss coefficients between sections 1 and 2, k_{12}, and between sections 3 and 4, k_{34}, are both 0.5 and they are based on the velocities in these sections. The nozzle velocity coefficient, $c_v = 0.95$, and the contraction coefficient is 1.00. The density of the water is to be taken as 1000 kg/m^3 and gravitational effects may be neglected.

(a) For a boat speed V_b, what is the mass flow rate through the pump in terms of ρ, d_5, V_b, and D?

(b) What is the pump head, in feet of fluid in terms of V_b, V_5, c_v, k_{1-2}, k_{3-4}, d_1, d_5?

(c) If the pump efficiency is 65 percent, what power is required to drive the pump for a boat speed V_b of 11 m/s?
Note: V_1 is not necessarily equal to V_b.

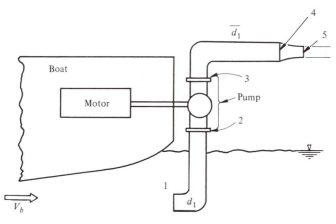

Problem 4.15

4.16 A body of mass m traveling at velocity V_0 has a hook that picks up the near end of a loose chain of linear density γ (kg/m) stretched out before the path of the body. Determine the subsequent motion of the body as a function of time. Neglect friction and assume all of the portion of the chain that has been picked up to be traveling at the instantaneous speed of the body.

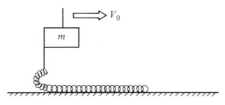

Problem 4.16

4.17 An incompressible, nonviscous liquid of height h and velocity V flows under the action of gravity through a sluice gate (see the accompanying figure). The depth downstream of the sluice gate is l. Determine the force per unit width R necessary to hold the plate in place in terms of ρ, g, h, and l.

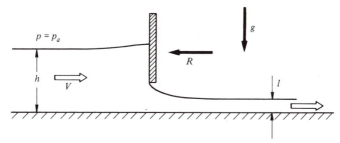

Problem 4.17

4.18 A weir discharges into a channel of constant breadth as shown. It is observed that a region of still water backs up behind the jet to a height a. The velocity and height of the flow in the channel are given as V and h, respectively, and the density of the water is ρ. Using the momentum theorem and the control surface indicated, determine a. Neglect the horizontal momentum of the flow which is entering the control volume from above and assume friction to be negligible. The gas pressure in the cavity below the crest of falling water is to be taken as atmospheric.

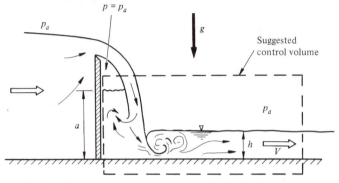

Problem 4.18

4.19 A pipe is arranged with a small slot along its length and fluid is allowed to drain out in a continuous sheet. In applications such as the production of paper, it is necessary to taper the pipe so that the vertical flow velocity is constant along the pipe length. The horizontal component of velocity at the slot is assumed to be a fixed fraction k of the average velocity in the pipe. The inlet flow rate is Q_0 and inlet diameter D_0, while the slot width is b and the overall pipe length is L. Frictional and gravity effects are negligible. If the pressure in the pipe is to be constant, show that k must be 1.0, and that $dU = 0$. Find the diameter $D(x)$.

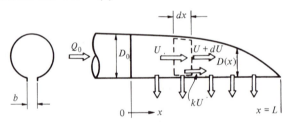

Problem 4.19

4.20 A two-dimensional body is mounted in a two-dimensional water tunnel of breadth h. The velocity V_0 far upstream is fixed. It is observed that when the upstream pressure p_0 is lowered a sufficient amount, the liquid behind the body boils and forms a very long cavity of vapor at the vapor pressure $p_v (p_v < p_0)$. The velocity across the liquid jets forming the boundary of the cavity can be assumed to be constant far downstream. Also assume that there is no friction, and that the density of the vapor and the effect of gravity are negligible. What is the drag force on the body per unit width in terms of V_0, p_0, p_v, h, and the constant density ρ of the liquid?

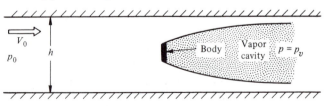

Problem 4.20

4.21 A plate of mass M is suspended between two opposing jets, each of area A, velocity V_0, and density ρ. The plate is suddenly given a velocity V_0 parallel to one of the jets. Determine the subsequent motion of the plate [i.e., $x(t)$] and the maximum displacement of the plate. Neglect any mass of liquid adhering to the plate and consider the flow at any instant as quasi-steady. Neglect friction and assume that motion takes place only along the x-direction.

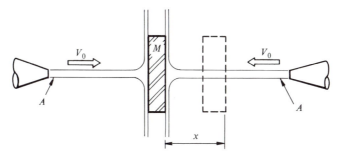

Problem 4.21

4.22 Rework Example 4.2 with the control surface *stationary.*

Hint: It will be necessary to account for the increase of momentum within the control volume due to the lengthening jet.

4.23 The mass of a rocket is $(m_0 - \dot{m}t)$, where m_0 is the original mass of the rocket, $\dot{m}$ the constant rate of propellant consumption, and t the time. Relative to the accelerating rocket the velocity of the escaping exhaust gas is V_e. Determine the acceleration of the rocket and the dependence of the rocket velocity V on time. For this purpose use a control volume that moves at a constant velocity V, where V is equal to the velocity of the rocket at the instant being considered. Let $V = 0$ at $t = 0$.

4.24 A horizontal flow with an inlet diameter of 1 m is divided and has equal flow rates from each outlet. The outlets have diameters of 0.5 m. The inlet pressure is 200 kPa and the inlet flow rate is 5 m³/s.
   (a) Calculate the centerline pressure at the outlets assuming no friction losses.
   (b) Calculate the reaction forces on the divided flow that must be absorbed by a support system.
   (c) Calculate the reaction forces on the divided flow if the friction loss between the inlet and either outlet is given by $h = 1.5V^2/2g$, where V is the inlet velocity.

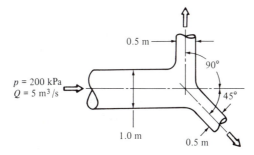

Problem 4.24

4.25 A bucket of mass m_0 rests on a set of scales with a spring constant k. A stream of BBs is dropped toward the bucket at a steady rate of $\dot{m}_b$ kg/s. At the point where they hit the bucket they have a (constant) velocity *downward* of magnitude V_b relative to ground. Let the position of the bucket relative to the ground be denoted by y and derive the differential equation for any position of the bucket $y(t)$.

Note: Clearly define the *control volume* used and *do not* try to solve the resulting equation. Take the mass of BBs contained in the bucket at the instant considered to be m_b. Indicate how m_b is to be found.

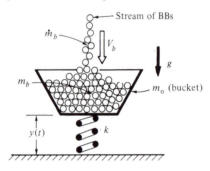

Problem 4.25

4.26 A piston in a tank of cross-sectional area A forces an ideal, incompressible fluid of density ρ out of the tank into the atmosphere through a reentrant pipe of cross-sectional area a. The jet in the pipe has a cross-sectional area Ca. Show that the contraction coefficient is

$$C = \frac{1}{2 - a/A}.$$

Note: Neglect gravitational acceleration.

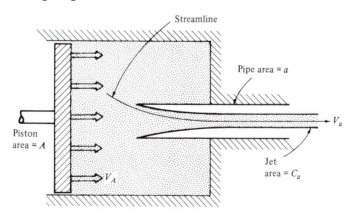

Problem 4.26

4.27 A conveyor belt moves at 3 m/s. A load of sand is dumped onto the belt at a rate that results in 100 kg being placed per meter of belt.
(a) What force is required to drive the belt?
(b) What power is required?

4.28 The RB 207 Turbofan engine used in commercial transport "jumbo" aircraft produces a static thrust (i.e., in still air) of 47,500 lbf with a mass flow of 1383 lbm/sec. The inlet to the turbofan has an area of 53.3 ft². The inlet air density is 0.075 lbm/ft³.

(a) Calculate the exit velocity in ft/sec.

(b) Calculate the inlet velocity to the fan assuming constant density.

4.29 A jet of area $A_j = 1$ m² discharges flow at $V = 50$ m/s into a duct of constant diameter $A_o = 5$ m² having an inflow velocity of 5 m/s. The pressure at this section is 100 kPa. The two streams mix thoroughly at a downstream station 2, where $A_o = 5$ m² (still).

(a) What is the velocity there?

(b) Assume no friction on the duct sidewalls and calculate the pressure rise if $\rho = 1000$ kg/m³ for both streams.

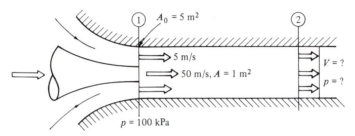

Problem 4.29

4.30 A penstock of 1.5 m² area feeds a nozzle of 0.07 m² area under a head of 300 m ($g = 9.81$ m/s²). Friction is to be neglected and $\rho = 1000$ kg/m³.

(a) Calculate the jet velocity V_j and the mass flux kg/s.

(b) The nozzle is connected to the penstock at 45°, as shown. Set up an appropriate control volume and write (in symbols) equations that define the external reactions R_x and R_y in terms of velocities and pressures, areas, and so on.

(c) Solve for numerical values of R_x and R_y.

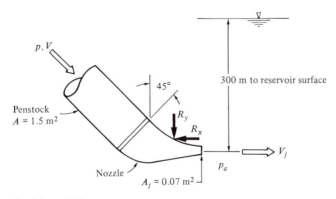

Problem 4.30

4.31 An axial-flow compressor can be approximated by the three rows of blades shown in the figure. Two of the rows are fixed (stator rows), while the intermediate row moves at a steady velocity u. The blades are shaped so that the angle the flow makes at the blade exit α_1, β_2, and α_3 are the true angles relative to the blades. Assume frictionless incompressible flow. Standard air flows in with axial velocity $V_0 = 100$ ft/sec. The other parameters have the following values: $u = 200$ ft/sec, $\alpha_1 = 45°$, $\beta_2 = 70°$, and $\alpha_3 = 80°$.

(a) Find the magnitudes of the (absolute) velocities V_1, V_2, and V_3.

(b) Find the force that must be applied to the moving rotor blades (for a unit area normal to the flow).

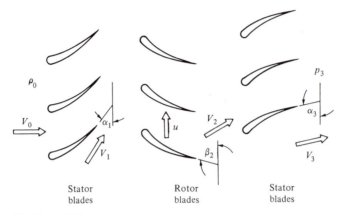

| Stator blades | Rotor blades | Stator blades |

Problem 4.31

4.32 An airfoil is tested in a large wind tunnel. The velocity upstream is uniform over the cross section and equal to V_0. At a cross section downstream, Pitot tube measurements show a distorted velocity profile, as shown in the figure. The pressure at both cross sections is the same and constant. Find the drag of the airfoil (per unit depth normal to the surface of the paper). The air may be considered incompressible.

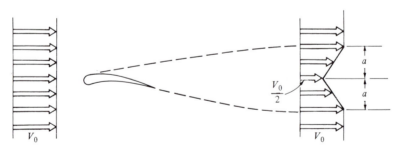

Problem 4.32

†4.3 Transformation of Volume Integral

It may have occurred to the reader that the transformation from the material volume integral to the surface and volume integrals for a fixed control volume is of more widespread utility than we have mentioned. This is indeed the case, and the derivation leading to Eq. (4.4a) yields a generally useful result. For example, consider a quantity ξ (a vector or a scalar), which is to represent some property that

---

† This section may be omitted without loss of continuity.

refers to a unit mass. Then from the discussion following Eqs. (4.1) to (4.4), we obtain

$$\frac{D}{Dt} \int_{\mathcal{V}'} \rho \xi \, d\mathcal{V} = \frac{\partial}{\partial t} \int_{\mathcal{V}_0} \rho \xi \, d\mathcal{V} + \int_{S_0} \rho \xi \mathbf{v} \cdot \mathbf{n} \, dS, \tag{4.6}$$

where, as before, $\mathcal{V}'$ is a material volume consisting of the volume of a given amount of mass, $\mathcal{V}_0$ is a control volume that coincides initially with $\mathcal{V}'$, and S_0 is the surface of $\mathcal{V}_0$. The rate of change of momentum is obtained by setting $\xi = \mathbf{v}$, since $\mathbf{v}$ is the momentum of the fluid per unit mass. The integral form of the conservation of mass principle is obtained by setting $\xi = 1$, so that

$$0 = \frac{D}{Dt} \int_{\mathcal{V}'} \rho \, d\mathcal{V} = \int_{\mathcal{V}_0} \frac{\partial \rho}{\partial t} \, d\mathcal{V} + \int_{S_0} \rho \mathbf{v} \cdot \mathbf{n} \, dS.$$

In this last example ξ has the meaning of "mass per unit mass," which, of course, is just equal to unity.

4.4 Alternative Derivation of the Momentum Theorem

In order to obtain a still more complete grasp of the momentum theorem, it is instructive to derive this theorem directly from the basic equations of motion. This can be done by integrating these equations over a fixed volume. *The momentum equation may therefore be described as the volume integral of the equations of motion, whereas the extended Bernoulli theorem was the result of integrating the equations of motion along a streamline.*

For this alternative way of deriving the momentum theorem, let us start with the equation as given in Eq. (2.22), which is a very general form. Taking the first of this set of equations and multiplying through by ρ (which may be a variable), we obtain

$$\rho \frac{\partial u}{\partial t} + \rho u \frac{\partial u}{\partial x} + \rho v \frac{\partial u}{\partial y} + \rho u \frac{\partial u}{\partial z} = -\frac{\partial p}{\partial x} + \rho f_x + \rho f_{fx},$$

in which all the viscous terms are lumped into f_{fx}. It is now desired to integrate this equation over a fixed volume $\mathcal{V}_0$. To make the integration possible, we add the continuity equation after having first multiplied it by u. The integral may then be written as

$$\int_{\mathcal{V}_0} \left[\frac{\partial(\rho u)}{\partial t} \, dx \, dy \, dz + \frac{\partial(\rho u u)}{\partial x} \, dx \, dy \, dz + \frac{\partial(\rho u v)}{\partial y} \, dx \, dy \, dz + \frac{\partial(\rho u w)}{\partial z} \, dx \, dy \, dz \right]$$

$$= -\int_{\mathcal{V}_0} \frac{\partial p}{\partial x} \, dx \, dy \, dz + \int_{\mathcal{V}_0} \rho f_x \, dx \, dy \, dz + \int_{\mathcal{V}_0} \rho f_{fx} \, dx \, dy \, dz.$$

The corresponding equations may be derived for the other two directions. The volume integral for the second term may now be transformed into a surface integral. Let us first carry out the integration along the x-direction. This means that the quantity $(\partial \rho u u / \partial x) \, dx \, dy \, dz$ must be summed throughout a cylinder of cross section $dy \, dz$, starting at the left-hand surface of the volume and ending at the right-hand surface (see Fig. 4.6). This integration is a definite integral and the result is

$$\int \frac{\partial(\rho u u) \, dy \, dz}{\partial x} \, dx = \int [(\rho u u)_{\text{II}} - (\rho u u)_{\text{I}}] \, dy \, dz,$$

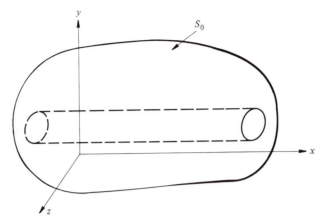

FIG. 4.6 General control volume indicating the element of volume involved in the integration along the x-axis.

where the subscripts I and II refer to the two sides of the volume. The quantity $(\rho u u)_\text{I}$ is evaluated at one surface, and the quantity $(\rho u u)_\text{II}$ is evaluated at the other surface. When the integration indicated on the right-hand side of the equation above is completed, the quantity $(\rho u u)$ will have been evaluated at every element of the surface. This term may also be written as

$$\int [(\rho u u)_\text{II} - (\rho u u)_\text{I}] \, dy \, dz = \int_\text{surface} (\rho u u) \, dy \, dz.$$

If this is done, however, one now has to think of the element $dy \, dz$, a surface element, and that the term $(\rho u u) \, dy \, dz$ has to be treated as positive or negative depending on whether the outward normal of the surface points in the direction of the velocity. The two integrals corresponding to the third and fourth terms, as well as the integral involving the pressure term, are transformed analogously.

The foregoing integral expression may then be rewritten as follows:

$$\int_{\mathscr{V}_0} \frac{\partial(\rho u)}{\partial t} \, dx \, dy \, dz + \int_{S_0} u(\rho u \, dy \, dz + \rho v \, dx \, dz + \rho w \, dy \, dx)$$

$$= -\int_{S_0} p \, dz \, dy + \int_{\mathscr{V}_0} f_x \rho \, dx \, dy \, dz + \int_{\mathscr{V}_0} f_{fx} \rho \, dx \, dy \, dz.$$

The meaning of these various integrals may now be determined. In the first integral $dx \, dy \, dz$ is a fixed volume element that is not a function of time, and we may therefore reverse the order of differentiation and integration and write

$$\frac{\partial}{\partial t} \int_{\mathscr{V}_0} \rho u \, dx \, dy \, dz$$

without changing its value. The quantity $\rho u \, dx \, dy \, dz$, however, is the momentum in the x-direction contained in the volume $dx \, dy \, dz$. The meaning of the integral is therefore the x-momentum contained in the volume $\mathscr{V}_0$, and the derivative consequently gives the time rate of change of x-momentum in the fixed volume $\mathscr{V}_0$.

In the second integral the quantity in parentheses describes the mass rate of flow through an element of surface. When multiplied by u, it represents the x-momentum flowing through the element of surface. The integral represents, therefore, the net rate of x-momentum passing through the surface of the given volume $\mathscr{V}_0$. The integral $\int p \, dz \, dy$ yields the x-component of the pressure force normal to the surface of the volume. The second integral on the right-hand side simply represents the sum of all body forces in the x-direction acting on the volume $\mathscr{V}_0$. The third integral finally gives the sum of the friction forces in the x-direction. This result may then be stated in words as follows: *The rate of change of x-momentum contained in a fixed volume plus the net flow of x-momentum through the surfaces of this volume is equal to the sum of all the forces in the x-direction acting on the volume.* A similar statement can, of course, be derived for the y- and z-directions, which then leads directly to the momentum theorem, as stated previously in Eq. (4.5).

†4.5 Momentum Theorem Referred to Moving Coordinates

The momentum theorem, Eq. (4.4), is valid in any uniformly translating coordinate system. Suppose that the velocities expressed in such a coordinate system are given by $\mathbf{v}_r$. Then, from Eq. (4.4),

$$\mathbf{F} = \frac{\partial}{\partial t} \int_{\mathscr{V}_0} \rho \mathbf{v}_r \, d\mathscr{V} + \int_{S_0} \rho \mathbf{v}_r \mathbf{v}_r \cdot \mathbf{n} \, dS.$$

Let $\mathbf{v}$ denote the velocities in another inertial coordinate system and let the two systems differ by a uniform velocity of translation $\mathbf{v}_0$, where

$$\mathbf{v} = \mathbf{v}_0 + \mathbf{v}_r.$$

The conservation of mass principle requires that

$$0 = \int_{\mathscr{V}_0} \frac{\partial \rho}{\partial t} \, d\mathscr{V} + \int_{S_0} \rho \mathbf{v}_r \cdot \mathbf{n} \, dS.$$

Multiplying this equation by $\mathbf{v}_0$ and adding the result to the expression for $\mathbf{F}$ above gives an alternative form of the momentum theorem

$$\mathbf{F} = \frac{\partial}{\partial t} \int_{\mathscr{V}_0} \rho \mathbf{v} \, d\mathscr{V} + \int_{S_0} \rho \mathbf{v} \mathbf{v}_r \cdot \mathbf{n} \, dS. \tag{4.7}$$

In this form, if one chooses $\mathbf{v}_0$ to be equal to the instantaneous absolute velocity of a moving body, the body will appear to be stationary in respect to the control volume. In many problems use of this form results in considerable simplification.

EXAMPLE 4.3 _____

As an example of the use of Eq. (4.7), let us consider again the problem of the hydraulic catapult already treated in Example 4.2. We take the control surface S_1 to enclose completely the airplane, bucket, and jet as well as a small bit of the jet and the return stream. The velocities $\mathbf{v}_r$ of Eq. (4.7) are measured relative to the moving airplane and the velocities $\mathbf{v}$ are measured in respect to earth.

Solution: The total force on the control surface is zero, so that we have

$$0 = M \frac{dV}{dt} + \int_{S_1} \rho \mathbf{v} \mathbf{v}_r \cdot \mathbf{n} \, dS,$$

where dV/dt is the acceleration of the airplane. Now with reference to Fig. 4.5a and noting that the velocity of the jet leaving the bucket measured in respect to the earth

† This section may be omitted without loss of continuity.

is $-(V_j - 2V)$, we have

$$\rho \mathbf{v} \mathbf{v}_r \cdot \mathbf{n} \, dS = \rho A_j V_j [-(V_j - V)] - \rho A_j (V_j - 2V)(V_j - V).$$

After simplification this becomes

$$\frac{dV}{dt} = \frac{2\rho A_j}{M} (V_j - V)^2,$$

as before.

PROBLEM

4.33 Rework Problem 4.23 using Eq. (4.7), with $\mathbf{v}_r$ being measured in respect to the moving rocket.

4.6 Angular Momentum Theorem

The momentum theorem developed in Section 4.2 gives the force acting on a fixed volume in terms of linear momentum flux through the surface of the volume. In many situations we are interested in the moment or torque on the volume. For this purpose we may adapt the angular momentum law of mechanics to the flow of fluids. Our starting point is the familiar law

$$\mathbf{F} = \frac{d}{dt} (m\mathbf{v}),$$

where m, $\mathbf{F}$, and $\mathbf{v}$ refer to a single particle. The torque exerted by the force $\mathbf{F}$ about a fixed point is

$$\mathbf{T} = \mathbf{r} \times \mathbf{F},$$

where $\mathbf{r}$ is the radius vector from the fixed point to the point of application of $\mathbf{F}$. The symbol $\times$ signifies, as usual, that the vector cross-product shall be taken. Then, from Newton's law of motion,

$$\mathbf{T} = \mathbf{r} \times \frac{d}{dt} (m\mathbf{v}).$$

We now define a vector $\mathbf{H}$ as the vector product of the radius vector to the particle and the linear momentum, that is,

$$\mathbf{H} = \mathbf{r} \times m\mathbf{v}. \tag{4.8}$$

The quantity $\mathbf{H}$ is called *angular momentum*. Upon differentiating $\mathbf{H}$ with respect to time, we find that

$$\frac{d\mathbf{H}}{dt} = \frac{d\mathbf{r}}{dt} \times m\mathbf{v} + \mathbf{r} \times \frac{d}{dt} (m\mathbf{v}).$$

However, $d\mathbf{r}/dt = \mathbf{v}$, and the cross-product of a vector parallel to itself is zero. The first term in the right-hand side therefore vanishes and we have the result that

$$\mathbf{T} = \frac{d\mathbf{H}}{dt}. \tag{4.9}$$

Equation (4.9) states that the rate of change of angular momentum of a particle about a fixed point is equal to the torque applied to the particle.

We now seek to modify the law as expressed by Eq. (4.9) to be suitable for a fixed volume. The development of this equation will therefore follow almost exactly that leading to the momentum theorem of Section 4.2. The torque on a material volume $\mathscr{V}'$ is then

$$\mathbf{T} = \frac{D}{Dt} \int_{\mathscr{V}'} \rho \mathbf{r} \times \mathbf{v} \, d\mathscr{V}.$$

This is readily transformed into a control volume integral as before, or by the application of Eq. (4.6) upon setting $\xi = \mathbf{r} \times \mathbf{v}$. We have, therefore,

$$\mathbf{T} = \frac{\partial \mathbf{H}}{\partial t} + \int_{S_0} \rho \mathbf{r} \times \mathbf{v}(\mathbf{v} \cdot \mathbf{n}) \, dS, \tag{4.10a}$$

where

$$\mathbf{H} = \int_{\mathscr{V}_0} \rho \mathbf{r} \times \mathbf{v} \, d\mathscr{V} \tag{4.10b}$$

is the angular momentum contained within the control volume. Equation (4.10a) represents the *angular momentum theorem*, corresponding to the linear momentum theorem given by Eq. (4.4a).

EXAMPLE 4.4

A liquid of density ρ flows through the nozzle shown below and leaves at a uniform velocity V_j. Calculate the restoring torque T_R required at point A assuming that there is no friction or gravity.

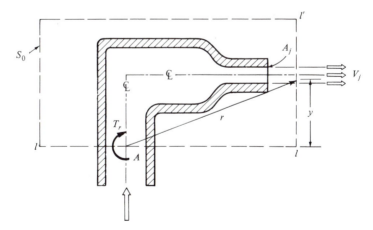

Example 4.4

Solution: The flow is steady, so that we have

$$\mathbf{T} = \int_{S_0} \rho \mathbf{r} \times \mathbf{v} \mathbf{v} \cdot \mathbf{n} \, dS$$

with the control surface shown. Point A is symmetrically located in the pipe so that there are no contributions to $\mathbf{T}$ from either the pressure forces or the angular momentum flux across $l - l$. Across $l - l'$, however, we have $\mathbf{r} \times \mathbf{v} = V_j y$, with the sign being determined by the "right-hand rule." Thus

$$T = T_R = \rho V_j^2 \int_{A_j} y \, dA,$$

since V_j is assumed constant over the area A_j of the jet. Hence

$$T_R = \rho V_j^2 A_j y_c = \dot{m} V_j y_c$$

in the clockwise sense. The length y_c determines the position of the centroid of the jet area above the line $l - l$ and is defined by

$$y_c = \int_{A_j} y \, dA \bigg/ \int_{A_j} dA.$$

EXAMPLE 4.5

A section of pipe is formed into an angle, as shown, and is rotated at a constant angular velocity Ω in the positive sense about the Oz-axis. Fluid of density ρ at a flow rate Q passes through the pipe. The x-, y-, and z-axes are fixed in space and at the instant shown the pipe angle lies in the xz-plane. We wish to calculate the components of the torque vector about the Ox-, Oy-, Oz-axes. For the purpose of this example we shall assume that the fluid approaching the angle has no angular momentum, that the discharge velocity is uniform over the discharge area A, and that the diameter of the discharge is small compared to the distance from the axis of rotation, $R \sin \theta$. With this assumption all of the fluid may be considered to leave the pipe at the same distance from the axis of rotation.

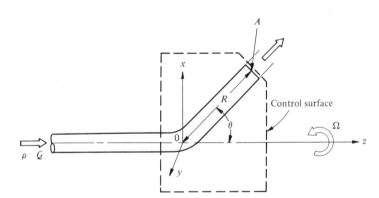

Example 4.5

Solution: We select a control volume as indicated by the dashed lines of the figure. In the present case both terms of Eq. (4.10a) contribute to the torque. To evaluate these terms we need to express the velocity of the fluid contained within the control volume in a manner suitable for the problem. In the present case we see that the velocity of fluid particles within the pipe is expressed by

$$\mathbf{v} = \mathbf{i}_R \frac{Q}{A} + \mathbf{i}_y r\Omega \sin \theta,$$

where $\mathbf{i}_R$ is a unit vector in the direction of the pipe angle, $\mathbf{i}_y$ is a unit vector parallel to the y-axis, and r is a radial coordinate along the pipe leg in the θ direction. At the location of the discharge, $r = R$. Inspection of Eqs. (4.10) shows that we need to evaluate the vector product $\mathbf{r} \times \mathbf{v}$. We do this by writing

$$\mathbf{r} = \mathbf{i}_R r = (\mathbf{i}_z \cos \theta + \mathbf{i}_x \sin \theta)r.$$

Noting that the cross-product of a vector with itself is zero, we obtain

$$\mathbf{r} \times \mathbf{v} = r^2 \Omega \sin \theta(-\mathbf{i}_x \cos \theta + \mathbf{i}_z \sin \theta).$$

It is now easy to evaluate the surface integral in Eq. (4.10) as

$$\int_{S_0} \rho \mathbf{r} \times \mathbf{v}(\mathbf{v} \cdot \mathbf{n}) \, dS = \rho Q R^2 \Omega \sin \theta(-\mathbf{i}_x \cos \theta + \mathbf{i}_y \cdot 0 + \mathbf{i}_z \sin \theta).$$

The instantaneous angular momentum is

$$\mathbf{H} = \Omega \sin \theta(-\mathbf{i}_x \cos \theta + \mathbf{i}_z \sin \theta) \int_{V_0} \rho r^2 \, dV.$$

The volume integral may be thought of as the polar moment of inertia I_0 of the fluid contained within the pipe about point O. We must now determine how H changes with time. After a little thought we can see that the angular momentum of the fluid within the pipe rotates with the pipe at angular velocity Ω. For example, when the axis has turned one quarter of a revolution, the x angular momentum component vanishes and a similar quantity appears instead along the y-axis (remember that the x-y-z axes are fixed); the z-component does not, however, change. We can then see that the x and y angular momenta are the components of a vector in the xy-plane rotating at angular velocity Ω. Thus we can write by inspection

$$\mathbf{H} = I_0 \Omega \sin \theta(-\mathbf{i}_x \cos \theta \cos \Omega t - \mathbf{i}_y \cos \theta \sin \Omega t + \mathbf{i}_z \sin \theta).$$

Therefore, at time $t = 0$ (corresponding to the given position of the pipe angle)

$$\frac{\partial \mathbf{H}}{\partial t} = -\mathbf{i}_y I_0 \Omega^2 \sin \theta \cos \theta$$

and finally,

$$T_{Ox} = -\rho Q R^2 \Omega \sin \theta \cos \theta,$$

$$T_{Oy} = -I_0 \Omega^2 \sin \theta \cos \theta,$$

$$T_{Oz} = \rho Q R^2 \Omega \sin^2 \theta.$$

PROBLEMS

4.34 Two straight pipes are connected by an offset (see illustration). If A is the cross-sectional area of the pipe, V the velocity of flow, and ρ the fluid density, determine the moment required about the point O to prevent rotation.

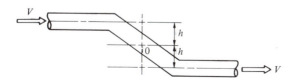

Problem 4.34

4.35 A two-dimensional, frictionless, steady, incompressible flow with a uniform radial velocity of v_{r_1} and tangential velocity v_{θ_1} enters a set of stationary guide vanes. The guide vanes align the flow so that $v_{\theta_2} = v_{r_2} \cot \alpha$, where v_{r_2} is the leaving radial velocity. Assume that v_θ and v_r do not depend on the angle θ, either at station 1 or station 2, and neglect the thickness of the vanes. The fluid density is ρ.

(a) Find the torque exerted on the guide vanes per unit axial length.

(b) Determine the equation of the streamlines outside of the guide vanes.

(c) Would your answer in part (a) change if there were friction? In part (b)?

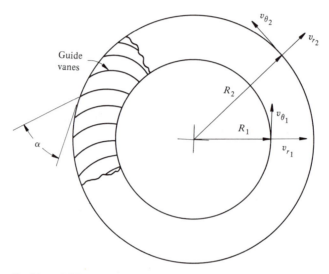

Problem 4.35

4.36 A pipe branches symmetrically into two legs each of length R and the whole system rotates at angular velocity Ω about its axis. Each branch is inclined at an angle α to the axis of rotation. A steady rate of fluid flow Q (m³/s) of constant density ρ enters the end of the pipe with no angular momentum. The flow is perfectly guided by the pipe and the pipe is of constant diameter d, where d is much less than R. Determine the torque required to turn the pipe.

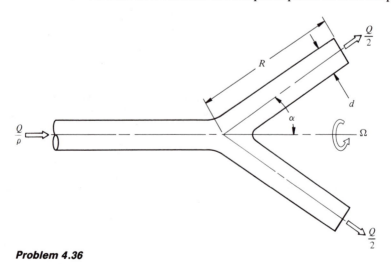

Problem 4.36

4.37 A steady, incompressible, frictionless, two-dimensional jet of fluid with density ρ, breadth h, velocity V, and unit width impinges on a flat plate held at an angle α to its axis. Gravitational forces are to be neglected.

  (a) Determine the total force on the plate, and the breadths a, b of the two branches.

  (b) Determine the distance l to the center of pressure (c.p.) along the plate from the point 0. (The center of pressure is the point at which the plate can be balanced without requiring an additional moment.)

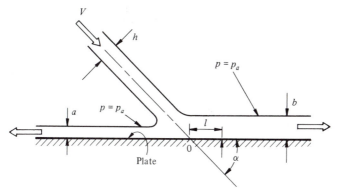

Problem 4.37

4.38 With the aid of the angular momentum theorem, Bernoulli's equation, or any of the other results obtained thus far, devise an instrument that will measure the *mass* rate of flow (i.e., kg/s).

4.39 Assume that the set of guide vanes in Problem 4.35 rotates at constant angular velocity Ω in the clockwise direction. And consider the velocities v_{θ_1} and v_{θ_2} to be measured with *respect to the rotating wheel.*

  (a) Determine the torque required to turn the wheel for a flow rate of Q (m³/s).

  (b) In part (a) it is seen that a torque is required to turn the rotating wheel. Therefore, an amount of power equal to the torque times the angular velocity is imparted to the flow, so that the total head of the flow must increase. Determine the ideal total head rise M across the wheel as a function of Q by equating the hydraulic power output to the input power. Depending on the sign of M, the device just described is a simplified pump or turbine and the rotating wheel is termed an impeller, or rotor.

4.40 The "Saint Catherine's wheel" consists of a toy wheel around which is wound a tube containing a combustible mixture, such as black gun powder. When ignited, the exhaust gases are rapidly expelled and the wheel is forced to rotate about its axis. The mass of the wheel m may be considered to have its center of gyration a distance r from the axis of rotation. The velocity of the ejected gases is V_e relative to a point on the rotating wheel at radius r, and the rate at which the gases are ejected, $\dot{m}$, is a constant. The initial mass of powder is m_0 and it may be assumed to be located at the same radius r. Determine the angular speed Ω of the wheel as a function of time after the powder is ignited in terms of m, m_0, $\dot{m}$, r, V_e, and t.

4.41 A water drop is falling through a vapor cloud and the drop accumulates additional mass at rate $\dot{m}$. The instantaneous velocity of the drop is V and its instantaneous mass is M. The drag force on the drop is D and the force due to gravity is, of course, gM.

(a) Find the expression for the instantaneous acceleration (dV/dt) of the drop.

(b) Assume the drop to be spherical with a radius r, and assume that the rate of accumulation may be approximated by

$$\dot{m} = \alpha S V,$$

where α is a constant and S is surface area of the drop. Considering the instantaneous acceleration of the drop in part (a), find the ratio of the term due to accumulation to the gravity term. Make a qualitative statement about the drop size and the velocity of the drop for which the mass accumulation term would be important compared to the gravity term.

4.42 A jet of velocity V_j, area A_j, and density ρ is shot into a tank of initial mass m_0. The tank retains all the fluid intercepted. It is mounted on a frictionless support, and from the force of the jet acquires a velocity V which is a function of time. By using the momentum principle, obtain an equation for the acceleration $\dot{V}$ of the tank. [Do not try to solve this equation to obtain $V(t)$.] Be *very clear* in your choice of control volume and state whether it is moving or stationary.

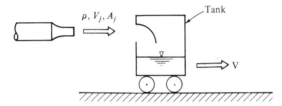

Problem 4.42

4.43 A total discharge of 0.02 ft³/sec (water) divides evenly between two nozzles in a sprinkler. The cross-sectional area of each nozzle is 0.05 in.², and the nozzles are inclined at 45° to the sprinkler arms, one of length 6 in. and the other, 12 in.

(a) Determine the torque that must be applied to the sprinkler arms to hold them from rotating.

(b) Determine the angular speed ω (rad/sec) if the arms are free to rotate and there is no friction.

(c) Determine the angular speed ω if there is a constant frictional torque (Coulomb friction) of 0.5 ft-lb resisting rotation of the arms.

(d) Determine the absolute velocity of the fluid leaving each nozzle for cases (b) and (c), and sketch the velocity vector diagrams.

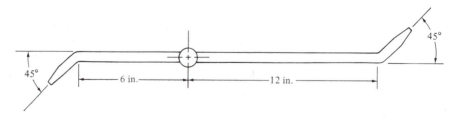

Problem 4.43

4.44 A rotating pipe arrangement as shown is turning at speed ω_0. The discharge from the two pipes is purely radial and the total mass discharge rate is $\dot{m}$. To maintain the speed ω_0, a torque is required. *No* torque is, however, applied and the unit is allowed to slow down. The flow rate is artificially maintained constant at the value $\dot{m}$. The cross-sectional area of the pipe is A, the density ρ, and the mass of the pipe per unit length is σ. The radial length of the two pipe ends is R. Find an expression for ω as a function of time.

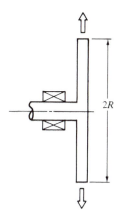

Problem 4.44

4.45 The rotor of a simple gas turbine has very short blades. Curvature effects may therefore be neglected and the blades may be considered to form a linear cascade as shown in the sketch. The blades move at a constant velocity u (which corresponds to a constant speed of rotation of the rotor). The gas is directed against the blades by a set of nozzles. The gas velocity in respect to a stationary frame of reference is V_1, and the direction is given by the angle α_1. The blades change the direction of the gas velocity so that the gas leaves the blades at angle β_2 *relative* to the moving blades. The blades are assumed to be *frictionless* and the pressure is *constant* throughout. To simplify matters, we may also assume that both α_1 and β_2 are very small so that $\cos \alpha_1 \approx \cos \beta_2 \approx 1.0$.

(a) Find the force on the blades in the direction of motion for a mass flow rate of $\dot{m}$.

(b) What is the power the turbine wheel can deliver for the conditions above?

(c) What is the turbine velocity u (in terms of V_1) for maximum power, and what is this maximum power?

(d) What is the numerical value for this power for $V_1 = 300$ m/s and $\dot{m} = 1.0$ kg/s? (The actual power will, of course, be somewhat less because of losses.)

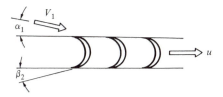

Problem 4.45

4.46 A pump, mounted on a boat, transmits M_P of work to each kilogram of water passing through it. The pump then discharges through a nozzle at a jet velocity c relative to the nozzle as shown. The pump suction line is so arranged that the pump may take water either from the sea (as shown), or from a large reservoir tank inside the boat. The boat is traveling at a speed V.

(a) Neglecting losses and elevation changes, derive an expression for the jet velocity relative to the boat, *for each of the two cases.*

(b) For a mass flow rate m through the nozzle, find the thrust on the boat *for each of the two cases,* assuming the jet velocity to be in a direction opposite to that of the boat velocity.

Note: Reference system leading to a steady state will be most convenient in both parts (a) and (b).

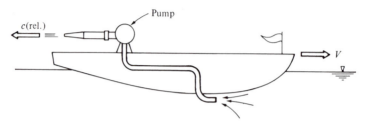

Problem 4.46

CHAPTER 5

Similitude

5.1 Introduction

In previous chapters several problems were solved by assuming that the flow was frictionless. These results were modified in certain instances to include the effects of friction by introducing experimentally determined coefficients. Orifice and Venturi meter coefficients are examples of such coefficients. An alternative approach to these problems would consist of solving the Navier–Stokes equations. From such solutions complete information about the flow could, in principle, be obtained and the above-mentioned coefficients could be determined analytically. Generally, however, the task of solving the Navier–Stokes equations can only rarely be accomplished. As a next best approach one might then attempt at least to determine in a more general way how the experimentally determined coefficients might depend on the other variables that enter into the problem, such as the velocity, the density, the viscosity, and the physical dimensions. To illustrate how one might proceed for such a purpose, let us consider the flow through an orifice. We would expect that the velocity coefficient of the orifice would depend in some way on the average jet velocity V, the density ρ, the viscosity μ, and the diameter of the orifice d. If a free surface is present, the surface tension σ and the acceleration of gravity g will probably also play a role. Symbolically, this expectation may be expressed as

$$c_v = c_v(\rho, V, d, \mu, \sigma, g).$$

Now c_v is a dimensionless number, and it therefore cannot depend only on the viscosity μ, say, independently of the other variables, since if it did, c_v would have the same "dimensions" as μ and would not be dimensionless. We must therefore conclude that the variables μ, V, d, etc., will have to occur in dimensionless groups, so that c_v remains dimensionless.

The procedure of sorting out the combinations of the several variables occurring in the problem in such a way as to have each of the combinations dimensionless is called *dimensional analysis*. Once the important variables have been selected, dimensional analysis becomes an entirely mathematical procedure that may be carried out without any further recourse to physical reasoning. When the differential equations governing the basic phenomenon to be examined are not known, dimensional analysis is helpful in determining the various dimensionless combinations of variables that might occur. The functional relationships between the various dimensionless groups must then, of course, still be found by experiment. However, when the basic differential equations are available, they may be used to derive the dimensionless groups in a more natural way, which at the same time allows more physical insight than does dimensional analysis alone. In this way the differential equations become most useful even when they cannot be solved. A more extensive discussion of dimensional analysis may be found in References 1 to 7.

In the next section we use the differential equations of fluid mechanics to derive the appropriate dimensionless groups, as indicated above. We then show how these dimensionless parameters may be used to classify and generalize experimental results. On this basis we also discuss under what conditions the experimental results obtained for a given flow are applicable to another flow, which is confined by boundaries that are geometrically similar, but are of different physical size. These considerations in turn lead to the concept of model testing.

5.2 Derivation of Similarity Parameters from Basic Equations

For simplicity we shall limit ourselves to an incompressible fluid. The continuity equation may then be written

$$\frac{\partial u}{\partial x} + \frac{\partial v}{\partial y} + \frac{\partial w}{\partial z} = 0, \tag{5.1}$$

and a typical component of the Navier–Stokes equation, including the gravitational term (assuming this term to be along the x-direction), is

$$\frac{\partial u}{\partial t} + u\frac{\partial u}{\partial x} + v\frac{\partial u}{\partial y} + w\frac{\partial u}{\partial z} = -\frac{1}{\rho}\frac{\partial p}{\partial x} + v\left(\frac{\partial^2 u}{\partial x^2} + \frac{\partial^2 u}{\partial y^2} + \frac{\partial^2 u}{\partial z^2}\right) + g. \tag{5.2}$$

In addition to the differential equations the boundary conditions also have to be specified for a complete statement of the problem. Two types of boundary conditions are of particular importance: In the first the velocities of the fluid are given at all surfaces. In the second the velocities are specified only at some of the boundaries, and the remaining surfaces are free surfaces at which the pressure is specified, although the exact location of the surface is not. The flow through a Venturi meter or the flow around a cylinder are examples of the first, whereas the flow of water in an open channel with a free surface is an example of the second kind of boundary condition. Symbolically, the first type may be expressed as

$$\left.\begin{aligned} u &= u_b \\ v &= v_b \\ w &= w_b \end{aligned}\right\} \quad \text{at } f(x_b, y_b, z_b) = 0, \tag{5.3}$$

where $f(x_b, y_b, z_b) = 0$ is the equation that completely defines the position of the boundaries. In the second type of boundary condition the specifications for the solid boundaries may be given in the same way as above. For the free surface neglecting surface tension, however, one has to write $p = p_b$ at $F(x_f, y_f, z_f) = 0$, where the function F is initially unknown.

In making Eqs. (5.1) and (5.2) dimensionless, it is first necessary to select a reference quantity for each of the variables. We shall call the reference velocity u_0, the reference length l_0, and the reference pressure p_0. For the reference time we shall use the ratio of the reference length to the reference velocity l_0/u_0. The reference quantities may be selected arbitrarily, but they must be well-defined quantities of the problem. For example, for the flow through the Venturi meter one might select the diameter of the throat as the reference length, the average velocity at the throat as the reference velocity, and the upstream static pressure as the reference pressure.

We shall next measure each of the variables in terms of the proper reference quantity by defining dimensionless variables in the following manner:

$$u = u_0 u^*, \quad v = u_0 v^*, \quad x = l_0 x^*, \quad t = \frac{l_0}{u_0} t^*, \quad p = p^* p_0, \quad \text{etc.,}$$

where the starred quantities are dimensionless.

The continuity and Navier–Stokes equations now become

$$\frac{\partial u^*}{\partial x^*} + \frac{\partial v^*}{\partial y^*} + \frac{\partial w^*}{\partial z^*} = 0 \tag{5.4}$$

and

$$\frac{\partial u^*}{\partial t^*} + u^* \frac{\partial u^*}{\partial x^*} + v^* \frac{\partial u^*}{\partial y^*} + w^* \frac{\partial u^*}{\partial z^*}$$

$$= -\frac{p_0}{\rho u_0^2} \frac{\partial p^*}{\partial x^*} + \frac{\nu}{u_0 l_0} \left(\frac{\partial^2 u^*}{\partial x^{*2}} + \frac{\partial^2 u^*}{\partial y^{*2}} + \frac{\partial^2 u^*}{\partial z^{*2}} \right) + \frac{g l_0}{u_0^2}. \tag{5.5}$$

Similarly, the boundary conditions become

$$\left. \begin{array}{l} u^* = u_b^* \\ v^* = v_b^* \\ w^* = w_b^* \end{array} \right\} \quad \text{at } f(x_b^*, y_b^*, z_b^*) = 0 \tag{5.6}$$

and

$$p^* = p_b^* \quad \text{at } F(x_f^*, y_f^*, z_f^*) = 0,$$

the latter condition being applicable only in case a portion of the boundary is a free surface, and provided that surface tension may be neglected at that boundary.

One must now acknowledge the fact that the above manipulations did not really help at all in arriving at a solution of the above set of differential equations. However, one important fact can be brought out. Let us consider *two* problems concerning the flow of an incompressible fluid for which the boundary conditions, when expressed in dimensionless terms, are identical.

Let us imagine that the two problems in question concern the flow around two cylinders—a large one and a small one. The boundary conditions require that the velocities at the cylindrical surface are zero and that the velocity at infinity is constant and we will call it the *approach velocity*. The region occupied by the fluid is essentially of infinite extent and there are no free surfaces. As a reference velocity we may here select the approach velocity; the diameter of the cylinder would be a convenient reference length, and the static pressure far upstream could serve as the reference pressure. Any other set of reference quantities would also be satisfactory, the only essential requirement being that the same reference quantities are selected for both problems.

If the boundary conditions are now expressed in dimensionless terms, they are seen to be identical for the two problems. As a consequence, the solutions to the two problems, when expressed in dimensionless terms, will be identical provided that the differential equations are also identical. Examining the differential equations, we see that the continuity equation will automatically satisfy this requirement, but that the Navier–Stokes equations will only be identical for the two cases if the three parameters $p_0/\rho u_0^2$, $\nu/u_0 l_0$, and $g l_0/u_0^2$ have the same values for both problems. If these three parameters are the same for the two problems, we can apply the solution for the one cylinder (say the smaller one) directly to the other. The two problems are then said to be *dynamically similar* in addition to being geometrically similar. One might add that the first solution could well be obtained experimentally. These experimental results would then be applicable to the flow over the second cylinder.

Let us now examine in detail the three parameters that appear in the Navier–Stokes equation. The first one $p_0/\rho u_0^2$ involves the ratio of the reference pressure to the velocity head based on the reference velocity. By analyzing this parameter from a physical rather than a mathematical point of view, it is seen that this ratio is of significance only when the absolute pressure level of the flow is of importance, that is, when the absolute value of p_0 is of significance. In many examples, however, the absolute value of p_0 does not influence the flow. Taking the flow of an incompressible fluid (water being a good approximation to such a fluid) over a cylinder as an example, we find that the flow pattern as well as all pressure differences cannot depend on the general pressure level. In flows of this type two simplifications are possible. First the quantity ρu_0^2 (or more usually $\frac{1}{2}\rho u_0^2$) may be taken as a measure of pressure instead of p_0. In this way the number of reference quantities is reduced, and the parameter $p_0/\rho u_0^2$ simply becomes unity (or $\frac{1}{2}$) and no longer has to be counted a separate parameter. Second, since the absolute pressure level does not influence the flow, pressure may be measured with respect to any convenient reference. In problems in which this simplification is possible the number of parameters reduces to two, namely $v/u_0 l_0$ and $g l_0/u_0^2$.

The simplification above is not always possible. In some flows, for example, the pressure at certain points may become so low that the vapor pressure of the liquid is reached and that vapor cavities form—a phenomenon known as *cavitation*. In these flows the pressure level is very important, since a high pressure level will tend to suppress cavitation, whereas a low pressure level will tend to promote it. At a certain point on the cylinder the pressure may be $p_\infty - \rho u_0^2$, where p_∞ is the static pressure at infinity. Depending on the value of p_∞, the pressure $(p_\infty - \rho u_0^2)$ may or may not correspond to the vapor pressure of the water. In examples of this kind the parameter $p_0/\rho u_0^2$ is an essential factor. Because of the importance of the vapor pressure in this type of problem, pressures are usually measured relative to this vapor pressure. Consequently, if the pressure at infinity is a suitable quantity for the reference pressure, one would select p_0 to be equal to

$$p_0 = p_\infty - p_v,$$

so that

$$\frac{p_0}{\rho u_0^2} = \frac{p_\infty - p_v}{\rho u_0^2}.$$

Since $\frac{1}{2}\rho u_0^2$ has a certain physical meaning, the parameter used more frequently is

$$\frac{p_0}{\frac{1}{2}\rho u_0^2} = \frac{p_\infty - p_v}{\frac{1}{2}\rho u_0^2},$$

which is known as the *Euler number*, Eu.

The remaining two parameters are $v/u_0 l_0$ and $g l_0/u_0^2$. By tradition the reciprocal of the first and the square root of the reciprocal of the second of these quantities are generally used and they are known as the *Froude number*

$$\mathrm{Fr} = \frac{u_0}{\sqrt{g l_0}}$$

and the *Reynolds number*

$$\text{Re} = \frac{u_0 l_0}{\nu}.$$

If the Euler number can be disregarded, then the Froude number and the Reynolds number remain as the parameters that determine the flow characteristics. This means that when two flows have the same Reynolds number and the same Froude number, the description of these two flows in terms of dimensionless quantities will be identical, provided, of course, that the dimensionless boundary conditions for the two problems are also the same.

There is now an additional very important class of problems in which a further simplification can be made. Consider the flow in a duct or more generally the flow of a fluid without a free surface, as illustrated in Fig. 5.1. The fluid is incompressible and let us assume further that the Euler number is unimportant for the problem at hand.

For a configuration of this type let us first write an equation for the pressure distribution that would exist if the duct contained fluid at rest. In dimensionless form the Navier–Stokes equation for this case simply becomes

$$\frac{1}{2}\frac{\partial p_r^*}{\partial x^*} = \frac{g l_0}{u_0^2},$$

where $\frac{1}{2}\rho u_0^2$ has been used as the reference pressure.

The subscript r has here been added to the pressure to indicate that this is the pressure existing in the fluid when at rest. Next, subtracting this equation from the Navier–Stokes equation, we obtain

$$\frac{\partial u^*}{\partial t^*} + u^* \frac{\partial u^*}{\partial x^*} + \cdots = -\frac{1}{2}\frac{\partial p_n^*}{\partial x^*} + \frac{\nu}{u_0 l_0}\frac{\partial^2 u^*}{\partial x^{*2}} + \cdots,$$

where $p_n^* \equiv p^* - p_r^*$. Let us call p_n^* the dimensionless "nongravitational" pressure. By introducing the concept of a nongravitational pressure, it is seen that the gravity term and hence the Froude number could be eliminated from the equation of motion. We next have to examine the boundary conditions. If these are of the type in which the velocities are specified at the boundaries all of which are fixed in space, then the boundary conditions remain unchanged by the substitution of the true pressure by the nongravitational pressure. Also, the continuity equation remains unaffected by this substitution. As a consequence, we now have the two differential equations and the boundary conditions in a dimensionless form with only a single dimensionless parameter, the Reynolds number. For problems in which the above simplification is permissible, only the Reynolds number, therefore, has to be kept

FIG. 5.1   Fluid flowing in a conduit. The fluid fills the duct and does not form a free surface.

constant to ensure dynamic similarity. Although this simplification applies only under the conditions mentioned, it is applicable for very many physical problems, such as flow through closed ducts of any kind, flow in turbomachinery, flight of objects through the air at speeds at which compressibility effects can be neglected, and so on. The Reynolds number is, therefore, a most essential parameter in fluid mechanics.

Let us consider now the class of problems in which a free surface exists—that is, a boundary, the shape of which depends on the motion. Then the introduction of a nongravitational pressure no longer leads to any simplification, as the correspond-ing pressure distribution for the fluid at rest (p_r) would have to be obtained for the free surface shape as it exists for the moving fluid. The determination of this shape, however, would have to come again from the solution of the complete dynamic equations, including the gravity term. The gravity term, therefore, cannot be elimi-nated from the basic equations in this case, and the Froude number has to be carried as a separate parameter. This condition exists, for example, for the flow in open channels, the propagation of surface waves, and the flow around ships.

In summary, it may be said that the dynamic similarity of the flow of an incom-pressible fluid under the influence of gravity and viscosity is in general governed by the three parameters, the Euler number $(p_\infty - p_v)/\frac{1}{2}u_0^2$, the Froude number $u_0/\sqrt{gl_0}$, and the Reynolds number $u_0 l_0/\nu$. If cavitation is of no concern, the Euler number may be disregarded. If, in addition, there is no free surface, the Froude number may be disregarded, leaving the Reynolds number as the only essential parameter for dynamic similarity. For flows in which a free surface exists, both the Froude number and the Reynolds number will generally have to be considered.

One may add here, however, that in some flows similarity depends essentially on the Froude number only. This occurs when the model as well as the prototype are in a range in which the flow conditions are insensitive to changes in Reynolds number. It is difficult to establish general rules as to when this condition is applic-able. However, dynamic similarity for the flow in rivers and over dams, as well as for the wave motion in harbors, is almost always controlled by the Froude number alone.

The analysis above has been limited to an incompressible fluid with constant viscosity. The Navier–Stokes equation together with the boundary conditions also did not include all possible forces that could act on the fluid. Among the forces omitted are those due to surface tension and those due to electromagnetic effects. Any additional forces have to be added to the Navier–Stokes equation for problems in which they play an important role. The enlarged equations may then also be made dimensionless and additional parameters will appear. If, furthermore, the fluid properties such as the density and viscosity depend on the temperature of the fluid, a third differential equation governing the temperature distribution has to be intro-duced. This differential equation is called the energy equation and is based on the first law of thermodynamics as well as on Fourier's law of heat conduction. The dimensional form of this third equation with its corresponding boundary conditions will also give rise to new dimensionless parameters. The method of analysis and the procedural steps in deriving the dimensionless parameters will, however, be the same as those followed for the incompressible fluid with constant viscosity. As a matter of incidental interest, a parameter of major importance when compressibility of the fluid is taken into account is the ratio of the reference fluid velocity, u_0, to the speed of sound referred to some reference state, a_0. This ratio, u_0/a_0, is called the *Mach number*, and it arises as a natural result of the similarity analysis of this section. (We have already seen an example of this ratio in Section 3.9.) A detailed study of the effects of compressibility and the additional parameters that arise will be taken up in Chapters 9 and 10 and will not be pursued further here.

5.3 Alternative Method of Deriving Similarity Parameters

(a) Balance of Forces

The foregoing method of discussing the role of the Reynolds number and the Froude number is perhaps the most straightforward. There is a simpler way of arriving at such dimensionless parameters, which does not require the derivation of the complete differential equations. It does, however, require a deeper physical insight. We shall illustrate this approach by considering the viscous flow through a Venturi meter as in Example 3.1. In this case one may guess that the important forces in this flow are the viscous shear forces and the inertia forces. The condition of dynamic similitude is that the ratio of these forces must be constant.

Let us suppose that we have a cube of fluid of side d in a fluid flow of velocity V. To accelerate this body requires a force equal to the mass times the acceleration. As a measure of the acceleration we take $dV/dt \propto V/t \propto V^2/d$, and with the mass equal to ρd^3, we obtain an inertial force

$$F_{in} \propto \rho d^2 V^2.$$

The viscous shear forces are proportional to the viscosity, cross-sectional area, and the velocity gradient; that is,

$$F_{vis} \propto \mu V d$$

and the ratio

$$\frac{F_{in}}{F_{vis}} \propto \frac{\rho V d}{\mu} = \text{Re}.$$

The Froude number and other similarity parameters may be derived in a similar manner.

(b) The Pi Theorem

There is still another method of developing dimensionless parameters, which also avoids the necessity of deriving the full differential equations. Again, a thorough physical understanding of the problem is required, sufficient to identify all of the significant variables that govern the problem to be solved. To illustrate this step, let us consider an example of the problem of determining the drag on a sphere in a large body of water. The water flows over the sphere with an approach velocity, V. The drag is the force exerted by the fluid on the sphere in a direction parallel to the flow. Based on intuition, it seems very plausible that the drag would depend on the velocity, V, the diameter of the sphere, d, and the density of the liquid, ρ. It is easy to imagine that an increase in each one of these variables might lead to an increase in the drag. In addition, we have previously seen that a fluid can exert shear forces and we should include, therefore, the viscosity, μ, which relates to these forces, in our set of variables.

We may then write in symbolic form

$$D = \phi(V, d, \rho, \mu),$$

where ϕ indicates a functional relationship. Both sides of this equation must, of course, have the same dimensions. Using the force–length–time (F-L-T) system, we find that the dimensions of the variables above are

$$[D] = F,$$

$$[V] = \frac{L}{T},$$

$$[d] = L,$$

$$[\rho] = \frac{F \times T^2}{L^4},$$

$$[\mu] = \frac{F \times T}{L^2}.$$

(The brackets indicate that a dimensional relationship only is shown.)

The equation for the drag may be rearranged to read

$$\frac{D}{\phi(V, d, \rho, \mu)} - 1 = 0$$

and in this form each term is clearly dimensionless. The same expression may also be written as

$$f(D, d, V, \rho, \mu) = 0, \tag{5.7}$$

where f designates a new functional relationship, which, of course, is still dimensionless. For this to be true, however, the variables D, V, d, ρ, and μ, appearing in this relationship, must be combined in such a way as to form groups which themselves are dimensionless. This simple consideration holds the key to the present approach to developing dimensionless parameters.

To illustrate the point in a less abstract manner, let us recall Eq. (b) of Example 3.2, in which the jet velocity, V_j, is given as

$$V_j^2 = 2\left[gy + \frac{1}{\rho}(p_t - p_a)\right]\frac{1}{1 - (A_j/A_t)^2}.$$

In this equation, y is a length, ρ is a density, p_t and p_a are pressures, g is the gravitational constant, and A_j/A_t is an area ratio. We may rearrange this equation into the form

$$\frac{2gy}{V_j^2} + \frac{p_t - p_a}{\rho V_j^2} - \left[1 - \left(\frac{A_j}{A_t}\right)^2\right] = 0.$$

Each term in this arrangement is dimensionless, representing a dimensionless group, as anticipated.

Returning to the problem of the drag on the sphere and the most general formulation of Eq. (5.7), we now try to form the appropriate dimensionless groups. The number of variables in Eq. (5.7) is five. The number of dimensions available to describe these variables is three: force (F), length (L), and time (T). The groups must be formed so that these three dimensions cancel out in each group. This requirement specifies three conditions that have to be met, which in turn provides three equations that determine the relations between the variables. It is plausible, then, to expect that the number of groups that can be formed from the variables will be three less than the original number of variables, or two groups in our case. This conclusion is a special case of a general theorem that can be proven rigorously and states that if a functional relationship exists between s variables and these variables are characterized by r fundamental dimensions, then it is possible to express the relationship in terms of $(s - r)$ dimensionless groups. This statement is known as the Pi theorem, and very careful and complete derivations of the theorem have been given (e.g., see Reference 4).

The next task concerns the actual formation of the dimensionless groups. To arrive at meaningful parameters, certain rules have to be followed. Continuing with our example in which we have three fundamental dimensions, we find that it is necessary to select three variables that among themselves contain the three fundamental dimensions; however, no two of them should have exactly the same dimensions. We shall call these variables the "recurrent" ones. A good choice might be the length d, the velocity V, and the density ρ. Having selected these three, we derive the first parameter P_1, by combining the three together with one of the remaining variables in the form

$$P_1 = d^{x_1} V^{x_2} \rho^{x_3} D^{x_4},$$

where the drag D was taken as the first of the remaining variables and the exponents x_1, x_2, and so on, are dimensionless numbers. Writing the dimensions corresponding to the expression for P_1, we see that the product

$$(L)^{x_1} \left(\frac{L}{T}\right)^{x_2} \left(\frac{FT^2}{L^4}\right)^{x_3} F^{x_4}$$

has to be dimensionless, and that the exponents have to be selected in such a way that all the dimensions cancel out. Considering the exponent associated with force first, we obtain

$$x_3 + x_4 = 0.$$

For time we obtain

$$-x_2 + 2x_3 = 0,$$

and for length,

$$x_1 + x_2 - 4x_3 = 0.$$

Solving these equations, we find that

$$x_3 = -x_4,$$
$$x_2 = -2x_4,$$

and

$$x_1 = -2x_4.$$

The parameter P_1 may then be written as

$$P_1 = \frac{D^{x_4}}{d^{2x_4} V^{2x_4} \rho^{x_4}} = \left(\frac{D}{\rho V^2 d^2}\right)^{x_4}.$$

The exponent x_4 is, so far, undetermined. It may be seen, however, that no matter what value it takes, the fact that the combination $D/\rho V^2 d^2$ is a parameter remains unchanged. After all, if a function depends on $D/\rho V^2 d^2$, it also depends on $(D/\rho V^2 d^2)^2$ or any power of $(D/\rho V^2 d^2)$. For convenience we let $x_4 = 1$ and define

$$P_1 = \frac{D}{\rho V^2 d^2}.$$

To develop P_2, we let

$$P_2 = d^{y_1} V^{y_2} \rho^{y_3} \mu^{y_4};$$

that is, P_2 has been formed from the recurrent variables that we selected (d, V, and ρ), and from the second of the remaining ones (μ). Letting y_4 be equal to unity right at the start, we may write

$$P_2 = d^{y_1} V^{y_2} \rho^{y_3} \mu.$$

Substituting the dimensions for each variable and proceeding as before, we find that $y_1 = y_2 = y_3 = -1$ and

$$P_2 = \frac{\mu}{\rho V d}.$$

To satisfy the dimensional requirement, we see that the five variables may be combined into two parameters, and instead of

$$F(D, V, d, \rho, \mu) = 0,$$

we may write

$$\phi(P_1, P_2) = \phi\left(\frac{D}{\rho V^2 d^2}, \frac{\mu}{\rho V d}\right) = 0.$$

This functional relationship may also be expressed as

$$\frac{D}{\rho V^2 d^2} = \phi_1\left(\frac{\mu}{\rho V d}\right),$$

where ϕ_1 simply indicates another functional relationship. Many different ways exist to present the same relationship. Using the reciprocal of the second parameter, we see that

$$\frac{D}{\rho V^2 d^2} = \phi_2\left(\frac{\rho V d}{\mu}\right);$$

or, if P_1 is a function of P_2, then P_1/P_2^2 is also a function of P_2, which would result in the new expression

$$\frac{D\rho}{\mu^2} = \phi_3\left(\frac{\rho V d}{\mu}\right).$$

By forming combinations of this kind, many different parameters can be formed, but they are all just combinations of the first two. Furthermore, different parameters would have been obtained had we selected D, V, and μ as our basic variables instead of D, V, and ρ. The resulting parameters, however, would again have been among the combinations derivable from the first set.

The most suitable set of parameters cannot be determined by the present method but will depend on other information or on experience. For the example of the drag on a sphere, the parameter P_1 is directly related to a commonly defined drag coefficient

$$C_D = \frac{D}{(\rho V^2/2)(\pi d^2/4)} = P_1 \frac{8}{\pi}$$

and P_2 is the inverse of the Reynolds number,

$$\mathrm{Re} = \frac{\rho\, V d}{\mu} = \frac{1}{P_2}.$$

For that case C_D and Re form the commonly used set of parameters and we may conclude that C_D is a function of Re. Of course, C_D and Re are very simply related to P_1 and P_2.

Having discussed a particular example, we may now outline the method of finding dimensionless parameters in a more general way:

1. A certain problem is to be solved in the sense that one unknown quantity is to be expressed in terms of several other quantities that are essential for a description of the problem. A very clear understanding of the physical aspects of the problem is required in order to select these quantities properly and to omit quantities that are extraneous to the problem.

2. Say that all of the quantities selected together with the unknown one form q variables, which will be denoted as $S_1, S_2, \ldots, S_q$. The unknown relationship between these variables may then be written symbolically as

$$f_1(S_1, S_2, S_3, \ldots, S_q) = 0.$$

3. Each of the variables S has certain dimensions. All the basic dimensions required to describe the set of variables have to be listed and their total number is designated by r.

4. The Pi theorem now states that it will be possible to simplify the function f_1 so that instead of a relationship between q dimensionless variables, one obtains a new relationship f_2 between $(q - r)$ dimensionless parameters. This result may be written symbolically as

$$f_2(P_1, P_2, P_3, \ldots, P_{q-r}) = 0.$$

5. To find the set of parameters, we first select r of the variables S. These have to be selected in such a way that among them they contain all the fundamental dimensions, and no two of them should have exactly the same combination of dimensions. Let us call these r variables the "recurring" variables.

6. To find a dimensionless parameter, we write

$$P_i = S_1^{x_1} S_2^{x_2} S_3^{x_3} \cdots S_r^{x_r} S_j,$$

where S_1 to S_r are the recurring variables and S_j is one of the remaining variables. We may now solve for the various exponents so that the parameters, P_j, will be dimensionless. The number of different variables S_j will be $(q - r)$, and consequently, we shall be able to obtain the predicted number of $(q - r)$ dimensionless parameters P.

An elaborate and elegant theoretical structure has been developed that supports the foregoing method of dimensional analysis. The development also contains rules for the selection of what we called the "recurrent" variables.

An unfortunate initial choice of the recurrent variables could at times prevent the systematic development of a suitable set of dimensionless parameters. It will, however, never lead to an incorrect set. If, therefore, an attempt to find the dimensionless parameters should be unsuccessful, a different set of recurrent parameters should be selected and a new trial should be attempted.

The method just described is based on the Pi theorem stated by Buckingham (see Reference 7). We have only provided a sketchy background of the theorem and some practical rules for its application. The theorem does allow the derivation of a complete set of dimensionless parameters for a given problem. The parameters greatly simplify the presentation of the data that result from solution of the problem. The theorem could then be a very powerful tool in the analysis of any engineering problem and a particularly valuable tool, as it is rather simple to use.

In practice, however, the task of selecting the key variables for a problem that is not yet understood is often quite difficult. Because of uncertainty as to the relative importance of the several variables, one may be forced to retain too many, which leads to a large set of dimensionless parameters. As was seen, these parameters can be combined to form many different sets, and it is often very difficult to arrive at a set of parameters that are of help in simplifying the problem.

5.4 Application to Flow Measurement

The concepts of dimensional analysis and dynamic similarity are used frequently in engineering. As an example, let us consider the calibration of a certain design of

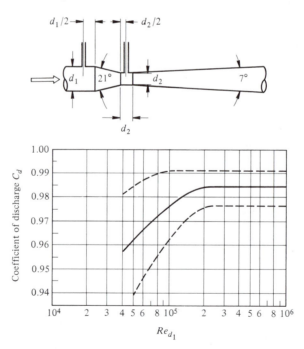

FIG. 5.2 Discharge coefficient c_d as a function of the Reynolds number. The Reynolds number is based on the average velocity in the pipe and the diameter of the pipe. The coefficients apply to the particular type of Venturi meter shown. The dashed lines show the range of values that may be encountered and the range of diameter ratios is 0.3 to 0.75. (Data from Chapter 4 of *Flow Measurement*, Publication PTC 19.5; 4-1959, American Society of Mechanical Engineers.)

Venturi meters. In Section 3.8 it was shown that for an incompressible fluid, the flow rate through a Venturi meter could be expressed in terms of the Venturi dimensions, pressure drop, and discharge coefficient as

$$Q_{\text{actual}} = \frac{c_d A_2 \sqrt{2(p_1 - p_2)/\rho}}{\sqrt{1 - (A_2/A_1)^2}}.$$

Suppose that we have a number of geometrically similar Venturi meters, each of different size. They are to be used for different flow rates and for different fluids, and we are to find the correct discharge coefficient for each application. In view of the previous discussions, this can be accomplished by testing any one of the meters, using only one fluid, and covering a range of Reynolds numbers simply by varying the flow rate.

The results can be presented as a plot of c_d versus Re, and they are applicable to any fluid of any viscosity and with any rate of discharge (as long as the fluid remains incompressible). A graph containing this kind of information may be found in Fig. 5.2, and similar information for a particular kind of orifice is given in Fig. 5.3. The virtue of such graphs is that the number of experiments necessary to obtain the discharge coefficient has been reduced tremendously, compared to the number that would have to be carried out if we had not taken advantage of the similarity considerations. In that event we would have had to find the effect of V, μ, ρ, and d separately.

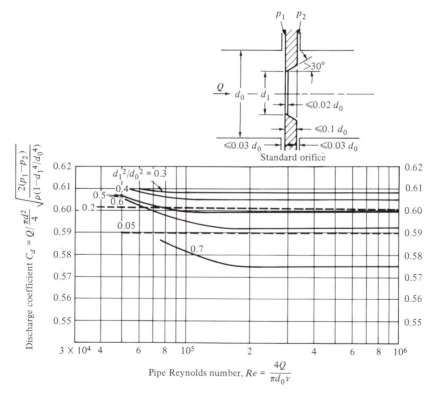

FIG. 5.3 Discharge coefficient as a function of Reynolds number of various ratios of $(d_1/d_0)^2$ for the type of orifice shown. The Reynolds number is based on the average velocity in the pipe and the pipe diameter. (Based on data from Report T.M. 952, Nat. Aero. and Space Adm., formerly NACA.)

PROBLEMS

When not otherwise specified, the properties of the fluids are to be evaluated at 20°C and 1 atmosphere pressure.

5.1 In some cases (e.g., studies on ship hulls) it is necessary to retain both the gravity and the viscosity term to describe the flow adequately. Under what conditions can a model be built in such a case?

5.2 By direct consideration of the inertial and gravitational forces derive the Froude number. By the same method derive the *Weber number*, $V_0/\sqrt{\sigma/\rho l_0}$, which is a measure of the ratio between inertial and surface tension forces.

5.3 **(a)** What is the merit of a wind tunnel in experimental work?
(b) Why is it desirable to pressurize wind tunnels?

5.4 A model of a harbor is made on the length ratio of 280 : 1. Storm waves of 1.5 m amplitude and 9 m/s velocity occur on the breakwater of the prototype harbor.
(a) Neglecting friction, what should be the size and speed of the waves in the model?
(b) If the time between tides in the prototype is 12 hr, what should be the tidal period in the model?

5.5 A research group was interested in designing an axial-flow air compressor with a rotor diameter of 0.6 m and an angular speed of 8000 rpm. Before deciding on the blade shapes, model tests were proposed. The model finally chosen was 0.3 m in diameter and was to pump water rather than air. The properties of the air are $\rho = 1.0 \text{ kg/m}^3$ and $v = 1.7 \times 10^{-5} \text{ m}^2/\text{s}$.
(a) For similar flows, what has to be the speed of the model?
(b) For similar flows in the two machines, what are the ratios of the head produced and the power required? Any possible effects of compressibility should be neglected.

5.6 A hydraulic model of the spillway for the Grand Coulee dam was made with a scale ratio of 180 : 1. Assume the flows in model and prototype to be insensitive to changes in Reynolds number.
(a) State under what conditions the model may be expected to predict adequately the behavior of the prototype.
(b) The velocity measured in the model at the bottom of the model spillway was 4 ft/sec and this value corresponds to a flow rate of 1 ft³/sec. What were the corresponding prototype quantities?

5.7 The propeller of a ship operates in close proximity to the free surface. The propeller of an ordinary cargo ship 170 m long is about 2.5 m in diameter and turns at 100 rpm. If a 1 : 100 scale model is to be tested of the ship and propeller in a towing tank:
(a) What should be the speed of the model propeller?
(b) What is the ratio of the horsepower required by the model to that of the ship? Assume that the effects of viscosity are negligible.

5.8 In the testing of aircraft-launched torpedoes, it is common to study the water-entry phenomenon by means of model studies. Suppose that a representative prototype torpedo is 8 ft long and enters the water at approximately 1200 ft/sec. A 1/10 scale model projected into a laboratory water tank is to be used.

Assume the flow to be insensitive to changes in Reynolds number in this range and neglect any influence of compressibility.

(a) What is the proper launching velocity for the model?

(b) If the model experiences a maximum instantaneous force, F_m, of 2 lbf upon entering the water at the velocity found in part (a), what force F_p may be expected in the full-scale torpedo?

(c) In studying water-entry phenomena, it is also important to model the cavitation characteristics of the flow. If the water is at 80°F ($p_v = 0.507$ psia) in both prototype and model, and the ambient atmospheric pressure is 15 psia for the prototype, find the desired atmospheric pressure p_{0m} for the model test.

5.9 At a temperature of 24°C and Reynolds number of 2.5×10^6 (based on length) the lift force of a model hydrofoil 0.15 m in length was tested in a water tunnel and found to be 11,000 N. The prototype, to be used in water at the same temperature, has a length of 0.6 m. What will be the force on the prototype at the same Reynolds number? What is the fluid velocity for the prototype?

5.10 A centrifugal pump is being designed for the cooling system of a nuclear reactor. The coolant is to be liquid sodium at 400°C, at which temperature it has a density of 850 kg/m^3 and dynamic viscosity of 0.269×10^{-3} Pa s. The pump is to circulate 0.03 m^3/s at a total pumping head of 2 m. It will be driven by a motor turning at 1760 rpm.

A model of this pump is to be constructed at four times life size and tested with water at 20°C. In modeling, the Reynolds number is to be considered the significant parameter. In computing this parameter, the diameter and peripheral velocity of the impeller may be taken as the characteristic length and velocity, respectively. Find (a) the appropriate speed, (b) the head, (c) the discharge of the model pump.

5.11 A model of a water siphon is to be made and tested. The siphon is to be operated at sea level ($p_a = 100$ kPa). At the summit it will lift the fluid to a maximum height of 9 m above the reservoir. The model is to be one-tenth full size and will be operated with water. The vapor pressure of water may be assumed to be 3 kPa in each case. Assuming that the effects of viscosity are negligible in both model and prototype, determine the ambient pressure to be used for the model test to ensure similitude and the ratio of the flow rates of the model to that of the prototype.

5.12 A 1-in. sharp-edged orifice is installed in a 2-in. pipeline. The fluid flowing in the line is water. Find the discharge through the line if a mercury manometer indicates a deflection of 1.5 in. Use the data of Fig. 5.3.

5.13 A model ship propeller is to be tested in water at the same temperature and pressures that would be encountered by a full-scale propeller. Over the speed range considered, it is assumed that there is no dependence on the Reynolds or Euler numbers, but only on the Froude number (based on forward velocity V and propeller diameter d). In addition, it is thought that the ratio of forward to rotational speed of the propeller must be constant (the ratio V/Nd, where N is propeller rpm).

(a) With a model 0.4 m in diameter, a forward speed of 2.5 m/s and a rotational speed of 450 rpm are recorded. What are the forward and rotational speeds corresponding to an 2.4-m-diameter prototype?

(b) A torque of 20 Nm is required to turn the model, and the model thrust is measured to be 250 N. What are the torque and thrust for the prototype?

5.14 A cylinder of diameter D is observed to oscillate with frequency f cycles/sec when held in a uniform flow of U_0 ft/sec (an unsteady flow situation).
 (a) By direct examination of the equations of motion, show that a measure of the ratio of unsteady force to "inertial force" is given by the *Strouhal number* $S = fD/U_0$.
 (b) A flagpole 0.15 m in diameter and 8 m high is observed to oscillate in its second mode at a frequency of 10 Hz in air ($v = 1.6 \times 10^{-5}$ m^2/s). Estimate the wind speed at this condition (trial and error).

 Wind tunnel test results for the Strouhal number at various Reynolds numbers follow:

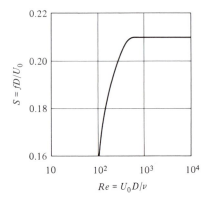

Problem 5.14

5.15 The equations of motion for a jet of hot gas at a temperature T_H (abs.) moving through a colder fluid at T_0 (abs.) are

$$\frac{D\mathbf{v}}{Dt} = \frac{\nabla p}{\rho_0} + v_0 \nabla^2 \mathbf{v} + \mathbf{g}\left(\frac{T_H}{T_0} - 1\right).$$

 (a) Show that the ratio of gravity (buoyancy) to inertial forces acting on a fluid element is

$$\frac{Lg}{V_0^2}\left(\frac{T_H}{T_0} - 1\right),$$

 where L and V_0 are reference lengths and velocity, respectively.
 (b) A 1/100 scale model of a stack for discharging hot gases is to be tested in a wind tunnel. The prototype stack is 100 ft high and discharges 10^7 ft^3/day of a 150°F gas to atmosphere ($T_0 \sim 60$°F). The model gas is limited to 80°F, and also discharges to atmospheric conditions. What is the model discharge rate for dynamic similarity?

5.16 A large pump receives water at an inlet pressure of 2×10^5 N/m^2 and discharges the water at 10^6 N/m^2. A half-scale model of this pump is to be tested. Gravity effects are assumed to be negligible, but cavitation is important. The model is to be tested with water at the same temperature as that used for the prototype (the full-scale unit). The vapor pressure under this condition is 3×10^3 N/m².

(a) For accurate scale experiments, which parameters have to be maintained constant between model and prototype?

(b) If the velocity to the inlet of the prototype is 2 m/s, what must the velocity be at the inlet of the model?

(c) What is the proper inlet pressure (in N/m²) for the model?

5.17 If a ship's propeller is well submerged, the thrust F and the torque T depend upon

d = propeller diameter,

ω = angular speed of the propeller,

V = forward speed of the propeller,

ρ = density of fluid in which the propeller operates,

μ = absolute viscosity of fluid in which the propeller operates.

By dimensional analysis, obtain the form of expressions for F and T. Select the appropriate quantities to combine with F and T to form dimensionless groups.

5.18 The streamline patterns in two geometrically similar centrifugal pumps, with impeller diameters D_1, D_2 and operating speeds N_1, N_2, can be made geometrically similar by suitably adjusting the discharges Q_1 and Q_2.

(a) Determine the ratio Q_1/Q_2 in terms of D_1, D_2, N_1, and N_2 necessary for kinematic similitude (streamline patterns geometrically similar).

(b) If all the energy in the flow discharging from the pumps is converted to kinetic energy by nozzles, what is the ratio of nozzle velocities V_1 and V_2 under conditions of kinematic similitude, in terms of D_1, D_2, N_1, and N_2?

(c) The unit weights of the fluids pumped are γ_1 and γ_2. What is the ratio of power P_1/P_2 added to the flow by the pumps operating under conditions of kinematic similitude, in terms of D_1, D_2, N_1, N_2, γ_1, and γ_2?

5.19 Suppose that a large amount of energy E is suddenly released in air (a point explosion). Experimental evidence suggests the radius R of the high-pressure blast wave depends on time, as well as the energy E and the density of the ambient air ρ.

(a) Using the Pi theorem, find the equation for R as a function of t, ρ, and E.

(b) Show that the speed of the wave front V decreases as R increases (i.e., $V \sim 1/R^{3/2}$).

(c) If the pressure rise Δp across the blast wave also is known to depend only on t, E, and ρ, show that $\Delta p \sim 1/R^3$.

5.20 A very long vertical plate is withdrawn at a steady speed V from a bath of liquid of density ρ as shown. The fluid is viscous (viscosity μ), so that a thin layer of fluid adheres to the plate, but gravity g also tends to make it drain back. The local film thickness t depends on the vertical location y, as well as ρ, μ, V, g, and the surface tension σ at the liquid–vapor interface. Determine the dimensionless groups governing such a flow, and indicate the functional dependence of t on these groups.

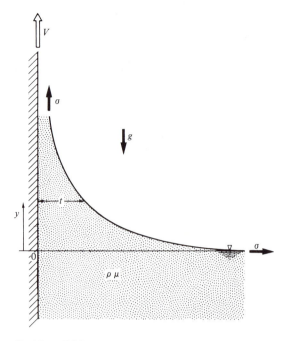

Problem 5.20

5.21 A thin, inextensible but perfectly flexible film (something like cellophane) rests on a thin layer of viscous fluid (density ρ, viscosity μ), which in turn covers a large flat surface. If a tension T is applied to the film, it splits and peels from the surface at some angle θ and at a steady speed V. This process can be visualized as a steady process when observed at the peeling point as shown in the sketch. Gravity forces are negligible at high-enough peeling speed, but surface tension σ is important in splitting the film. Find a relationship for the peeling angle θ in terms of two independent dimensionless groups.

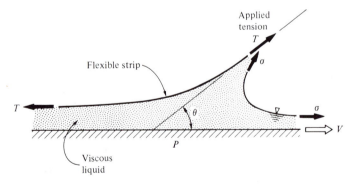

Problem 5.21

5.5 Application to Pipe Friction

If a fluid flows through a length of pipe and the pressure is measured at two stations along the pipe, one will find that the pressure decreases in the direction of flow. This pressure decrease is caused by the friction of the fluid against the pipe wall. The prediction of this friction loss is one of the important problems in fluid mechanics. It is a very complicated problem and only in a special case, to be discussed in a subsequent chapter, can the friction factor be computed analytically.

The problem of friction in pipes must generally be solved experimentally. To approach this question, let us consider the following experimental situation. A large tank of fluid is connected to a long pipe of constant diameter and a pressure is applied to the tank so that a steady flow results. Near the entrance of the pipe the flow has not been subjected to the influence of viscosity for long, so that the velocity profile is reasonably constant across the pipe, except for a region near the walls, where the velocity will have been decreased due to the wall shear. This region will grow as the flow proceeds downstream, and the velocity profile at the same time will undergo changes. Eventually, however, the distribution of velocity will no longer change with distance, and an *equilibrium profile*, expressible as $u^* = u^*(r^*)$, will become established. The function $u^*(r^*)$ will, in the light of our previous discussion, still depend on an appropriate Reynolds number characterizing the flow in the pipe. Similar to the velocity profile, the pressure gradient will also change in the entrance region of the pipe. Whenever the velocity distribution becomes independent of distance, however, the gradient dp^*/dx^* should also be expected to reach a constant value. This gradient will, of course, also depend on the characteristic Reynolds number, that is,

$$\frac{dp^*}{dx^*} = f(\text{Re}). \tag{5.8}$$

For fully established flow, this equation may then be integrated in respect to x^* to obtain

$$\Delta p^* = \Delta x^* f(\text{Re}).$$

The characteristic length of the problem is chosen as the pipe diameter d, so that $x^* = x/d$, and the dimensionless pressure is taken as $p^* = 2p/\rho V^2$, where V is the average velocity in the pipe. The pressure drop Δp over a length of pipe l can now be expressed in terms of the head loss h_f as

$$\frac{\Delta p}{\rho g} = h_f = f \frac{l}{d} \frac{V^2}{2g}, \tag{5.9}$$

where f has been written for $f(\text{Re})$. The function f is called the *friction factor*.† The Reynolds number Re is generally computed with the average velocity V, the diameter of the pipe d, and the kinematic viscosity. The actual dependence of f on Re will have to be determined by experiment. Having determined the curve of f versus Re, the friction loss in any pipe at any flow rate can be computed from Eq. (5.9). The present discussion is limited to "long" pipes in which the velocity distribution has become fully established or uniform along its length. It should be apparent that friction factors determined for this case do not apply near the entrance portion of a pipe, where the flow changes fairly quickly from one cross section to the next, or to any other flow in which the acceleration terms are not negligible.

There is one complication we have not considered so far. Due to manufacturing methods and corrosion, different pipe walls may be of different *roughness*. If the roughness for two pipes of the same diameter is different, or if the roughness does

---

† The friction factor f defined here is called the *Darcy friction factor*, common in hydraulics. Another friction factor, the *Fanning factor*, often used in heat transfer and aerodynamics, is one-quarter as large. Care should be exercised when using results from the literature to make sure that one has the right coefficient.

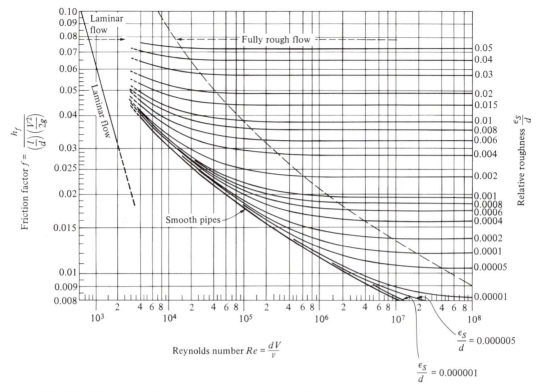

FIG. 5.4 Friction coefficient as a function of Reynolds number for round pipes of various relative roughness ratios ε_s/d. (L. F. Moody, "Friction Factors for Pipe Flow," *Trans. ASME*, vol. **66**, no. 8, 1944, p. 671.)

not increase in the same geometrical proportions as the diameter, we really violate the conditions of geometrical similarity. We then must expect an influence on f. To take the effect of roughness into account, it will be necessary to plot a set of curves of f versus Re, each for a different "relative roughness." The definitions of "roughness" and "relative roughness" are still incomplete, since they do not take the shape or distribution of the roughness elements into account.

The most systematic work with rough pipes has probably been done with pipes in which the surfaces were coated with sand grains of a definite size. The relative roughness in that case was defined as ε_s/d, where ε_s was the sand grain diameter and d the pipe diameter. It was then possible to present the friction factor as a function of Reynolds number for each ratio ε_s/d. The results also showed that at sufficiently high Reynolds numbers the curves approach horizontal lines, indicating that f becomes independent of Re. The flow conditions in this region are sometimes called "fully rough." The curves of f versus Re for pipes with natural roughnesses differ in some details from those with sand grain roughness; however, the curves also approach a constant value at high Reynolds number and exhibit the phenomenon of "fully rough" flow. Fairly good correlation for naturally rough tubes is attained by selecting as an *effective* relative roughness the ε_s/d value of a sand-grain roughness, which matches the friction factor of the naturally rough pipe in the fully rough region. Typical values for the equivalent roughness might be 3 mm for a riveted steel pipe, 1 mm for a concrete pipe, 0.3 mm for a wood stave or cast iron pipe, 0.05 mm for a commercial steel pipe, and 0.001 mm for a smoothly drawn steel tube. A graph giving f versus Re for various effective values of ε_s/d for naturally rough

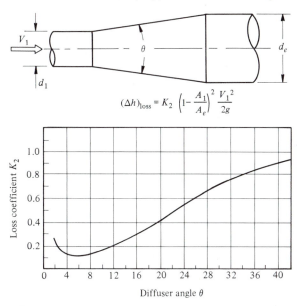

$$(\Delta h)_{\text{loss}} = K_2 \left(1 - \frac{A_1}{A_e}\right)^2 \frac{V_1^2}{2g}$$

FIG. 5.5   Loss coefficient for a conical diffuser as a function of the divergence angle. The loss coefficient is defined by the equation shown. These results are applicable for relatively high Reynolds numbers (say $> 10^5$). (Data obtained from Report ARR-L4F26, by J. R. Henry, Nat. Aero. and Space Adm., formerly NACA, 1944.)

pipes is shown in Fig. 5.4. With this graph and the appropriate value for ε_s we are now in a position to calculate with fair accuracy the pressure drop due to friction for a large variety of naturally rough pipes.

Experiments have also verified that the friction factor for a pipe of any given relative roughness is a function of Re only, no matter what the individual values of V, d, ρ, and μ that make up Re.

It should be pointed out that the curves in Fig. 5-4 are for pipes of circular cross section. For pipes of different cross sections, equivalent sets of graphs will have to be prepared from experimental data. In the absence of such graphs, however, estimates of the pressure drop may be obtained from the graphs for circular pipes by assigning an effective diameter to the noncircular pipe. This effective diameter is equal to four times the cross-sectional area divided by the wetted perimeter, which quantity becomes equal to the actual diameter if the cross section is circular. The approximation is the better, the closer the cross section resembles a circle. As a matter of record it should also be mentioned that the ratio of the cross-sectional area divided by the perimeter is, somewhat misleadingly, called the *hydraulic radius*.

5.6 Application to Losses in Pipeline Components

In the first part of Section 5.4 we analyzed the losses due to pipe friction by inspection of the governing differential equation. Similar results may be expected to occur for pipe fittings and the like. If, for example, we were to investigate the friction loss in a 90° elbow we could express this loss in the form

$$h_f = k \frac{V^2}{2g}, \tag{5.10}$$

which is analogous to Eq. (5.9).

A geometrical factor (such as the length/diameter ratio in the equation for pipe friction) does not have to appear separately here, as this factor would be the same for all geometrically similar elbows. This factor is therefore included in k. Of course, k would in general still be a function of Reynolds number.

Instead of expressing the pressure loss as a function of k as defined by Eq. (5.10), equivalent expressions are also possible. In Fig. 5.5, for example, a loss coefficient k_2 for a conical pipe is given which is defined by the equation

$$\Delta h_{\text{loss}} = k_2 \left(1 - \frac{A_1}{A_e}\right)^2 \frac{V_1^2}{2g},$$

the quantities of the equation being indicated in the sketch accompanying the figure. Such a duct in which the flow velocity is decreased and in which, therefore, the static pressure is increased is called a *diffuser*.

In Table 5.1 the loss coefficient k of Eq. (5.10) is expressed somewhat indirectly in terms of an *equivalent length* of straight pipe such that

$$k = f\left(\frac{l}{d}\right)_{\text{equiv}}.$$

The factor f is to be determined for the Reynolds number of a straight pipe made in the same way and being of the same diameter as the pipe component in question. In most practical applications, the variation of loss coefficient k with Reynolds number is slight. (The flow can be thought of as in the fully rough region in Fig. 5.4.)

Referring back to Fig. 5.5, we should point out that k_2 is represented as independent of Reynolds number. This is, of course, not strictly true, and the data in Fig. 5.5 should be interpreted as to mean that the influence of Reynolds number on k_2 is negligible, provided that the Reynolds number is "sufficiently large." In this example a Reynolds number $V_1 d_1/\nu$ greater than about 10^5 would satisfy this condition. Another frequently encountered situation in which a head loss is experienced occurs when fluid flows from a large reservoir into a pipeline. A few such arrangements are shown in Fig. 5.6. The loss associated with such a flow is termed an *entrance loss* and is usually expressed in terms of a coefficient

$$\Delta h_{\text{loss}} = k \frac{V^2}{2g},$$

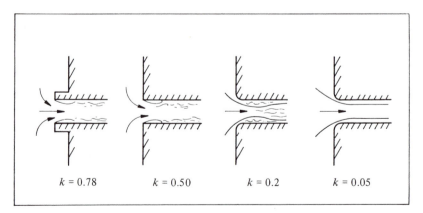

$k = 0.78 \qquad k = 0.50 \qquad k = 0.2 \qquad k = 0.05$

FIG. 5.6 Pipe entrance loss coefficients. (Adapted from E. Brater and H. King, *Handbook of Hydraulics*, 6th ed., 1976, McGraw-Hill Book Company.)

TABLE 5.1 Representative Equivalent Lengths in Pipe Diameter (l/d) of Various Valves and Fittings[a]

	Description of Product			Equivalent Length in Pipe Diameters (l/d)
Globe valves	Conventional	With no obstruction in flat, bevel, or plug-type seat	Fully open	340
		With wing or pin-guided disc	Fully open	450
	Y-pattern	No obstruction in flat, bevel, or plug-type seal		
		With stem 60° from run of pipe line	Fully open	175
		With stem 45° from run of pipe line	Fully open	145
Angle valves	Conventional	With no obstruction in flat, bevel, or plug-type seat	Fully open	145
		With wing or pin-guided disc	Fully open	200
Gate valves	Conventional wedge disc, double disc, or plug disc		Fully open	13
Check valves	Conventional swing	0.5[b]	Fully open	135
	In-line ball	2.5 vertical and 0.25 horizontal[b]	Fully open	150
Foot valves with strainer		With poppet lift-type disc 0.3[b]	Fully open	420
Butterfly valves (6-in. and larger)			Fully open	20
Fittings	90° standard elbow			30
	45° standard elbow			16
	90° long radius elbow			20
	90° street elbow			50
	45° street elbow			26
	Square corner elbow			57
	Standard tee	With flow through run		20
		With flow through branch		60
	Close pattern return bend			0

[a] Data from *Flow of Fluids*, Crane Co., Technical Paper 410, 1969. (Courtesy of the Crane Co., Chicago, Ill.)

[b] Minimum calculated pressure drop (psi) across valve to provide sufficient flow to lift disc fully.

TABLE 5.2 Loss Coefficients for Sudden
Contractions in Pipes[a, b]

$\dfrac{A_2}{A_1}$	0.0	0.1	0.3	0.5	0.7	0.9
k	0.5	0.46	0.36	0.24	0.12	0.02

[a] Data from V. L. Streeter, ed., *Handbook of Fluid Dynamics*,
McGraw-Hill Book Company, New York, 1961.
[b] The loss coefficient is defined by the equation $\Delta h_{loss} = k(V_2^2/2g)$. The data apply to relatively high Reynolds numbers
only ($> 10^5$).

where k is now called the entrance loss coefficient. This coefficient depends upon the inlet flow geometry and Reynolds number. The values listed, as for the case of the diffuser, are limited to large Reynolds numbers. When the orifice edge is relatively sharp, as in the first three sketches of Fig. 5.6, the flow forms a vena contracta somewhat similar to that of an orifice. The mixing process downstream of the vena contracta is responsible for most of the loss of total head. When the edge is well rounded, as for the "bell-mouthed" inlet of the last sketch, the flow follows the contour of the inlet and only a relatively slight loss is experienced. A loss in head also takes place when the transition from one pipe diameter to a smaller one is abrupt and the corners are square-edged. One then speaks of a *sudden contraction*.

The loss may, as usual, be expressed in terms of a coefficient k, where k is defined by the relation

$$\Delta h_{loss} = k \frac{V_2^2}{2g}$$

and V_2 represents the average velocity in the smaller pipe. In general, k should again be regarded as a function of Reynolds number Re. At high values of Re, however, the coefficient k becomes nearly independent of Re. A tabulation of the loss coefficient for sudden contractions at high Reynolds numbers is given in Table 5.2. A significant reduction in this loss may be brought about by rounding the sharp transition.

PROBLEMS

When not otherwise specified, the fluid is assumed to be water at 20°C.

5.22 (a) What discharge will produce cavitation in the Venturi tube for the configuration shown if the indicated dimensions are $d_1 = 2.5$ cm, $d_2 = 10$ cm, $d_3 = 5$ cm, and $l = 310$ m? The friction coefficient for the pipe of length l is $f = 0.04$, the velocity coefficient of the nozzle at the discharge is $c_v = 0.9$, and the diffuser angle $\alpha = 7°$. The atmospheric pressure in terms of "head of water" is 10.4 m and the pressure for incipient cavitation is 0.3 m in the same terms.

(b) What is the power in the jet (corresponding to its kinetic energy) for the rate of discharge determined in part (a)?

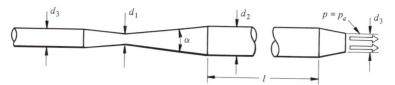

Problem 5.22

5.23 Crude oil having a kinematic viscosity of 7.5×10^{-5} m²/s and density of 880 kg/m³ is pumped through a pipe of 0.75-m diameter at an average velocity of 1.2 m/s. The roughness of the pipe is equivalent to that of a "commercial steel" pipe. If pumping stations are 330 km apart, find the head loss (in meters of oil) between pumping stations and the power required to overcome friction.

5.24 The cold water faucet of a kitchen sink is fed from a water main through the following simplified piping system.
   (1) 30 m of commercial steel pipe with a diameter $d = 12.5$ mm leading from the main line to the base of the faucet.
   (2) An entrance at the main which produces entrance conditions equivalent to those created by a sharp-edged entrance.
   (3) Six 90° standard elbows.
   (4) One wide-open angle valve (with no obstruction).
   (5) The faucet. Consider the faucet to be made up of two parts: (a) a conventional globe valve and (b) a nozzle with a velocity coefficient of $c_v = 0.70$. The nozzle spout has a cross-sectional area of 8×10^{-5} m².
The pressure in the main line is 650 kPa (practically independent of flow) and the velocity there is negligible.

   Find the maximum rate of discharge from the faucet. As a first try, assume for the pipe that $f = 0.035$. Any possible changes in elevation throughout the piping system may be neglected. It will be advantageous first to write down the Bernoulli equation in *symbols*, keeping each loss term separate.

5.25 A water tunnel is shown schematically in the figure. The four bends at the corners may be assumed to have the same loss as "long radius" elbows. The combined length $l_1 + l_2 + 2l_3 + l_4 + l_5 = 125$ ft, and the material of the pipe is

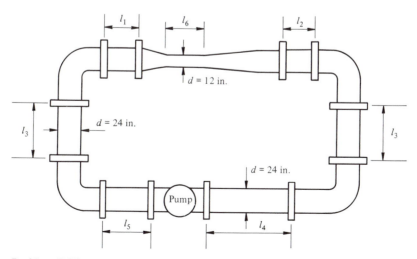

Problem 5.25

commercial steel. The diameter of the elbows and all straight pipe sections is 24 in. The length of the working section $l_6 = 4$ ft, and the material is also commercial steel. The divergence angle of the diffuser section is 8°. If the velocity in the working section is to be 100 ft/sec, find:

(a) The head (in feet of water) to be delivered by the pump.
(b) The power to be delivered to the pump, if the pump efficiency is 80 percent. Losses in the gradually convergent section may be neglected.

5.26 A piece of glass tubing having the shape of a total head tube is immersed in a stream of water at 20°C with a uniform velocity as shown. The tube diameter is d, the total tube length is l, and the contraction coefficient at the top of the tube is unity.

(a) Find the free stream velocity V necessary to obtain a volume discharge rate Q from the top of the tube. (Give the answer in terms of Q, d, h, l, f, and g.)
(b) Find the numerical value of V if

$$l = 0.6 \text{ m}, \qquad d = 9 \text{ mm},$$
$$h = 0.3 \text{ m}, \qquad Q = 8.5 \times 10^{-5} \text{ m}^3/\text{s}.$$

Neglect entrance losses, additional losses due to the bend, and capillary effects. Also assume the tube to be hydraulically smooth.

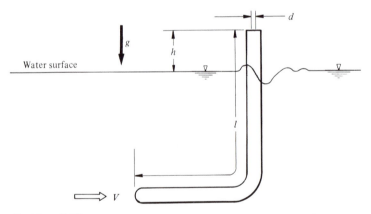

Problem 5.26

5.27 The West Texas Gulf "Biggest Inch" Pipe Line, of 26-in.-diameter (actual inside diameter of 25.375 in.) commercial steel pipe, is one of the largest pipe lines in the United States. The sketch shows the westernmost section of the line from the western terminus pumping station at Colorado City (elevation 2255 ft) to the booster pumping station at Ranger (elevation 1440 ft), where the pressure in the line at the pump suction is 25 psig. The crude oil pumped through the line has a density of 1.5 slugs/ft³ and a kinematic viscosity of 0.0000565 ft²/sec at the operating temperature. For the design flow of 330,000 bbl/day (21.4 ft³/sec), find (a) the discharge pressure p_1 required at the Colorado City pumping station, and (b) the net horsepower required at Ranger to restore this discharge pressure in the line.

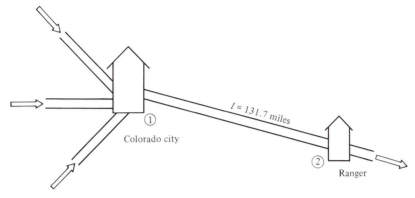

Problem 5.27

5.28 A reinforced concrete pipeline 1500 m long is to be laid as shown to carry by gravity flow the sewage effluent from a treatment plant out into the ocean for ultimate disposal at a point 30 m below mean sea level (MSL) (measured to center of the pipe). (Such a pipe is commonly called an outfall sewer.) There is to be a gate valve for use in emergencies between the treatment plant and the ocean. The maximum tidal range is MSL $\pm$ 1.2 m and the density of seawater is 1020 kg/m³. Sewage effluent has the same density and viscosity as fresh water for all practical purposes.

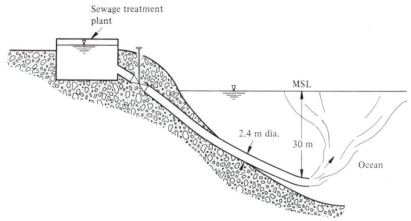

Problem 5.28

The plant is so constructed that if the level of the sewage in the collecting basin at the pipe entrance exceeds 6.7 m MSL the sewage will overflow the sides of the basin and flood the plant. Can a 2.4-m pipe carry the design peak flow of 10 m³/s without flooding the plant for any possible tide level? If so, how much is the smallest freeboard that could occur? (Freeboard is the distance the actual level is below the overflow level.) At the entrance to the pipe the head loss coefficient is 0.3. After aging, the equivalent sand roughness of the pipe is expected to be about 3 mm. Assume that the temperature is 20°C.

5.29 Flow takes place from a reservoir through a rough pipe ($\varepsilon_s/d = 0.0001$) 300 m long and 0.15 m in diameter, and this pipe is connected to another pipe with the same ε_s/d that is 150 m long and 0.075 m in diameter. The outlet of the small pipe is 30 m below the surface of the reservoir. Determine the *outlet* velocity (i.e., of the 0.075-m pipe). As a first trial take $f = 0.015$. Neglect any entrance or transition losses.

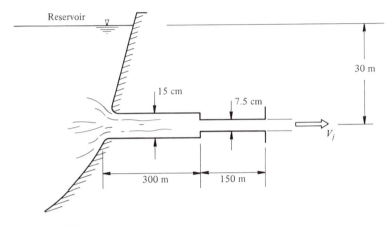

Problem 5.29

5.30 Two water reservoirs of height $h_1 = 60$ m and $h_2 = 30$ m are connected by a pipe that is 0.4 m in diameter. The exit of the pipe is submerged a distance $h_3 = 7.6$ m from the reservoir surface.
(a) Determine the flow rate in m^3/s through the pipe if the pipe is 76 m long and the friction factor $f = 0.016$. The pipe inlet is set flush with the wall.
(b) Assume that a pipe with an equivalent roughness of $\varepsilon_s/d = 0.004$ is used. Check the value of the friction factor used above and suggest a more correct value if necessary.

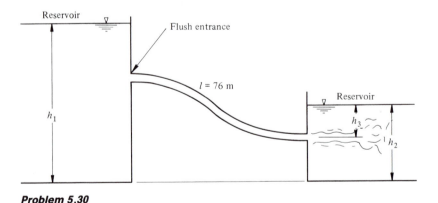

Problem 5.30

5.31 A 6-in. dia. steel pipe 5000 ft long is connected to a reservoir of 300-ft elevation. The pipe is led over a slight rise of 200-ft elevation, the distance to this point being 2500 ft. The discharge is at an elevation, of 0 ft. However, it is necessary that the flow rate through the pipe be controlled by a valve at the end to limit the minimum pressure within the pipe to atmospheric pressure.

Neglect any entrance loss. Calculate the maximum permissible velocity in the pipe and the pressure just upstream of the valve.

5.32 Three reservoirs with surfaces at elevations 300, 360, and 390 m are connected by means of three lengths of smooth pipe leading to a common junction. Each length is 300 m long and 8 cm in diameter. Find the flow rate in each pipe and the head at the junction. Neglect all losses other than pipe friction.

Note: This problem must be done by trial and error.

5.33 A $\frac{1}{2}$-in.-I.D. garden hose has a length of 50 ft. The average roughness element is 0.0005 in. The pressure at the inlet to the hose is 80 psi above atmospheric pressure. If water is allowed to discharge directly from the hose without a nozzle, find the velocity and discharge rate. Neglect entrance losses.

5.34 A wind tunnel is constructed primarily of 6-m-diameter piping arranged with four 90° elbows as shown in the sketch. The working section is 3 m in diameter, is preceded by a nozzle, and is followed by a diffuser. A fan is installed to create an air velocity of 80 m/s in the working section. Find the power that must be provided to the fan. Assume that the following losses occur in the tunnel:

(1) A loss in each of the four corner bends equivalent to a length of 20 diameters of the large piping.

(2) A friction factor, f, of 0.02 in the 138 m of 6-m-diameter pipe.

(3) A total loss in the nozzle, working section, and diffuser equivalent to one-fifth of the velocity head in the working section.

Air at these speeds can be assumed incompressible with a density of 1.2 kg/m³.

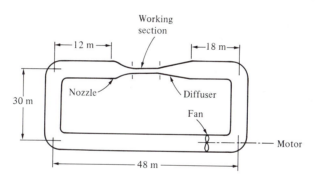

Problem 5.34

5.35 Two pipes of unequal length and diameter are connected to a water reservoir of height h above the level of discharge. The two pipes are joined together at their ends. The two flows mix and the resultant stream is discharged into the atmosphere. The total flow rate is measured at this point and is found to be 10 cubic feet per second. The following information is known about the two pipes:

$$L_1 = 3000 \text{ ft}, \qquad L_2 = 4000 \text{ ft},$$

$$D_1 = 12 \text{ in.}, \qquad D_2 = 16 \text{ in.},$$

$$\varepsilon_1 = 0.0002 \text{ ft}, \qquad \varepsilon_2 = 0.003 \text{ ft}.$$

Neglecting all minor losses such as entrance loss and possible loss at the common junction, find the velocity V in each pipe.

Hint: By appropriate use of the Bernoulli equation, show that the head loss in each of the pipes is the same. The resulting equation must be solved by trial and error.

5.36 A pipe 0.019 m in diameter carries water at 3 m/s. The roughness of the pipe is such that $\varepsilon_s/d = 0.002$.

(a) Find the friction loss in a length of 20 m, expressed in "height of water."

(b) If the pipe were to be replaced by a perfectly smooth one, what would be the percentage decrease in the friction loss?

5.37 An irrigation system for a small orchard is to be installed, and a particular pipeline leading from a reservoir to the orchard is 500 ft long, and has to deliver 0.1 ft^3/sec. The pressure at the reservoir is 20 psig and that at the discharge end is atmospheric. Compute the pipe size for a relative roughness $\varepsilon/d = 10^{-3}$.

5.7 Application to Flows over Bodies Moving Through Fluids

The foregoing considerations of dynamic similarity may also be applied to other flow situations. We now know that in an incompressible, viscous flow (in the absence of any body forces or cavitation), the flow through or around any object depends only on the Reynolds number. Dynamic similarity (the ratios of pressures and forces) and kinematic similarity (similar shape of the streamlines) for two geometrically similar objects is then assured when the Reynolds number is the same for the two flows.

We can expect, therefore, that the drag force on a sphere (for example) when expressed as a dimensionless coefficient will depend only on the Reynolds number. Thus if D is the drag force, r_0 the radius of the sphere, and V_0 and ρ the velocity and density of the approaching stream, then the drag coefficient

$$C_D = \frac{2D}{\pi r_0^2 \rho V_0^2}$$

is a function only of the Reynolds number. If the sphere is rough, then the same remarks concerning relative roughness made about pipe friction also apply here. A graph of the drag coefficients for a sphere and a cylinder is given in Fig. 5.7.

It should be noted that for either the sphere or cylinder, surface roughness has a small effect on drag coefficient for Reynolds numbers below approximately 100,000. Above this value, the drag coefficient drops sharply at a critical Reynolds number, which in turn depends on surface roughness. This interesting effect, due to changes in the nature and subsequent separation of the viscous boundary layer at the surface of the body, will be discussed briefly at the beginning of the next chapter and in more detail in Chapter 8.

Similarly, the force coefficients of other shapes depend on the Reynolds number. For example, the lift force experienced by an airplane wing at a given inclination to an oncoming wind stream when expressed in coefficient form depends only upon the Reynolds number. Again, the Reynolds number can be based on any length as long

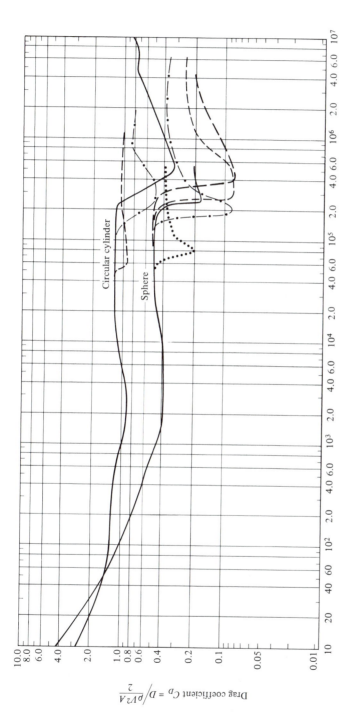

FIG. 5.7   Drag coefficients for a sphere and a cylinder as a function of Reynolds number. The Reynolds number is based on the diameter, and A is the projected area normal to the flow. The data are taken from References [9], [10], and [11]. The solid lines are for smooth surfaces. For the cylinder, the dashed line corresponds to a roughness ratio $k/d = 9 \times 10^{-3}$; the dash-dot line to $k/d = 1 \times 10^{-3}$. For the sphere, the dotted line corresponds to $k/d = 17.5 \times 10^{-3}$; the dash-dot line to $k/d = 1.5 \times 10^{-3}$; and the long dash line to $k/d = 0.25 \times 10^{-3}$. The short dash lines are a second set of experimental results for a smooth sphere, indicating the range of variation of available data.

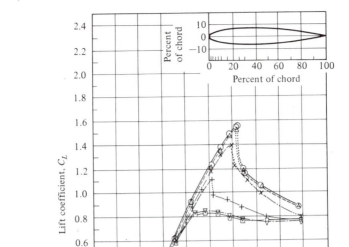

FIG. 5.8 Lift coefficient as a function of the angle of attack for various values of the Reynolds number, Vc/v. The data are for an NACA 0012 airfoil, which is shown in the insert. ○—3,180,000; △—2,380,000; ×—1,340,000; +—660,000; ▽—330,000; and ☐—170,000. (Data from Report TR 586 by E. N. Jacobs and A. Sherman, Nat. Aero. and Space Adm., formerly NACA, 1957.)

as it is chosen consistently. The usual definition of the lift coefficient is

$$C_L = \frac{L}{(\rho V_0^2/2)A},$$

where A is the planform area of the wing. Figure 5.8 shows the lift coefficient of a typical airfoil as a function of the angle of attack for several Reynolds numbers. Once again the sharp change in lift coefficient beyond a certain angle of attack can be explained by the separation of the viscous boundary layer from the upper surface of the airfoil (called *stall*). This is also discussed briefly at the beginning of the next chapter.

Naturally, if additional physical phenomena become important, they too will enter into the similarity parameters and must be considered. Such was seen to be the case in pipe roughness. If the fluid is compressible, or if there exists a free surface with gravitation and surface tension, more similarity parameters are required. The force coefficients (and similarly, the dimensionless pressure coefficients) may then depend on these additional terms.

For example, when the Mach number approaches or exceeds unity, the effect of compressing the fluid in front of a body produces an additional drag. At supersonic speeds, this is called *wave drag*. Even if the body is very slender, as in a thin wing or

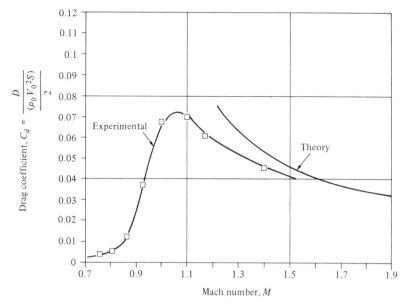

FIG. 5.9 Drag coefficient for a unit length of a circular airfoil, showing the effect of the Mach number. The ratio of maximum thickness to length of the airfoil is 0.1 and the drag coefficient c_d is based on the length S. The theory is explained in Chapter 9; the experimental results are from R. Michel, F. Marchaud, and J. Le Gallo, "Étude des écoulements transoniques autour des profiles lenticulaires, à incidence nulle," ONERA Publication no. 65, 1953.

airfoil, the wave drag may be much larger than the drag due to viscous effects. In this case the drag coefficient is very dependent on Mach number, as shown in Fig. 5.9.

In a somewhat similar way, the free surface waves formed at the bow of a ship or other surface craft produce an additional drag which depends on the Froude number. The *total* drag coefficient then depends in a complicated way on both the Froude and Reynolds numbers. The relative importance of the Froude and Reynolds numbers must be found by performing experiments on models, such as those described briefly in the next section.

The present section is not intended to be an exhaustive description of the role of Reynolds number or other parameters in the flow of real fluids. Its purpose is only to point out that the basic principles of dynamic similarity apply to these (and many other) flows, and that one should *expect* the dependence of force coefficients and similar quantities on various dimensionless groups. The problem of determining the drag coefficient of various bodies and the dependence of the drag coefficient upon the various similarity parameters of the particular flow situation is a recurring problem in fluid mechanics. Reference 8 provides a useful summary of much data of this type.

PROBLEMS

5.38 Electrical transmission towers are stationed at 400-m intervals and a smooth conducting cable 1.3 cm in diameter is strung between them. If a 30-m/s wind is blowing transversely across the wires, compute the total drag force on one such wire in between two towers (see Fig. 5.7). The drag coefficient C_D for a cylinder is defined by the equation

$$C_D = \frac{\text{drag per meter}}{\frac{1}{2}\rho V_0^2 d},$$

where d is the diameter in meters. For air assume that the kinematic viscosity $v = 1.5 \times 10^{-5}$ m²/s and that the specific weight $\rho = 1.2$ kg/m³.

5.39 A sphere with a density of 1.5 times that of ocean water is dropped into the ocean. Compute the terminal velocity if the diameter is 0.15 m. The kinematic viscosity of ocean water may be assumed to be that of fresh water at 22°C.

5.40 A man bails out of an airplane at 12,000 ft elevation (above sea level) but cannot make his parachute work at first.
   (a) Determine the approximate rate at which he will fall (after an initial brief period of acceleration). Make any reasonable assumptions, but state them clearly. (Base calculations on properties of air at 10,000 ft, where $\rho = 1.75 \times 10^{-3}$ slug/ft³.)
   (b) By what percentage will his falling speed have changed by the time he reaches the 1000-ft level, where $\rho = 2.31 \times 10^{-3}$ slug/ft³?

5.41 A sphere of 0.25-in. diameter is dropped into water at 20°C. After an initial period of acceleration it is observed to sink steadily at a velocity of 0.5 m/s. Determine the ratio of the density of the sphere to that of water.

5.42 A toy balloon is filled with hydrogen and inflated to a diameter of 0.3 m. The balloon is then allowed to escape and after a few seconds has reached a steady rise velocity. The height at which this occurs is such that normal atmospheric conditions may be assumed. The weight of the balloon (including the skin as well as the gas inside) is equal to one-tenth of the displaced atmospheric air. Find the steady rise rate of the balloon.

5.8 Engineering Model Studies

As mentioned earlier, the principles of dynamic similarity also form the basis of model studies and testing. If we want to find the characteristics of a certain prototype flow, it is in principle possible to build a geometrically similar model and to adjust the velocity, viscosity, and so on, to ensure that the model flow has the same similarity parameters as the flow to be investigated. There are, however, a few practical aspects encountered in such model studies, which we mention briefly here.

Two of the important ratios for dynamic similitude that have been introduced so far are Reynolds number and Froude number. The first ensures the proper ratio of viscous to inertia forces and the second ratio the proper relationship between gravity and inertia forces. If, in some model problem, we expect that more than two types of forces are important in determining the flow, more than one ratio has to be kept unchanged from model to prototype. It sometimes is possible to satisfy the conditions for two parameters by using a different fluid in the model than in the prototype. Generally, however, if more than two important parameters are involved, the construction of a proper model is usually not feasible. A simple illustration of this occurs during drag tests on a model ship, where we would expect both Reynolds and Froude numbers to be important. The most practical model fluid (water)

FIG. 5.10   A 13:1 scale model of a fishing boat (a singlechine seiner) being tested in a towing tank. The drag on the model is measured by instruments on the overhead carriage, which pulls the model along the tank. The pronounced effect on the free surface of the water can easily be seen. (Courtesy of S. Calisal, University of British Columbia.)

is the same as that for the prototype. As a consequence, Reynolds number similarity implies that the product Vl_0 remains constant, while Froude number similarity requires constant V^2/l_0. Clearly, both requirements cannot be met at the same time.

Fortunately, the results of model tests (e.g., that on the fishing boat shown in Fig. 5.10) suggest that in practice it is sometimes possible to relax the strict conditions of simultaneously matching Froude and Reynolds numbers in model and prototype. The actual drag test results shown in Fig. 5.11 indicate that the portion of the drag coefficient due to friction remains constant over a practical range of model speeds, while the *total* drag coefficient increases sharply at higher speeds due to the creation of the free surface wave structure shown in Fig. 5.11. Since Froude number effects are thus the more significant in describing the flow, ship model tests are usually conducted at matching Froude numbers, while Reynolds number effects are estimated by other methods.

A further aspect that must be kept in mind is that if a small geometrical model is to be made of a certain prototype, forces that may have been entirely negligible in the prototype may be of importance in the model. The effect of such forces must be taken into account or minimized if possible. For example, if a small model is made of a harbor, and water is used as a fluid in both cases, surface tension forces are entirely negligible in the prototype but they may be of major importance in the model. Therefore, steps have to be taken to eliminate the effect of surface tension in the model, since otherwise strict similitude would not exist. In other words, the ratio of inertial to surface tension forces, called the Weber number (see Problem 5.2), has to be kept at a value so that in both cases the surface tension forces are insignificant compared to the other forces.

In addition to the difficulty of strictly maintaining dynamic similarity in both model and prototype, in many cases geometric similarity may also be difficult to achieve. For example, in a model study of a river it is difficult to the point of impossibility to reproduce the detailed structure of the rocks and sand grains that form the channel bed. In this event one must rely on experience to evaluate the usefulness of the model results. Nevertheless, engineers often find experiments with

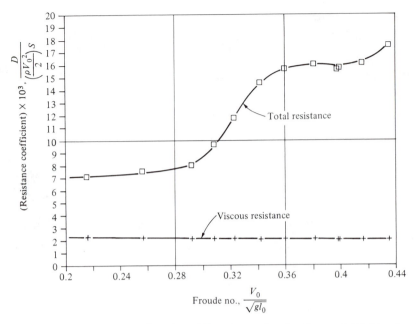

FIG. 5.11 Drag coefficients as a function of Froude number for the model fishing boat shown in Fig. 5.10. The *total* drag is measured, while the friction drag has been estimated from the results of tests where the free surface is not present. The length l_0 used in the Froude number is the length of the hull measured on the waterline, while the area S in the drag coefficient is the area of wetted hull surface. The photograph in Fig. 5.10 corresponds to a Froude number Fr = 0.36.

such imperfect models to be of value. The model study in such cases is intended to exhibit at least the most important and salient features of the flow. The discrepancy between the results of the imperfect model and prototype is usually attributed to a so-called "scale" effect. To obtain some estimate of the scale effect, models of several different sizes may be constructed. By comparing the behavior of the various models, conclusions may sometimes be drawn as to the relative importance of such dimensionless parameters as the Reynolds number, Froude number, and Weber number. Often, and particularly for river and harbor models, the model is distorted to alleviate the effect of forces that may become predominant in the model but that can be neglected in the prototype. For example, in a harbor model the ratio of model to prototype depth will usually be different from the ratio of the horizontal distances to avoid unwarranted effects of surface tension. In such cases one must verify that the model is behaving in the appropriate way by comparison either with observations made on the original harbor or with other model studies on a larger scale.

For another type of limitation that arises in model studies, let us consider experiments in a wind tunnel. In wind tunnel testing a wing profile or airplane model is held stationary in a moving stream. However, this situation is not equivalent to the same object moving through still air unless the turbulence (see Chapter 7) level in the tunnel is negligible. Special provisions must be made in a wind (or water) tunnel to reduce any velocity fluctuations originating from, for example, the fan or the turning vanes. It is not surprising that different results are sometimes obtained from various experimental facilities because of differences in the level of fluctuations in the incoming stream. Let us emphasize again, however, that all of the discrepancies discussed above are due to imperfect matching of the geometric and kinematic boundary conditions of the flows, and not to the basic theory of dynamic similitude.

REFERENCES

1. H. Rouse, *Engineering Hydraulics*, Wiley, New York, 1948, Chap. 2.
2. L. I. Sedov, *Similarity and Dimensional Methods in Mechanics*, Academic Press, New York, 1959.
3. H. L. Langhaar, *Dimensional Analysis and the Theory of Models*, Wiley, New York, 1951.
4. T. Z. Fahidy and M. S. Quaraishi, *Encyclopedia of Fluid Mechanics*, Vol. 1, Chapter 12, N. P. Cheremisionoff, ed., 1986.
5. P. W. Bridgeman, *Dimensional Analysis*, McGraw-Hill, New York, 1922.
6. S. J. Kline, *Similitude and Approximation Theory*, McGraw-Hill, New York, 1965.
7. E. Buckingham, "Model Experiments and the Forms of Empirical Equations," *Trans. ASME*, vol. **37**, 1915.
8. S. F. Hoerner, *Fluid-Dynamic Drag*, Dr. Ing. S. F. Hoerner (publisher), Midland Park, N.J., 1965.
9. A. Roshko, "Experiments on the Flow Past a Circular Cylinder at Very High Reynolds Number," *J. Fluid Mech.*, vol. **10**, 345 (1961).
10. E. Achenbach, "Influence of Surface Roughness on the Cross-Flow Around a Circular Cylinder," *J. Fluid Mech.*, vol. **46**, 321–335 (1971).
11. E. Achenbach, "Experiments on the Flow Past Spheres at Very High Reynolds Number," *J. Fluid Mech.*, vol. **54**, 565–575 (1972).

6

CHAPTER

Elements of Potential Flow

6.1 Introduction

In this chapter and several that follow we shall be concerned with the calculation of detailed flow patterns for several different kinds of situations that arise in engineering applications. In all cases the Navier–Stokes equations apply, but because complete solutions of these equations can rarely be found, it is necessary to make certain approximations. In this chapter we make the simplest approximation of all, namely, that the fluid is in a sense perfect and possesses no viscosity whatsoever. This is certainly a drastic simplification, for we know that practically all fluids are viscous to some degree. We then have to ask how good the perfect fluid approximation is, and when, if ever, it can be made. It is difficult to give a precise answer to these questions, although some qualitative statements may be made that help clarify the physical processes that take place. First, from our preceding discussion on similitude, we expect that in the absence of other important forces the flow will be governed by the Reynolds number, and, as we have seen, the Reynolds number is a measure of the relative importance of the inertial forces to the viscous forces. If Re is very large, we may anticipate that the inertial effects will outweigh the viscous effects, so that the latter may be neglected except, possibly, in regions of high shear such as may occur near the surface of a body. We have already seen an example of this behavior in Example 2.7 in connection with the nonsteady viscous flow of a moving flat plate. In this case the most important effects of viscosity were confined within a distance perpendicular to the plate $y \sim 4\sqrt{vt}$, or expressing t in terms of the distance traveled by the plate,

$$y \sim 4\sqrt{\frac{vl}{U_0}},$$

where U_0 is the plate velocity and l the length traveled. When y is greater than this value, the fluid is essentially unaffected by viscosity. If we express this result in nondimensional form, the limit of the viscous effect for this problem is

$$\frac{y}{l} \sim 4\sqrt{\frac{v}{U_0 l}}.$$

The governing parameter is seen to be the ratio $U_0 l/v$ or a Reynolds number, as we expected. As Re becomes larger, the region where viscosity is of importance becomes relatively *smaller*. Outside this region the flow is essentially a perfect fluid flow. Although these conclusions have been deduced from a single simple example, it is

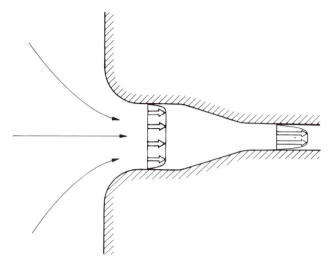

FIG. 6.1a Flow in a contracting nozzle. Note the reduction in velocity near the wall for the downstream velocity profile. The flow closely conforms to the boundaries and there is no separation.

generally correct to state that the relative extent of the region in which viscosity is of importance decreases with increasing Reynolds number, as will be seen in more detail in Chapter 8.

The foregoing discussion indicates that for many purposes the flow can be divided into two types of regions. First, there are the regions in which the viscosity has a marked influence on the flow. The fluid in these regions consists of the fluid in the immediate neighborhood of solid surfaces (as discussed above), as well as of the fluid that has *been* in the neighborhood of a solid boundary. The corresponding regions are generally called *boundary layers* and *wakes*, respectively. Second, there are the regions outside the boundary layers and wakes in which viscosity can be neglected. The latter regions are the ones of special interest in the present chapter. Before discussing the flow in these regions further, however, it is instructive to study a few situations in which we may observe the development of boundary layers and wakes and their effects on the flow field.

The first of these, Fig. 6.1a, is an idealized sketch of the flow through the contracting nozzle of a Venturi tube at a large Reynolds number. A boundary layer develops on the walls as the flow accelerates through the duct. In this case the

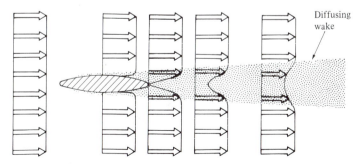

FIG. 6.1b Flow past a slender body without separation. The fluid whose energy has been reduced by friction forms the wake, which gradually spreads out by diffusion.

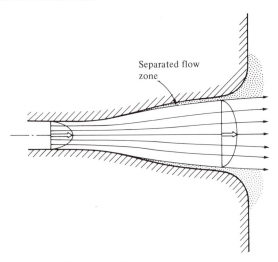

FIG. 6.1c Flow through a rapidly expanding duct, in which the flow separates from the walls.

pressure decreases continually in the direction of flow and the boundary layer conforms closely to the walls of the tube. A similar situation is shown in Fig. 6.1b, where the flow past a slender strut at high Reynolds number is sketched. The boundary layer gradually develops and is then shed into a wake, where it is diffused gradually as shown by successive velocity profiles. The third and fourth examples show a flow through an enlarging duct and around a relatively blunt object. In Fig. 6.1c the flow takes place through a rapidly expanding duct—the reverse of the flow in Fig. 6.1a. Here, because of the large and rapid increase of pressure required of the low-energy boundary layer (it has a lower kinetic energy than the main stream), it cannot regain pressure together with the main stream to fill the downstream duct completely. Instead, the main flow, while remaining more or less unaffected by viscosity, does not follow the wall but continues into the channel as a jet. The region between the jet and the wall is filled with fluid of lower velocity that churns and eddies in an irregular way. In this way the boundary layer has brought about a rather severe alteration of the flow picture. When the main stream (that portion unaffected by viscosity) does not follow the walls of the surrounding duct or adjacent solid surface, it is said to *separate* or *break away*. A similar situation is observed for the flow around a cylinder at sufficiently high Reynolds number (Fig. 6.1d). The main flow separates at a position near the maximum height; behind the cylinder

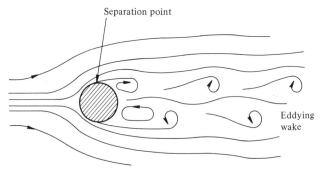

FIG. 6.1d Separation behind a cylinder. In this particular instance the wake formed is nonsteady.

there is an eddying wake that extends for many diameters downstream. On the forward portion of the cylinder a relatively thin boundary layer develops (at fairly large Reynolds numbers), and the flow exterior to this and to the wake may be considered to be inviscid.

These examples illustrate two features of recurring importance in fluid mechanics: the development of viscous regions and the occurrence of the separation phenomenon. Because of the possibility of separation one cannot always assume that the inviscid portion of the flow will conform to the shape of solid boundaries in the flow. In the absence of separation and at sufficiently high Reynolds numbers, however, the actual boundaries may be taken as the boundaries for the inviscid flow, because of the thinness of the boundary layer. In general these conditions are fulfilled in the forward portions of blunt bodies or along airfoil-like shapes, particularly in regions where the pressure decreases in the direction of flow.

We now turn our attention to the essentially inviscid flow external to any boundary layers and wakes. The flow in these regions may often be assumed to be such that the quantity

$$\boldsymbol{\omega} = \text{curl } \mathbf{v}$$

is equal to zero. The quantity $\boldsymbol{\omega}$ which was introduced in Section 2.7 was called the *rotation*, or the *vorticity*. A flow for which $\boldsymbol{\omega} = 0$ is consequently called an *irrotational* flow. In view of our discussion on angular momentum (Section 4.6), we may now add that the angular momentum of an infinitesimal particle is proportional to $\boldsymbol{\omega}$ and that changes in $\boldsymbol{\omega}$ can, therefore, only be brought about by the action of a torque. The fact that $\boldsymbol{\omega}$ may be set equal to zero in certain regions of flow is important because it leads to significant simplifications in the equations of motion.

Before examining the mathematical aspects of irrotational flows, let us consider at least qualitatively some of the conditions under which such flow may occur. Many flows of engineering importance, for example, may be considered to originate from a reservoir where the fluid has zero velocity or from a duct in which the field velocity is axial and constant. In both cases the fluid is initially irrotational, that is, the vorticity is equal to zero. We now postulate that the force field is a conservative one (such as the gravitational field), that the fluid has a constant density, and that all viscous forces and the corresponding friction losses are negligible. It may then be seen from Eq. (3.8b) that the Bernoulli constant is the same for all streamlines. It may further be shown that the subsequent flow will remain irrotational. Although this may be demonstrated in general, we shall for simplicity limit ourselves to two-dimensional flow for this illustration. We may then refer to Eq. (3.9):

$$\frac{\partial B}{\partial \psi} = -\omega_z.$$

Since we have shown that B is the same for all streamlines under consideration, this equation verifies the fact that the vorticity in the subsequent flow remains zero. The same conclusion can also be reached by a more physical reasoning. Again we postulate an incompressible fluid, a conservative force field, and the absence of vorticity initially. In that case rotation of fluid elements can only be produced by shear forces, and if such forces are negligible, it follows that the flow will be irrotational. As a consequence, the main body of a flow may often be considered irrotational, whereas the flow in boundary layers and wakes will have to be considered rotational. The remarkable success of analytical fluid mechanics in the last six decades has in large part been made possible by the careful "patching" together of solutions for an external irrotational flow to solutions that apply to the regions in which viscosity cannot be neglected and which therefore have to be considered rotational.

The conclusions above are also obtained for a fluid that has a barotropic $[\rho = \rho(p)]$ density distribution instead of a constant density. The results are, however, not generally true for variable density. Flows in which heating takes place and where the fluid density is a function of both pressure and temperature are usually rotational.

6.2 Definition of Potential Flow

Let us now turn to the mathematical treatment of flows for which the vorticity vector $\boldsymbol{\omega}$ may be set equal to zero such that

$$\boldsymbol{\omega} = \text{curl } \mathbf{v} = \nabla \times \mathbf{v} = 0.$$

A function whose curl is equal to zero can always be represented by the gradient of a scalar function because of the vector identity curl (grad ϕ) = 0. In our particular case the velocity vector may therefore be written as the gradient of a scalar function $\varphi(x, y, z, t)$, that is,

$$\mathbf{v} = \text{grad } \varphi = \nabla \varphi. \tag{6.1}$$

The function φ is called the *velocity potential*, in analogy to the force potential of mechanics, and because of the existence of this potential function, irrotational flow is often called *potential flow*. The Bernoulli equation for an incompressible fluid [Eq. (3.8c)] in potential flow becomes

$$\frac{\partial \varphi}{\partial t} + \frac{1}{2} (\nabla \varphi)^2 + \frac{p}{\rho} - U = \text{const.} \tag{6.2a}$$

For a barotropic fluid we have

$$\frac{\partial \varphi}{\partial t} + \frac{1}{2} (\nabla \varphi)^2 + \int \frac{dp}{\rho} - U = \text{const.} \tag{6.2b}$$

6.3 Basic Equations for Incompressible Potential Flow

We shall now derive the differential equation for the potential function from the continuity equation and the condition of irrotationality [Eq. (6.1)]. If the fluid is incompressible, the continuity equation yields

$$\nabla \cdot \mathbf{v} = 0.$$

Thus in view of Eq. (6.1), φ must satisfy the well-known relation

$$\nabla \cdot (\nabla \varphi) = \nabla^2 \varphi = 0, \tag{6.3}$$

which is called *Laplace's equation*. The symbolic operator ∇^2 is called the *Laplacian*. In Cartesian coordinates Eq. (6.3) becomes

$$\frac{\partial^2 \varphi}{\partial x^2} + \frac{\partial^2 \varphi}{\partial y^2} + \frac{\partial^2 \varphi}{\partial z^2} = 0 \tag{6.3a}$$

and in polar coordinates (r, θ, z) it is

$$\frac{1}{r}\frac{\partial}{\partial r}\left(r\frac{\partial\varphi}{\partial r}\right) + \frac{1}{r^2}\frac{\partial^2\varphi}{\partial\theta^2} + \frac{\partial^2\varphi}{\partial z^2} = 0. \tag{6.3b}$$

It is interesting to note that the equations of motion are not involved in determining the governing differential equation for the velocity potential for an incompressible fluid. The kinematics of the flow are fully determined by Eq. (6.3). The dynamic quantities of the flow may now be found with the aid of the Bernoulli relation Eq. (6.2), which is, of course, an integral of the equations of motion.

It is also possible for the flow of compressible fluids to be irrotational. This subject is discussed in Chapter 10.

6.4 Two-Dimensional, Incompressible Potential Flow

We have seen that for incompressible potential flow, the velocity potential satisfies Laplace's equation. In two dimensions this equation becomes

$$\frac{\partial^2\varphi}{\partial x^2} + \frac{\partial^2\varphi}{\partial y^2} = 0.$$

The stream function as we saw in Chapter 2 is related to the velocity components by the equations

$$u = \frac{\partial\psi}{\partial y},$$

$$v = -\frac{\partial\psi}{\partial x}. \tag{6.4}$$

But for a potential flow we know that u and v are related to the velocity potential by the equations

$$u = \frac{\partial\varphi}{\partial x},$$

$$v = \frac{\partial\varphi}{\partial y}. \tag{6.5}$$

Comparison of these two sets of equations gives the relation between the velocity potential and stream function as

$$\frac{\partial\varphi}{\partial x} = \frac{\partial\psi}{\partial y}, \qquad \frac{\partial\varphi}{\partial y} = -\frac{\partial\psi}{\partial x}. \tag{6.6}$$

These equations are known as the *Cauchy–Riemann* equations. From them, if either the potential or stream function is known, the other may be computed. Using steps analogous to the above, we find that in two-dimensional polar coordinates the relations corresponding to Eqs. (6.6) are

$$\frac{\partial\varphi}{\partial r} = \frac{\partial\psi}{r\,\partial\theta}, \qquad \frac{1}{r}\frac{\partial\varphi}{\partial\theta} = -\frac{\partial\psi}{\partial r}, \tag{6.7}$$

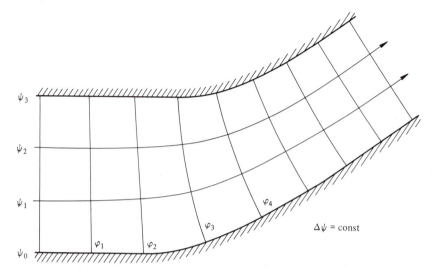

FIG. 6.2 Orthogonal flow net for the flow in a two-dimensional reducing elbow. The net is formed by lines of constant φ and constant ψ.

and the velocity components are

$$v_r = \frac{\partial \varphi}{\partial r},$$

$$v_\theta = \frac{\partial \varphi}{r \, \partial \theta}.$$

If Eqs. (6.4) are substituted into the irrotationality condition, we find that the stream function also satisfies Laplace's equation in two-dimensional Cartesian coordinates, that is,

$$\frac{\partial^2 \psi}{\partial x^2} + \frac{\partial^2 \psi}{\partial y^2} = 0 \tag{6.8}$$

for incompressible, irrotational, two-dimensional flow. (However, the stream function does not satisfy Laplace's equation in axisymmetric flow.) Since ψ and φ satisfy the same equation, in two-dimensional flow their roles may be interchanged to obtain different flows.

In two-dimensional potential flows, lines of constant stream function are perpendicular to lines of constant velocity potential. To show this result, we note that for φ constant

$$d\varphi = 0 = \frac{\partial \varphi}{\partial x} \, dx + \frac{\partial \varphi}{\partial y} \, dy,$$

and for ψ constant

$$d\psi = 0 = \frac{\partial \psi}{\partial x} \, dx + \frac{\partial \psi}{\partial y} \, dy.$$

From the first equation

$$\left. \frac{dy}{dx} \right|_{\varphi \text{ const}} = -\frac{\partial \varphi / \partial x}{\partial \varphi / \partial y} = -\frac{u}{v},$$

and for ψ constant

$$\left.\frac{dy}{dx}\right|_{\psi \text{ const}} = -\frac{\partial\psi/\partial x}{\partial\psi/\partial y} = +\frac{v}{u}.$$

Thus

$$\left[\left.\frac{dy}{dx}\right|_{\varphi \text{ const}}\right]\left[\left.\frac{dy}{dx}\right|_{\psi \text{ const}}\right] = -1,$$

which is the requirement that lines of constant φ and ψ be orthogonal.

Lines of constant φ and ψ therefore form an orthogonal network. Figure 6.2 shows such a network for irrotational, incompressible, two-dimensional flow in a reducing elbow. Note that when the flow is parallel, equally spaced φ and ψ lines form small squares. When the velocity is not constant, the φ and ψ lines form small curvilinear squares. From the spacing of the ψ and φ lines velocities can be computed, and the pressure may then be determined from Bernoulli's equation. Since there is no flow through any of the streamlines, such as ψ_1 or ψ_2, any one of them could also be considered to be a possible solid boundary.

6.5 Superposition

The governing equation of potential flow is a linear, partial differential equation. The fact that the equation is linear has the important consequence that various solutions of Laplace's equation may be superimposed to obtain new solutions. In this way complicated flow patterns may be built up from simple elements.

The proof of the superposition principle is as follows: Let φ_1 and φ_2 represent two different solutions of Laplace's equation. Then if $\varphi_3 = \varphi_1 + \varphi_2$, φ_3 is also a solution, since

$$\frac{\partial^2(\varphi_3)}{\partial x^2} + \frac{\partial^2(\varphi_3)}{\partial y^2} = \frac{\partial^2(\varphi_1)}{\partial x^2} + \frac{\partial^2(\varphi_1)}{\partial y^2} + \frac{\partial^2(\varphi_2)}{\partial x^2} + \frac{\partial^2(\varphi_2)}{\partial y^2} = 0.$$

Similarly, the velocities given by φ_1 and φ_2 can be added (vectorially) to give the velocities due to φ_3. However, the pressures corresponding to φ_1 and φ_2 cannot be added to give the pressure from φ_3, since Bernoulli's equation is nonlinear in the velocity terms.

We now consider a few simple but useful solutions of Laplace's equation.

EXAMPLE 6.1

Let us consider a two-dimensional velocity potential of the form $\varphi = Ax + By + Cx^2 + Dxy + Ey^2$. This function satisfies Laplace's equation if $C + E = 0$.

(a) *Uniform Flow*

Suppose that $B, C, D, E = 0$. Then $\varphi = Ax$, and the velocity components are

$$u = \frac{\partial\varphi}{\partial x} = A,$$

$$v = \frac{\partial\varphi}{\partial y} = 0.$$

The stream function corresponding to the potential can be found by utilizing the Cauchy–Riemann equations (6.6). Thus

$$A = \frac{\partial \psi}{\partial y},$$

or

$$\psi = Ay + f(x),$$

where $f(x)$ is an arbitrary function of x, since the integration was with respect to y alone. But we have

$$\frac{\partial \varphi}{\partial y} = 0 = -\frac{\partial \psi}{\partial x} = \frac{df}{dx}.$$

Hence f is at most a constant. Let us take it to be zero for convenience, so that

$$\psi = Ay,$$

$$\varphi = Ax.$$

This flow is entirely parallel to the x-axis with the constant velocity A and is called a *uniform flow* for that reason. In polar coordinates the velocity potential and stream function are, respectively,

$$\varphi = Ar \cos \theta,$$

$$\psi = Ar \sin \theta.$$

A sketch of the flow is shown in Fig. 6.3.

(b) *Stagnation Flow*

Let us take $C = -E = 1$ and let $A = B = D = 0$, so that $\varphi = x^2 - y^2$. The velocity components therefore are $u = \partial \varphi / \partial x = 2x$, $v = \partial \varphi / \partial y = -2y$. With this information it is possible to plot the flow pattern; however, for this purpose it is convenient

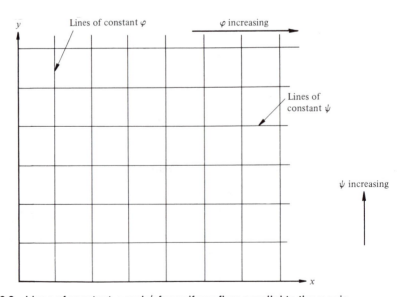

FIG. 6.3 Lines of constant φ and ψ for uniform flow parallel to the x-axis.

to find the stream function ψ. Since

$$u = \frac{\partial \psi}{\partial y} = 2x,$$

we can integrate this equation to get

$$\psi = 2xy + f(x),$$

where as before $f(x)$ is an arbitrary function of x, since the integration was with respect to y only. We also know that

$$v = -\frac{\partial \psi}{\partial x} = -2y,$$

but from the above

$$v = -\frac{\partial \psi}{\partial x} = -2y + \frac{df}{dx}.$$

By comparison of the two expressions for v, we see that $f(x)$ is again only a constant. Setting this constant equal to zero, one has

$$\varphi = x^2 - y^2,$$

$$\psi = 2xy.$$

The lines $\psi = 0$ are given by $x = y = 0$, and lines of constant stream function are rectangular hyperbolas. A sketch of this flow appears in Fig. 6.4. This flow is sometimes called *flow in a corner*, or *stagnation flow*.

EXAMPLE 6.2 _____

A slightly more complicated example in polar coordinates is afforded by the potential function $\varphi = r^{1/2} \cos \theta/2$. The radial velocity corresponding to this potential is

$$v_r = \frac{\partial \varphi}{\partial r} = \frac{1}{2} r^{-1/2} \cos \frac{\theta}{2}$$

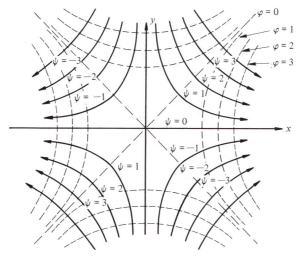

FIG. 6.4 Lines of constant φ and ψ for the flow in a 90° corner formed by two infinite plates.

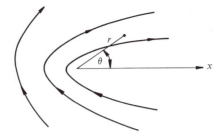

FIG. 6.5 Flow about a semi-infinite flat plate. Several streamlines are shown.

and the tangential velocity is

$$v_\theta = \frac{1}{r}\frac{\partial \varphi}{\partial \theta} = -\frac{1}{2}r^{-1/2}\sin\frac{\theta}{2}.$$

The tangential velocity v_θ is zero for $\theta = 0$, 2π, for all values of r. Therefore, the line $\theta = 0$ extending from the origin to infinity must be a streamline. Similarly, the streamlines must cross $\theta = \pi$ perpendicularly, since $v_r = 0$ there. Noting that v_r is positive for $\pi \le \theta \le 0$ and negative for $\pi < \theta < 2\pi$, we are led to the idea that the flow proceeds generally from the lower right-hand quadrant flowing around the positive half of the x-axis to the upper right quadrant. This flow is sketched in Fig. 6.5. The stream function can be determined as in the preceding examples and is

$$\psi = r^{1/2}\sin\frac{\theta}{2}.$$

When transformed to Cartesian coordinates, ψ becomes

$$\psi = 2y\sqrt{x^2 + y^2},$$

which is recognizable as the equation of a parabola.

PROBLEMS

6.1 The stream function of an incompressible, steady, two-dimensional flow is

$$\psi = x^2 + y^2.$$

Does this flow have a velocity potential?

6.2 In Section 6.4 it was shown that lines of constant φ and ψ form an orthogonal network in two-dimensional, incompressible potential flow. Show that in two-dimensional, *compressible* steady flow the same result follows.
   Hint: Use the result of Problem 2.15.

6.3 Obtain the stream function and plot the streamlines of the velocity potentials:
   (a) $\varphi = xy$;
   (b) $\varphi = x^3 - 3xy^2$;
   (c) $\varphi = x/(x^2 + y^2)$;
   (d) $\varphi = (x^2 - y^2)/(x^2 + y^2)^2$.
   Note: It may be convenient to write some of these expressions in polar coordinates before proceeding.

6.4 Plot the streamlines for the three-dimensional velocity potential

$$\varphi = x^2 + y^2 - 2z^2.$$

6.5 Show that

$$\varphi = Ar^{\pi/\alpha} \cos \frac{\pi}{\alpha} \theta$$

(A and α are constants) satisfies Laplace's equation and determine the stream function ψ. Sketch the streamlines for two cases, $\alpha = \pm \pi/3$.

6.6 Sources in Incompressible Fluids

The conservation of mass principle requires that the net mass flow or volume flow of an incompressible fluid through the surface of an arbitrary volume be zero, as long as no fluid is produced within the volume. This condition may be written in vector form as

$$\int_{S_0} \mathbf{v} \cdot \mathbf{n} \, dS = 0.$$

Application of the divergence theorem to this surface integral results in the continuity equation already derived for an incompressible fluid

$$\nabla \cdot \mathbf{v} = 0.$$

Let us imagine now a situation in which fluid somehow is produced within the volume under consideration. The surface integral above will then be equal to

$$Q = \int_{S_0} \mathbf{v} \cdot \mathbf{n} \, dS, \tag{6.9}$$

where Q is the volume flow rate produced within the volume. Q is said to be the "source strength" and has units of volume per second.

As an example of a flow in which the concept of a source is useful, consider the flow through the pipe and annular space between two parallel plates in Fig. 6.6. Assume that there is no friction and that the angular momentum of the approaching

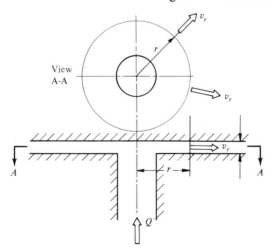

FIG. 6.6 A possible way of simulating the flow from a two-dimensional source.

flow is zero. Then far from the inlet pipe the velocity v_r will be purely radial. Let us assume further that v_r is independent of angle. If the integral in Eq. (6.9) is now evaluated for a surface consisting of two flat circular areas of radius r and the annular strip of width b and radius r between the plates, then evidently

$$\int_{S_0} \mathbf{v} \cdot \mathbf{n} \, dS = 2\pi r b v_r = Q = \text{const},$$

provided that Q does not change with time. Thus

$$v_r = \frac{Q}{b} \cdot \frac{1}{2\pi r} = \frac{q}{2\pi r}, \tag{6.10}$$

where q is the source strength per unit distance between the plates. The inlet pipe can be taken to be infinitely small without changing Eq. (6.10). The velocity at the origin becomes infinite, of course, since fluid is being injected there. Although no real fluid flow can follow Eq. (6.10) near the origin, the concept of a source is most useful. The velocity distribution of Eq. (6.10) is that of a potential flow—since it is irrotational everywhere and since the continuity equation is satisfied everywhere, except at the source itself. The velocity potential obtained from Eq. (6.10) is

$$\varphi = \frac{q}{2\pi} \ln r + \text{const.} \tag{6.11}$$

The point $r = 0$, where φ becomes logarithmically infinite and v_r becomes infinite, is said to be a *singularity* of the solution. It is readily verified that Eq. (6.11) is a solution of Laplace's equation, for in polar coordinates neglecting any variation in angle Eq. (6.3b) is

$$\frac{1}{r} \frac{\partial}{\partial r} \left(r \frac{\partial \varphi}{\partial r} \right) = \frac{1}{r} \frac{\partial}{\partial r} \left(r \cdot \frac{q}{2\pi r} \right) = 0.$$

The stream function corresponding to the potential of Eq. (6.11) is

$$\psi = \frac{q}{2\pi} \theta + \text{const.} \tag{6.12}$$

To ensure that ψ has only a single value within the flow, θ is restricted to lie within the sector $0 \le \theta \le 2\pi$. The source strength may be negative, so that fluid is being drained into the origin. Rather than a source, we now have a *sink*.

In this chapter we are concerned primarily with two-dimensional flow. When the words *source* or *sink* are used in this context they will generally refer to a source or sink *per unit length*. Similarly, any *forces* that may occur in a two-dimensional problem are to be regarded as *forces per unit length*.

6.7 Circulation and Vorticity

We have seen that the velocity of a potential flow is given by

$$\mathbf{v} = \nabla \varphi.$$

Thus if $\mathbf{v}$ is known, φ can be calculated by the integration

$$\varphi - \varphi_A = \int_A \mathbf{v} \cdot d\mathbf{r}, \tag{6.13}$$

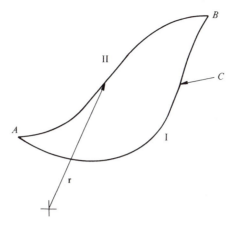

FIG. 6.7 Possible paths of integration between points A and B.

where the subscript A denotes a definite point and $\mathbf{r}$ is the radius vector from some fixed point. The difference of potential between any two points AB is obtained by integrating Eq. (6.13) along some suitable path, say I of Fig. 6.7, to get

$$(\varphi_B - \varphi_A)_{\mathrm{I}} = \int_A^B \mathbf{v}_{\mathrm{I}} \cdot d\mathbf{r}. \tag{6.14}$$

The subscript I indicates the path chosen for integration. However, some alternative path II could just as well be chosen so that

$$(\varphi_B - \varphi_A)_{\mathrm{II}} = \int_A^B \mathbf{v}_{\mathrm{II}} \cdot d\mathbf{r}. \tag{6.15}$$

Now the velocity in a physical problem is, of course, a single-valued function of position and time, and one might expect that the potential function ϕ is therefore also single valued. In many of the previous examples this is actually the case, and the two integrals above would then be the same. Nevertheless, we may not conclude that φ is always single valued, and the two integrals may differ from each other in certain cases. Therefore, if we proceed from A to B by I and then back to A from B by II, we shall in certain cases obtain a difference or discontinuity $\Delta\varphi$ in the value of φ_A of the amount

$$\Delta\varphi = \int_A^B \mathbf{v}_{\mathrm{I}} \cdot d\mathbf{r} + \int_B^A \mathbf{v}_{\mathrm{II}} \cdot d\mathbf{r} = \oint_C \mathbf{v} \cdot d\mathbf{r}. \tag{6.16}$$

The last integral symbol on the right denotes the contour integral traversed in the counterclockwise direction around the contour C formed by paths I and II. The magnitude $\Delta\varphi$, the discontinuity in velocity potential obtained after one traverse of the contour, is called the *circulation*. In the subsequent paragraphs we shall examine in more physical terms just when such a discontinuity $\Delta\varphi$ exists and when $\Delta\varphi = 0$.

Fortunately, $\Delta\varphi$ can be expressed in terms of physically meaningful quantities. We transform the contour integral of Eq. (6.16) into a surface integral by Stokes' theorem (Appendix 3). Limiting ourselves now to a two-dimensional, Cartesian coordinate system, we obtain

$$\oint_C \mathbf{v} \cdot d\mathbf{r} = \oint_C (u \, dx + v \, dy) = \int_S \left(\frac{\partial v}{\partial x} - \frac{\partial u}{\partial y} \right) dx \, dy,$$

in which S denotes the surface enclosed by the contour C. The integrand of the right-hand side is the vorticity ω_z of a two-dimensional flow. Thus

$$\Delta\varphi = \int_S \omega_z \, dS, \tag{6.17}$$

so that the circulation around any contour is the integral of the vorticity over the area enclosed by the contour. In a purely irrotational flow where ω_z is *everywhere* zero, the circulation about every path is zero. This means that in a purely irrotational flow φ is *single valued*. That is, at any given point φ can have only one value independent of the manner in which the point is approached. However, if the contour encloses a region in which the mean value of ω_z is not zero, φ will be *multiple valued*, increasing by the amount $\Delta\varphi$ for each circuit enclosing the region in question.

As an example of a flow exhibiting these features let us consider one that is called the Rankine *combined vortex*. It is defined by the velocity components

$$v_r = 0,$$
$$v_\theta = kr, \qquad r \le r_0,$$

and

$$v_\theta = k\frac{r_0^2}{r}, \qquad r \ge r_0, \tag{6.18}$$

where k is a constant and r_0 is a reference radius.

The vorticity ω_z expressed in polar coordinates is

$$\omega_z = \frac{1}{r}\frac{\partial(rv_\theta)}{\partial r} - \frac{1}{r}\frac{\partial v_r}{\partial\theta},$$

so that

$$\omega_z = 2k, \qquad r \le r_0,$$
$$\omega_z = 0, \qquad r > r_0.$$

Hence the flow is irrotational outside the reference radius r_0 and has uniform vorticity inside r_0. Thus for *any* circuit completely enclosing the circle r_0,

$$\Delta\varphi = 2\pi kr_0^2.$$

The inner velocity distribution of Eq. (6.18) is the same as that of a rotating solid body, and the corresponding flow is sometimes called a *forced vortex*. The outer distribution is also a circulatory flow but is irrotational and is usually called a *free vortex*. Now if r_0 is made to approach zero and k is increased to keep the product kr_0^2 constant, the entire flow field is irrotational, except at the origin $r = 0$. Then

$$v_r = 0,$$
$$v_\theta = \frac{\Gamma}{2\pi r},$$

where Γ is a constant and equal to $\Gamma = 2\pi kr_0^2$. The corresponding potential is

$$\varphi = \frac{\Gamma}{2\pi}\theta, \tag{6.19a}$$

and in accordance with our previous discussion, it has the discontinuity of Γ for one revolution around the origin. Γ is called the *vortex strength*. The strength of the vortex Γ and the circulation $\Delta\varphi$ are therefore the same.

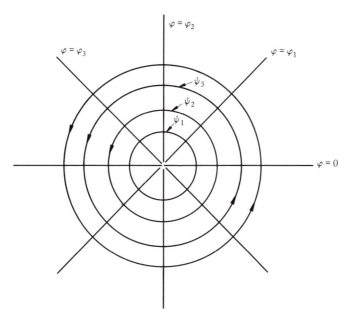

FIG. 6.8   Flow in a free vortex. Lines of constant φ and ψ are shown.

The stream function of the free vortex is

$$\psi = -\frac{\Gamma}{2\pi} \ln r, \qquad (6.19b)$$

and the streamlines, of course, are circles around the origin, as shown in Fig. 6.8. A free vortex, like a source, is a singularity of the flow. These singularities, together with some of the elementary flows discussed in the examples and problems to follow, will enable us to construct by superposition flows that are of engineering importance.

EXAMPLE 6.3

As an example of the method of superposition, let us select one that involves the concept of a source and a sink. In particular, we shall consider a flow which consists of that due to a source and a sink with a uniform velocity parallel to the line joining the source and sink. Let the source and sink be separated by a distance $2a$ with the origin midway between (Fig. 6.9). At a general point P the complete potential is

$$\varphi = Vx + \frac{q}{2\pi} \ln r_1 - \frac{q}{2\pi} \ln r_2 = Vx + \frac{q}{4\pi} \ln \frac{(x+a)^2 + y^2}{(x-a)^2 + y^2}.$$

From this expression the velocity components and stream function can be found. The stagnation points are those points for which $u = v = 0$. At the stagnation point the streamline divides into two branches that enclose the source and sink. This streamline can be considered to be an oval-shaped body. (For the body to be closed, the source strength must be equal to the sink strength.)

Ovals generated by a source and sink are called *Rankine ovals*. It is evident that more complicated shapes can be obtained (in three dimensions as well) by distributing sources and sinks along the axis or even off the axis. Such shapes are useful for representing the flow around struts, bridge piers, airship hulls, and so on.

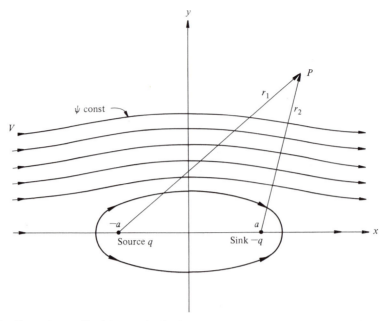

FIG. 6.9 Flow about a Rankine oval, obtained by superposition of a uniform flow and a source–sink pair.

An important special case of the Rankine oval described in Example 6.3 occurs when the source and sink combination are brought close together. As might be expected, the oval shape tends toward that of a circle; however, the strength of the source–sink pair must be increased in a special way so that they do not simply cancel each other.

Consider once again the expression for the velocity potential φ in Example 6.3, but allow the separation distance $2a$ to tend to zero. For very small values of a, the argument of the logarithm, then the logarithm itself, may be expanded in a suitable series. Upon retaining only first-order terms of a, the velocity potential becomes

$$\varphi = Vx + \frac{qa}{\pi}\frac{x}{x^2 + y^2}.$$

If the strength, q, of the source–sink pair is kept constant while the separation $2a$ tends to zero, the pair simply cancel each other and only the uniform flow remains. If, however, q is correspondingly increased so that the *product qa remains constant*, the resulting potential function φ represents that of a cylinder of radius $r_0 = (qa/\pi V)^{1/2}$ in a uniform flow (see Problem 6.18). This special source–sink pair represented by the second term in the equation above is sometimes called a *doublet*, or *dipole*.

6.8 The Method of Images

With the aid of the superposition principle many methods can be devised to synthesize flows that satisfy various boundary conditions. One useful approach of this type is frequently called the *method of images*. The type of problem in which this approach is best used is illustrated in the following example.

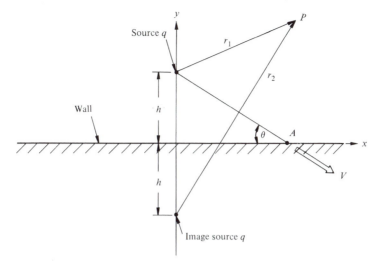

FIG. 6.10 Method of images as applied to the problem of finding the flow field produced by a source near an infinite flat plate.

EXAMPLE 6.4 _____

Consider the flow due to a two-dimensional source of strength q located a distance h from a wall, as shown in Fig. 6.10. It is desired to determine the potential flow due to this source in the region of space above the straight wall. The wall is solid and therefore will be a streamline of the flow since the velocity perpendicular to the wall is zero. The source in question induces a radial velocity at point A on the wall that has a vertical component $V \sin \theta$. Therefore, it can represent only part of the complete potential. Evidently, the simplest potential we can add that cancels this perpendicular component is that due to an additional source of the same strength at the point at which the image of the given source would appear if the wall had the properties of a mirror. It is because of this manner of devising the complementary flow that the method receives its name. The two sources together now make the wall a streamline and the total velocity potential at a general point P is

$$\varphi = \frac{q}{2\pi} \ln r_1 + \frac{q}{2\pi} \ln r_2.$$

With the coordinate system shown,

$$\varphi = \frac{q}{4\pi} \ln [x^2 + (y - h)^2][x^2 + (y + h)^2]$$

is the potential for the flow. The streamlines and potential lines are sketched in Fig. 6.11. The geometrical shapes traced by the equipotential lines are sometimes called *ovals of Cassini*. The origin is a stagnation point of this flow and in the vicinity of the stagnation point the flow is similar to that of Example 6.1b.

_____

PROBLEMS

6.6 Determine the stream function and plot several streamlines for the potential $\varphi = -(\Gamma/2\pi)\theta + (q/2\pi) \ln r$. The resulting flow is called a *spiral vortex*.

6.7 Tornado damage to structures is often caused by an excess interior pressure on walls and roofs when the atmospheric pressure drops suddenly. Judging from

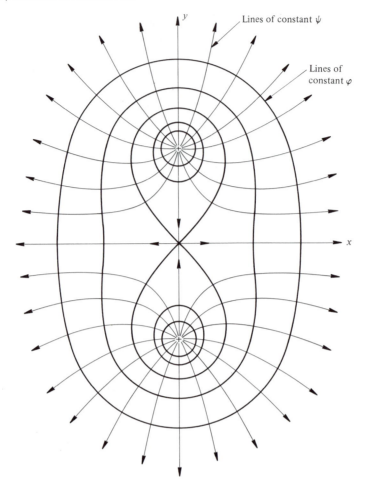

FIG. 6.11   Flow field produced by a source near an infinite flat plate. Both the source and its mirror image are shown.

the structural damage caused, we find that this temporary pressure differential may have been of the order of 1.5 kPa. Idealize a tornado as a two-dimensional free vortex (axis perpendicular to ground) superimposed on a uniform wind velocity, which is responsible for the translation of the vortex. (A steady flow is obtained by superimposing a velocity equal and opposite to the uniform wind.) Neglect all frictional effects for radial distances r greater than 30 m. (Inside this "eye," the flow cannot be assumed irrotational.)

(a) At $r = 30$ m, what would the wind velocity have to be to cause a drop in pressure of 2 kPa below the atmospheric pressure (at infinity)? Assume incompressible flow.

(b) For such a tornado moving at 30 m/s ground speed, how long would it take the pressure at a point on the ground directly in the path to drop from 0.5 kPa below atmospheric pressure to 2 kPa below? Assume the ground to be an infinite plane.

6.8   A two-dimensional free vortex is located near an infinite plane at a distance h above the plane. The pressure at infinity is p_0 and the velocity at infinity is V_0

parallel to the plane. Find the total force (per unit depth normal to the paper) on the plane if the pressure on the back side of the plane is p_0. The strength of the vortex is Γ. The fluid is incompressible and perfect. To what expression does the force simplify if h becomes very large?

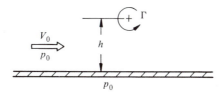

Problem 6.8

6.9 A two-dimensional source of strength q per unit length is located in a rectangular corner formed by two infinite walls. The distance of the source from each wall, h, is given. The fluid is perfect, incompressible, and the flow is steady.

(a) Find the configuration of sources that will make the corner a streamline.
(b) Find the velocities at the wall as a function of the distance x along the wall.
(c) Set up the equation for finding the pressure along the wall and sketch the pressure distribution. Take the pressure at infinity, p_0, as given.
Hint: Work with velocities directly.

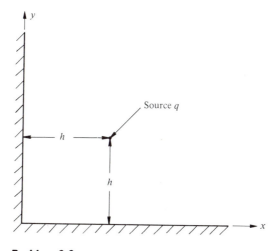

Problem 6.9

6.10 A perfect, incompressible irrotational fluid is flowing past a wall with a sink of strength q per unit length at the origin. At infinity the flow is parallel and of uniform velocity V_0. Find the pressure distribution along the wall as a function of x. Taking the free-stream static pressure at infinity to be p_0, make a sketch of the pressure distribution along the wall.

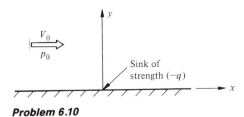

Problem 6.10

6.11 If a two-dimensional source flow of strength q per unit length is superposed on a uniform straight flow of velocity V_0, it will be found that the central streamline separates into two branches at a stagnation point and that the branches continue on to infinity. The shape defined by the two streamlines just described is known as a *half-body*, since its length is infinite. The fluid is incompressible.

(a) The stagnation streamline corresponds to $\psi = q/2$. Plot this streamline on graph paper and sketch the other streamlines outside the body. Determine the breadth of the body far downstream and at the location of the source. Also find the distance between the source and the stagnation point.

(b) Sketch neatly the pressure distribution along the surface of the body.

6.12 A source of strength q is superposed on a uniform stream of velocity V_0, pressure p_0, and density ρ as in Problem 6.11. With the use of the momentum theorem calculate the force on the surface of the resulting half-body.

Hint: Consider a control surface consisting of the surface of the half-body and the portion of a circle of very large radius not intersected by the body. It will be necessary to account for both the momentum and pressure terms on the circular portion of the control surface.

6.13 For Problem 6.11 find the locus of all points in the flow field for which the direction of the local flow differs by one degree from that of the uniform approaching flow. Show that this locus is a circle and give the coordinates of the center and radius in terms of V_0, the speed of the uniform flow, and q the source strength.

6.14 At a certain point at the beach the coastline makes a rectangular bend as shown. The flow of the salt water in this bend can be approximated by the potential flow of an incompressible fluid in a rectangular corner. The potential for this case is equal to

$$\varphi = Ar^2 \cos 2\theta,$$

where A is a given constant.

A freshwater underground reservoir is located near the corner. The salt water is to be kept away from that reservoir to avoid any possible seepage of the salt water into the fresh water. For this purpose a *source* of fresh water is located at the corner of the bend and fresh water is discharged from this source at the rate of q ft^3/sec per foot of depth.

(a) *Carefully* sketch the streamlines and indicate in particular the line dividing the fresh water flow from the saltwater flow.

(b) Find the amount of fresh water q that has to be discharged through the source per unit depth to keep the salt water from the coast for at least a distance b from the corner.

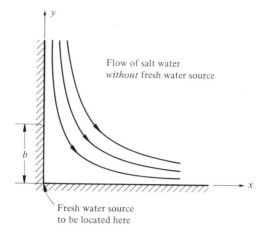

Flow of salt water
without fresh water source

Fresh water source
to be located here

Problem 6.14

6.15 A two-dimensional source of strength q per unit length is located midway between the walls of a channel of width h. The fluid velocity far upstream in the channel is V_0, and the pressure p_0 there may be taken as zero. The central streamline splits to form a half-body about the source as shown. Assume the flow to be frictionless and of constant density ρ.

(a) Calculate the width b of the half-body. It will help to recognize that far downstream the velocity is uniform across the channel.

(b) Calculate with the aid of the momentum theorem the total force on the surface of the half-body. Express your answer only in terms of ρ, V_0, q, and h. To what expression does the result simplify when h approaches infinity?

Note: In parts (a) and (b) do not try to set up the potential for this flow.

(c) Indicate by means of a sketch the location of the image sources necessary to make the walls of the channel a streamline.

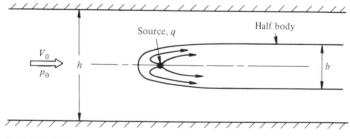

Problem 6.15

6.16 Proceeding similarly as in Problem 6.11, determine the diameter of a three-dimensional half body produced by a three-dimensional source of strength Q and a uniform stream of velocity V_0. Also find the distance from the source to the stagnation point.

6.17 Use the results of Example 6.3 to determine the approximate length and breadth of a Rankine oval that is so slender that the maximum thickness is much less than the spacing $2a$. Determine the pressure at the central section of the oval in terms of the length-to-breadth ratio, the velocity of the free stream V_0, and the density ρ.

Hint: The velocity at the center section of the oval may then be assumed constant over the thickness.

6.18 Consider the flow represented by the potential

$$\phi = V_0 r\left(1 + \frac{r_0^2}{r^2}\right) \cos \theta - \frac{\Gamma}{2\pi} \theta,$$

where V_0, r_0, and Γ are constants. Show that this represents the flow past a stationary circular cylinder of radius r_0 due to a uniform flow of velocity V_0 at infinity. Determine the location of the stagnation points and carefully sketch the streamlines *outside* the cylinder for the cases $\Gamma = 0$, $\Gamma/r_0 = 2\pi V_0$, and $\Gamma/r_0 = 4\pi V_0$.

6.19 Consider the following potential in a three-dimensional *axisymmetric* flow: $\varphi = V_0 r(1 + r_0^3/2r^3) \cos \theta$, where θ is the polar angle from the axis of symmetry and r is the radius from the origin.

   (a) Show that this is the potential for the uniform flow of an incompressible fluid past a sphere.

   (b) Calculate the maximum velocity on the sphere and location of the point where the static pressure is equal to the free-stream value and compare these with those for two-dimensional flow around a cylinder.

6.20 (a) A cylindrical tube with three radially drilled orifices can be used as a flow direction indicator. Whenever the pressure on the two side holes is equal, the center hole (located halfway between the side holes) will point in the direction of the flow. The pressure at the center hole is then the stagnation pressure. Such an instrument is called a direction-finding Pitot tube, or a cylindrical yaw probe. If the orifices of a direction-finding Pitot tube were to be used to measure free-stream static pressure, where would they have to be located? Assume incompressible potential flow. In the case of real fluid flow, the static pressure point is at a somewhat greater angle.

   (b) For a direction-finding Pitot tube with orifices located as calculated in part (a), what is the sensitivity? Let the sensitivity be defined as the pressure change per unit angular change (i.e., $\partial p/\partial \theta$).

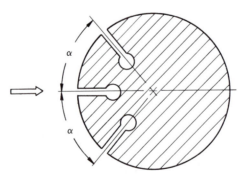

Problem 6.20

6.21 An arctic hut in the shape of a half-circular cylinder has a radius R_0. A wind of velocity V_0 batters the hut and threatens to raise it off its foundations due to the *lift* of the wind. This lift is partly due to the fact that the entrance to the hut is at ground level at the location of the *stagnation* pressure. A clever occupant sized up the situation quickly and relocated the entrance at an angle θ_0 from the ground level, which caused the net force on the hut to vanish. What is the angle θ_0? For the purpose of this problem the opening will be assumed to be very small compared to the radius R_0. Assume incompressible potential flow and observe that the static pressure inside the hut p_s will depend on θ_0.

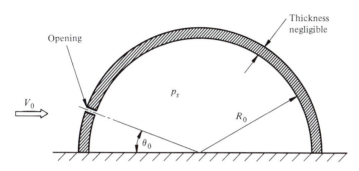

Problem 6.21

6.22 Using Cartesian coordinates (x, y, z), show that the velocity components u, v, w of a potential flow each satisfy Laplace's equation. Thus if φ is a solution to Laplace's equation, so is $\partial\varphi/\partial x$, $\partial^2\varphi/\partial x^2$, and so on.

6.23 A two-dimensional source of strength q per unit length is separated from a sink of the same strength by a distance $2a$. Show that the streamlines are all circles passing through the source and sink. The fluid is incompressible.

6.24 An x-y coordinate system is used to describe a potential flow. The flow is caused by:
(1) A source of strength Q located at the origin,
(2) A second source, also of strength Q, located on the x-axis at $x = f$, and
(3) A sink of strength Q also located on the x-axis at $x = a^2/f$. Find the x-component of velocity on the x-axis at $x = a$, and also the y-component of velocity on the y-axis at $y = a$.

6.25 A hurricane can be visualized as a planar incompressible flow consisting of a rotating circular core surrounded by a potential flow. A particular hurricane has a core of radius 40 m and air is sucked into this core at a volume flow rate per meter depth perpendicular to the diagram of 5000 m^3/s. Furthermore, the pressure difference between the air far away from the hurricane and the air at the edge of the core is 1.5 kPa. The velocity of the air far from the core is assumed to be negligible. The density of the air is assumed uniform and constant at 1.2 kg/m^3. Find the angular rate of rotation of the hurricane.

6.26 Consider the flow that results from the superposition of a source of strength q and a vortex of strength Γ. Both are located at the origin and the flow is two-dimensional.
(a) At any given radius, what is the angle between the velocity and the radial direction? (Call this angle α.)
(b) If the pressure at $r \to \infty$ is equal to p_0, write an expression for the pressure, p, as a function of the radius, r.

6.27 A competent engineer observes the flow of air over a cylinder. Some flow visualization method (e.g., smoke injection) is used and she notices that the flow separates at $\theta = 60°$ (i.e., $120°$ from the stagnation point). From this she concludes that the pressure distribution from the forward stagnation point to the separation point is given quite accurately by the pressure distribution in the corresponding potential flow. The pressure in the separated region she assumes to be equal to that *at* the separation point. Using these assumptions, calculate the drag coefficient for this cylinder. (Assume that air may be considered to be incompressible for the given test conditions.) (The answer does *not* match the experimental data very well.) For uniformity, integrate counterclockwise from 0 to 60° and from 60 to 180°. The flow over the lower half is, of course, symmetrical. Take the direction $\theta = 0$ as the positive one.

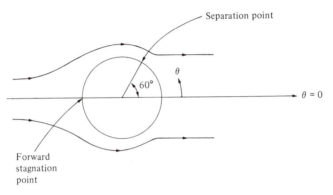

Problem 6.27

6.28 A three-dimensional source with a total volume discharge rate Q is located at a distance b from an infinite flat plate.
(a) Determine the potential for a single three-dimensional source in an infinite fluid in the absence of the plate.
(b) Obtain an expression for the velocity at the plate for the configuration given.
(c) Find the pressure at any point R along the plate, where R is measured from the point on the plate that is closest to the source.
(d) Find the distance R_m for which the pressure is a minimum.
(e) Sketch the pressure distribution in a p versus R graph.
(f) Letting the pressure at infinity be equal to zero, derive the integral expression from which the total force on the plate may be computed. (Do not evaluate the integral.) The fluid is frictionless and incompressible.

6.29 The velocity potential for a two-dimensional "doublet" (source–sink combination) is given by

$$\varphi = u_s \frac{r_0^2}{r} \cos \theta.$$

Find the total kinetic energy in the entire flow field $(r > r_0)$.

6.30 Consider the potential flow of a perfect incompressible fluid around a cylindrical "direction-finding" Pitot tube as in Problem 6.20. The central pressure orifice is pointing in the upstream direction $(\theta_1 = 180°)$ and the side orifices are located at $\theta_2 = \theta_1 \pm 30°$. In this way the pressure difference between the central orifice and any one side orifice is proportional to the velocity head of the approaching flow, provided that the flow is steady. In the case to be considered here, however, the upstream velocity u_0 is uniform (the same for all $r \to \infty$), but changes with time [i.e., $u_0 = u_0(t)$]. For this flow find an expression for the pressure difference $(p_1 - p_2)$, where p_1 is the pressure of the centrally located orifice and p_2 is the pressure at either of the side orifices.

6.9 Potential Flows Around Moving Bodies

The methods developed in previous sections make it possible to visualize the streamlines during steady flow past fixed objects and next to walls whose shapes are specified. In some cases these methods can also be used for the description of the *unsteady* flow formed when a body moves through a stationary fluid. Knowing the analytic formulas for steady streamlines and velocities can make the computation of an unsteady flow somewhat easier.

As may be recalled from Chapter 2, in an unsteady flow, *pathlines* identify the trajectories of individual "particles" of fluid within the flow, while *streaklines* identify the collection of particles that have all passed through the same point at different times. In general, the instantaneous streamlines, pathlines, and streaklines do not coincide in an unsteady flow.

To illustrate the general approach for finding the flow field formed by a moving body, let us consider as an example the case of a cylinder with radius r_0 moving steadily at a speed U along the x-axis from left to right through a fluid that is stationary everywhere far away from the cylinder. To begin with, we first consider a uniform flow from right to left (a flow speed $-U$) past a *stationary* cylinder of radius r_0 located at the origin. If we use the methods of Section 6.8, the velocity potential for such a flow is

$$\varphi = -Ur_0^2 \frac{\cos \theta}{r} - Ur \cos \theta.$$

For this case the oncoming fluid can be brought to rest (as seen by a stationary observer) by moving the entire coordinate system to the right at a velocity U. This would also make the previously stationary cylinder move to the right at the speed U into the now stationary fluid. In this way we have obviously created exactly the flow we originally set out to study.

The correct velocity potential for this *unsteady* flow (at the instant the cylinder is at the origin) can be obtained by adding the velocity potential for a uniform flow to the equation above, with the result

$$\varphi = -Ur_0^2 \frac{\cos \theta}{r}$$

(note that this is once again the velocity potential of a dipole). As a check that this is the right result, note that on the boundary of the circle $r = r_0$,

$$v_r = U \cos \theta, \qquad v_\theta = U \sin \theta,$$

which expresses the kinematic boundary condition that there be no flow *through* the surface of the moving cylinder.

At the instant the moving cylinder is at the origin, the lines of constant velocity potential are concentric circles; their centers lie on the x-axis and the circles are

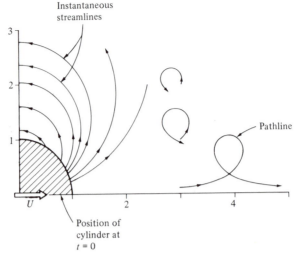

FIG. 6.12 The flow around a *moving* cylinder. The cylinder is shown at the instant it passes the origin. The instantaneous streamlines are circles; they pass through the origin and their centers are on the *y*-axis. The streamline pattern remains attached to the cylinder as it moves. Three different pathlines are shown. The initial particle positions correspond to the cylinder as shown. As the cylinder moves to the right, the pathline is generated as shown.

tangent to the *y*-axis at the origin. The streamlines at this instant are also a set of circles, centered on the *y*-axis and tangent to the *x*-axis, as shown in Fig. 6.12. The fact that the streamlines intersect the surface of the cylinder is a consequence of this being an *unsteady* flow; the fluid is pushed away from the front of the cylinder and must return to fill in behind as the cylinder passes by. The complete network of instantaneous potential lines and streamlines described moves with the moving cylinder, so that at any other time we need only to find the center of the moving cylinder and suitably shift the pattern of lines.

As a further illustration of the features of this particular unsteady flow, let us calculate the pathline of a fluid particle whose coordinates we shall call $X(t)$ and $Y(t)$. The pathline coordinates are related to the instantaneous fluid velocity components by the definitions

$$u = \frac{\partial X}{\partial t}, \qquad v = \frac{\partial Y}{\partial t}.$$

These equations can now be integrated with respect to time to obtain the pathline. (This is a generalization of the step-by-step procedure in Example 2.1, using infinitesimally small steps.) We first need to express the velocity potential in Cartesian coordinates,

$$\varphi = -\frac{U r_0^2 (x - x_0)}{(x - x_0)^2 + y^2},$$

where x_0 is the instantaneous center of the cylinder, so that the velocity components u and v are given by

$$u = \frac{U r_0^2 [(x - x_0)^2 - y^2]}{[(x - x_0)^2 + y^2]^2}, \qquad v = \frac{2(U r_0^2)y(x - x_0)}{[(x - x_0)^2 + y^2]^2}.$$

The instantaneous velocities are now integrated with respect to time to obtain the trajectory coordinates. For this purpose let us start at $t = 0$ with the initial locations X_i and Y_i of a particular fluid particle; then the trajectory equations are

$$X(t) - X_i = U r_0^2 \int_0^t \frac{(X - x_0)^2 - Y^2}{[(X - x_0)^2 + Y^2]^2} \, dt,$$

$$Y(t) - Y_i = U r_0^2 \int_0^t \frac{2y(X - x_0)}{[(X - x_0)^2 + Y^2]^2} \, dt.$$

It should be noted that the integrands must be evaluated at the current particle position. Since the center of the circle x_0 has the position

$$x_0(t) = x_0(i) + Ut,$$

where $x_0(i)$ is the initial position of the circle at $t = 0$, this expression for $x_0(t)$ must be inserted into the integrand.

Although it is possible to write out the equations for the trajectories of pathlines as outlined above, these equations are quite complicated to integrate analytically. It is possible, however, to do so numerically (at least approximately). Many such methods have been developed to integrate just these kinds of equations; these are fully described in both References 6 and 7. In the present example we shall illustrate one of these methods, the *improved Euler method*. For this we write the trajectory equations in the symbolic form for the x equation

$$X(t) - X_i = \int_0^t f(X, Y, t) \, dt,$$

where the function f represents, in the present case, the u velocity. As in all numerical methods, the intervals of time are discretized into finite amounts; here these are indicated as t_n and the index numbering these increments is $n = 0, 1, 2, \ldots$. The difference in the function $X(t)$ between a time t_n and t_{n+1} is

$$X_{n+1} - X_n = \int_{t_n}^{t_{n+1}} f(X, Y, t) \, dt.$$

If X and Y were known at the upper limit, we could evaluate the integral approximately (with $\tau = t_{n+1} - t_n$) as

$$X_{n+1} - X_n = \frac{[f(X_n, Y_n, t_n) + f(X_{n+1}, Y_{n+1}, t_{n+1})]\tau}{2}.$$

But usually X_{n+1} and Y_{n+1} are not known and must be replaced by estimates. A simple estimate is just

$$X^*_{n+1} = X_n + f(X_n, Y_n, t_n)\tau$$

plus an equivalent statement for Y_{n+1}. Then, approximately,

$$X_{n+1} - X_n = \frac{[f(X_n, Y_n, t_n) + f(X^*_{n+1}, Y^*_{n+1}, t_{n+1})]\tau}{2}.$$

By repeating this incremental step, the equation may be integrated as a function of time. This process is readily carried out using a simple spreadsheet program on a personal computer. The results of such a calculation for a cylinder of radius r_0 equal to unity moving to the right are shown in Fig. 6.12, along with the instantaneous streamlines when the cylinder is at the origin (taken as $t = 0$). Three different pathlines are shown; they correspond to particles initially at different heights on the line

$x = 3$ at time $t = 0$. The motion of the particles is clearly seen to be away and upward initially, ending with a following of the cylinder as it moves far away to the right. Each of the particles experiences a permanent displacement to the right as a result of the moving cylinder.

Many other examples of unsteady flow may be studied by the methods outlined above. A few of these are described in the following problems.

PROBLEMS

6.31 Consider a cylinder with unit radius moving to the right along the x-axis. *Sketch* the streakline for particles passing through $x = 3$, $y = 1$, beginning from a time when the moving cylinder is at the origin. Devise a simple way to compute the streakline.

6.32 Determine the instantaneous kinetic energy of the flow due to the motion of a circular cylinder of radius r_0 and velocity V_0 through an otherwise undisturbed perfect incompressible fluid of density ρ.

6.33 Find the stagnation pressure at the front of the moving cylinder described in Problem 6.32 by evaluating the instantaneous velocity field and using the *unsteady* form of the Bernoulli equation, Eq. (3.7).

6.34 At a particular instant, a two-dimensional free vortex of strength $-\Gamma$ is located on the y-axis at a height h_0, while an equal and opposite free vortex is located on the y-axis at $-h_0$.
  (a) First consider the vortex pair *fixed* at these locations (i.e., the flow is steady). Find the expressions for the x and y velocity components throughout the flow.
  (b) Returning now to the original problem statement, assume that the vortex pair is *not* fixed, but is free to move in the x-direction under the influence of the induced velocity of the neighboring vortex with the total separation distance $2h_0$ remaining constant. What is the speed of the moving vortex pair?

6.35 Problems 6.11, 6.12, and 6.13 deal with flow past a half-body formed by imposing a point source of strength q on a uniform flow. Suppose that the same body is made to move to the right at a speed V_0 through a stationary fluid.
  (a) Sketch the instantaneous streamlines when the nose of the body is at the origin.
  (b) Calculate and plot the pathline of a particle located at $x = y = q/2V_0\pi$ at the same time that the nose of the body is at the origin. These calculations are most easily done in tabular form, or by using a simple spreadsheet program on a personal computer.

†6.10 General Solution for Two-Dimensional, Incompressible, Potential Flow

Thus far we have exhibited solutions of Laplace's equation in two dimensions and have shown what physical flows they represented. A truly vast body of mathematics

† This section may be omitted without loss of continuity.

has been devoted to finding solutions of this equation for various given boundary conditions. We shall not explore this mathematical problem to any great extent. However, it must be mentioned that in two dimensions the extremely powerful methods of complex variable theory provided practical and useful means of obtaining solutions to Laplace's equation. We introduce the basic concepts of these methods and discuss a few examples for illustration.

Instead of considering φ or ψ to be a function of x and y, let us consider as a new variable the combination $z = x + iy$, where $i = \sqrt{-1}$. The new variable z is said to be *complex* when y is not zero. The variable x is said to represent the *real* part and y the *imaginary* part of the complex number z. Let us examine now a general function F which has z as its argument [i.e., $F(z)$]. In general, we expect that F will have a real part and another part multiplied by i, the imaginary part. We shall denote these as follows:

$$F(z) = F(x + iy) = h(x, y) + ig(x, y). \tag{6.20}$$

For example, if $F(z) = z^2 = (x + iy)^2 = x^2 - y^2 + 2ixy$, then $h = x^2 - y^2$ and $g = 2xy$. The reason for introducing the complex variable z is that it generally leads to the two functions h and g, which both satisfy Laplace's equation. This is easily proved, since

$$\nabla^2 F = \frac{\partial^2}{\partial x^2} F(x + iy) + \frac{\partial^2}{\partial y^2} F(x + iy)$$

$$= F''\left(\frac{\partial z}{\partial x}\right)^2 + F''\left(\frac{\partial z}{\partial y}\right)^2 = F'' - F'' = 0,$$

as $(i \times i) = -1$, and F' denotes differentiation of $F(z)$ with respect to its argument z. Thus F satisfies Laplace's equation, but since F has real and imaginary parts, we have

$$\nabla^2 F = 0 = \frac{\partial^2 h}{\partial x^2} + \frac{\partial^2 h}{\partial y^2} + i\left(\frac{\partial^2 g}{\partial x^2} + \frac{\partial^2 g}{\partial y^2}\right).$$

The real and imaginary parts must each separately equal zero, so that

$$\nabla^2 h = \nabla^2 g = 0. \tag{6.21}$$

There is, however, a further requirement upon F that we already have tacitly assumed in the foregoing equations. We are interested in functions that are "smooth" and change in a continuous way from point to point. Therefore, we shall require F to be continuous and that at each point the derivative $F' = dF/dz$ can be computed. For such functions the value of F' at each point should not depend on the particular direction in the (x, iy)-plane which dz takes. For example, if we take dz parallel to the x-axis so that $dz = dx$,

$$F' = \frac{\partial F}{\partial x} = \frac{\partial h}{\partial x} + i\frac{\partial g}{\partial x}.$$

However, if we take dz to be $i\,dy$,

$$F' = \frac{\partial F}{i\,\partial y} = \frac{1}{i}\frac{\partial h}{\partial y} + \frac{\partial g}{\partial y} = -i\frac{\partial h}{\partial y} + \frac{\partial g}{\partial y}.$$

These equations must be the same, so that upon equating real and imaginary parts, we have

$$\frac{\partial h}{\partial x} = \frac{\partial g}{\partial y},$$

$$\frac{\partial h}{\partial y} = -\frac{\partial g}{\partial x}. \tag{6.22}$$

Equations (6.22) are the Cauchy–Riemann equations [Eq. (6.6)] satisfied by the velocity potential and stream function. We have shown then that both h and g satisfy Laplace's equation [Eq. (6.21)] and the Cauchy–Riemann equations [Eq. (6.22)], just as do the velocity potential and stream function. Without further delay we can therefore put

$$F(z) = \varphi + i\psi$$

and

$$z = x + iy$$

with F being any function of z, which, however, must be twice differentiable. F is called the *complex velocity potential*. Equation (6.20) generates an infinite variety of solutions and it is up to the skill and ingenuity of the analyst to determine the correct ones for the particular problem. All of the preceding examples and problems are special cases of the functions

$$F(z) = \frac{1}{2\pi} (q + i\Gamma) \ln z + \sum_{-\infty}^{\infty} (a_n + ib_n)z^n, \tag{6.23}$$

where a_n, b_n are constants and n is an index.

In polar coordinates (r, θ) we have

$$z = x + iy = re^{i\theta} = r(\cos \theta + i \sin \theta)$$

and we may derive from Eq. (6.23) [noting that $\ln (re^{i\theta}) = \ln r + i\theta$]

$$\varphi = \frac{q}{2\pi} \ln r - \frac{\Gamma}{2\pi} \theta + \sum_{-\infty}^{\infty} r^n(a_n \cos n\theta - b_n \sin n\theta)$$

$$\psi = \frac{q}{2\pi} \theta + \frac{\Gamma}{2\pi} \ln r + \sum_{-\infty}^{\infty} r^n(a_n \sin n\theta + b_n \cos n\theta). \tag{6.24}$$

Other functions may, of course, also lead to the description of pertinent flow fields. For illustration the example below involves the use of another elementary function, the exponential function.

EXAMPLE 6.5

Let us consider the complex function

$$F(z) = V_0[z + y_0 \, e^{i2\pi z/\lambda}] = V_0[(x + iy) + y_0 \, e^{[2\pi i(x + iy)/\lambda]}],$$

with V_0, y_0, and λ being constants. Now

$$\varphi + i\psi = V_0\left[x + iy + y_0 e^{-(2\pi y/\lambda)}\left(\cos \frac{2\pi x}{\lambda} + i \sin \frac{2\pi x}{\lambda}\right)\right],$$

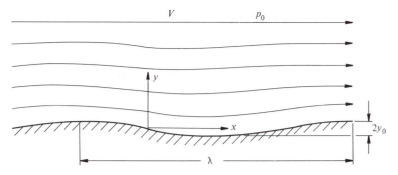

FIG. 6.13   Streamlines past a sinusoidally corrugated wall.

so that the stream function is

$$\psi = V_0 \left(y + y_0 e^{-(2\pi y/\lambda)} \sin \frac{2\pi x}{\lambda} \right).$$

For large positive values of y, $\psi = V_0 y$, which is the stream function of a uniform flow of velocity V_0 parallel to the x-axis. The exponential term may be expanded, that is,

$$\psi = V_0 \left[y + y_0 \sin \frac{2\pi x}{\lambda} - 2\pi \frac{y_0}{\lambda} y \sin \frac{2\pi x}{\lambda} + \frac{y_0}{2} \left(\frac{2\pi y}{\lambda} \right)^2 \sin \frac{2\pi x}{\lambda} + \cdots \right],$$

and when $2\pi y/\lambda$ is small compared to unity we have (approximately)

$$\psi = V_0 \left(y + y_0 \sin \frac{2\pi x}{\lambda} \right).$$

The streamline $\psi = 0$ is then approximately given by

$$y = -y_0 \sin \frac{2\pi x}{\lambda},$$

and the function ψ may therefore be regarded as representing the flow over a sinusoidally corrugated wall, where y_0 is the amplitude of the corrugation and λ is the wavelength (Fig. 6.13).

In most, if not all, problems, the use of the complex variable notation leads to less algebra, and in addition the stream function corresponding to the desired velocity potential is found automatically. This alone would be sufficient reason for its use. However, we have touched on only the most superficial aspects of complex variable theory. For a detailed account of this theory as applied to fluid mechanics, several of the references at the end of the chapter may be consulted.

PROBLEMS

6.36 The flow against an infinite flat plate is often called stagnation point flow, or corner flow. If the flow is incompressible and irrotational, the corresponding potential function is $\varphi = A(x^2 - y^2)$, where A is a constant. Suppose that such a flow is directed against a plate that is flat except for a circular bump of unit radius centered on the origin as shown.

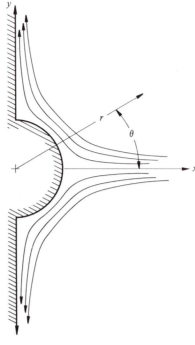

Problem 6.36

(a) Write down the appropriate boundary conditions to be satisfied on the solid boundaries and for r approaching infinity.

(b) Evaluate the coefficients a_n, b_n of Eq. (6.24) and determine the maximum velocity on the circular bump.

6.37 A circular cylinder of radius r_0 is concentric with a larger stationary cylinder of radius R_0. At the instant of observation, the inner cylinder has a velocity V_0 to the right. The flow is inviscid, incompressible, and irrotational.

(a) State the conditions to be satisfied by the velocity at the inner cylinder and at the stationary outer cylinder.

(b) Find the potential for this flow starting with Eq. (6.24).

6.38 Determine the location of the maximum and minimum pressures on the wavy wall of Example 6.5 and show that approximately

$$\frac{p_{max} - p_0}{\frac{1}{2}\rho V_0^2} = \frac{4\pi y_0}{\lambda},$$

where p_0 is the pressure far from the wall. By examining the result, note that to measure p_0 by means of an orifice located in the side of a wall, the wall must be quite smooth.

6.39 If $F(z)$ is the complex velocity potential show that:

(a) In two-dimensional Cartesian coordinates $dF/dz = u - iv$, where u and v are the velocities parallel to the x- and y-axes, respectively.

(b) In two-dimensional polar coordinates (r, θ), $dF/dz = (v_r - iv_\theta)e^{-i\theta}$.

Hint: Let $dz = d(re^{i\theta})$ when determining dF/dz and use the relations of Eq. (6.7).

6.40 Write the potential functions corresponding to Examples 6.1 and 6.2 and Problem 6.5 in terms of complex variables in the form of functions $F(z)$.

6.41 Show that the uniform flow of magnitude V_0 inclined to the x-axis at the angle α is $F(z) = V_0 e^{-i\alpha} z$.

6.42 Write the potential function for the flow of Problem 6.18 in complex form by specifying the appropriate function $F(z)$.

6.43 Suppose that a complex potential is given by the function

$$F(z) = F(x + iy) = A(z) + iB(z),$$

where A and B are functions of z but A is that part of F that is real when z is real and B is that part of F that is imaginary when z is real. Show that the flow corresponding to the potential function F about a circle of radius r_0 whose center is at $z = 0$ is

$$G(z) = F(z) + A\left(\frac{r_0^2}{z}\right) - iB\left(\frac{r_0^2}{z}\right),$$

where $A(r_0^2/z)$, and so on, means that the variable z is to be replaced by r_0^2/z.
  Hint: Let $z = r_0 e^{i\theta}$ and show that $G(z)$ is only real on the circle. (This result is known as the Milne–Thompson circle theorem.)

6.44 Derive the potential function for the flow around a circle of radius r_0 without circulation due to a uniform flow V_0 parallel to the x-axis using the method of Problem 6.43.

6.45 Derive the potential function for the flow around a circle of radius r_0 without circulation due to a uniform flow of speed V_0 inclined to the x-axis at an angle α using the method of Problem 6.43.

6.46 A source is given by the complex potential $F = (q/2\pi) \ln (z - b)$. Using the method of Problem 6.43, show that the flow around a circle of radius r_0 due to the source is that of a source located at the point r_0^2/b inside the circle, a sink of strength q per unit length at the center of the circle, and the original source at b.

6.47 A two-dimensional, incompressible potential flow consists of two vortices of equal and opposite strength Γ, located on the y-axis at $+a$ and $-a$, respectively (the vortex at $+a$ has a strength $-\Gamma$).
  (a) Express the complex velocity potential for the two vortices in terms of the complex variable $z = x + iy$, Γ, and a.
  (b) Show that streamlines are coaxial circles centered on the y-axis.
  (c) What uniform velocity parallel to the x-axis must be superposed to reduce the force on an individual vortex to zero? Locate all of the stagnation points for this case and sketch the resulting flow.
  (d) Suppose that a is made to approach zero in such a way that the product Γa is kept constant. To what potential does this flow reduce?

6.48 Potential flow within a 45° wedge is given by the potential function

$$\varphi = u_0 r^4 \cos 4\theta,$$

where u_0 is a reference velocity. A source with a strength q is now placed at the origin.

(a) May the walls of the original wedge still be regarded as walls?

(b) Find the location (r_s, θ_s) of the stagnation point.

Hint: This point must be along a line where *both* original flows had radial components of velocity only.

6.49 A circle of radius a is at a certain instant *concentric* with another circle of larger radius b. The inner circle is stationary, but the larger one is moving with a velocity U_0 in the negative x-direction (an unsteady flow).

(a) What are the boundary conditions on the velocity at $r = b$ and $r = a$?

(b) Noting that the velocity potential satisfies Eq. (6.24), find the velocity potential for this problem. It is assumed that there are no sources or vortices present. (See also Problem 6.37.)

6.50 A flow in the ζ-plane is given by $F = u_0\,\zeta$. The lines $\theta = 0$, $r > 0$ and $\theta = \pi$, $r > 0$ are seen to be streamlines. A corresponding flow in the z-plane is to be investigated. The relation between ζ and z is given by

$$z = \zeta^{1/2}.$$

(a) What are the streamlines in the z-plane that correspond to the lines mentioned above?

(b) Find the velocities in the z-plane by forming the derivative dF/dz.

(c) Find an expression for the streamlines in the z-plane.

6.51 (a) By making the substitution $m = i(n)$ into Eq. (6.23), where n is a real integer, show that a family of solutions to Laplace's equation for the velocity potential which are alternative to that in Eq. (6.24) are

$$\varphi = \frac{q}{2\pi}\ln r - i\frac{\Gamma\theta}{2\pi} + \sum_{m=-\infty}^{\infty} e^{-m\theta}[a_n \cos\,(m\,\ln r) - b_n \sin\,(m\,\ln r)].$$

Note: The form of this solution indicates that the velocity potential decreases (or increases) exponentially in annular channel flow.

(b) Find the stream function that corresponds to the solution of part (a).

6.52 Consider the complex velocity potential

$$F(z) = i\sqrt{z^2 - 1}.$$

(a) Show that when $|z| \gg 1$, the flow is a uniform flow of speed unity parallel to the y-axis.

(b) Sketch the streamline $\psi = 0$ [i.e., when $F(z)$ is purely real]. Pay special attention to the line segment $-1 \le Rl(z) \le 1$.

Hint: Identify and locate any stagnation points and any points where the velocity may be infinite.

†6.11 Aerodynamic Lift

Consider a vortex of strength Γ located at the origin in a uniform flow of velocity V_0 shown in Fig. 6.14. We shall compute the force per unit length that is necessary to maintain the vortex in equilibrium by means of the momentum theorem applied to a circular control volume as shown in Fig. 6.14. The direction of the vortex is shown in the diagram. From the momentum theorem the momentum flux through the

---

† This section may be omitted without loss of continuity.

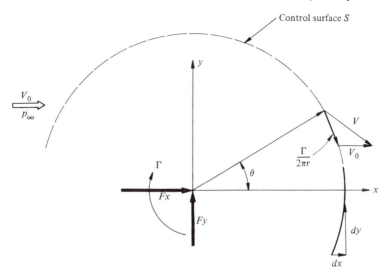

FIG. 6.14 Control volume for computing the force required to hold a vortex in place in a uniform flow field.

surface S of the control volume is equal to all the forces acting on or within the control volume. The force can be separated into two terms, of which one is due to the pressure on S. Thus if we resolve the force into x- and y-components, we can write (with positive increments in dx and dy as shown)

$$\dot{M}_x = x \text{ momentum flux} = -\int p \, dy + F_x,$$

$$\dot{M}_y = y \text{ momentum flux} = +\int p \, dx + F_y,$$

where F_x and F_y are the forces to be determined. For the x momentum flux through S, we obtain (see Fig. 6.14)

$$\dot{M}_x = \int_0^{2\pi} \rho\left[\frac{\Gamma}{2\pi r} \sin\theta + V_0\right] V_0 \cos\theta \, r d\theta = 0$$

and

$$\dot{M}_y = -\int_0^{2\pi} \rho \frac{\Gamma}{2\pi r} V_0 \cos^2\theta \, r d\theta = -\frac{\rho V_0 \Gamma}{2},$$

with Γ positive (clockwise) as shown in the diagram, and V_0 representing the magnitude of the approaching velocity. The first result is due to the periodicity of the sine and cosine functions and the second is obtained with the use of the double-angle formula. The pressure may be computed from the Bernoulli equation

$$p = p_\infty - \frac{\rho}{2}\left[V^2 - V_0^2\right]$$

$$= p_\infty - \frac{\rho}{2}\left[\left(V_0 + \frac{\Gamma}{2\pi r}\sin\theta\right)^2 + \left(\frac{\Gamma}{2\pi r}\cos\theta\right)^2 - V_0^2\right].$$

Now $dy = rd\theta \cos \theta$ and $dx = -rd\theta \sin \theta$, so that

$$
F_x = \oint p \, dy + \dot{M}_x
$$

$$
= \int_0^{2\pi} r \cos \theta \, d\theta \left\{ p_\infty - \frac{\rho}{2} \left[\left(V_0 + \frac{\Gamma}{2\pi r} \sin \theta \right)^2 + \left(\frac{\Gamma}{2\pi r} \cos \theta \right)^2 - V_0^2 \right] \right\} + 0
$$

$$
= -\frac{\rho}{2} \int_0^{2\pi} r \cos \theta \, d\theta \left[\frac{V_0 \Gamma \sin \theta}{\pi r} + \left(\frac{\Gamma}{2\pi r} \right)^2 \right]
$$

$$
= 0.
$$

The force component in the y-direction is

$$
F_y = -\oint p \, dx + \dot{M}_y = -\frac{\rho}{2} \int_0^{2\pi} r \sin \theta \, d\theta \left[\frac{V_0 \Gamma \sin \theta}{\pi r} + \left(\frac{\Gamma}{2\pi r} \right)^2 \right] - \frac{\rho V_0 \Gamma}{2}
$$

or

$$
F_y = -\frac{\rho V_0 \Gamma}{2\pi} \int_0^{2\pi} \sin^2 \theta \, d\theta - \frac{\rho V_0 \Gamma}{2} = -\rho V_0 \Gamma.
$$

Thus the force that must be applied to the vortex is $\rho V_0 \Gamma$ directed downward. The force per unit length experienced by the vortex from the fluid is, therefore, upward and is called the *lift*. If L denotes the lift force, we have

$$
L = \rho V_0 \Gamma. \tag{6.25}
$$

This important result is frequently known as the *Kutta–Joukowsky theorem*. It is interesting to observe that the resultant force on the vortex is perpendicular to the upstream flow (i.e., there is no "drag" force). As a consequence, no work is expended in maintaining a lift force.† Although Eq. (6.25) is applicable for an isolated vortex, it is also true if the vortex is created by a body of finite length enclosing the vortex. The truth of this statement can be indicated as follows. Since the body is not of infinite length there is no net source strength, as reference to the results of Example 6.3 will show. Thus outside the body the potential from Eq. (6.24) can be expressed in terms of functions such as $\varphi = 1/r^n \sin (n\theta)$ or $\varphi = 1/r^n \cos (n\theta)$, since in polar coordinates these are the only functions that do not give rise to a disturbance at infinity. The radial and tangential velocities then vary as $1/r^{n+1} \sin (n\theta)$ or $1/r^{n+1} \cos (n\theta)$. When these components are introduced into the pressure and momentum integrals, it is found that there is no contribution to the force for all values of $n \geq 1$.

The problem of determining the lift on a two-dimensional body is now reduced to finding the vortex strength Γ supplied by the body. It is just this problem that gave difficulty in the early history of aerodynamics.

PROBLEMS

6.53 A stream of velocity V_0 flows past a circular cylinder of radius r_0 (see the sketch) containing a vortex Γ of such strength that the lift coefficient, C_L, is

† It will be noticed that this result is analogous to the force on a wire carrying a current in a magnetic field.

equal to one-half, where

$$C_L = \frac{2L}{2r_0 \rho V_0^2}.$$

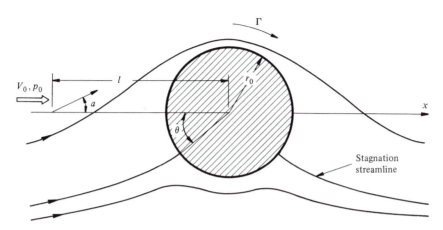

Problem 6.53

The fluid is incompressible.
(a) Obtain the magnitude of the vortex strength Γ.
(b) Find the angle θ from the horizontal axis to the stagnation point.
(c) To what distance (l) on the x-axis must one go before the local flow inclination α is less than $0.5°$ to the horizontal?
(d) What is the value of the pressure at that point? Express the pressure in terms of a pressure coefficient defined as

$$C_p = \frac{p - p_0}{(\rho/2)V_0^2}.$$

Note: Four-figure accuracy will be needed.

†6.12 The Kutta Condition

When a slender body with a sharp trailing edge, such as an airfoil or hydrofoil, is immersed in a real fluid which is in motion, it is found experimentally that the flow streams smoothly off the trailing edge as shown in Fig. 6.15a. A boundary layer is formed around the body and it is discharged into the wake at the trailing edge, point B. In general, a lifting force is exerted on the body by the flow, and if one were to measure the distribution of velocity on a contour outside the surface boundary layers, it would be found that a circulation would be present also. Figure 6.15b shows the inviscid potential flow over the same body without circulation. The stagnation streamline divides near the nose of the body, and the two branches meet at the rear stagnation point located at A. As it proceeds around the body, the flow passes over the sharp trailing edge at point B where, because of the infinite curvature at the trailing edge, the ideal velocity is infinite at this point. Such a flow

† This section may be omitted without loss of continuity.

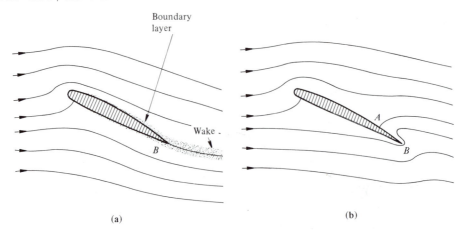

FIG. 6.15   Streamlines for (a) the flow of a real fluid about a slender airfoil and (b) the potential flow without circulation about the same airfoil.

picture, however, does not resemble that which is produced by any real fluid. It is not possible even for a fluid of very low viscosity to flow into regions of large positive pressure gradient as exists between points B and A because of the friction loss that would have been caused by the high velocities about point B. The rear stagnation point A of the ideal potential flow, however, can be moved by adjusting the circulation about the body. The potential flow will most closely resemble the actual flow in Fig. 6.15a, when the magnitude of the circulation is selected so that the rear dividing streamline streams smoothly from the trailing edge, thereby avoiding infinite velocities.

Thus, when treating the flow about bodies with sharp trailing edges, the circulation of the potential flow simulating the actual flow is selected so that the velocity at the trailing edge is finite. This procedure was first suggested by Kutta, and the condition that infinite velocities may not exist at the sharp trailing edge is known as the *Kutta condition*. The actual pressures and velocities generally agree well with those calculated on the basis of potential flows that meet the Kutta condition.

†6.13 The Lifting Flat Plate

In this section we treat the problem of the flow past a flat plate of zero thickness and length c held in a stream of velocity V_0 at an angle of incidence α^* to the plate as shown in Fig. 6.16. Several methods can be used to determine the flow past the plate, but we shall use the method of superposition of vortices to represent this flow. Suppose that we place on the plate an infinitesimal vortex of strength $d\Gamma$ at a distance from the center as shown, with positive Γ being in the clockwise sense. The complex velocity potential at any complex point $z = x + iy$ due to the vortex of strength $d\Gamma$ located at $x = \xi$ is

$$dF = i\,\frac{d\Gamma}{2\pi}\,\ln\,(z - \xi),$$

† This section may be omitted without loss of continuity.

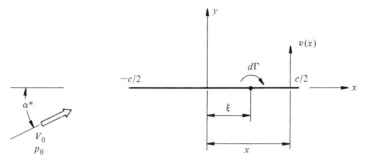

FIG. 6.16 System of notation for computing the flow over a flat plate at an angle of attack α^*.

as may be derived from Eq. (6.23). From Problem 6.39, the (complex) velocity at z is

$$d(u - iv) = i \frac{d\Gamma}{2\pi} \frac{1}{z - \xi}.$$

The incremental symbols d are used, since we expect to calculate the total contribution to the velocity from a continuous distribution of infinitesimal vortices along the plate. The total velocity may be written

$$u - iv = \frac{i}{2\pi} \int_{-c/2}^{c/2} \frac{d\Gamma}{z - \xi},$$

where the magnitude of $(z - \xi)$ is the distance from an elemental vortex at which the velocity is to be found. The integration has to be carried out to include all of the vortices along the plate and for this reason the integration is carried out from $-c/2$ to $+c/2$. The quantity z is constant during the integration. We can express the strength $d\Gamma$ of the elemental vortices $d\Gamma$ as

$$d\Gamma = \frac{d\Gamma}{d\xi} d\xi = \gamma(\xi) \, d\xi,$$

in which γ is the intensity or "density" of the vortices distributed along the axis. Such a distribution is called a vortex *sheet*. Thus

$$u - iv = \frac{i}{2\pi} \int_{-c/2}^{c/2} \frac{\gamma(\xi) \, d\xi}{z - \xi}, \tag{6.26}$$

and ξ is the integration variable. The denominator of the integrand can be rationalized to separate the real and imaginary parts. In this way we obtain the velocity components

$$u = \frac{1}{2\pi} \int_{-c/2}^{c/2} \gamma(\xi) \, d\xi \, \frac{y}{(x - \xi)^2 + y^2},$$

$$v = -\frac{1}{2\pi} \int_{-c/2}^{c/2} \gamma(\xi) \, d\xi \, \frac{(x - \xi)}{(x - \xi)^2 + y^2}. \tag{6.27}$$

These results, of course, could have been obtained also without the use of the complex variable notation. On the plate itself, $y = 0$ and the vertical, and hence normal velocity is†

$$v = -\frac{1}{2\pi} \int_{-c/2}^{c/2} \frac{\gamma(\xi)\, d\xi}{x - \xi}. \tag{6.28}$$

The velocity u is harder to determine when y approaches zero. By changing variables to

$$x - \xi = y \cot \theta$$

the integral for u becomes

$$u = \pm \frac{1}{2\pi} \int \gamma(x - y \cot \theta)\, d\theta, \tag{6.29}$$

with the limits on θ (corresponding to $\xi = \pm c/2$) not shown. The positive sign applies for y positive and the negative sign for y negative. When y is made to approach zero, the limits on θ become 0 and π, so that we have on the plate

$$u(x) = \pm \frac{\gamma(x)}{2}.$$

We see that the horizontal velocity on such a vortex sheet is discontinuous and that the magnitude of the discontinuity in velocity is equal to the local vortex sheet strength.

Equations (6.28) and (6.29) give the velocity contribution resulting from the vortex distribution $\gamma(x)$. To these must be added the velocity of the uniform flow $V_0 \cos \alpha^*$ in the x-direction and $V_0 \sin \alpha^*$ in the y-direction. The complete velocity distribution on the plate is therefore

$$v(x) = V_0 \sin \alpha^* - \frac{1}{2\pi} \int_{-c/2}^{c/2} \frac{\gamma(\xi)\, d\xi}{x - \xi}, \tag{6.29a}$$

$$u(x \pm 0) = V_0 \cos \alpha^* \pm \frac{\gamma(x)}{2}, \tag{6.29b}$$

where the $(+)$ sign refers to the upper surface and the $(-)$ to the lower. For Eq. (6.29a) to represent the flow around the flat plate, $v(x)$ must be zero, otherwise the plate would not be a streamline. Hence we have the equation

$$V_0 \sin \alpha^* = \frac{1}{2\pi} \int_{-c/2}^{c/2} \frac{\gamma(\xi)\, d\xi}{x - \xi} \tag{6.30}$$

to determine the unknown function $\gamma(\xi)$. Once γ has been determined the lift force from the Kutta–Joukowsky theorem is

$$L = \rho V_0 \Gamma = \rho V_0 \int_{-c/2}^{c/2} \gamma(\xi)\, d\xi, \tag{6.31}$$

and it is perpendicular to V_0.

---

† It should be noted that the integrand becomes infinite when $x = \xi$. The proper value of Eq. (6.28) is taken as the limiting case as $\delta \to 0$, where $\delta = |x - \xi|$ is the portion of the range of integration excluded on either side of the point $x = \xi$.

Equation (6.30) is said to be an *integral* equation since the unknown function is under the integral sign. Formal procedures are available for obtaining solutions to such equations, but it would take us too far beyond our present objectives to describe them. Instead, we shall merely observe that both the functions

$$\gamma = 2V_0 \sin \alpha^* \sqrt{\frac{c/2 - \xi}{c/2 + \xi}} \tag{6.32a}$$

and

$$\gamma = 2V_0 \sin \alpha^* \sqrt{\frac{c/2 + \xi}{c/2 - \xi}} \tag{6.32b}$$

satisfy Eq. (6.30). That is, either or both of Eqs. (6.32) make the plate a streamline. As the plate lies on the x-axis the point $x = c/2$ may be said to be the trailing edge. If this flow is to satisfy the Kutta condition at that point, that is, the velocity must be finite at $x = c/2$, Eq. (6.32a) is the permissible solution since the distribution expressed by Eq. (6.32b) would lead to an infinite velocity at the trailing edge as seen from Eq. (6.29b).

With the function γ known, the lift force becomes

$$L = 2\rho V_0^2 \sin \alpha^* \int_{-c/2}^{c/2} d\xi \sqrt{\frac{c/2 - \xi}{c/2 + \xi}} = \pi \rho V_0^2 c \sin \alpha^* \tag{6.33}$$

and

$$\Gamma = \pi V_0 c \sin \alpha^*.$$

This is usually expressed in coefficient form

$$C_L = \frac{2L}{\rho V_0^2 c} = 2\pi \sin \alpha^*, \tag{6.34}$$

where L is the lift force per unit length of the plate. The coefficient C_L is called the *lift coefficient*.

The lift coefficient formula for a flat plate [Eq. (6.34)] serves to estimate the force on a flat plate airfoil or hydrofoil, such as the rudder of a ship. Strictly speaking, it is limited to a flat plate. However, the expression for the lift coefficient for a slender curved plate is nearly the same—except that the angle of attack for zero lift force is changed. For example, the lift coefficient of a shallow, circular arc at a small angle of attack is

$$C_L = 2\pi \left(\alpha^* + 2\frac{h}{c} \right), \tag{6.35}$$

where h is defined in Fig. 6.17.

The camber (h/c) of the shape does not affect the slope of the lift curve. The effect of thickness is even smaller and usually amounts to less than 8 to 10 percent for a shape of that thickness-to-length ratio. Equation (6.35) affords a rapid estimate, therefore, of the lift of an airfoil or hydrofoil if its mean line (or skeleton line) does not depart appreciably from a circular arc.

In the foregoing discussion we have shown how Eq. (6.29a) leads to an exact solution of the potential flow past a flat plate. Based on the same approach, satisfactory approximate solutions for flows past slightly curved plates may also be obtained provided that the shape does not depart appreciably from that of a flat

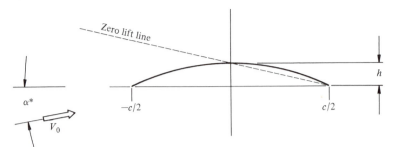

FIG. 6.17   A thin circular arc as an airfoil.

plate and that the angle of attack is small. It may then be assumed that the local changes in velocity brought about by the object are small compared to the approaching velocity and that the condition of flow tangency may then be applied to the line joining the leading and trailing edge. The condition of tangency (i.e., no flow through the surface of the airfoil) may be expressed approximately as

$$\frac{dy}{dx} = \frac{v}{V_0 \cos \alpha^*}, \tag{6.36}$$

where y is the ordinate of the airfoil and V_0 is the undisturbed velocity parallel to the x-axis.

Various airfoil configurations may then be investigated. It is easiest to adopt an indirect approach by selecting arbitrary distributions of γ and computing v from Eq. (6.29a). The corresponding airfoil shapes may then be determined by means of Eq. (6.36). For further details of this approach Reference 5 may be consulted.

Let us finally point out that other and much simpler potential flows than the one corresponding to the flat plate may be used to discuss two-dimensional lifting surface profiles. For example, a primitive airfoil could consist of the superposition of a source, a sink, and one or more discrete vortices. In such a case the potential and stream function are easily written down, but the coordinates of the streamline boundary will usually be difficult to find. The use of numerical methods, discussed briefly in the following section, along with modern high-speed computers can make this an attractive approach for some problems.

PROBLEMS

6.54 Verify that Eq. (6.32a) satisfies Eq. (6.30).

Hint: Use the result from p. 92 of Reference 5, that is,

$$\int_0^\pi \frac{\cos n\theta}{\cos \theta - \cos \theta'} \, d\theta = \pi \frac{\sin n\theta'}{\sin \theta'}.$$

6.55 With the aid of Eqs. (6.29b) and (6.32a) calculate the distribution of the pressure coefficient $c_p = (p - p_0)/(\rho/2)V_0^2$ along the upper and lower surfaces of the flat plate for $a^* = 5.7°$. Plot this distribution as a function of position.

6.56 Consider a lifting surface with a vortex strength

$$\gamma = k\left[1 - \left(\frac{2x}{c}\right)^2\right]^{1/2}$$

(where k is a constant) immersed in a uniform flow V_0 parallel to the x-axis. Using Eqs. (6.29a) and (6.36), determine the shape of the surface and the lift coefficient. (Use the integral given in Problem 6.54.)

6.57 Complete all the steps leading from Eq. (6.27) to Eqs. (6.29).

6.58 The complex potential for a source of strength q located at a point $z = \xi$ is $F(z) = (q/2\pi) \ln (z - \xi)$. A continuous distribution of infinitesimal sources of strength $m(x) = dq/dx$ may be made analogous to the continuous vortex sheet distribution of Eq. (6.26). By following similar steps as in the development of Eqs. (6.29), show that (a) the vertical velocity normal to a straight-line distribution of source strength density $m(x)$ is

$$v(x, \pm 0) = \pm \frac{m(x)}{2}$$

and (b) the horizontal velocity is

$$u(x) = \frac{1}{2\pi} \int_{-c/2}^{c/2} \frac{m(\xi)\, d\xi}{x - \xi}.$$

†6.14 Numerical Solutions

In the previous sections, analytic solutions of the potential flow equations were derived. These provide valuable insight into many useful flows, but they are limited in representing the flow past a *complex* geometrical shape as commonly occurs in engineering. Instead of formulating an analysis, it is possible to solve the potential flow equations numerically (indeed, as mentioned in the closing remarks of Chapter 2, it is even possible to solve the much more complex Navier–Stokes equations numerically).

There are a number of different approaches that may be used to obtain a numerical solution of the flow field. As mentioned at the end of Chapter 2, the flow can be imagined as being made up of many, very small, adjoining fluid elements. The conservation of mass and momentum equations are then applied to each element and the entire set of equations for all the elements solved simultaneously. This approach is called the finite-element or finite-volume method. The elements need not be rectangular; triangular and other shapes are frequently used. The size and shape of the elements chosen is dictated by the boundaries and nature of the flow, as well as by the computational efficiency and accuracy required.

In previous sections we have seen that by distributing source and sink combinations (or perhaps continuous-line vortex elements) through the fluid, the flow field over many different bodies or walls can be generated. This leads to a very elegant method, called the boundary element method, of distributing singularities of varying strength on the boundaries of a shape for which the flow field is desired. The nature and strength of singularity is chosen to obtain the proper body or wall shape and the remaining flow calculated by summing the contributions of all the boundary elements at any point in the flow. For some special problems the boundary element method may require less computation than the finite element method, particularly if only a certain region of the flow is of interest.

By far the simplest and most widely used numerical procedure for solving fluid

---

† This section may be omitted without loss of continuity.

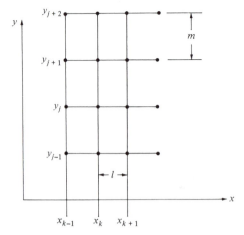

FIG. 6.18 Diagram showing finite difference mesh for numerical solution of the Laplace equation in Cartesian coordinates.

flow and other problems is the finite-difference method, in which the flow is calculated only at certain points forming a regular grid throughout the flow. The continuous functions and their derivatives (as in Laplace's equation) are then replaced by approximations based on the values of the functions at the grid points and the grid spacing or mesh size. Because of its widespread use and great facility in dealing with many flows of interest, we shall present only the finite-difference method here. The references at the end of the chapter give further information on finite- and boundary element methods; our intent here is only to outline one of the simplest and easiest methods for performing numerical analysis of fluid flows.

(a) Finite-Difference Approximation

To develop the finite-difference method, consider the representation of the velocity potential or stream function at discrete points x_k, y_j, and so on, where k and j are indices. Such an array of discrete points is shown in Fig. 6.18 in Cartesian coordinates. The spacing l between points along the x-axis is uniform, and similarly, the spacing m on the y-axis is also uniform but not necessarily the same as l. The discrete point x_k, y_j has the numerical value $x = kl$, $y = jm$ and the stream function there will be denoted by $\psi_{k,j}$. Values of the stream function at neighboring points $(k + 1, j)$, $(k, j + 1)$, and so on, are obtained by the truncated Taylor series expressions

$$\psi_{k+1, j} = \psi_{k, j} + \left.\frac{\partial \psi}{\partial x}\right|_{k, j} l + \frac{1}{2} \left.\frac{\partial^2 \psi}{\partial x^2}\right|_{k, j} l^2 + \cdots,$$

$$\psi_{k-1, j} = \psi_{k, j} - \left.\frac{\partial \psi}{\partial x}\right|_{k, j} l + \frac{1}{2} \left.\frac{\partial^2 \psi}{\partial x^2}\right|_{k, j} l^2 + \cdots.$$

Using these two expressions, we can represent the second derivative of ψ with respect to x at the point k, j, in terms of values of ψ at adjacent points and the grid spacing

$$\left.\frac{\partial^2 \psi}{\partial x^2}\right|_{k, j} = \frac{\psi_{k+1, j} + \psi_{k-1, j} - 2\psi_{k, j}}{l^2}.$$

Similarly,

$$\frac{\partial^2 \psi}{\partial y^2}\bigg|_{k,j} = \frac{\psi_{k,j+1} + \psi_{k,j-1} - 2\psi_{k,j}}{m^2},$$

so that the Laplace equation [Eq. (6.8)] becomes

$$\frac{\partial^2 \psi}{\partial x^2} + \frac{\partial^2 \psi}{\partial y^2} = \frac{\psi_{k+1,j} + \psi_{k-1,j}}{l^2} + \frac{\psi_{k,j+1} + \psi_{k,j-1}}{m^2} - 2\psi_{k,j}\left(\frac{1}{l^2} + \frac{1}{m^2}\right) = 0.$$

Solving for the stream function, we get

$$\psi_{k,j} = \frac{1}{2(l^2 + m^2)}\left[m^2(\psi_{k+1,j} + \psi_{k-1,j}) + l^2(\psi_{k,j+1} + \psi_{k,j-1})\right]. \qquad (6.37)$$

Often an expression for a spatial derivative is necessary. From the foregoing development there are three choices (using the x-direction as an example),

$$\frac{\partial \psi}{\partial x}\bigg|_{k,j} = \frac{\psi_{k+1,j} - \psi_{k,j}}{l}, \qquad (6.38a)$$

$$\frac{\partial \psi}{\partial x}\bigg|_{k,j} = \frac{\psi_{k,j} - \psi_{k-1,j}}{l}, \qquad (6.38b)$$

$$\frac{\partial \psi}{\partial x}\bigg|_{k,j} = \frac{\psi_{k+1,j} - \psi_{k-1,j}}{2l}. \qquad (6.38c)$$

The first of these is known as "forward" difference, the second as "backward" difference, and the last as the "central" difference. The first two represent the derivative but with an error term proportional to the spacing, whereas the central difference formula has a much smaller error, proportional to the *square* of the spacing.

If the "mesh" or spacing is the same in the two directions, then the finite-difference formula [Eq. (6.37)] is particularly simple; namely,

$$\psi_{k,j} = \frac{\psi_{k-1,j} + \psi_{k+1,j} + \psi_{k,j+1} + \psi_{k,j-1}}{4}, \qquad (6.39)$$

or succinctly in words: "The stream function at a given location is the average of its neighbors." The use of this equation for solving a complete flow field is best illustrated by studying a particular flow, as in the example that follows.

EXAMPLE 6.6

The square whose corners are located at $(x, y) = (0, 0), (1, 0), (0, 1), (1, 1)$ has the distribution of stream function ψ on its boundaries given in the table.

This flow is constructed from a uniform flow in the x-direction having a velocity of 100 dimensionless units and two sinks each of strength (-320) dimensionless units located at $x = 1.5$ and $y = +0.5$ and -0.5. The streamlines tend to converge as they approach the sinks and thus have the effect of the flow in a converging nozzle. We could, of course, use the methods of the previous sections to solve for the flow field in this case. Our purpose, however, is to show how the finite-difference method is used.

We now apply Eq. (6.39) to the region described in the table; the spacing in both directions is 0.1, but it does not enter directly into Eq. (6.39), as seen. Each of the indices k, j for the unknown values of the function ψ will run from 1 to 9 to cover

TABLE For Example 6.6[a]

x	0	0	0	0	0	0	0	0	0	0	0
y	0	0.1	0.2	0.3	0.4	0.5	0.6	0.7	0.8	0.9	1.0
ψ	0	16.1	32.2	48.2	64.1	80.0	95.6	111.1	126.4	141.5	156.4
x	0	0.1	0.2	0.3	0.4	0.5	0.6	0.7	0.8	0.9	1.0
y	1	1	1	1	1	1	1	1	1	1	1
ψ	156.4	159.2	162.3	165.2	169.5	173.7	178.3	183.5	189.4	196.0	203.6
x	1	1	1	1	1	1	1	1	1	1	1
y	0.0	0.1	0.2	0.3	0.4	0.5	0.6	0.7	0.8	0.9	1
ψ	0	20.3	40.9	62.2	84.1	106.4	128.3	149.3	168.8	186.9	203.6

[a] On $y = 0$, $\psi = 0$ for all x.

the entire region of the square; it will be recalled that the function is prescribed on the boundaries of the square for which the indices each have the values 0 and 10. Thus Eq. (6.39) can be written somewhat more directly as

$$\psi(k, j) = \frac{[\psi(k - 1, j) + \psi(j + 1, j) + \psi(k, j + 1) + \psi(k, j - 1)]}{4} \tag{6.40}$$

for $k, j = 1, 2, \ldots, 9$. There are thus $9 \times 9 = 81$ unknowns, and of course just that many equations are available to determine them. These equations are *linear*, so they can be solved by the methods of linear algebra discussed in the references at the end of the chapter. One of the most important of these (long used by engineers) is the solution of simultaneous equations by iteration. In this method an initial guess is made of unknown values (they can be zero, for example) and an approximation to the correct solution is found by substitution into the right-hand side of Eq. (6.40). This process is then repeated. Of course, for this to be a useful process it must converge to a unique answer. It is shown in References 6 and 7 that equations of the type shown in Eq. (6.40) do actually converge in this iteration process and that the rate of convergence is speeded up if the updated terms are used in the right-hand side of Eq. (6.40) when they become available. Fortunately, this is the case in even the simplest of programs in which each of the indices is repetitively cycled through all their values. This process was carried out for the data given in the table, and after 100 iterations the value of the stream function at the center of the square $x = 0.5$, $y = 0.5$, or $k = j = 5$ was found to be $\psi(5, 5) = 90.01$. The exact value is 90.00, so that the accuracy is more than adequate for most engineering purposes. In the present case this accuracy level is assured by the smoothness of the function distribution on the boundaries of the region considered; that is, there are no singularities there. Often, however, this may not be the case.

(b) Irregular Boundaries

The finite-difference formation in Cartesian coordinates is perfectly suitable to represent potential flow past bodies whose boundaries are square or at 45° to the grid axis. Such is rarely the case, unfortunately, for most practical engineering geometries. Then the boundaries of the flow may intersect a Cartesian mesh as indicated in Fig. 6.19 as an example. In this example the mesh lengths to points A and B on the irregular boundary are $a \times l$ and $b \times l$, respectively, from the node at k, J (i.e., $x_k = kl$, $y_J = Jl$; here we take the mesh spacing to be equal). The developments

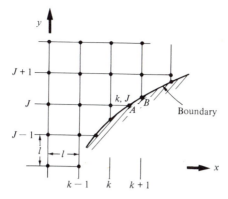

FIG. 6.19 Diagram showing a finite-difference mesh with an irregular boundary node (k, J). The length of the mesh for point A on the boundary is al and that of B is bl.

leading to Eq. (6.37) clearly have to be altered; specifically, the Taylor series expansions for the node (k, J) must reflect the new mesh length values, that is,

$$\psi_B = \psi_{k, J} + \left.\frac{\partial \psi}{\partial x}\right|_{k, J} bl + \frac{1}{2} \left.\frac{\partial^2 \psi}{\partial x^2}\right|_{k, J} (bl)^2 + O(bl)^3$$

and

$$\psi_A = \psi_{k, J} - \left.\frac{\partial \psi}{\partial y}\right|_{k, J} al + \frac{1}{2} \left.\frac{\partial^2 \psi}{\partial y^2}\right|_{k, J} (al)^2 - O(al)^3.$$

The expressions for the regular mesh lengths are unaltered, so that we can proceed to find, for example, the expression for the second partial derivative in the x-direction. First we divide the expression for ψ_B by bl and the previous equation for ψ_{k-1}, J by l and add as before to eliminate the first derivative. After simplification we find

$$\left.\frac{\partial^2 \psi}{\partial x^2}\right|_{k, J} = \frac{2}{l^2(b + 1)} \left[\frac{\psi_B}{b} + \psi_{k-1, J} - \psi_{k, J}\left(\frac{1 + b}{b}\right) \right] + O\left[l\left(\frac{b - 1}{b}\right) \right].$$

Here we have been a little more careful about the next term in the expansion; the O symbol means "order of," or simply proportional to the quantity in parentheses. It may be seen that if $b = 1$ (for a regular mesh size), this term disappears. If $b \neq 1$, an error proportional to the spacing appears in this representation of the second derivative. This error can be eliminated by using the information at an additional neighboring node but we shall not do so here. Neglecting this term and carrying out a similar development for the y-direction, we obtain the following expression for the Laplace equation at the irregular node:

$$\nabla^2 \psi_{J, k} = \frac{2}{l^2} \left[\frac{1}{b + 1}\left(\frac{\psi_B}{b} + \psi_{k-1, J}\right) + \frac{1}{a + 1}\left(\frac{\psi_A}{a} + \psi_{k, J+1}\right) - \psi_{k, J}\left(\frac{1}{b} + \frac{1}{a}\right) \right], \quad (6.41)$$

which reduces to Eq. (6.39) when $a = b = 1$.

EXAMPLE 6.7

In this example incompressible potential flow circulates around a circle of radius unity centered inside a square whose sides are 4 units in length. The stream function on the circle is set at a value of $\psi = 0$. The flow is to be analyzed numerically with a *square* mesh of spacing *l* and the velocity on the circle at a point nearest the wall is to be calculated. Because of symmetry, only one-eighth of the pattern need be analyzed; the resulting figure is shown below with a mesh of distance $l = \frac{1}{3}$. There are two irregular nodes at 5 and 8, where in the notation of Eq. (6.41) we have

$$\text{Node 5:} \quad a = 1, \quad b = 0.1716$$

$$\text{Node 8:} \quad a = 1, \quad b = 0.7638$$

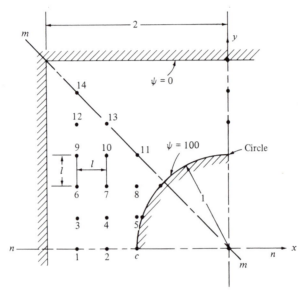

Example 6.7 Lines *n–n* and *m–m* are lines of symmetry. The mesh spacing here is $l = \frac{1}{3}$; the unknown nodes are numbered 1 though 14. Nodes 5 and 8 are *irregular* nodes.

The governing equation for these irregular nodes is still $\nabla^2\psi = 0$, or

$$\psi_{k,J} = \frac{ab}{a+b}\left[\frac{1}{b+1}\left(\frac{\psi_B}{b} + \psi_{k-1,J}\right) + \frac{1}{a+1}\left(\frac{\psi_A}{a} + \psi_{k,J+1}\right)\right]. \qquad (6.42)$$

Equation (6.42) is applied to the irregular nodes, while if $a = b = 1$, the node is regular and Eq. (6.39) is applicable. The lines of symmetry $n - n$, $m - m$ need a brief comment; here the stream function is symmetric or "even" in respect to these lines, so that, as an example, for node 13

$$\psi_B = \frac{\psi_{10} + \psi_{10}^* + \psi_{12} + \psi_{12}^*}{4},$$

where the starred quantities are the images or the symmetric counterparts of the variables indicated. These are in the present case $\psi_{10}^* = \psi_{10}$, $\psi_{12}^* = \psi_{12}$, and so on. Thus

$$\psi_{13} = \frac{\psi_{10} + \psi_{12}}{2},$$

and similarly,

$$\psi_2 = \frac{2\psi_4 + 100}{4}.$$

In this fashion equations are written for each of the 14 nodes, two of which are for the irregular nodes 5 and 8. The resultant set of equations can be solved by iteration or methods of linear algebra to get

$$
\begin{aligned}
&\psi_1 = 29.05, &\psi_6 = 23.64, &\quad \psi_{11} = 56.84, \\
&\psi_2 = 61.05, &\psi_7 = 48.72, &\quad \psi_{12} = 12.28, \\
&\psi_3 = 27.56, &\psi_8 = 76.55, &\quad \psi_{13} = 24.70, \\
&\psi_4 = 57.58, &\psi_9 = 18.26, &\quad \psi_{14} = 6.14. \\
&\psi_5 = 92.98, &\psi_{10} = 37.12
\end{aligned}
$$

The mesh in the present case is large, so that these values cannot be expected to be too accurate.

The velocity at point c is given by

$$
v(c) = -\frac{\partial \psi}{\partial x}(c),
$$

but the gradient there is not known. Instead, a Taylor series may be used to estimate it, namely,

$$
\left. \frac{\partial \psi}{\partial x} \right|_{(c)} = \left. \frac{\partial \psi}{\partial x} \right|_{(2)} + l \left. \frac{\partial^2 \psi}{\partial x^2} \right|_{(2)}.
$$

The derivative, $\partial \psi / \partial x|_{(2)}$ can be expressed by the central difference formula

$$
\left. \frac{\partial \psi}{\partial x} \right|_{(2)} = \frac{1}{2l} [\psi(c) - \psi(1)]
$$

and the central difference formula for the second derivative may be used to obtain the result

$$
\left. \frac{\partial \psi}{\partial x} \right|_{(c)} = \left[2\psi(2) - \frac{\psi(1)}{2} - \frac{3}{2}\psi(c) \right].
$$

In the present example,

$$
v(c) = -\left. \frac{\partial \psi}{\partial x} \right|_{(c)} = -127.3
$$

with the mesh chosen.

The simple finite-difference method outlined above is useful for studying a great many flow field problems. Techniques for solving the resulting linear equations are readily adapted to a personal computer (see Reference 11), so that fairly complex flow fields are now routinely solved by engineers in practice using this method. The flow over very complicated shapes, or flow field problems where the solutions must be known with great accuracy, are best solved by the more advanced boundary element or finite-element methods described in the references.

PROBLEMS

6.59 Use the data given in the table for Example 6.6.
  (a) Using a finite-difference method, calculate the values of the stream function for y varying from 0 to 1 and $x = 0.2, 0.4, 0.6,$ and 0.8.
  (b) Sketch the streamline for a stream function value of 100.
  (c) Calculate and plot the x-component of velocity along $y = 0$.
    Note: In parts (a) and (c) you should *check* your results by using the exact solution for at least one point in the flow.

6.60 Verify Eq. (6.41).

6.61 Flow circulates around a square annulus as shown.
  (a) Set up a finite-difference calculation procedure using a grid with at least six *spaces* across the gap and determine the stream function. Because of symmetry only one-eighth of the region need be considered.
  (b) Find the velocity at the midpoint along the surface of the inner wall. Can you calculate the tangential velocity at the corner along the inner wall? (Explain briefly.)

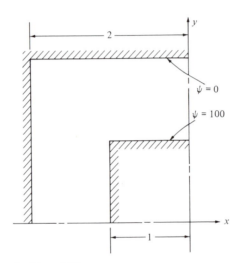

Problem 6.61

6.62 Consider again the flow described in Example 6.7, but imagine that rather than the annulus, it is a *bend* as shown. We want to find the flow field when the length of channel entering and leaving the bend is very long. In this case the flow far away from the bend is uniform, with the streamlines equally spaced across the channel. To limit the number of grid points the finite-difference computation is stopped at some distance along the channel, with the hope that reasonable accuracy is still maintained.
  (a) Begin by assuming that the value of the stream function is evenly spaced across the channel along a line $x = 0$, $y = 0$. Find the fluid velocity along the inner surface at the midpoint of the bend.
  (b) Repeat the calculation in part (a), but assume that the uniform flow condition is imposed at $x = 2$, $y = -2$. Can you guess from this if the grid size is large enough? Where along the channel do you think the uniform flow condition can reasonably be applied?

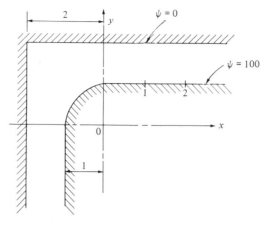

Problem 6.62

REFERENCES

1. A. H. Shapiro, *Shape and Flow*, Doubleday, New York, 1961.
2. L. Prandtl and O. G. Tietjens, *Fundamentals of Hydro- and Aero-mechanics*, McGraw-Hill, New York, 1934.
3. L. M. Milne-Thompson, *Theoretical Aerodynamics*, 4th ed., Macmillan, London, 1960.
4. H. Valentine, *Applied Hydrodynamics*, Butterworth, London, 1959.
5. H. Glauert, *The Elements of Aerofoil and Aeroscrew Theory*, 2nd ed., Cambridge Press, Cambridge, 1948.
6. B. Carnahan, H. A. Luther, and K. E. Wilkes, *Applied Numerical Methods*, Wiley, New York, 1969.
7. J. H. Ferziger, *Numerical Methods for Engineering Application*, Wiley, New York, 1981.
8. A. J. Chapman, *Heat Transfer*, 3rd ed., Macmillan, New York, 1986 (for application of numerical methods to the Laplace equation).
9. O. C. Zienkiewicz, *The Finite Element Method*, McGraw-Hill, New York, 1977.
10. C. A. Brebbia and S. Walker, *Boundary Element Techniques in Engineering*, Newnes—Butterworth, London, 1980.
11. W. J. Orvis, *1-2-3 for Scientists and Engineers*, Sybex, San Francisco, 1987.

CHAPTER 7

Analysis of Flow in Pipes and Channels

7.1 Introduction

In Chapter 2 we developed the Navier–Stokes equations, which are the general equations of motion for a Newtonian fluid. In the same chapter a few restricted problems were solved to illustrate the application of these equations. In Chapter 5 the Navier–Stokes equations were cast into dimensionless form, and the significant parameters governing the flow of a viscous fluid were determined, without, however, attempting to describe the flow in detail. We now investigate further the detailed behavior of the flow and such characteristics as the velocity distribution and the shear forces. Since the flows are going to be quite complicated, we restrict ourselves to simple geometrical boundaries, and examine principally flows through circular pipes and channels with walls parallel to the flow direction. To become acquainted with a very general viscous flow phenomenon, we again consider the flow in a circular pipe.

7.2 Laminar and Turbulent Flow

In the analysis of Chapter 5 it was predicted that for a circular pipe the friction coefficient f would be a function of Re. This fact was confirmed by experimental results such as shown in Fig. 5.4. Nothing at all, however, could be said as to the shape of the curve. Nevertheless, it is somewhat surprising that the curves of f versus Re for any of the pipes in Fig. 5.4 show a rather sudden variation in slope at a Reynolds number of about 3000.

To investigate the cause of this sudden change, it is necessary to observe the flow directly. For this purpose one might devise an experiment in which water flows through a transparent pipe. The Reynolds number can be varied by varying the flow rate. To make the flow visible, dye may be injected along the center of the pipe as shown in Fig. 7.1. For simplicity we shall assume in the present discussion that the flow is being observed at a position sufficiently far downstream of the entrance, so that no further changes in the flow profile with distance are to be expected. The flow is then said to have "reached equilibrium" or to be *fully developed*.

If one starts at a very low flow rate, one finds that the dye stream will follow a well-defined straight line parallel to the axis of the pipe. The dye streak remains straight as the flow rate is slowly increased. Suddenly, however, beyond a certain flow rate the line will become wavy, and if the flow is increased still further, the distinct line will disappear completely and the dye will be spread uniformly throughout the pipe. This experiment was first conducted by Osborne Reynolds, and he demonstrated by it that there were two different possible modes of flow in a pipe. In

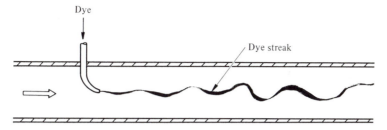

FIG. 7.1   Injection of dye into a pipe for detecting whether the flow is laminar or turbulent. The behavior of the dye streak as shown approximates that of a streak in transition to a turbulent flow.

the first mode the water particles follow along straight lines parallel to the axis of the pipe, but in the second mode each water particle seems to follow a random path throughout the pipe, only the average motion being along the pipe axis. This first mode of flow is called *laminar* flow and the second mode, *turbulent*. The transition from laminar to turbulent flow is, of course, again a function of the Reynolds number rather than of the velocity alone, as can be demonstrated by experiment. Furthermore, the transition coincides with the Reynolds number at which the sudden increase in friction coefficient occurs, and this change in mode of flow may, therefore, be taken as the cause for this effect.

The transition occurs because above a certain Reynolds number the laminar flow becomes unstable: If small disturbances are imposed on the fluid, these disturbances continue to increase with time. A stable flow is one in which the disturbances are damped out. It turns out that below a certain Reynolds number laminar pipe flow is stable for all small disturbances and remains, therefore, laminar. As the Reynolds number is increased, laminar pipe flow becomes unstable for a disturbance of a certain frequency and ultimately for all small disturbances. For these higher Reynolds numbers the disturbances grow and interact with each other to result in the irregular, fluctuating motion characteristic of turbulent flow.

Because the transition is dependent on disturbances either imposed externally (by vibration or other means) or from roughness elements in the surface, it may occur in a range of Reynolds numbers. In very carefully controlled experiments it has been possible to maintain laminar flow in smooth pipes up to Re = 40,000 based on pipe diameter. Under normal engineering conditions transition occurs in pipes at Reynolds numbers of 2000 to 3000. Below 2000 the flow is completely stable and will always be laminar.

Turbulent flow and turbulence is not restricted to flow in pipes. It occurs as well in flows over surfaces or objects; in fact, it may occur in every type of flow provided that the appropriate Reynolds number is sufficiently high. Turbulence is also a feature of many flows that are encountered in daily life. For example, the atmosphere is in an almost constant turbulent state. Winds, breezes, the formation of thunderheads, and so on, are all turbulent processes. The scale of these phenomena is, however, much larger than, say, those in a small water pipe but nonetheless the phenomena are of the same type. A good appreciation of the chaotic nature of turbulence (and therefore of the intrinsic difficulty in studying turbulent flow) is to be had by observing the foam of an ocean breaker or the motion resulting from rapidly filling a glass of water from a tap.

It is apparent that turbulent flows cannot be considered strictly steady; furthermore, the flow path of individual portions of the fluid cannot be predicted a priori even for simple geometrical boundary conditions. At best the *time average* of the flow can be considered steady and the *average* direction of the flow can be specified. We tacitly have already used the term "steady" in this manner in applying the Bernoulli equation and the momentum theorem in some of the previous problems. Whenever we simplify the Navier–Stokes equations, however, by assuming a strictly steady flow or by specifying the path of the fluid particles a priori, then we automatically restrict ourselves to laminar flow. Flows such as those of Examples 2.6 and 2.7 are, therefore, laminar flows only.

7.3 Shear Distribution in Circular Pipes

We shall now return to the analysis of the flow in a circular pipe. Rather than apply the Navier–Stokes equation directly, it is instructive first to develop a general relationship between the shear, the pressure drop, and the radius. To do this, consider a section of circular pipe of constant cross section, carrying a fluid of constant density. We shall assume that the flow has become fully developed, and that the velocity distribution as well as the pressure gradient does not change any further. Let the length and radius of the section be l and r, respectively, and let Δp be the pressure drop in the length l as shown in Fig. 7.2. The pressure distribution is the same over each cross section, since the pipe is straight and the average flow is therefore along straight parallel lines. For the cylindrical section outlined by dashed lines in Fig. 7.2, we can write the following condition of equilibrium:

$$\Delta p \pi r^2 = 2\pi r \tau l, \tag{7.1}$$

where τ is the shear stress along the cylindrical shell. If r_o is the outside radius of the pipe, then Eq. (7.1) becomes

$$\Delta p \pi r_o^2 = 2\pi r_o \tau_0 l, \tag{7.2}$$

where τ_0 is the shear stress at the wall.

From Eqs. (7.1) and (7.2) it is seen that

$$\tau = \tau_0 \frac{r}{r_o}, \tag{7.2a}$$

which shows that the shear stress must vary linearly with the radius. The wall shear τ_0 is of course related to the friction factor f, and from the definition of this factor the relation is

$$\tau_0 = \frac{1}{4} f \rho \frac{V^2}{2}, \tag{7.3}$$

where V is the average velocity in the pipe.

Equations (7.1) and (7.3) are quite general in that they have not been restricted to laminar or turbulent flow. If we knew τ as a function of the velocity and its deriv-

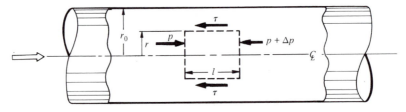

FIG. 7.2 Force equilibrium on a cylindrical portion of the fluid inside a circular pipe.

atives, Eq. (7.1) could be integrated and the friction factor could be obtained. In the case of laminar motion the shear stress τ is known in terms of the velocity gradients, but, as we shall see later, for turbulent flow the relationship between the shear and the characteristics of the average velocity profile is only poorly understood.

7.4 Steady Laminar Flow in Pipes and Channels

(a) Flow in a Circular Pipe

For laminar flow we have seen from Reynolds' experiment that each fluid particle moves along a straight line parallel to the pipe axis. The equation of motion can then be greatly simplified, and from Eq. (2.25) it can be shown that the shear stress is just equal to

$$\tau = \mu \frac{du}{dr}. \tag{7.4}$$

From this relationship between the shear forces and the velocity profile for the case of laminar flow in a straight pipe, together with Eq. (7.1), one obtains

$$\frac{du}{dr} = \frac{\Delta p}{2l\mu} r.$$

This equation is limited to laminar flow, because it contains the assumption that the flow is steady and that the streamlines are straight lines parallel to the axis. Making the same assumptions, this equation could also have been derived starting directly from the Navier–Stokes equation (see Problem 7.1). Upon integrating, the velocity is seen to be

$$u = \frac{\Delta p}{2l\mu} \frac{r^2}{2} + \text{const.}$$

The fluid velocity at the boundary ($r = r_o$) is equal to the velocity of the boundary, which is zero in this case. The integration constant can then be evaluated to obtain the following formula for the velocity profile in the pipe:

$$u = -\frac{\Delta p}{4l\mu} (r_o^2 - r^2). \tag{7.5}$$

The flow distribution for laminar flow in a pipe is seen to be parabolic. This example shows how from the general condition of equilibrium and the shear stress law, the velocity profile can be found.

With the use of Eq. (7.5) the flow rate through the pipe is found to be

$$Q = \int_0^{r_o} u 2\pi r \, dr = -\frac{\Delta p}{2l\mu} \pi \int_0^{r_o} (r_o^2 r - r^3) \, dr = -\frac{\Delta p \pi r_o^4}{8\mu},$$

and the average velocity is therefore

$$V = \frac{Q}{\pi r_o^2} = -\frac{\Delta p r_o^2}{8l\mu}, \tag{7.6a}$$

which happens to be equal to one-half of the maximum velocity that occurs at the center of the pipe. Equation (7.6a) may also be rearranged into the form

$$\frac{|\Delta p|}{\rho g} = h_f = \frac{64l\mu}{\rho V d^2} \frac{V^2}{2g} = \frac{64}{\text{Re}} \frac{V^2}{2g} \frac{l}{d}. \tag{7.6b}$$

This last equation can be compared to the general expression for the friction loss in a pipe. For the special case of laminar flow it is seen that

$$f = \frac{64}{\text{Re}},$$

a simple function of Reynolds number. Experiments verify this value of f very closely in the first part of the f versus Re curve (i.e., before transition). Furthermore, the influence of roughness is not detectable in the laminar flow range, so that in this range the line $f = 64/\text{Re}$ is applicable for all pipes.

It is interesting to observe that for laminar flow it was possible to determine f analytically without recourse to pipe friction experiments. In simplifying the equation of motion, we did, however, use the experimental information that in laminar pipe flow the fluid particles *steadily* follow straight paths parallel to the pipe axis. This condition does not apply to turbulent flow.

The laminar flow with a parabolic velocity distribution in a circular tube is known as *Poiseuille* flow and Eq. (7.6a) is sometimes called the *Hagen–Poiseuille* law. It should also be pointed out again that the velocity distribution [Eq. (7.5)] could also have been obtained directly by integration of the Navier–Stokes equations.

PROBLEMS

7.1 Starting with the Navier–Stokes equation, derive the velocity profile for the steady laminar flow in a pipe of radius r_o and a uniform drop of $\Delta p/l$ per unit length.

7.2 A very viscous fluid of density ρ is flowing in a tube of small diameter. The Reynolds number based on the tube diameter and the average flow velocity is 100. The tube is inclined at an angle of 30° with the horizontal, gravity being present. Assume the flow to be a fully developed laminar flow.
(a) Set up the differential equation for the flow for the case that $\Delta p = 0$.
(b) Find the velocity as a function of radius.
(c) Find μ as a function of the discharge Q.
The setup indicated represents one method of measuring viscosity.

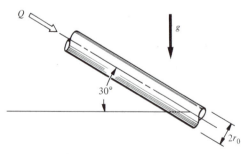

Problem 7.2

7.3 Calculate the velocity profile for fully developed laminar flow in the annulus formed by two concentric circular pipes of radii a and b. The pressure drop per unit length is Δp and the density is constant. Give your answer in terms of a, b, r, Δp, and μ.

7.4 A shaft of radius r_i is held concentrically in a cylindrical case of radius r_o. The shaft is kept stationary and the case is moved at a constant velocity V_0. There is no pressure gradient and the flow is laminar and incompressible.
  (a) Determine the distribution of shear stress τ between the inner and outer surfaces and integrate the resulting equation to obtain the velocity distribution.
  (b) What force (per unit length) is needed to move the outer cylinder?

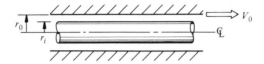

Problem 7.4

7.5 A common type of viscosimeter for liquids consists of a relatively large reservoir with a very slender outlet tube, the rate of outflow being determined by timing the fall in the surface level. If oil of constant density flows out of the viscosimeter shown at the rate of 3×10^{-7} m³/s, what is the kinematic viscosity of the fluid? Assume laminar, quasi-steady flow.

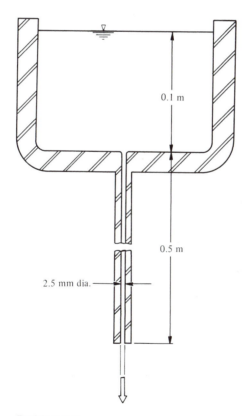

Problem 7.5

7.6 Consider laminar, steady flow of a gas in a circular pipe. The density is very low but the Mach number (velocity) is also low. Because of the latter, all properties are constant. Because of the former, the phenomena of "wall slip" must be considered. This means that at the tube wall,

$$u(r_o) \neq 0, \quad \text{but} \quad u(r_o) = -a \left. \frac{\partial u}{\partial r} \right|_{r_o},$$

where the tube radius is r_o and $a = $ constant. Find the flow rate Q in the pipe for a specified pressure gradient $(\Delta p / L)$.

7.7 Consider steady, laminar flow (incompressible, constant properties) in a very long, round pipe of radius r_o, driven by an axial pressure gradient $\Delta p / \Delta x$.

(a) By considering the Navier–Stokes equation directly, show that the shear stress varies linearly with radius across the pipe

$$\tau_{rz} = \tau_w \frac{r}{r_o},$$

where

$$\tau_w = \frac{r_o}{2} \left(\frac{\Delta p}{\Delta x} \right).$$

(b) A "pseudoplastic" fluid (see Problem 1.1) has a shear stress given by

$$\tau_{rz} = \mu_0 \left(\frac{\partial u}{\partial r} \right)^m$$

$$m = \text{const} < 1.$$

Show that the velocity profile in the pipe $u(r)$ is given by

$$\frac{u(r)}{u_0} = \left[1 - \left(\frac{r}{r_o} \right)^{(m+1)/m} \right],$$

where $u_0 = u$ at $r = 0$ (pipe centerline).

7.8 A simple model for the flow of blood in a circular pipe of diameter d relates shear stress to velocity gradient by

$$\tau = \alpha \sqrt{\frac{du}{dr}},$$

where α is a constant.

(a) Find the velocity profile $U(r)$.

(b) Determine the shear stress at the wall and thus the axial pressure gradient $\Delta p / L$.

(c) A further refinement of this model considers that blood will not flow but will remain as a solid plug unless the shear stress is greater than some critical value τ_c. Sketch the resulting velocity profiles when $\tau_c \ll \tau_w$ and $\tau_c \sim \tau_w$.

7.9 In a heat exchanger, oil flows through circular tubes 5×10^{-3} m in diameter. The tubes are 0.2 m long. The average velocity of the oil is 4.0 m/s and the kinematic viscosity is 1.0×10^{-4} m^2/s. Assuming laminar flow and neglecting entrance effects, calculate the power required to pump the oil through such an exchanger at a rate of 0.1 kg/s.

7.10 Both the mammalian respiration system and the mammalian blood circulation system are networks of tubes in which the flow from one large tube (respectively, the trachea and the aorta) branches into parallel flows in tubes of smaller size. This branching continues through a number of stages as shown in the illustration.

(a) If, for each stage, the number of tubes is denoted by n and the cross-sectional area of each and every tube in that stage is denoted by A_n, find the relation between A_n and n such that the pressure gradient $\partial p/\partial x$ is the same for each stage. How does the average velocity depend on n? (Assume steady, fully developed Poiseuille flow in all tubes.)

(b) If the diameter of the aorta is 3 cm and the diameter of the microcirculation (the smallest tubes) is 8×10^{-6} m, calculate the number of tubes at the microcirculation stage required for the property above to hold. The actual number is a great deal less than this.

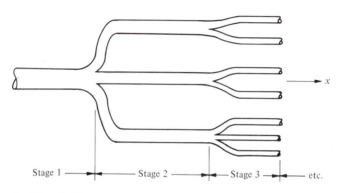

Stage 1 ——————— Stage 2 ——————— Stage 3 —— etc.

Problem 7.10

(b) Flow in a Duct or Channel

Now instead of a circular pipe we have a channel of infinite width, and gap h separating the two stationary surfaces. Again, as in the case of the circular pipe, the flow is assumed to be fully developed; the pressure decrease in any length l is a constant, say Δp; the flow is parallel to the axis of the channel; and the velocity u is a function of distance across the gap only. We can proceed as in the case of the pipe, or integrate the Navier–Stokes equations directly. In the present case the latter approach will be followed, and for that purpose we let the streamwise direction be denoted by x and the distance normal to the x-axis across the gap, by y, with $y = 0$ being the lower surface of the channel. The flow is fully developed, so that no changes in the x-component of velocity occur with distance x; the velocity is then a function of y only. The velocity v perpendicular to the channel is zero everywhere. These are the same conditions as Example 2.6, in which the Navier–Stokes equations were solved to obtain the Poiseuille flow in a channel

$$u(y) = -\frac{1}{2\mu}\frac{dp}{dx}(hy - y^2).$$

Another type of flow in such a channel is one that has *no* pressure drop along the flow; however, the upper plate must have a velocity U_0 for this case. This flow was described in Example 1.1 and is called Couette flow; it has a linear velocity profile

$$u(y) = U_0\frac{y}{h}.$$

The Poiseuille and Couette flows occur together frequently in situations of great practical interest. The flow in certain types of pumps or bearings, where one surface moves adjacent to another and a pressure gradient is superimposed, is an example of combined Couette–Poiseuille flow. By using the results of Example 2.6, the flow rate per unit width in a channel for such a combined case is

$$q_x = U_0 \frac{h}{2} - \frac{1}{12} \frac{h^3}{\mu} \frac{dp}{dx}.$$

Let us now suppose that the channel gap is *not* a constant but varies slightly along the distance x [i.e., $h = h(x)$]. We can imagine that in the case of a very slight variation of gap thickness with distance, the result above for the flow rate is still approximately correct at any location x along the channel. The gap height h and pressure gradient dp/dx are then interpreted as *local* values. The flow is still assumed to be only in the x-direction, so the expression for flow rate above must remain constant, that is, $dq_x/dx = 0$. This result leads to the equation

$$\frac{d}{dx}\left(\frac{h^3}{\mu}\frac{dp}{dx}\right) = 6U_0 \frac{dh}{dx},$$

which governs the distribution of pressure (approximately) in a channel of variable gap $h(x)$ if the rate of change, dh/dx, is very small. As mentioned, such variable gaps are a good representation of many types of *slider* (or journal) bearings; a few of these are described in some of the problems at the end of this section.

A more complicated bearing flow is one for which the bearing has finite width in the z-direction, causing a flow out the side of the bearing as well. For this case we can also imagine a combined Couette and Poiseuille flow in the z-direction, with the component of velocity of the upper plate in the z-direction being W_0. Then by analogy we can write

$$q_z = W_0 \frac{h}{2} - \frac{1}{12} \frac{h^3}{\mu} \frac{dp}{dz},$$

except, of course, that if the two flows (one parallel to the x-axis and the other parallel to the z-axis) occur simultaneously, partial derivatives should be introduced for the changes in the two directions. Conservation of volume flow (the continuity equation) is

$$\frac{\partial q_x}{\partial x} + \frac{\partial q_z}{\partial z} = 0,$$

so we are led to the equation

$$\frac{\partial}{\partial x}\left(\frac{h^3}{\mu}\frac{\partial p}{\partial x}\right) + \frac{\partial}{\partial z}\left(\frac{h^3}{\mu}\frac{\partial p}{\partial z}\right) = 6\left(U_0 \frac{\partial h}{\partial x} + W_0 \frac{\partial h}{\partial z}\right).$$

This famous result is called the *Reynolds lubrication equation*, and the approximations made above lead to what is known as *hydrodynamic lubrication theory* (for laminar flow).

An interesting special case of this equation occurs for constant viscosity and gap h. Then the Reynolds equation simply becomes

$$\frac{\partial^2 p}{\partial x^2} + \frac{\partial^2 p}{\partial z^2} = 0;$$

that is, the pressure in such a viscous channel flow satisfies the Laplace equation. This somewhat surprising result is the basis of a method of flow visualization devised by Hele–Shaw for potential flows.

PROBLEMS

7.11 The device illustrated is known as a *viscosity* pump. It consists of a stationary case inside of which a drum is rotating. The case and the drum are concentric. Fluid enters at A, flows through the annulus between the case and the drum, and leaves at B. The pressure at B is higher than that at A, the difference being Δp. The length of the annulus is l. The width of the annulus h is very small compared to the diameter of the drum, so that the flow in the annulus is equivalent to the flow between two flat plates. Assume the flow to be laminar and the fluid to be of density ρ and viscosity μ. Find the characteristics of this pump (i.e., pressure rise, torque, power input, and efficiency) as a function of the flow rate per unit depth.

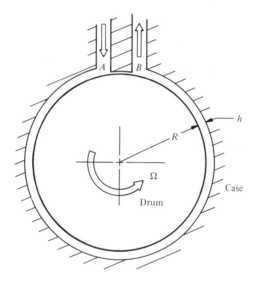

Problem 7.11

7.12 A continuous belt passes upward through a chemical bath at velocity V_0 and picks up a liquid film of thickness h, density ρ, and viscosity μ. Gravity tends to make the liquid drain down, but the movement of the belt keeps the fluid from running off completely. Assume that the flow is a well-developed laminar flow with zero pressure gradient, and that the atmosphere produces no shear at the outer surface of the film.
  (a) Set up the equilibrium of forces on a small element of the fluid, one side being bounded by the belt at $y = 0$ and the other by the variable y as shown.
  (b) State clearly the boundary conditions at $y = 0$ and $y = h$ to be satisfied by the velocity.
  (c) Calculate the velocity profile.

(d) Determine the rate at which fluid is being dragged up with the belt in terms of μ, ρ, g, h, and V_0.

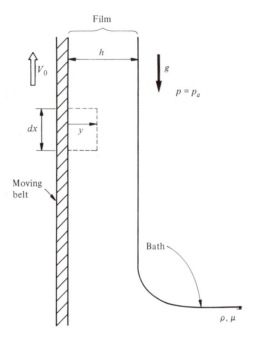

Problem 7.12

7.13 A *stepped bearing* is shown in the accompanying figure. The lower plates move in respect to the bearing shoe and drags oil through the clearance space. The bearing is of infinite extent in a direction normal to the paper, and all values have to be computed per unit width. The pressure can be assumed to increase linearly from the entrance to the step and then to decrease again linearly from the step to the discharge section. Assume laminar incompressible flow.

(a) Find an expression for the discharge q in the first part of the bearing (from the entrance to the step). Give your answer in terms of p_m, V_0, l, h_1, and μ, where p_m is the pressure at the step. (The flow in this section corresponds to that between two flat plates in the presence of a pressure gradient.)

(b) Find the discharge in the second section of the bearing (between the step and the discharge). Note that the pressure is now decreasing in the direction of flow. Give the answer in terms of p_m, V_0, l, h_2, and μ.

(c) From the two expressions for q [answers to parts (a) and (b)] eliminate q and find p_m in terms of V_0, h_1, h_2, l, and μ.

(d) Find the load that the bearing can carry per unit width.

(e) What is the load capacity if $h_2 = h_1/2$?

Note: The flow rate per unit width between two plates, one of which is moving with a velocity V_0, is given by the equation

$$q = -\left(\frac{dp}{dx}\right)\frac{h^3}{12\mu} + \frac{V_0 h}{2},$$

where dp/dx is the pressure rise per unit length and h is the separation between the plates.

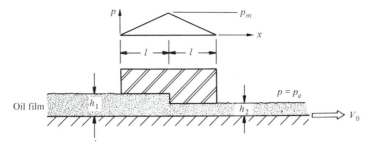

Problem 7.13

7.14 A simplified form of the "stepped" thrust bearing consists of a stationary plate held parallel to a moving surface as shown. The plate has a "step" that prevents flow through the gap. The height of the gap is h, and consider the bearing to be of unit width. The fluid is of density ρ and viscosity μ, and the flow is fully laminar. The velocity of the moving surface is V_0.
(a) Calculate the load F supported by the bearing in terms of μ, V_0, l, and h.
(b) Calculate the ratio of the drag force D on the moving surface to the load F.
(c) If $V_0 = 15$ ft/sec, $h = 10^{-3}$ ft, $l = 0.5$ ft, and we use SAE 30 oil at a temperature of $150°F$ (specific gravity $= 0.85$), calculate F and D/F.

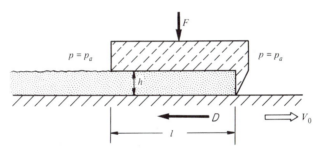

Problem 7.14

7.15 A simple slider bearing has a uniform taper from the inlet, where the gap is b_1, to the exit, where the gap is b_2. The bottom surface moves at a steady speed V as shown. The local channel height is h.
(a) Show that the pressure gradient along the slider can be written as

$$\frac{dp}{dx} = \frac{12\mu V}{h^3}\left(\frac{h}{2} - \frac{b_1 b_2}{b_1 + b_2}\right).$$

(b) If the pressure at each open end of the bearing is p_0, sketch the pressure profile in the fluid. Find the location and value of the maximum pressure.

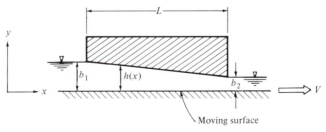

Problem 7.15

7.16 A slider bearing has an exponentially varying gap, so that

$$h(x) = h_0 e^{kx},$$

where k is a positive constant and the coordinate system is shown in the illustration. If the pressure is p_0 at each end of the slider, find the pressure profile $p(x)$, the total lift on the stationary part, and the total drag force on the moving slider.

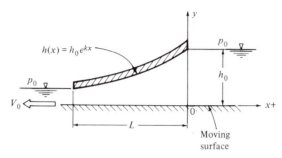

Problem 7.16

7.17 (a) Show that the mean velocity V for incompressible laminar flow between two parallel plates separated by a small distance h due to a pressure gradient $\Delta p/l$ is

$$V = \frac{\Delta p h^2}{12\mu l}.$$

(b) Consider now the flow in a channel for which h is not constant but varies with z, the direction transverse to the flow (see sketch). The flow is in the x-direction. If the width of the channel B in the z-direction is great compared to $h = h(z)$, then $\partial^2 u/\partial z^2$ may be neglected compared to $\partial^2 u/\partial y^2$. Also neglect the effect of the vertical sides. Write an integral expression for the discharge Q as a function of Δp, μ, l, h, and B, assuming that $B \gg h$.

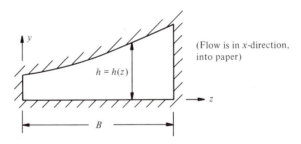

Problem 7.17b

(c) Finally, consider an eccentric plunger of radius R working in a cylinder with mean radial clearance h_0 and eccentricity εh_0. The variable radial clearance is then $h = h_0(1 + \varepsilon \cos z/R)$. Letting Q_ε represent the leakage flow in a direction normal to the paper past the eccentric plunger, *due to pressure only*, and Q_0 represent the same leakage with the plunger centered, derive an expression for Q_ε/Q_0 as a function of ε only. (Assume that $\varepsilon \ll 1$.)

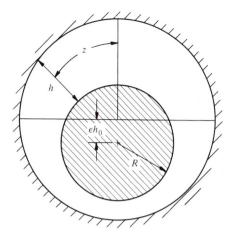

Problem 7.17c

7.18 A circular plate of radius r_o is immersed in a fluid of density ρ and viscosity μ and it is forced down with a constant force F onto a flat parallel surface as shown. The motion is sufficiently slow so that the acceleration and kinetic energy of the fluid can be neglected. The outward viscous flow between the plates at any radial position can then be assumed to have the same velocity distribution as a two-dimensional, fully developed laminar flow.

(a) Use the conservation of mass principle and derive an expression for the distribution of pressure as a function of radius assuming that $p = 0$ when $r = r_o$. Express your answer in terms of r, μ, V_0 (the vertical velocity of the plate), and h (the gap).

(b) Calculate the force F in terms of the same variables.

(c) Integrate this expression to get the time t necessary for the plate to move to within a distance h of the surface (assuming for calculation purposes that h is infinite for $t = 0$).

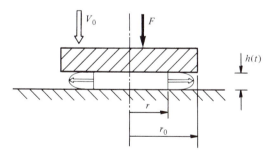

Problem 7.18

(c) Fully Developed Laminar Flow in Noncircular Pipes

The cross section of a pipe need not be circular; triangular, square, and other shapes are common in heat exchangers and other engineering flows. The shear distribution is not as easily determined as for the circular pipe and the Navier–Stokes equations must be solved directly. In fully developed laminar flow, the velocity is parallel to the pipe axis and the pressure gradient is along the direction of flow. (The effect of

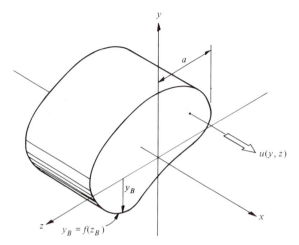

FIG. 7.3   Steady, laminar, fully developed flow in a duct of arbitrary cross section.

gravity may be accounted for as a hydrostatic pressure as shown in Chapter 5.) Let us take the pipe axis to be in the x-direction and assume the flow to be incompressible with constant viscosity. Then Eq. (2.22) becomes

$$\nabla^2 u = \frac{1}{\mu}\frac{dp}{dx} = \text{const}, \tag{7.7}$$

where the Laplacian function is only in the coordinates of the cross section of the pipe (y, z), as shown in Fig. 7.3.

For simple cross-sectional shapes it is possible to solve Eq. (7.7) analytically; the round pipe and infinitely wide parallel channel are, of course, special cases for which analytical solutions can be obtained easily. While other simple shapes, such as the ellipse in Example 7.1, can also be solved analytically, many practical shapes require numerical methods of solution.

EXAMPLE 7.1

Consider the steady laminar flow in a duct with an elliptical cross section in the yz-plane normal to the flow direction. We wish to find the velocity profile as well as the pressure drop for a given flow rate. First, Eq. (7.7) becomes

$$\frac{\partial^2 u}{\partial y^2} + \frac{\partial^2 u}{\partial z^2} = \frac{1}{\mu}\frac{dp}{dx}$$

and on the boundaries of the pipe the velocity $u(y, z)$ must be zero. The equation of the ellipse is

$$\frac{y_b^2}{a^2} + \frac{z_b^2}{b^2} = 1,$$

where y_b and z_b denote boundaries of the pipe cross section; a is the semimajor axis and b is the semiminor axis of the pipe. A solution can be tried following Example 6.1, letting,

$$u(y, z) = U_0(Ay^2 + Bz^2 + C).$$

Then

$$2U_0(A + B) = \frac{1}{\mu}\frac{dp}{dx}.$$

On the boundary we must have

$$0 = Ay_b^2 + Bz_b^2 + C,$$

which is achieved if $A = -1/a^2$, $B = -1/b^2$, and $C = +1$. Finally,

$$2U_0\left(\frac{1}{a^2} + \frac{1}{b^2}\right) = -\frac{1}{\mu}\frac{dp}{dx}, \tag{7.8}$$

so that the velocity distribution is

$$u(y, z) = U_0\left(1 - \frac{y^2}{a^2} - \frac{z^2}{b^2}\right), \tag{7.9}$$

where U_0 is the velocity at the pipe center. The approach we tried was indeed successful and the solution for the velocity distribution has been obtained.

Let us consider next the equation for the friction factor f, which can be written as

$$f = \frac{(-dp/dx)D_e}{\rho V^2/2}, \tag{7.10}$$

where D_e, the effective diameter, was previously defined to be four times the cross-sectional area divided by the duct perimeter P. For the elliptical duct, the pressure gradient can be taken from Eq. (7.8), while the average velocity V must be found by integrating the velocity distribution, Eq. (7.9), across the duct and dividing by the duct area, πab. After considerable (but straightforward) algebra, we can express f for an elliptical duct in terms of a factor multiplying the result already obtained for round pipes.

$$f = \frac{64}{\text{Re}}\left[\frac{1 + (b/a)^2}{2}\left(\frac{2\pi a}{P}\right)^2\right],$$

where

$$\text{Re} = \frac{VD_e}{v}$$

and

$$D_e = \frac{4\pi ab}{P}.$$

The perimeter of the ellipse P can be expressed in terms of the complete elliptic integral of the second kind (see Reference 1) and is approximately

$$P = 2\pi a\left(1 - \frac{k^2}{4} - \frac{3}{64}k^4 - \frac{15}{2304}k^6 - \cdots\right),$$

where

$$k^2 = 1 - \left(\frac{b}{a}\right)^2$$

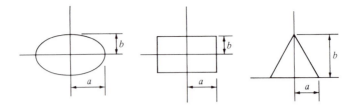

	Ellipse			Rectangle			Isoceles triangle		
b/a	A/a^2	P/a	fRe	A/a^2	P/a	fRe	A/a^2	P/a	fRe
10.00	31.42	4.06	77.26	40.00	44.00	84.68	10.00	22.10	50.17
5.00	15.71	4.20	74.41	20.00	24.00	76.28	5.00	12.20	51.62
2.00	6.28	4.84	67.29	8.00	12.00	62.19	2.00	6.47	53.28
1.00	3.14	6.28	64.00	4.00	8.00	56.91	1.00	4.83	52.61
0.90	2.83	5.97	64.09	3.60	7.60	57.04	0.90	4.69	52.22
0.80	2.51	5.67	64.39	3.20	7.20	57.51	0.80	4.56	51.84
0.70	2.20	5.38	64.98	2.80	6.80	58.42	0.70	4.44	51.45
0.60	1.88	5.11	65.92	2.40	6.40	59.92	0.60	4.33	51.06
0.50	1.57	4.84	67.29	2.00	6.00	62.19	0.50	4.24	50.49
0.40	1.26	4.60	69.18	1.60	5.60	65.47	0.40	4.15	49.81
0.30	0.94	4.39	71.58	1.20	5.20	70.05	0.30	4.09	49.12
0.20	0.63	4.20	74.41	0.80	4.80	76.28	0.20	4.04	48.63
0.10	0.31	4.06	77.26	0.40	4.40	84.68	0.10	4.01	48.31

TABLE For Example 7.1

(this result is good only for small values of k). The area, perimeter, and friction factor are tabulated for a few values of the ratio b/a in the table.

We now return to the case of steady laminar flow due to a pressure gradient in a duct of arbitrary, but unchanging, cross section, as shown in Fig. 7.3. For practical purposes we need to develop f versus Re relationships for such ducts.

We first define a dimensionless velocity

$$u^* = \frac{u}{(a^2/\mu)(dp/dx)}$$

and dimensionless coordinates

$$y^* = \frac{y}{a},$$

$$z^* = \frac{z}{a},$$

where a is a dimension typical of the duct size, as shown in Fig. 7.3. Then Eq. (7.7) becomes

$$\frac{\partial^2 u^*}{\partial y^{*2}} + \frac{\partial^2 u^*}{\partial z^{*2}} = 1.$$

With the further transformation

$$\varphi^* = u^* - \frac{y^{*2} + z^{*2}}{4},$$

we obtain

$$\frac{\partial^2 \varphi^*}{\partial y^{*2}} + \frac{\partial^2 \varphi^*}{\partial z^{*2}} = 0;$$

that is, φ^* satisfies Laplace's equation in dimensionless coordinates y^* and z^*.

If the equation of the duct boundary is written in a general way as

$$y_B^* = f(z_B^*),$$

the value of φ^* on the boundary of the duct must be

$$\varphi_B^* = -\frac{y_B^{*2} + z_B^{*2}}{4}.$$

Laplace's equation for φ^* can now be solved by the numerical techniques outlined in Chapter 6, along with the known value of φ_B^* on the boundary for any shape of duct. The friction factor f is obtained from Eq. (7.10); we first note that the average velocity is

$$V = \frac{a^4}{\mu A} \left(\frac{dp}{dx}\right)\left(\iint u^* \, dy^* \, dz^*\right),$$

where A is the total cross-sectional area of the duct and the velocity profile u^* is known from the numerical solution for φ^*. The friction factor f then becomes

$$f = \frac{2(A/a^2)(D_e/a)^2/[\iint (-u^*) \, dy^* \, dz^*]}{\text{Re}}.$$

The numerator in the equation above is a number that can be evaluated once a duct shape is specified. In this way the friction factor can be evaluated for many shapes of interest in engineering practice. A few of these are given in Table 7.1.

PROBLEM

7.19 Consider the triangular duct shown. The base has a dimension $2a$ and the angle at the apex is 90°. A fluid of viscosity μ flows under an imposed pressure gradient dp/dx.

(a) Find the volumetric flow rate $Q = V/A$.

(b) Find the location and magnitude of the maximum velocity.

(c) Verify that $f = (52.61)/\text{Re}$, as shown in the table for Example 7.1.

Note: A grid equal in both z- and y-directions with a spacing smaller than $a/50$ should be used.

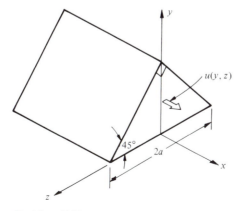

Problem 7.19

7.5 Turbulent Flow in Pipes

To determine the turbulent velocity profile in a circular pipe, it would seem logical to proceed in the same manner as we did for the laminar case. The reason that this approach has not been successful is that the flow is now no longer steady and the path of the fluid particles is erratic, which makes it impossible to predict the streamlines.

There are a few superficial things we can say about the velocity distribution: The time average of the velocity must be directed along the axis of the pipe, the profile of the average velocity must be symmetrical about the axis; and the velocity must be zero at the wall, because of the no-slip condition. The general shape of the velocity profile must be as shown in Fig. 7.4.

We have seen from Reynolds experiments that in the case of turbulent flow, elements of the fluid (much larger than a molecule) move along the pipe at random. If a particle moves normal to the direction of the average flow (say from A to B in Fig. 7.4), it is moving from a region of lower average x-momentum to a region of higher average x-momentum. The particle will therefore exert a drag on the fluid around B. Similarly, if a particle moves away from the pipe axis, it will tend to accelerate the fluid around its new position. These forces are the result of the turbulent cross motion of fluid particles and are in large part responsible for the shear force (or apparent shear force) within the fluid. It is of interest to recall here that the viscous forces in a perfect gas may be explained on the basis of random *molecular* motion; namely, that shear forces result from the transfer of momentum by the thermal motion of the molecules. An analogous phenomenon occurs in turbulent motion on a much larger scale, but it must be remembered that fluid particles are constantly sheared and distorted, and unlike individual molecules, quickly lose their identity.

In a purely formal way one might, for a circular pipe, express the shear in turbulent flow in terms of the average velocity as

$$\frac{\tau}{\rho} = v \frac{d\bar{u}}{dr} + v_T \frac{d\bar{u}}{dr}, \tag{7.11}$$

where $\bar{u}$ is an appropriate time average of the axial velocity component. For fully established flow the force due to this total shear must again balance the forces due to the pressure, and this shear stress must therefore obey the linear relationship, $\tau = \tau_0(r/r_o)$, which was developed in Section 7.3.

The expression for the shear in Eq. (7.11) differs from the expression for laminar flow by the second term on the right-hand side. The coefficient v represents the kinematic viscosity of the fluid, and v_T is sometimes called the *eddy viscosity* and results from the turbulent momentum exchange. But whereas v is a well-defined property of the fluid, v_T depends on the fluid as well as on the type of flow and will vary from the wall to the center of the pipe. Near the wall it will also depend on the surface roughness. For a smooth surface the cross-motion of the fluid particles near

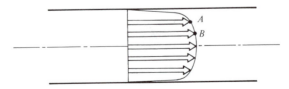

FIG. 7.4 Typical velocity profile in a circular pipe for turbulent flow.

the wall must approach zero because of the presence of the wall, and v_T should therefore be practically zero at this surface. The fluctuations of the molecules themselves, on the other hand, are so small that the wall does not constitute a solid barrier against their motion, so that the viscosity μ (and therefore the kinematic viscosity) is not explicitly dependent on the distance from the wall. (It is assumed that the fluid is Newtonian, that is, the viscosity does not depend on the shear stress.) In the region near the wall, then, the flow approaches laminar motion, and this region has been appropriately called the *viscous sublayer*. In the central region of the pipe the random motion of the particles should be most pronounced. Because of the large masses involved in this motion, it should be expected to contribute more to the shear than the molecular motion. In the central region, therefore, the first term should be negligible compared to the second term. Of course, there must also be a region in which both terms of Eq. (7.11) are about equally important. This region has been called the *transition region*, or *buffer layer*.

As may be inferred from the foregoing, the analysis of turbulent flow is exceedingly difficult. Although the governing equations (the Navier–Stokes and the continuity equations) are still valid, little can be done theoretically because of the nonlinearity and complexity of these equations and because the lack of a priori information on the unsteady paths of the fluid particles. For this reason it has become customary to abandon the study of the detailed flow and regard turbulent flow as consisting of an average flow with certain fluctuating velocity components superposed on it. Thus, in Eq. (7.11), $\bar{u}$ refers to the average velocity with respect to time at a given point. With this simplification semiempirical theories of turbulent pipe flow based on assumptions about the behavior of the eddy viscosity v_T have been developed, originally by Ludwig Prandtl and Theodore von Kármán. The development of computational fluid dynamics has led to the eddy viscosity being calculated directly from additional differential equations representing the turbulence itself. Unfortunately, empirical information is still needed in these equations as well, but considerable understanding has been gained in recent years about such flows. Both the earlier semiempirical as well as the more recent computer-based methods will be discussed briefly in later sections. First, however, we shall review certain basic experimental findings concerning turbulent flow.

7.6 Law of the Wall

A large part of the experimental work on turbulent flow has been done in circular pipes, mainly because of the well-defined geometry, the ease of controlling the flow conditions, and the flow symmetry. In the portion of the pipe near the entrance, the velocity profile will normally undergo some adjustments as described previously, until an equilibrium profile is finally established that will remain essentially unchanged as the flow proceeds further downstream. Depending on the exact flow conditions, a velocity profile closely approaching the fully developed one is established at a distance from about 50 to 300 diameters downstream of the entrance. The lower part of the range corresponds to turbulent and the higher one to laminar flow.

Velocity profiles in fully developed turbulent flow in smooth pipes have been measured by many investigators for a wide range of flow rates and pipe diameters, and for different fluids. From similarity considerations one may conclude that it should be possible to present the results in terms of a series of curves of velocity ratio $\bar{u}/V$ versus the radius ratio r/r_o, where V is the average velocity in the pipe and

r_o the pipe radius. Each curve would correspond to a different Reynolds number. This expectation is certainly borne out and experimental results have been successfully presented in this manner.

Another presentation, however, has been proposed for the velocity profile in the neighborhood of a wall. If we restrict ourselves to a region for which the radius ratio r/r_o is approximately equal to unity (i.e., near the wall), then this ratio or the radius itself might be expected to be no longer of great importance. It may then be reasoned that the velocity in the neighborhood of the wall must be determined by the conditions at the wall, the fluid properties, and the distance from the wall. The quantities that might be expected to be relevant are

$$y = \text{distance from the wall,}$$

$$\tau_0 = \text{wall shear,}$$

$$\rho = \text{density,}$$

$$\nu = \text{kinematic viscosity.}$$

For the dimensionless representation of the velocity profile near the wall, one may then construct from the quantities above a reference velocity $\sqrt{\tau_0/\rho}$ and a reference length $\nu/\sqrt{\tau_0/\rho}$. One may then attempt to present the experimental results in terms of a functional relationship between

$$\frac{\bar{u}}{\sqrt{\tau_0/\rho}} \quad \text{and} \quad \frac{y\sqrt{\tau_0/\rho}}{\nu}, \tag{7.12}$$

and any such relationship is called a *law of the wall*. For convenience the velocity $\sqrt{\tau_0/\rho}$ is often written as u_τ and is called the *friction velocity*. The group $\bar{u}/u_\tau$ is abbreviated as u^*, and the group $y u_\tau/\nu$ is often written as y^*, where y^* has the form of a Reynolds number.

Data from many experiments involving smooth pipes of many different sizes and flows at various pipe Reynolds numbers (Vd/ν) have been plotted on a graph of u^* versus y^*, and some of the data points are reproduced in Fig. 7.5. Purely fortuitously, the dimensionless graph of Fig. 7.5 gives a fairly accurate representation of the velocity profile in a pipe also when r/r_o differs considerably from unity, although small derivations occur near the center of the pipe. For this reason the graph in this figure has sometimes been regarded as one of rather universal applicability. Whenever r/r_o differs much from unity the representation cannot, of course, be expected to be exact. Depending on the pipe diameter, the pipe center will correspond to different values of y^*, and the universal curve, for example, will indicate a finite value for the velocity slope $d\bar{u}/dy$, whereas a zero slope should be expected for the actual profile.

In addition to the measurements in pipes, the velocity profiles in the neighborhood of the flat plates have also been measured. The same type of presentation of results in terms of the variables $\bar{u}/u_\tau$ and $y u_\tau/\nu$ has been found possible. Again purely fortuitously, the general curve of $\bar{u}/u_\tau$ versus $y u_\tau/\nu$ for the velocity profiles near a flat plate has been found to be quite similar to the pipe profiles—at least near the walls.

Let us next examine the curve in Fig. 7.5 more closely. In normal mathematical terms the functional relationship represented by the curve may be written as

$$\frac{\bar{u}}{u_\tau} = f\left(\frac{y u_\tau}{\nu}\right). \tag{7.13}$$

It is also possible to fit the curve—or portions of it—by suitable algebraic expressions. If this is done, it is seen that at low values of y^* ($y^* < 5$) the experimental

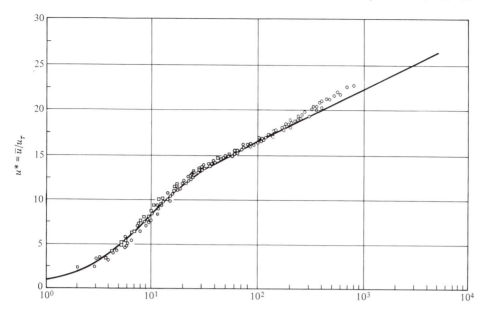

FIG. 7.5 Dimensionless velocity profile u^* versus y^*. This type of representation could have been expected for the region near the wall. It actually gives a fair representation even at distances where r/r_o differs significantly from unity. In addition, the profile for the flow near a flat plate is also well approximated by the graph in the figure. This velocity profile has, therefore, often been called a universal one. (Adapted from Report 1210 by R. G. Deissler, Nat. Aero. and Space Adm., formerly NACA, 1955.)

curve is well approximated by the linear relationship $u^* = y^*$. Since this is the relationship that would be obtained for laminar flow, it suggests that a laminar-like flow exists close to the wall. Measurements of velocity fluctuations as well as some heat transfer measurements support the hypothesis that the flow in the immediate neighborhood of the wall is dominated by viscous effects. This is the region then in which $v(d\bar{u}/dy) \gg v_T(d\bar{u}/dy)$ [see Eq. (7.11)] and which has been termed the *viscous sublayer*. Since the linear approximation fits the experimental curve well below $y^* = 5$, this value has been arbitrarily set for the width of the viscous sublayer. Next, for values of $y^* > 30$, the experimental curve is well approximated by the equation

$$\frac{\bar{u}}{u_\tau} = 5.75 \log_{10} \frac{y u_\tau}{v} + 5.5. \tag{7.14}$$

This region is associated with the flow in the main portion of the core, in which v_T is expected to greatly exceed v. The region between $y^* = 5$ and $y^* = 30$ is then called the *transition region*, or *buffer layer*.

Instead of Eq. (7.14), alternative relationships are sometimes given, as, for example,

$$\frac{\bar{u}}{u_\tau} = 8.7 \left(\frac{y u_\tau}{v} \right)^{1/7}. \tag{7.15}$$

This relationship does not fit the experimental curve as well as Eq. (7.14); it is, however, more easily handled mathematically and is therefore used as an approximation in some computations.

7.7 Derivations from the Law of the Wall

For most cases of turbulent flow in a pipe, the thickness of the combined viscous sublayer and the transition layer is very small compared to the pipe diameter. For computing the total flow rate in the pipe one may, therefore, without incurring any significant error, use for the entire pipe an expression for $\bar{u}$ which applies to the central part only. The discharge rate may then be expressed as

$$Q = 2\pi \int_0^{r_o} \bar{u}(r_o - y)\,dy,$$

where the velocity $\bar{u}$ is given by Eq. (7.14) or the equivalent. It should be pointed out here again that even in the central region of the pipe the universal representation of the velocity profile as given by Fig. 7.5 or the equivalent mathematical expressions can only be regarded as an approximation. In particular, the condition that the velocity gradient be zero at the pipe centerline due to symmetry is not satisfied. In the calculation above, however, the velocity appears in the integrand, and for mathematical reasons the inaccuracies in the expression for the velocity will affect the result to a minor degree. The relations derived from this integration may therefore be used with some confidence. Carrying out the integration for the discharge rate and substituting for τ_0 in terms of the friction coefficient, one obtains the following expression relating the friction coefficient to the Reynolds number of the pipe:

$$\frac{1}{\sqrt{f}} = 2.0 \log_{10} \text{Re}\sqrt{f} - 0.9, \tag{7.16}$$

where $\text{Re} = Vd/\nu$ and V is the *average* velocity and d is the diameter of the pipe.

A relation for the ratio between the average and the maximum velocity $(V/\bar{u}_{max})$ may be obtained by combining the foregoing equation with Eq. (7.14), remembering that $\bar{u}_{max}$ occurs at the position $y = r_o$. The resulting expression is

$$\frac{V}{\bar{u}_{max}} = \frac{1}{1 + 1.33\sqrt{f}}. \tag{7.17}$$

One more derivation of interest can be made from the velocity profile in a circular pipe as given by Eq. (7.14). By evaluating the equation for $y = r_o$ and subtracting from it the general expression which is valid at all distances, one obtains

$$\frac{\bar{u}_{max} - \bar{u}}{u_\tau} = 5.75 \log \frac{r_o}{y}. \tag{7.18}$$

The importance of this equation lies in the fact that it shows that the dimensionless velocity difference and therefore the general shape of the profile is a function of the distance ratio only and does not depend on the viscosity of the fluid or the corresponding Reynolds number $u_\tau y/\nu$.

†7.8 The Form of the Velocity Profile in a Circular Pipe

A brief comment may be added as to the general form of the velocity profiles above, by following a line of reasoning quite independent of the equations discussed before.

---

† This section may be omitted without loss of continuity.

Considering the flow in the vicinity of the wall, we see that the relationship

$$\frac{\bar{u}}{u_\tau} = h\left(\frac{u_\tau y}{v}\right)$$ (7.19)

could be expected. It has been argued further that at a reasonably large distance from the wall the viscosity factor should no longer be predominant in determining the velocity profile. Velocity differences in this region should therefore only depend on geometrical factors, so that

$$\frac{\bar{u} - u_{max}}{u_\tau} = g\left(\frac{y}{r_o}\right),$$ (7.20)

a fact which has actually been demonstrated in Eq. (7.18). (The symbols h and g here represent simply a general functional relationship.) The general expression of Eq. (7.20) has to be limited to velocity differences because the simple geometrical relationship only holds in a region away from the wall and the zero velocity at the wall cannot be used as a reference. A velocity within the region in question has to be selected as a reference, and for convenience $\bar{u}_{max}$ has here been used. The argument is then continued by pointing out that there must be some region of overlap in which both Eqs. (7.19) and (7.20) are valid. If this hypothesis is accepted, it can be shown that at least in this region the function $h(u_\tau y/v)$ which appears in Eq. (7.19) must be of logarithmic form. As we have seen [Eq. (7.14)], the experimental results verify this conclusion. The reasoning above was first pointed out by C. B. Millikan (Reference 2).

7.9 The Velocity Profile in a Fully Rough Pipe

The flow profile in a rough pipe may be discussed in a manner very similar to that in a smooth pipe. By examining the parameters that could influence the velocity ratio $\bar{u}/u_\tau$ in the neighborhood of the wall, one concludes that the ratio y/ε has to be considered in addition to the parameter yu_τ/v, which was the only one necessary to describe the smooth tube profile. The symbol ε represents the so-called *characteristic length* of the roughness. This length is related to the average height of the roughness elements but is not identical to it, and the ratio between the characteristic length and the average varies with different types of roughnesses. The relationship has to be determined experimentally for each type, as will be described more precisely later. In terms of the two parameters the velocity profile near the wall of a rough tube will be of the form

$$\frac{\bar{u}}{u_\tau} = f\left(\frac{yu_\tau}{v}, \frac{y}{\varepsilon}\right).$$

One may further hypothesize that when ε becomes large compared to v/u_τ, the parameter y/ε becomes predominent, or in different words, that the length ε rather than the length v/u_τ is the significant one in determining the velocity profile. The velocity profile will then be a function of y/ε only, and a pipe in which this type of flow occurs is called *fully rough*. By following the steps presented in Section 7.7, one may further hypothesize that for a fully rough pipe the velocity profile in the neighborhood of the wall should again be a logarithmic function of the significant parameter. Both of these expectations are confirmed experimentally, and the velocity profile for a *fully rough* pipe is well approximated by the equation

$$\frac{\bar{u}}{u_\tau} = 5.75 \log_{10} \frac{y}{\varepsilon} + C_2. \tag{7.21}$$

Because of fortuitous circumstances this equation, like Eq. (7.14), gives a fair approximation to the velocity even near the center of the pipe. The first extensive experiments on velocity profiles in rough pipes were performed in pipes in which the walls were coated with sand grains. In comparing with Eq. (7.21), ε was set equal to ε_s, the average diameter of the grains, and the constant C_2 was then determined by means of the experimental results. Under these conditions C_2 was found to be equal to 8.5. For different types of roughnesses it has become customary to let C_2 again be equal to 8.5 and to define an *equivalent sand grain roughness* ε_s, so that Eq. (7.21) is satisfied. For roughness elements different from sand grains, ε_s, therefore, is no longer equal to the size of the element, but it is related to this size.

The velocity profile for the rough pipe may again be integrated so as to yield the average velocity, and from this expression an equation for the friction in terms of the roughness ratio may be developed. This resulting equation may also be used to determine the equivalent roughness, and it is in fact usually the more accurate way of establishing this parameter. The procedure of obtaining the relation for the friction coefficient is analogous to the one outlined in Section 7.7.

It is important to note that in the fully rough flow regime the velocity profile in turbulent flow as well as the friction factor are independent of viscosity.

Experimentally, it has been determined that the conditions for a fully rough pipe are well satisfied when

$$\frac{\varepsilon_s u_\tau}{\nu} > 80.$$

Just as there exists a fully rough flow regime for which the dimension ε_s is so large as to make the effect of the dimension ν/u_τ negligible, there is also a flow regime in which ε_s is negligible compared to ν/u_τ. This is the regime in which the flow is essentially equivalent to that in a smooth pipe. Experimentally, it has been found that the flow is unaffected by roughnesses when the surface roughnesses are such that

$$\frac{\varepsilon_s u_\tau}{\nu} < 5.$$

The walls are then said to be *hydraulically smooth*. It is interesting to note that a surface may be hydraulically smooth even though the roughness size ε_s is not zero. For the same reason, in producing a hydraulically smooth pipe surface it is only necessary to polish the walls to a finish such that $\varepsilon_s < 5\nu/u_\tau$.

Let us point out again that a fully rough flow is independent of viscosity. Accordingly, one should not expect a viscous sublayer or transition region. The velocity profile as given by Eq. (7.21) may, therefore, be expected to hold throughout the fluid right up to the roughness elements themselves.

PROBLEMS

7.20 Starting from the velocity profile for a *fully rough* pipe, derive a relationship between the friction factor and the roughness ratio ε_s/d for the case in which the walls can be considered fully rough.

7.21 Fluid is flowing in a 0.15-m-diameter smooth glass pipe at an average velocity of 15 m/s. Find the thickness of the viscous sublayer and the transition layer:
(a) If the fluid is water at 21°C.
(b) If the fluid is air at 101 kPa and 21°C and the density may be taken as constant.

7.22 If the Reynolds number does not exceed 10^5, the turbulent velocity profile in a circular pipe may be approximated by

$$\frac{\bar{u}}{\bar{u}_{max}} = \left(\frac{r_0 - r}{r_0}\right)^{1/7},$$

where r_0 is the radius. Air at 21°C and 1 atmosphere pressure flows in a tube of 0.15-m diameter with an average velocity of 8 m/s. The walls are smooth.
(a) Determine the shear stress at $r/r_0 = 0.95$.
(b) What proportion of this shear is due to viscosity rather than turbulent fluctuations?

7.23 Let the mean effective height of the roughness elements in a pipe of radius r_0 be ε_s. From measurements we know that pipe roughness has little or no effect on friction as long as the roughness heights do not exceed the thickness of the laminar sublayer. If the relative roughness ε_s/d in part (a) of Problem 7.21 were 0.0005, what is the Reynolds number at which roughness will start to become important?

7.24 When the Reynolds number is less than about 10^5, the velocity profile is well represented by Eq. (7.15). With this formula show that the friction factor in a pipe is given by the equation $f = 0.31(Vd/v)^{1/4}$, where V is the average velocity. Also show that the shear at the wall may be given by $\tau_0 = 0.0228\rho u_{max}^2/(u_{max} r_0/v)^{1/4}$.

7.25 Fluid flows in a long, smooth, circular pipe of $\frac{1}{2}$-in. inside diameter. The velocity is measured in the center of the tube and far away from the entrance. The measurement is 5 ft/sec. What is the discharge rate (in ft³/sec) through the pipe:
(a) If the fluid is an oil with a kinematic viscosity of 10^{-3} ft²/sec?
(b) If the fluid is water at usual room temperature?

7.26 Consider steady flow of a fluid in a fully rough pipe. Find an expression for $\bar{u}_{max}/V$ as a function of friction factor f alone.

7.27 For ordinary Newtonian fluids, the "universal" velocity profile can generally be approximated by the equation

$$u^* = 8.7(y^*)^{1/7}.$$

A fluid containing a small amount of certain additives behaves somewhat differently. It has been found that the corresponding profile for that fluid can be expressed by the equation

$$u^* = 8.7(y^*)^{1/7} + B,$$

where B is a constant.
(a) For such a fluid, find the pipe friction coefficient f. Give the answer in terms of an equation involving f, Re, and B, where Re $= Vd/v$. (It may not be possible to solve this equation explicitly for f.)

(b) Show that the expression reduces to the one derived for an ordinary Newtonian fluid when $B = 0$.

(c) Find the friction coefficient, f, first for ordinary water and then for a solution for which $B = 10$ for $\text{Re} = 10^5$. (A trial-and-error solution may be necessary.)

†7.10 Reynolds Stresses in Turbulent Flow

We have now examined some of the experimental results for turbulent flow in a pipe and over a flat plate. Mention has also been made of the difficulties that we must expect when analytical solutions for turbulent flow are attempted. A start toward an analytical approach may, however, be made by considering the Navier–Stokes equation. This equation, as was pointed out before, embraces turbulent flow and is, therefore, still applicable. Let us now look at the velocity in a turbulent flow as consisting of a time-averaged portion and a fluctuating part as was already suggested. The x-component of the velocity for example could be written accordingly as

$$u = \bar{u} + u',$$

where $\bar{u}$ is the averaged portion and u' is the fluctuating part. The average $\bar{u}$ is defined as

$$\bar{u} = \frac{1}{T} \int_t^{t+T} u \, dt$$

and has already been used in the preceding section. By the definition of u',

$$\bar{u}' = \frac{1}{T} \int_t^{t+T} u' \, dt = 0.$$

The time T represents the interval over which the averaging is to be carried out. The exact value of T is difficult to prescribe, and several possible choices, leading to different ways of describing a turbulent flow, have been proposed. In the simplest approach, T is chosen to be long compared to any of the frequencies of the fluctuations. Since u is a time-dependent quantity, $\bar{u}$ even then will vary slightly as a function of the exact value of T. For sufficiently large values of T, however, the variation will be entirely negligible, and $\bar{u}$ will consequently be essentially independent of T, provided that T has been selected as specified. The minimum required value of the time T is quite different for different flows and essentially has to be established experimentally. For the flow of water in a 0.15-m pipe at 7 m/s, for example, a few seconds would constitute a quite sufficient measuring interval, but an interval of several minutes may be required for a large stream or for winds in the atmosphere. As a rule of thumb, the ratio of a significant dimension of the problem (such as a pipe diameter) to the stream velocity will yield a time that is of the order of magnitude of the time interval required to establish the average velocity.

One may also imagine changes in the average components $\bar{u}$, $\bar{v}$, and $\bar{w}$ with time. Such changes would have to be slow compared to the averaging interval T, but with this stipulation we will include such terms as $\partial \bar{u} / \partial t$, and so on.

---

† This section may be omitted without loss of continuity.

Consider now the Navier–Stokes equation for an incompressible fluid in the form that is obtained after adding the continuity equation, that is,

$$\frac{\partial u}{\partial t} + \frac{\partial u^2}{\partial x} + \frac{\partial uv}{\partial y} + \frac{\partial uw}{\partial z} = -\frac{1}{\rho}\frac{\partial p}{\partial x} + v\left(\frac{\partial^2 u}{\partial x^2} + \frac{\partial^2 u}{\partial y^2} + \frac{\partial^2 u}{\partial z^2}\right).$$

(Only one component has to be given here, as all steps for the other two will be completely analogous.)

Next let us substitute for the velocities and the pressure in terms of the average and the fluctuating components. The resulting equation becomes

$$\frac{\partial \bar{u}}{\partial t} + \frac{\partial u'}{\partial t} + \frac{\partial \bar{u}^2}{\partial x} + 2\frac{\partial uu'}{\partial x} + \frac{\partial u'^2}{\partial x} + \frac{\partial \bar{u}\bar{v}}{\partial y} + \frac{\partial \bar{u}v'}{\partial y} + \frac{\partial u'\bar{v}}{\partial y} + \frac{\partial u'v'}{\partial y}$$

$$+ \frac{\partial \bar{u}\bar{w}}{\partial z} + \frac{\partial \bar{u}w'}{\partial z} + \frac{\partial u'\bar{w}}{\partial z} + \frac{\partial u'w'}{\partial z}$$

$$= -\frac{1}{\rho}\frac{\partial \bar{p}}{\partial x} - \frac{1}{\rho}\frac{\partial p'}{\partial x} + v\left(\frac{\partial^2 \bar{u}}{\partial x^2} + \frac{\partial^2 \bar{u}}{\partial y^2} + \frac{\partial^2 \bar{u}}{\partial z^2}\right) + v\left(\frac{\partial^2 u'}{\partial x^2} + \frac{\partial^2 u'}{\partial y^2} + \frac{\partial^2 u'}{\partial z^2}\right). \quad (7.22)$$

In a further operation we shall now average each term in this equation over a time interval T, where the meaning of the averaging procedure is the same as that discussed before. It is important to note here that for the type of averaging process under consideration, the average of the derivative of a function is equal to the derivative of the average. To verify this statement, let us consider first the special derivative $\partial F/\partial n$, where F is any time-dependent function (e.g., $\bar{u}$, u', v', etc.) and n is one of the space coordinates (e.g., x, y, or z). By definition

$$\overline{\frac{\partial F}{\partial n}} = \frac{1}{T}\int_t^{t+T} \frac{\partial F}{\partial n}\,dt,$$

and since the differentiation process is here independent of the integration one may write the integral also in the form

$$\frac{1}{T}\int_t^{t+T} \frac{\partial F}{\partial n}\,dt = \frac{\partial}{\partial n}\left(\frac{1}{T}\int_t^{t+T} F\,dt\right).$$

The expression in parentheses is the definition of $\bar{F}$ and therefore

$$\overline{\frac{\partial F}{\partial n}} = \frac{\partial \bar{F}}{\partial n}.$$

Similarly, by applying the rules of differentiation for an integral in which the limits depend on the variable of differentiation, we may write

$$\overline{\frac{\partial F}{\partial t}} = \frac{1}{T}\int_t^{t+T} \frac{\partial F}{\partial t}\,dt = \frac{1}{T}[F(t+T) - F(t)],$$

but also

$$\frac{\partial \bar{F}}{\partial t} = \frac{\partial}{\partial t}\left(\frac{1}{T}\int_t^{t+T} F\,dt\right) = \frac{1}{T}[F(t+T) - F(t)],$$

and therefore

$$\overline{\frac{\partial F}{\partial t}} = \frac{\partial \bar{F}}{\partial t}.$$

By averaging each of the terms in Eq. (7.22) and remembering that the average of any individual fluctuating quantity is zero, the equation becomes

$$\frac{\partial \bar{u}}{\partial t} + \frac{\partial \bar{u}^2}{\partial x} + \frac{\overline{\partial u'^2}}{\partial x} + \frac{\partial \bar{u}\bar{v}}{\partial y} + \frac{\overline{\partial u'v'}}{\partial y} + \frac{\partial \bar{u}\bar{w}}{\partial z} + \frac{\overline{\partial u'w'}}{\partial z} = -\frac{1}{\rho}\frac{\partial \bar{p}}{\partial x} + \nu\left(\frac{\partial^2 \bar{u}}{\partial x^2} + \frac{\partial^2 \bar{u}}{\partial y^2} + \frac{\partial^2 \bar{u}}{\partial z^2}\right).$$

(7.23)

If we treat the continuity equation in a similar way, it can be shown that

$$\frac{\partial \bar{u}}{\partial x} + \frac{\partial \bar{v}}{\partial y} + \frac{\partial \bar{w}}{\partial z} = 0.$$

By multiplying the continuity equation by $\bar{u}$, subtracting it from Eq. (7.23), and rearranging, we obtain

$$\frac{\partial \bar{u}}{\partial t} + \bar{u}\frac{\partial \bar{u}}{\partial x} + \bar{v}\frac{\partial \bar{u}}{\partial y} + \bar{w}\frac{\partial \bar{u}}{\partial z} = -\frac{1}{\rho}\frac{\partial \bar{p}}{\partial x} + \frac{1}{\rho}\frac{\partial}{\partial x}\left(\mu\frac{\partial \bar{u}}{\partial x} - \rho\overline{u'^2}\right)$$

$$+ \frac{1}{\rho}\frac{\partial}{\partial y}\left(\mu\frac{\partial \bar{u}}{\partial y} - \rho\overline{u'v'}\right) + \frac{1}{\rho}\frac{\partial}{\partial z}\left(\mu\frac{\partial \bar{u}}{\partial z} - \rho\overline{u'w'}\right).$$

(7.24a)

The corresponding equations for the y- and z-directions may be obtained by cyclic permutation. In cylindrical coordinates the same set of equations takes the form

$$\frac{\partial \bar{v}_z}{\partial t} + \bar{v}_z\frac{\partial \bar{v}_z}{\partial z} + \bar{v}_r\frac{\partial \bar{v}_z}{\partial r} + \frac{\bar{v}_\theta}{r}\frac{\partial \bar{v}_z}{\partial \theta}$$

$$= -\frac{1}{\rho}\frac{\partial \bar{p}}{\partial z} + \frac{1}{\rho}\frac{\partial}{\partial z}\left(\mu\frac{\partial \bar{v}_z}{\partial z} - \rho\overline{v_z'^2}\right) + \frac{1}{\rho r}\frac{\partial}{\partial r}\left(\mu r\frac{\partial \bar{v}_z}{\partial r} - \rho r\overline{v_z'v_r'}\right)$$

$$+ \frac{1}{\rho r}\frac{\partial}{\partial \theta}\left(\frac{\mu}{r}\frac{\partial \bar{v}_z}{\partial \theta} - \rho\overline{v_z'v_\theta'}\right)$$

$$\frac{\partial \bar{v}_r}{\partial t} + \bar{v}_r\frac{\partial \bar{v}_r}{\partial r} + \frac{\bar{v}_\theta}{r}\frac{\partial \bar{v}_r}{\partial \theta} + \bar{v}_z\frac{\partial \bar{v}_r}{\partial z} - \frac{\bar{v}_\theta^2}{r}$$

$$= \frac{1}{\rho}\frac{\partial \bar{p}}{\partial r} + \frac{\partial}{\partial z}\left(\mu\frac{\partial \bar{v}_r}{\partial z} - \rho\overline{v_z'v_r'}\right) + \frac{\partial}{\rho r\,\partial r}\left(\mu r\frac{\partial \bar{v}_r}{\partial r} - \rho r\overline{v_r'^2}\right) \qquad (7.24b)$$

$$- \left(\mu\frac{\bar{v}_r}{r^2} - \frac{\overline{v_\theta'^2}}{r}\right) + \frac{1}{\rho r}\frac{\partial}{\partial \theta}\left(\frac{\mu}{r}\frac{\partial \bar{v}_r}{\partial \theta} - \rho\overline{v_r'v_\theta'}\right) - \frac{\mu}{\rho}\frac{2}{r^2}\frac{\partial \bar{v}_\theta}{\partial \theta}$$

$$\frac{\partial \bar{v}_\theta}{\partial t} + \bar{v}_z\frac{\partial \bar{v}_\theta}{\partial z} + \bar{v}_r\frac{\partial \bar{v}_\theta}{\partial r} + \frac{\bar{v}_\theta}{r}\frac{\partial \bar{v}_\theta}{\partial \theta} + \frac{\bar{v}_r\bar{v}_\theta}{r}$$

$$= -\frac{1}{\rho r}\frac{\partial \bar{p}}{\partial \theta} + \frac{1}{\rho}\frac{\partial}{\partial z}\left(\mu\frac{\partial \bar{v}_\theta}{\partial z} - \rho\overline{v_z'v_\theta'}\right) + \frac{1}{\rho r}\frac{\partial}{\partial r}\left(\mu r\frac{\partial \bar{v}_\theta}{\partial r} - \rho r\overline{v_r'v_\theta'}\right)$$

$$- \left(\mu\frac{\bar{v}_\theta}{r^2} + \frac{\overline{v_r'v_\theta'}}{r}\right) + \frac{1}{\rho r}\frac{\partial}{\partial \theta}\left(\frac{\mu}{r}\frac{\partial \bar{v}_\theta}{\partial \theta} + 2\mu\frac{\bar{v}_r}{r} - \rho\overline{v_\theta'^2}\right).$$

Upon comparing Eq. (7.24a) to the Navier–Stokes equation in its original form [Eq. (2.22)], great similarities are observed. In fact, except for the terms $\rho\overline{u'^2}$, $\rho\overline{u'v'}$, and $\rho\overline{u'w'}$, and so on, in Eq. (7.24a), it is seen that the average velocity components and the average pressure satisfy an equation identical to the Navier–Stokes equation. In the case of laminar flow the fluctuating components are all equal to zero and the Navier–Stokes equation is exactly satisfied by the average velocities and pressure.

This fact is, of course, not surprising because in laminar flow there is no difference between the average and the instantaneous components. In turbulent flow, however, the equation for the average quantities contains extra terms such as $\overline{\rho u'^2}$, $\overline{\rho u'v'}$, and $\overline{\rho u'w'}$, and in the way in which the equation [Eq. (7.24a)] has been arranged these terms may be regarded as additional stresses. The additional terms have actually been called *turbulent stresses*, or *Reynolds stresses*, in honor of Osborne Reynolds. With the concept of turbulent stresses a turbulent flow may be analyzed by considering only the average velocity and the average pressure, but including the special Reynolds stresses in addition to the usual viscous stresses. It should be stated here, however, that in deriving Eqs. (7.23a) and (7.23b), we have only rearranged the Navier–Stokes equation and introduced some new notation, without adding any new physical information. Nevertheless, the concept of the Reynolds stresses, which is based on this rearranged equation, has proved fruitful in studying turbulent flows.

As an example, let us simplify Eq. (7.24b) for the case of an established flow in a circular pipe for which all average velocities have the same direction and are parallel to, say, the z-axis. For such a flow Eq. (7.24b) reduces to

$$\frac{\partial}{\partial r} r \left(\mu \frac{\partial \bar{v}_z}{\partial r} - \overline{\rho v'_z v'_r} \right) = \frac{\partial \bar{p}}{\partial z} r$$

and upon integration

$$r \left(\mu \frac{\partial \bar{v}_z}{\partial r} - \overline{\rho v'_z v'_r} \right) = \int \frac{\partial p}{\partial z} r \, dr + \text{const.}$$

By considering the force balance on an element of fluid as in Section 7.3, the local shear stress can be identified as

$$\mu \frac{\partial \bar{v}_z}{\partial r} - \overline{\rho v'_z v'_r} = \tau.$$

Here again the role of the turbulent fluctuations in producing an effective "turbulent shear stress" is evident.

In this form the equation above can easily be compared to the expression

$$\tau = \mu \frac{\partial \bar{v}_z}{\partial r} + \rho v_T \frac{\partial \bar{v}_z}{\partial r},$$

which was previously [Eq. (7.11)] written for the same flow. From this comparison it is seen that the eddy viscosity is related to the velocity fluctuations in the following way:

$$v_T = - \frac{\overline{v'_z v'_r}}{d\bar{v}_z / dr}. \tag{7.25}$$

This expression, of course, is not sufficient to obtain a value for v_T, but it gives us a better idea of the factors on which v_T depends. (The total derivative may now be used, since variations with r only are now being considered.)

It is now possible to obtain some additional insight into the mechanism by which the turbulent fluctuations produce shear. For this purpose consider a control volume within a fluid for a general one-dimensional flow, and let us apply the momentum theorem to this control volume. For convenience we shall here revert to Cartesian coordinates (see Fig. 7.6). The average normal velocity component is zero, so that the average net mass flow through each side of the control volume is zero; but it is still possible to have an average net momentum flow through each side. The

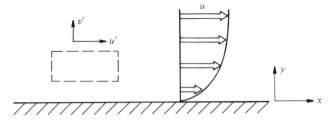

FIG. 7.6   Control volume near a wall to illustrate the origin of the turbulent stresses.

average net momentum flow through the side normal to the y-axis (which is equivalent to a force in the opposing direction) is

$$\frac{1}{T} \rho \int_0^T (\bar{u} + u')v' \, dt = \rho \overline{u'v'}.$$

But $-\rho \overline{u'v'}$ is precisely the term previously identified as the turbulent stress, and we therefore again conclude that the turbulent stress is caused by the momentum flux associated with the turbulent fluctuations. As was also mentioned earlier, it is interesting to note that in a gas, molecular fluctuations are in a very similar manner the cause for viscous shear.

A remark should finally be made regarding the sign of the quantity $\overline{u'v'}$. Consider a velocity profile with a positive slope $d\bar{u}/dy$, and let us suppose that the fluctuating components u' and v' are associated with more-or-less discrete portions of the fluid. The portions that have a positive v' will generally, because of the positive velocity gradient, have a lower x-component of velocity than the fluid into which they are moving. The fluctuation u' caused by the upward-moving fluid is therefore mostly negative, and the average product $\overline{u'v'}$ is expected to have a negative value. The term $-\rho \overline{u'v'}$ is therefore augmenting the viscous stress $\mu \, d\bar{u}/dy$. The same conclusion is, of course, reached for a negative slope $d\bar{u}/dy$. Direct measurements in turbulent flows have not only confirmed the expectations as to the sign of $\overline{u'v'}$, but have also verified that the magnitude of the average product of the velocity fluctuations properly accounts for the existing total shear stress. For illustration a graph is given (Fig. 7.7a) which shows the measured distribution of the turbulent shear term in a circular pipe. In the graph the velocity average $\overline{u'v'}$ is made dimensionless by means of the ratio of the wall shear τ_0 and the density ρ, and the radial position is given in

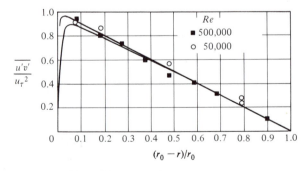

FIG. 7.7(a)   Experimental results of measurements for the average of the product $\overline{u'v'}$ as a function of radial position in a circular pipe. The product $\overline{u'v'}$ has been made dimensionless by means of $u_\tau^2 = \tau_0/\rho$. (Data from Report 1174 by J. Laufer, Nat. Aero. and Space Adm., formerly NACA, 1954.)

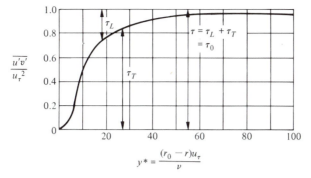

FIG. 7.7(b) The same experimental results as for Figure 7-7a (for *Re* = 500,000), with the region next to the wall enlarged. The laminar and turbulent portions of the total shear stress are defined in Eq. (7.11), while the total shear is assumed to be the same as the wall shear. Note that $y^* = 100$ is approximately the same position as $(r_0 - r)/r_0 = 0.01$.

terms of the pipe radius r_0. Two sets of data are shown for pipe Reynolds numbers of 50,000 and 500,000, respectively.

It should be remembered that the total shear stress in the pipe varies linearly with radius [i.e., $\tau = \tau_0(r/r_0)$]. Since the ordinate of Fig. 7.7a is the ratio of the turbulent shear stress to wall shear stress, it can also be seen that for most of the pipe the laminar component of the shear is negligible. Figure 7.7b shows the results of the same experiment, plotted against y^*, so that the region close to the wall is enlarged; in that region the total shear stress has been taken approximately equal to the shear stress at the wall.

A comment on notation should be made at this point. In the present and some of the previous sections, when discussing turbulent flows, we have marked the time average velocities by a bar ($\bar{u}$, $\bar{v}$, etc.) and the fluctuating components by a prime (u', v', etc.), and we have designated the complete instantaneous velocity components by plain letters such as u, v, and so on. In other chapters, however, the bars are often omitted for simplicity, and the plain letters then refer to the time-averaged velocity components. The context in each of these instances is such that this simplification is not likely to lead to any confusion.

†7.11 Models for Reynolds Stresses

As mentioned in Section 7.10, the Reynolds stresses can be represented in principle with equations like Eq. (7.25):

$$-\overline{\rho u'v'} = \rho v_T \frac{d\bar{u}}{dy};$$

however, we still need further information about the eddy viscosity v_T.

The various methods for deriving v_T in their increasing complexity involve the "models" referred to in modern computation of turbulent flows. The earliest work was that of Prandtl, who suggested that turbulent momentum exchange perpendicular to the main flow direction would occur over a "mixing length," l. He further reasoned that v_T was related to the mean flow by the expression

$$v_T = l^2\left(\frac{d\bar{u}}{dy}\right).$$

If we further assume, for example, that the mixing length varies in a prescribed way from zero near the wall (where fluctuations die out) to some known value in the mean flow, the time-averaged turbulent flow equations [Eq. (7.24a) or (7.24b)] contain only time-averaged (and not fluctuating) quantities. Thus they can be solved in principle, albeit a large computer may still be needed to make the calculations for flows of engineering interest.

Since this type of Reynolds stress modeling relates the turbulence to mean flow quantities and requires no *further* equations, it has frequently been called the "zero-equation" model. This type of model was used by von Kármán (along with additional physical reasoning) to derive the logarithmic form of the law of the wall (see Problem 7.29). Further refinements involve specifying more elaborate (but still empirical) variations of the mixing length l as a function of distance y from the surface. This information is usually based on experimental results and depends very much on the specific flow being studied.

The next level of complexity in describing Reynolds stresses involves use of the concept of kinetic energy associated with the turbulence (as if the kinetic energy of the mean flow were removed). Thus it is possible to define the turbulent kinetic energy k as

$$k = \frac{\overline{u'^2 + v'^2 + w'^2}}{2},$$

where the primes indicate the fluctuating components, as before. The turbulent kinetic energy itself must be derived from an additional partial differential equation which stems from further manipulation of Eq. (7.22) and the corresponding equations for the y- and z-direction motions. This equation (called a "transport" equation for the turbulent kinetic energy) must be solved along with the time-averaged Navier–Stokes equations. However, the Reynolds stress in this model still requires a mixing length, that is,

$$\nu_T = c_1 k^{1/2} l,$$

where c_1 is a constant and the variation of l is given by an empirical formula as in the zero-equation model. Since one additional equation (for the turbulent kinetic energy) is needed, this method is called the "one-equation" model. It should be pointed out, however, that in most cases the results are not much different from those of the zero-equation model, and the one-equation model is used only in certain classes of turbulent flows (e.g., thin shear layers).

The most frequently used model for Reynolds stresses adds the concept of rate of dissipation of turbulent kinetic energy, ε. The transport (involving k), production, and dissipation ε of turbulence are linked by two partial differential equations for k and ε; the derivation of these equations is involved and beyond the purposes of this discussion. The interested reader should consult some of the references listed at the end of this chapter. The eddy viscosity in the "two-equation," or k–ε model is given by

$$\nu_T = \frac{c_2 k^2}{\varepsilon},$$

where c_2 is another constant.

The two additional unknowns, k and ε, along with the two additional equations, form a complete set along with the time-averaged Navier–Stokes equations. Once again, however, considerable empirical information is still needed on the higher turbulent correlations which occur in this method, and great care must be taken when using the results of such calculations; some of the empirical correlations are obtained from very particular experimental conditions and may be entirely inappropriate in other applications. The flow near a wall or sharp corner (giving rise to free shear layers), or flows with significant secondary components, require special treatment, for example.

Finally, it should be mentioned that even if a suitable Reynolds stress model is available, the actual numerical computation involved can be formidable. Nevertheless, useful engineering results can be obtained, and many commercially available programs are now routinely in use to calculate turbulent flows in a variety of engineering applications. This emerging area of fluid mechanics research holds great promise for improved engineering flow calculation in the future.

PROBLEMS

7.28 It has been suggested that the eddy viscosity v_T near the wall of a pipe or flat plate can be approximated by $v_T = k\bar{u}^2/(d\bar{u}/dy)$, where k is a constant. The total shear stress near the wall then becomes

$$\tau = \mu \frac{d\bar{u}}{dy} + \rho k\bar{u}^2.$$

For a region close to the wall the shear stress may be considered constant so that $\tau = \tau_0$. For that case find $\bar{u}$ as a function of y, the distance from the wall.

Note:

$$\int \frac{dx}{a^2 + b^2 x^2} = \frac{1}{ab} \tan^{-1} \frac{bx}{a},$$

$$\int \frac{dx}{a^2 - b^2 x^2} = \frac{1}{ab} \tanh^{-1} \frac{bx}{a}.$$

7.29 For turbulent flow in a round pipe, von Kármán suggested that outside the viscous layer near the wall, the mixing length l depends only on the distance from the wall (i.e., $l = Ky$, where K is a constant).

(a) Show that if the turbulent shear stress is approximately constant and equal to the wall shear, the velocity profile becomes

$$u^* = \frac{1}{K} \ln y^* + c.$$

(b) Using the data on Fig. 7.5, estimate the constants K and c, and compare your estimates with the values in Eq. (7.14).

REFERENCES

1. M. Abramowitz and I. Stegun, *Handbook of Mathematical Functions*, Dover, New York, 1965.
2. C. C. Lin, ed., *Turbulent Flows and Heat Transfer*, Vol. 5, *High Speed Aerodynamics and Jet Propulsion*, Princeton University Press, 1959 (especially Section B).

3. L. Prandtl, *Essentials of Fluid Mechanics*, Hafner Press, New York, 1952 (especially Chapter 3).

4. *Collected Works of Theodore von Kármán*, Butterworth Scientific Publications, London, 1956.

5. B. E. Launder and D. B. Spalding, *Mathematical Models of Turbulence*, Academic Press, London, 1972.

6. P. Bradshaw, T. Cebeci, and J. Whitelaw, *Engineering Calculation Methods for Turbulent Flow*, Academic Press, London, 1981.

CHAPTER 8

Flow over External Surfaces

8.1 Introduction

The preceding chapter was concerned with fully developed flow in pipes or channels of constant cross section. A fully developed flow is one in which the velocity profile is the same at all locations along the pipe. However, the flow is *not* fully developed at the entrance to a pipe, say from a reservoir, as sketched in the introduction to Chapter 6. The flow in the main portion of the pipe away from the walls is uniform across a cross section. However, since a *real* fluid with nonzero viscosity must stick to the wall or bounding surface, a thin region near the wall is formed in which viscous effects are important. This thin region, as mentioned before, is called the *boundary layer*. As the flow proceeds downstream, the boundary layer grows through the action of viscosity and eventually fills the entire pipe to become a fully developed flow.

Unlike that of a pipe or channel, the flow over the *external* surface of a body such as a ship, airplane wing, or a lifting surface does not usually come to a fully developed shape with a constant-velocity profile. The region influenced by viscosity continues to grow as the flow proceeds downstream, although this region may remain "thin," so that the term *boundary layer* suitably describes this region of flow surrounding the body. Again, recalling the discussion of Section 6.1, the boundary layer may remain close to the surface or it may *separate* from the body, giving rise to a *wake*.

The boundary layer flow itself can be laminar, or like the pipe flow of Chapter 7, it may become turbulent at some distance from the leading edge of the body. The process of change from the laminar condition to the turbulent one is called *transition*. A photograph showing the flow on a flat plate with a laminar boundary layer undergoing transition to turbulence may be seen in Fig. 8.1. The flow near the leading edge of the boundary layer is seen to be smooth and regular, with fluid particles being sheared and distorted in an orderly way. In addition, there are no changes in the direction into the page (called the *spanwise* direction). The laminar flow becomes unstable and develops a wavy pattern that extends across the span. These waves grow and change into a more complicated pattern of twisted vortex lines that may also be parallel to the main flow. The three-dimensional vortex lines stretch and diffuse, with strong motion from the region nearest the wall out toward the edge of the boundary layer. At some point, the vortex twisting and mixing becomes sufficiently intense that turbulent spots can be noticed near the wall. These spots spread downstream quickly, and the entire boundary layer becomes turbulent. The mixing within the turbulent boundary layer is intense, the flow having regions reminiscent of those in fully developed turbulent pipe flow described in Chapter 7.

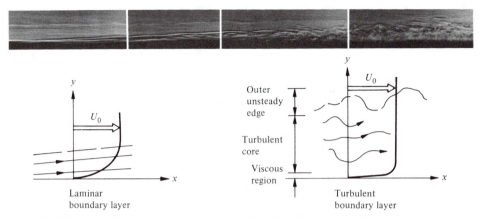

FIG. 8.1 Photograph of a boundary layer on a flat plate in a uniform flow approaching from the left. The fluid passing over the plate is liquid carbon dioxide with a little or no turbulence in the free stream. The plate is heated slightly in order to change the index of refraction of the fluid, thereby producing streaks when illuminated (called a *schlieren* photograph). The sketches below show the main features of the laminar and turbulent portions of the boundary layer.

Near the wall there is a viscous sublayer that is a small fraction of the overall thickness; the main portion of the boundary layer is very similar to that described by the law of the wall for pipe flow.

At the outer edge of a turbulent boundary layer the flow is highly irregular. Here the nonturbulent free-stream flow is mixed into the boundary layer at a distinct edge or interface, which is easily seen in the figure. The turbulent bursts or puffs within the boundary layer are intermittent; they grow and recede at the interface with the nonturbulent free stream. This behavior at the outer edge is unique to the turbulent boundary layer and one of the main differences when compared to turbulent pipe flow near the pipe center.

Although the foregoing description of transition to turbulence and the resulting turbulent flow in a boundary layer is somewhat incomplete, it points out the usefulness of visualization experiments in fluid mechanics. Much of our understanding of such complex flows, however, is still obtained from careful measurements of the velocity fluctuations throughout the boundary layer.

Of course, the boundary layer may still be thin whether or not the flow is turbulent. This notion of a thin boundary layer leads to a major simplification of the Navier–Stokes equations. This simplification makes it possible to obtain some "exact" solutions for laminar flows and provides a qualitative physical understanding of the flow. Turbulent boundary layers can be analyzed also, at least in principle, but as the results of Chapter 7 show, our knowledge of turbulent shear stress is still empirical. In the sections that follow, the boundary layer equations are derived for laminar and turbulent flow. They are then applied to the uniform flow past a flat plate. This classical problem is solved first for laminar flows in an exact sense, then in an approximate way. The chapter concludes with a brief discussion of flow around finite bodies and the contribution of friction to the total drag.

8.2 The Boundary Layer Equations

The Navier–Stokes equations are, of course, the governing equations (together with the continuity equation). For simplicity, the x-direction will be taken in the direction of the flow along the surface, and the y-direction will be normal to the surface.

If we neglect any effect of curvature, Eqs. (2.22) are (for steady flow in two dimensions with constant density and viscosity, and neglecting body forces): for the x-direction,

$$u \frac{\partial u}{\partial x} + v \frac{\partial u}{\partial y} = -\frac{1}{\rho} \frac{\partial p}{\partial x} + v\left(\frac{\partial^2 u}{\partial x^2} + \frac{\partial^2 u}{\partial y^2}\right)$$

and for the y-direction,

$$u \frac{\partial v}{\partial x} + v \frac{\partial v}{\partial y} = -\frac{1}{\rho} \frac{\partial p}{\partial y} + v\left(\frac{\partial^2 v}{\partial x^2} + \frac{\partial^2 v}{\partial y^2}\right).$$

The essential physical idea is that there is a small distance δ, the boundary layer thickness, normal to the surface of a body beyond which the viscous effect is negligible and that this distance is small compared to, say, the length L of the body (or perhaps the radius of curvature of the surface). We will now examine the various terms in these equations with a view toward introducing some acceptable simplifications. The basic assumption, $\delta \ll L$, implies that we may neglect the second derivatives in the x-direction compared to those in the y-direction, or in symbols,

$$\frac{\partial^2}{\partial x^2} \ll \frac{\partial^2}{\partial y^2}.$$

Next, we need to know v relative to u, and this may be obtained from the continuity equation [Eq. (2.5)]:

$$\frac{\partial u}{\partial x} + \frac{\partial v}{\partial y} = 0,$$

from which v is found by integration assuming that v is zero at $y = 0$:

$$v = -\int_0^y \frac{\partial u}{\partial x} \, dy.$$

We now estimate v as follows. As a measure of u we take a reference velocity U in the x-direction so that the value of $\partial u/\partial x$ is approximately

$$\frac{\partial u}{\partial x} \simeq \frac{U}{L}.$$

Then v at the edge of the boundary layer $y = \delta$ is

$$v(x, \delta) \approx U\left(\frac{\delta}{L}\right)$$

(the positive sign is taken because the x velocity is retarded in the flow direction). This result shows that $v/U \approx \delta/L$ and thus that every term in the y-direction of motion is less than corresponding terms in the x equation of motion by this same ratio, δ/L. One may conclude that the pressure gradient term in the y-direction, normal to the surface, is then proportional to δ/L times that of the x-direction. If δ/L is small enough, we will simply approximate the y-equation of motion as

$$0 \approx \frac{\partial p}{\partial y}. \tag{8.1a}$$

That is, the pressure is constant *across* the boundary layer. This result is one of the important consequences of the thin-boundary layer assumption. The other important result is that the derivative across the boundary layer is much greater than that along the boundary layer [e.g., $\partial u/\partial y \sim U/\delta$ and $\partial u/\partial x \sim U/x$ (or U/L)]. We have shown already that $v \sim U(\delta/L)$, so that the terms $u\,\partial u/\partial x \sim U^2/L$ and $v\,\partial u/\partial y$ $\sim U^2/L$ are seen to be of the same magnitude and, therefore, both must be retained in the left-hand side of the equation of motion. With these ideas in place, the final form for the boundary layer equation in the x-direction is (for incompressible flow)

$$u\,\frac{\partial u}{\partial x} + v\,\frac{\partial u}{\partial y} = -\frac{1}{\rho}\frac{dp}{dx} + v\,\frac{\partial^2 u}{\partial y^2}, \tag{8.1b}$$

which together with the continuity equation

$$\frac{\partial u}{\partial x} + \frac{\partial v}{\partial y} = 0, \tag{8.1c}$$

comprise the boundary layer equations. The present form of these equations, the boundary layer equations, is due to the arguments of Prandtl (Reference 1). The pressure gradient in Eq. (8.1b) is written as a total derivative because it is shown in Eq. (8.1a) that p is a function of x only [i.e., $p = p(x)$]. The pressure can therefore be determined from the Bernoulli equation, Eq. (3.2), for the flow just outside the boundary layer.

It should be noted that the boundary layer equations (8.1b) and (8.1c) have been derived directly from the Navier–Stokes equations, which of course are valid for either laminar or turbulent flow. However, the form of the boundary layer equations shown in Eqs. (8.1b) and (8.1c) are useful only for laminar flows, since the effects of turbulent stresses are not specifically identified as they were in Section 7.10, for example. It is not difficult to rearrange Eq. (8.1b) to make it apply to turbulent flows, however; the starting point is Eq. (7.24a), simplified for two-dimensional and steady average flow. The same arguments used previously for the order of magnitude of the velocity gradients still apply, so that for a turbulent boundary layer we may write Eq. (7.24a) as

$$\bar{u}\,\frac{\partial \bar{u}}{\partial x} + \bar{v}\,\frac{\partial \bar{u}}{\partial y} = -\frac{1}{\rho}\frac{\partial}{\partial x}\left(\bar{p} + \overline{\rho u'^2}\right) + \frac{1}{\rho}\frac{\partial}{\partial y}\left(\mu\,\frac{\partial \bar{u}}{\partial y} - \overline{\rho u'v'}\right).$$

The additional term $\overline{\rho u'^2}$ in the pressure gradient can be thought of as a correction to the pressure due to the *intensity* of the turbulence; for sufficiently low values of turbulence this can be neglected. While the remaining equation is reminiscent of the laminar form, the presence of the additional turbulent stress makes it very difficult indeed to solve the equations directly. The turbulence models described for *pipe* flows in Chapter 7 are sometimes used for boundary layers as well, but a great many additional approximations of the turbulence itself must be made, so that the approach is necessarily empirical. Further mention of turbulent boundary layers is reserved for a later part of this chapter.

Before trying to solve the laminar boundary layer equations, it is useful to discuss them in a qualitative way. The tangential velocity u should vanish at the boundary of a stationary solid surface. Similarly, the normal velocity v should also be zero at a solid surface; if the solid surface is *porous*, however, a value of $v(x, 0)$ may be prescribed. For the present we take v to be zero at a solid surface. The tangential velocity $u(x, y)$ changes from zero at the surface to the full external velocity $U(x)$ at the edge of the boundary layer. We shall assume that this velocity is known, perhaps by a potential flow calculation or by measurement of the surface pressure and hence by use of the Bernoulli equation. Other than these two limiting values, further information about the velocity profile may be found from the equation of motion (8.1b)

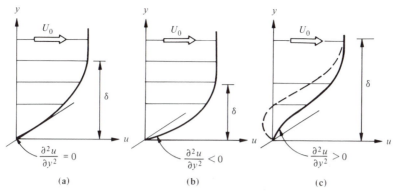

FIG. 8.2   Sketches of possible boundary layer velocity profiles based on the curvature of the velocity profile at the surface: (a) constant pressure flow (or flow past a flat plate); (b) flow with a *negative* pressure gradient that has "fuller" velocity profile than (a); (c) possible profiles for *positive* pressure gradients that have a positive curvature of the velocity profile and a positive pressure gradient. The curvature must increase for stronger pressure gradients and could conceivably result in the dashed curve, indicating a region of reverse flow. In this event the boundary layer flow would have become detached or *separated* from the body. For that reason, a positive pressure gradient is called an *adverse* pressure gradient.

right at the surface (where $u = v = 0$). Then

$$\left.\frac{\partial^2 u}{\partial y^2}\right|_{(x,\,0)} = \frac{1}{\mu}\frac{dp}{dx}(x),$$

and the left-hand side may be seen as the *curvature* of the velocity profile right at the surface itself. This curvature is zero for a zero pressure gradient, positive for a positive pressure gradient, and negative for a negative pressure gradient. Sketches of possible velocity profiles $u(x, y)$ developing as a result of different pressure gradients are shown in Fig 8.2. These sketches are purely schematic but do reflect the factors inherent in the equations of motion at a solid surface. A positive pressure gradient leads to a positive curvature of the velocity profile at the surface, which tends to decrease the velocity gradient and the velocity itself near the wall. This trend with increasing pressure gradient may actually result in a zero velocity gradient $\partial u/\partial y$ at the wall and thus zero shear stress. This condition is taken as one of impending reverse flow within the boundary layer and the flow is said to be on the verge of *separation*. Separation is caused by a positive or *adverse* pressure gradient such as that occurring at the rear of a body, for example. Some examples of how pressure gradient and separation influence the drag for different-shaped bodies are given at the end of this chapter.

8.3 The Flat Plate

A very simple body, which nevertheless illustrates the main features of boundary layer flow, is that of a vanishingly thin flat plate. In this case flow of uniform speed U_0 proceeds along the flat plate, which is placed parallel to the flow, rather like that in Fig. 8.1. In this case, the velocity external to the boundary layer, U_0, is constant, so that the velocity profile is similar to that of Fig. 8.2a. The flow is assumed to be laminar, so the equations to be solved are

$$u\,\frac{\partial u}{\partial x} + v\,\frac{\partial u}{\partial y} = \nu\,\frac{\partial^2 u}{\partial y^2} \tag{8.2}$$

and

$$\frac{\partial u}{\partial x} + \frac{\partial v}{\partial y} = 0, \tag{8.1c}$$

and we shall use the boundary conditions mentioned previously, namely,

$$u(x, 0) = v(x, 0) = 0$$

$$u(x, y \to \infty) = U_0. \tag{8.3}$$

This problem of the viscous flow past a flat plate was first solved by Blasius in 1908, and it is considered one of the classical achievements in fluid mechanics (an English version is available in Reference 2). Blasius proceeded following an inspiration that the velocity ratio u/U_0 (which in general is a function of x and y) might depend on a single dimensionless variable η of the form

$$\eta = \frac{Ay}{x^n},$$

where A and n are constants (a somewhat similar situation arose in Example 2.7, where the single variable is labeled z). The procedure is essentially to substitute η into Eqs. (8.2) and (8.1c) and then choose the constant n, so that the resulting equations depend only on η. The constant A may be selected rather arbitrarily so as to obtain simple coefficients in the resulting differential equation. This process is greatly facilitated by satisfying the continuity equation identically by introducing the stream function

$$u = \frac{\partial \psi}{\partial y},$$

$$v = -\frac{\partial \psi}{\partial x}.$$

The goal is to be able to express u/U_0 as a function of η only, say,

$$\frac{u}{U_0} = f(\eta),$$

where f is a function to be found. Then from the continuity equation, it follows that

$$u = \frac{\partial \psi}{\partial y} = U_0 f(\eta).$$

This relation can be integrated at a fixed position

$$\psi = U_0 \int f(\eta)\,dy.$$

From the definition of η we may then write

$$dy = \frac{x^n\,d\eta}{A},$$

so that

$$\psi = \frac{U_0 x^n \int f(\eta) \, d\eta}{A},$$

which is the relation that the stream function must satisfy. The integral is merely another function of η, say $F(\eta)$ or

$$f(\eta) = F'(\eta),$$

where the prime denotes differentiation with respect to variable η. Thus

$$\psi(x, y) = \frac{U_0 x^n F(\eta)}{A}.$$

It then follows that

$$u = \frac{\partial \psi}{\partial y} = U_0 F'(\eta),$$

$$\frac{\partial u}{\partial x} = -\frac{n U_0}{x} \eta F''(\eta),$$

$$\frac{\partial u}{\partial y} = \frac{U_0 A}{x^n} F''(\eta),$$

$$\frac{\partial^2 u}{\partial y^2} = \frac{U_0 A^2}{x^{2n}} F'''(\eta),$$

$$v = -\frac{\partial \psi}{\partial x} = -\frac{U_0}{A} [nx^{n-1} F(n) - n\eta x^{n-1} F'(\eta)],$$

where use has been made of the definition of η in the second and last expressions. Upon substitution into Eq. (8.2), one gets after simplification the equation

$$F''' + \frac{U_0 n x^{2n-1}}{\nu A^2} F F'' = 0$$

[dropping the parentheses for $F(\eta)$]. It can be seen now by inspection that this equation can be made to depend on η alone simply by setting $n = \frac{1}{2}$. The remaining group of terms must then just be a constant; that is,

$$\text{const} = \frac{U_0}{2\nu A^2}.$$

The value of the constant and hence that of A is arbitrary; a possible choice is to set the constant equal to $\frac{1}{2}$. Then $A = \sqrt{U_0/\nu}$,

$$\eta = y \sqrt{\frac{U_0}{\nu x}},$$

$$\psi = \sqrt{\nu U_0 x} \, F(\eta),$$

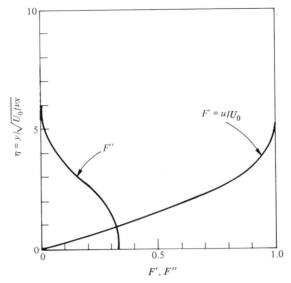

FIG. 8.3 Flat plate boundary layer velocity and shear profile corresponding to the equation $2F''' + FF'' = 0$.

and the equation satisfied by F is

$$2F''' + FF'' = 0, \tag{8.4}$$

with primes denoting differentiation in respect to η.†

The boundary conditions are that at the plate where $y = \eta = 0$, both u and v are zero and far from the plate, where y and η approach infinity, $u(x, y \to \infty) = U_0$. From these conditions and the relations between the stream function and the velocity components, we find

$$F'(0) = \frac{u(x, 0)}{U_0} = 0$$

$$F'(\infty) = \frac{u(x, \infty)}{U_0} = 1$$

and the normal velocity $v(x, 0)$ is made zero by requiring that

$$F(0) = 0.$$

The differential equation (8.4) cannot be solved in closed form despite its apparent simplicity. Instead, Eq. (8.4) must be solved either numerically or by series approximations as Blasius did.

It is not too difficult to integrate these equations numerically; an elementary numerical solution is sketched in Section 8.5. The results of a more accurate integration are tabulated in Table 8.1 and are shown graphically in Fig. 8.3.

The solution to the flat-plate boundary layer, Eq. (8.4), is called the Blasius solution. An important result of this solution is the surface shear stress,

$$\tau_0 \equiv \mu \left. \frac{\partial u}{\partial y} \right|_{y=0} = \mu U_0 \sqrt{\frac{U_0}{vx}} F''(0).$$

† In his original work, Blasius used $\eta = y\sqrt{U_0/4vx}$ and $\psi = \sqrt{vU_0 x}\,F(\eta)$, and the equation to be solved was $F''' + FF'' = 0$ with boundary conditions $F(0) = F'(0) = 0$, $F'(\infty) = 2$. We follow the usage of Reference 3.

TABLE 8.1   Numerical Values of the Solution of Eq. (8.4)[a]

$\eta = y\sqrt{\dfrac{U_0}{vx}}$	$\dfrac{u}{U_0} = F'(\eta)$	$F''(\eta)$
0	0	0.3321
1.0	0.3298	0.3230
2.0	0.6298	0.2668
3.0	0.8461	0.1614
4.0	0.9555	0.0642
5.0	0.9916	0.0059
6.0	0.9990	0.0024
7.0	0.9999	0.0002
8.0	1.0000	0.0001

[a] Adapted from L. Howarth, *Proc. Roy. Soc. London Ser. A*, Vol. **164**, 1938, p. 547.

In coefficient form this equation becomes

$$c_f \equiv \frac{\tau_0}{\rho U_0^2/2} = \frac{2}{\sqrt{U_0 x/v}}\, F''(0)$$

or simply (from Table 8.1)

$$c_f = \frac{0.664}{\sqrt{\mathrm{Re}_x}}. \tag{8.5}$$

The *drag* for one side of a plate of length L is

$$D = \int_0^L \tau_0\, dx$$

and in coefficient form the drag coefficient is

$$C_D = \frac{2D}{\rho U_0^2 L} = \frac{1.328}{\sqrt{\mathrm{Re}_L}}, \tag{8.6}$$

where

$$\mathrm{Re}_L = \frac{U_0 L}{v}.$$

Another interesting feature of this boundary layer flow is the velocity normal to the plate, v. From the development leading to Eq. (8.4) it is easy to see that

$$\frac{v}{U_0}\sqrt{\mathrm{Re}_x} = -\tfrac{1}{2}[F(n) - \eta F'(\eta)], \tag{8.7}$$

from which it may be verified that indeed at the wall, $y = 0$ and $v = 0$, as was required. The magnitude of v increases slowly to a maximum value of approximately

$$\left.\frac{v}{U_0}\sqrt{\mathrm{Re}_x}\right|_{y \to \infty} \approx 0.81.$$

From this result it can be seen that $v/U_0 \ll 1$ if the Reynolds number is large compared to unity, as was supposed in the development of the boundary layer equations. Of course, near the leading edge where $\text{Re}_x \ll 1$, the approximation is not valid.

8.4 Boundary Layer Parameters

An inspection of Table 8.1 or Fig. 8.3 reveals that the velocity ratio u/U_0 approaches unity gradually. As a practical matter one can say that the boundary layer "edge" has been reached at a value of η of about 5 or so, as the ratio u/U_0 differs by less than 1 percent from unity there. Nevertheless, the shear stress there is still about 5 percent of the wall value. The actual "edge" is, of course, infinitely far from the plate even though as a practical matter the boundary layer thickness, δ, may be approximated by

$$\frac{\delta}{x} \simeq \frac{5}{\sqrt{U_0 \, x/v}}.$$

Other boundary layer parameters have been developed that are free of this arbitrary specification, and these have important physical meanings. These are the *displacement thickness*, δ_d, and the *momentum thickness*, δ_m. The first, defined as

$$\delta_d = \frac{1}{U_0} \int_0^\infty (U_0 - u) \, dy, \tag{8.8}$$

gives the distance, δ_d, that the outer streamlines are shifted or displaced outward from the plate as a result of the retarded flow in the boundary layer. The second is defined as

$$\delta_m = \frac{1}{U_0^2} \int_0^\infty (U_0 - u)u \, dy. \tag{8.9}$$

It is evident from Eq. (8.9) that this length (multiplied by ρU_0^2) represents the reduced momentum flux in the boundary layer as a result of shear force on the surface of the plate.

For the Blasius solution these two lengths are

$$\delta_d = 1.72 \sqrt{\frac{vx}{U_0}} \tag{8.10}$$

and

$$\delta_m = 0.664 \sqrt{\frac{vx}{U_0}}. \tag{8.11}$$

PROBLEMS

8.1 A semi-infinite body of a viscous fluid is bounded by a flat porous plate. At time $t = 0$, the plate is suddenly set in motion at a velocity U_0 in the x-direction (in the plane of the plate). At the same time injection of fluid through the plate starts. The injection velocity is in the y-direction (normal to the plate). The flow is laminar at all times.

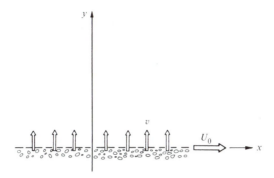

Problem 8.1

(a) What is the most general form of any solution for the velocity component v (in the y-direction)?

(b) Write the differential equation from which the velocity component u may be obtained.

(c) Suppose that the partial differential equation in (b) is to be converted into an ordinary differential equation for the single independent variable $z = y/\sqrt{2vt}$. How must the velocity component v be specified to make this possible?

8.2 Consider laminar, steady, constant-property flow over a semi-infinite flat plate. The free-stream velocity is constant, and a laminar boundary layer is formed on the plate as described in the preceding section. Now consider that the plate is slightly porous, and that a similar fluid is injected into the flow with a velocity at the plate of v_w, which should be considered a function of distance x along the plate.

(a) State the appropriate form of the boundary layer equations, and recast them as in the Blasius case in terms of a similarity variable η and a dimensionless stream function $F(\eta)$.

(b) Specify the boundary conditions at the wall and at the edge of the boundary layer.

(c) How must the injection velocity v_w vary with x to make $F(0) =$ constant? Why would this be necessary?

8.3 Consider laminar, steady flow of an incompressible fluid past an infinite flat plate. Now, however, fluid is withdrawn by a steady constant suction through the plate, which is slightly porous. In this case the boundary layer does not grow with distance along the plate but remains constant, so that $\partial u/\partial x = 0$.

(a) Starting with the boundary layer equations, show that

$$u(y) = u_0\left[1 - \exp\left(\frac{v_0\, y}{v}\right)\right],$$

where v_0 is the constant suction velocity at the plate.

(b) Find the displacement and momentum thickness.

(c) Find the drag on one side of the plate (the plate length is L).

8.4 Near the stagnation point on a cylinder in a uniform flow U_0, the outer free-stream velocity can be approximately represented by the potential flow solution

$$U_1 = U_0 \sin \theta \simeq \frac{2U_0 x}{R}.$$

Thus, for laminar flow the approximate boundary layer equations near the stagnation point may be written as

$$u \frac{\partial u}{\partial x} + v \frac{\partial u}{\partial y} = 4 \frac{U_0^2}{R} x + v \frac{\partial^2 u}{\partial y^2}.$$

Find a transformation of the form

$$\eta = \frac{Ay}{x^n}, \qquad \psi = Bx^m F(\eta)$$

that reduces the boundary layer equations to

$$\frac{d^3 F}{d\eta^3} + F \frac{d^2 F}{\partial \eta^2} - \left(\frac{dF}{d\eta}\right)^2 + 1 = 0.$$

8.5 A narrow two-dimensional jet issues into an infinite region of an undisturbed fluid of the same kind as the jet, as shown. The flow is laminar everywhere. Some of the undisturbed fluid will now be set in motion because of the shear forces produced by the jet. The region that is significantly affected by the jet is thin compared to the distance along the jet, so that the streamlines deviate very little from the axial direction of the jet. The equation of motion for the x-direction, therefore, takes the same form as that for a boundary layer on a flat plate. The boundary conditions, however, now are

$$\text{at } y = 0, \qquad \frac{\partial u}{\partial y} = 0$$

$$\text{at } y \to \infty, \qquad u \to 0.$$

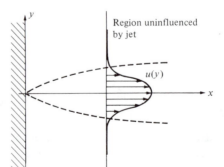

Problem 8.5

It is desired to convert the partial differential equation into an ordinary one by introducing the similarity variable

$$\eta = \frac{y}{x^m}$$

and letting $\psi = x^n f(\eta)$.

(a) Set up an integral for the total x-momentum flux across a vertical plane, and let η be the integration variable. The momentum flux should be the same through all vertical cross sections, as there are no lateral forces acting on the fluid. From this fact, determine the relation between m and n.

(b) Rewrite the differential equation in terms of f and its derivatives and determine the values of m and/or n, so that the resulting differential equation is an ordinary one. Then state the resulting equation.

†8.5 Integration of the Blasius Equation

In this section we are concerned with the numerical solution of the Blasius equation [Eq. (8.4)] with its boundary conditions, which in terms of the function F is written as

$$F(0) = F'(0) = 0,$$

$$F(\infty) = 1.$$

The process of numerical integration is made somewhat easier conceptually by replacing Eq. (8.4) with three first-order equations. Here, as before, $F(\eta)$ is the dimensionless stream function. Now let

$$G(\eta) = F'(\eta) \tag{8.12a}$$

and

$$H(\eta) = G'(\eta). \tag{8.12b}$$

Then the original Blasius equation becomes

$$2H'(\eta) + F(\eta)H(\eta) = 0. \tag{8.12c}$$

Equations (8.12) now replace the original Blasius equation; the conditions on the new functions G and H are

$$G(0) = 0, \tag{8.13a}$$

$$G(\infty) = 1, \tag{8.13b}$$

and, as before,

$$F(0) = 0. \tag{8.13c}$$

Integration can now proceed in a symbolic way, for example,

$$G(\eta) = G(0) + \int_0^\eta H(\eta)\, dy$$

$$F(\eta) = F(0) + \int_0^\eta G(\eta)\, dy$$

from Eqs. (8.12a) and (8.12b). Again in a symbolic way one could write the solution of Eq. (8.12c) as

$$H(\eta) = H_0 \exp\left[-\tfrac{1}{2} \int_0^\eta F(\eta) \right] d\eta,$$

† This section may be passed over without loss of continuity.

but it is apparent that in all these expressions intermediate values of F, H, and G are needed, but these are not available beforehand. Similarly, the term $H(0) \equiv F''(0)$ is not known in advance and must be found as a part of the solution. As in all numerical integrations, the range of integration is divided into discrete steps and the variables F, G, and H are evaluated as Example 6.6 at discrete locations η_k, $k = 0, 1, 2, \ldots, N$, where N is the maximum index and the variables there are denoted by F_k, F_{k+1}, G_k, G_{k+1}, and so on. The spacing between steps will here be assumed to be a constant value, l. Then the discrete analogs to the symbolic equations above are approximately

$$G_{k+1} = G_k + \frac{l(H_{k+1} + H_k)}{2},$$

$$F_{k+1} = F_k + \frac{l(G_k + G_{k+1})}{2},$$

in which use has been made of the *trapezoid* approximation for the area under a curve. Also, from Eq. 8.12c,

$$H_{k+1} = H_k - \frac{l(F_k H_k + F_{k+1} H_{k+1})}{4}$$

with the same approximation. Here again, not all the values at the $(k + 1)$ position are available to enable the solution to be stepped from k to $k + 1$. These values must be estimated. Also, $H(0)$ must be guessed and then subsequently improved to achieve the requirement of Eq. (8.13b). One possible scheme (of many) that is readily implemented is to let

$$G_{k+1}^* = G_k + lH_k;$$

then

$$F_{k+1} = F_k + \frac{l(G_k + G_{k+1}^*)}{2}, \tag{8.14a}$$

$$H_{k+1}^* = H_k - \frac{lF_k H_k}{2},$$

and

$$H_{k+1} = H_k - \frac{l(F_k H_k + F_{k+1} H_{k+1}^*)}{4}. \tag{8.14b}$$

Here the starred quantities are approximate or estimated values of the variables at the next spatial position, which can then be used to obtain "corrected" values in Eqs. (8.14a) and (8.14b). Then finally,

$$G_{k+1} = G_k + \frac{l(H_k + H_{k+1})}{2}, \tag{8.14c}$$

to complete one cycle of computation. This cycle is repeated until the maximum value N (or $\eta \doteq \infty$) is reached. From Table 8.1 this (infinitely far away) location can be seen to be about $\eta = 7$ or 8. It is at this location that $H(0)$ is to be determined so that Eq. (8.13b) is satisfied. Reference 4 describes a simple easy-to-program scheme based on interval halving that systematically converges on the correct value of $H(0)$.

The foregoing system of equations are readily programmed on a hand calculator or in any of the current languages on a microcomputer and requires, in fact, less space to do so than the discussion above. Four-figure accuracy is easily achievable with small step sizes ($l \simeq 0.02$), and relatively few iterations are required to achieve this accuracy. These equations described provide one of the simplest examples of the *Runge–Kutta* methods for integration of differential equations. The present method is accurate through l^2; Reference 4 describes more accurate schemes. These are applied in Reference 5 to the numerical treatment of the Blasius and other boundary layer equations.

EXAMPLE 8.1

In this example an even simpler method is used to illustrate the integration of the Blasius equation. The equations are, as before,

$$F = F,$$

$$G = F',$$

$$H = G',$$

and

$$2H' + FH = 0.$$

TABLE 8.2(a)

k	η	F	G	H	$c_f\sqrt{Re}$
0	0	0	0	0.325	0.65
1	0.02	0	0.0065	0.325	0.65
2	0.04	0.00013	0.013	0.325	0.65
3	0.06	0.0039	0.0195	0.324999	0.649999
4	0.08	0.00078	0.025999	0.324998	0.649996
—	—	—	—	—	—
397	7.94	6.145571	0.992708	1.42^{-5}	2.85^{-5}
398	7.96	6.165425	0.992708	1.34^{-5}	2.67^{-5}
399	7.98	6.185279	1.992708	1.25^{-5}	2.51^{-5}
400	8	6.205134	0.992709	1.18^{-5}	2.35^{-5}

TABLE 8.2(b)

$c_f(0)\sqrt{Re}$	η	F	G	H	$c_f\sqrt{Re}$
0.655	8	6.241437	0.997811	1.11^{-5}	2.22^{-5}
0.660	8	6.277661	1.002901	1.05^{-5}	2.09^{-5}
0.665	8	6.313805	1.007978	9.87^{-6}	1.97^{-5}
0.670	8	6.349871	1.013042	0.31^{-6}	1.86^{-5}

The *Euler* approximation to an integral is made in which the integrand is evaluated at the lower limit of the integral. Let the step size be l; then the equations are

$$F_{k+1} = F_k + lG_k,$$

$$G_{k+1} = G_k + lH_k,$$

and

$$H_{k+1} = H_k - \frac{F_k H_k l}{2},$$

where the index $k = 0, 1, 2$ to the maximum value N_{max}. The equations are to be integrated from $\eta = 0$ (where $k = 0$) to a value η_{max}. In the present case η_{max} is taken as 8, and the step size $l = 0.02$. Both F and G are zero at the origin. The value of $F''(0) \equiv H(0)$ is unknown, so that guesses of $F''(0)$ are successively tried in order to calculate the value of $u(\eta = 8)/U_0 \equiv G(\eta = 8)$, which should of course just equal unity.

Table 8.2(a) lists a portion of the calculated results for a trial value of $2F''(0) = c_f \sqrt{Re_x} = 0.65$. Such a table is easily prepared using a spreadsheet program on a personal computer and can be repeated for different initial guesses of $F''(0)$; the results of such guesses are shown in Table 8.2(b). From these results the shear stress coefficient may be seen to be $c_f \sqrt{Re_x} \doteq 0.66$, which is sufficiently accurate for most engineering purposes. The last column is the dimensionless shear stress at $\eta = 8$ and it is seen to be very small. Further calculations can be tried with different values of l and η_{max}, which will affect these results slightly.

8.6 The Boundary Layer Momentum Integral

We shall next obtain an integral relation between the viscous shear stress on a plate, the pressure gradient, and the momentum flux. This relationship may subsequently be used to formulate approximate solutions of the boundary layer equations which are consistent with the momentum principle of Chapter 4. In formulating this relationship, it is convenient to visualize the boundary layer as ending at some definite distance, δ, which changes with x [i.e., $\delta(x)$]. This is in agreement with the finding in Section 8.4. We now apply the momentum theorem to the control volume of Fig. 8.4. Outside the boundary layer the velocity is assumed to be that of the undisturbed flow U_0; we will derive the momentum integral for a general case where the pressure is not constant, and therefore the external velocity U_0 also depends on x [i.e., $U_0(x)$]. The forces acting on the control volume are due to normal pressure and the viscous shear stress along the plate. The normal viscous stresses are neglected. The present discussion will also be applicable to laminar as well as turbulent flows. In the latter case all velocities should be considered as time-averaged values, but for convenience we omit the bar over the symbols. We also neglect any normal Reynolds stresses, in keeping with the boundary layer approximation.

The mass flow through the left-hand vertical boundary is

$$\dot{m} = \int_0^\delta \rho u \, dy,$$

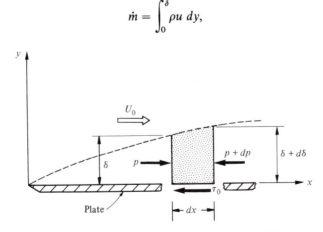

FIG. 8.4 Control volume for momentum integral on a plate with a pressure gradient.

and through the right-hand boundary is

$$\dot{m} + \frac{\partial \dot{m}}{\partial x} dx = \int_0^\delta \rho u \, dy + \frac{\partial}{\partial x} \left(\int_0^\delta \rho u \, dy \right) dx.$$

The difference is simply

$$\frac{\partial \dot{m}}{\partial x} dx = \frac{\partial}{\partial x} \left(\int_0^\delta \rho u \, dy \right) dx,$$

which by the continuity equation must be equal to the mass flow through the top boundary of the control volume, a fact we shall presently use. The x-momentum flow through the left-hand boundary is

$$\dot{M}_x = \int_0^\delta \rho u^2 \, dy,$$

and through the right-hand side it is

$$\dot{M}_x + \frac{\partial \dot{M}_x}{\partial x} dx = \int_0^\delta \rho u^2 \, dy + \frac{\partial}{\partial x} \left(\int_0^\delta \rho u^2 \, dy \right) dx,$$

the difference being

$$\frac{\partial \dot{M}_x}{\partial x} dx = \frac{\partial}{\partial x} \left(\int_0^\delta \rho u^2 \, dy \right) dx.$$

The x-momentum inflow over the top boundary is

$$(\dot{M}_x)_{\text{top}} = U_0 \frac{\partial}{\partial x} \left(\int_0^\delta \rho u \, dy \right) dx,$$

because this fluid has the external velocity U_0. The total flux of momentum is thus

$$\frac{\partial}{\partial x} \left(\int_0^\delta \rho u^2 \, dy \right) dx - U_0 \left(\frac{\partial}{\partial x} \int_0^\delta \rho u \, dy \right) dx,$$

while the force on the control volume is

$$-\tau_0 \, dx + p(\delta + d\delta) - (\delta + d\delta)\left(p + \frac{dp}{dx} dx \right),$$

and by the momentum principle these must be equal. Neglecting second-order quantities, we then obtain the equation

$$\frac{\partial}{\partial x} \int_0^\delta \rho u^2 \, dy - U_0 \frac{\partial}{\partial x} \int_0^\delta \rho u \, dy = -\tau_0 - \delta \frac{dp}{dx},$$

which can be brought into a simpler form with the use of Bernoulli equation for the external flow,

$$p + \frac{\rho}{2} U_0^2 = \text{const.}$$

Then, after some manipulation, the boundary layer momentum integral in its final form is

$$\frac{\partial}{\partial x} \int_0^\delta \rho(U_0 - u)u \, dy - \frac{dU_0}{dx} \int_0^\delta \rho(U_0 - u) \, dy = \tau_0. \tag{8.15}$$

This famous equation is due to von Kármán† and it is the starting point for many approximate methods of solving the boundary layer equations. Equation (8.15) may readily be put in terms of the displacement and momentum thickness parameters, that is,

$$\frac{\partial}{\partial x}\left(\rho U_0^2 \delta_m\right) - \rho U_0 \delta_d \frac{dU_0}{dx} = \tau_0. \tag{8.16}$$

The case of a flat-plate flow, $U_0 = \text{const}$, is particularly easy to interpret, for then Eq. (8.16) simply states that the rate of increase of the momentum thickness is equal to the wall shear stress (divided by ρU_0^2). In fact, for the flat plate, Eq. (8.16) shows that the shear stress coefficient is simply

$$c_f = \frac{\tau_0}{\rho U_0^2/2} = 2\,\frac{d\delta_m}{dx}, \tag{8.17}$$

so that if there were a known relation between δ_m and c_f, Eq. (8.17) could be used to determine both quantities. This approach is, in fact, one of the methods for analysis of turbulent boundary layers.

8.7 The Laminar Boundary Layer: Approximate Solutions

The momentum integral, Eq. (8.15), is the basis of many different approximate methods of solution of the laminar boundary layer equation, which we now derive for the flat plate, $U_0 = \text{const}$. In this method it is assumed that

$$\frac{u}{U_0} = f\!\left(\frac{y}{\delta}\right);$$

that is, the velocity profiles, u/U_0, along the plate are similar to each other. Equation (8.15) can then be simplified into the form

$$\frac{d}{dx}\left[\rho U_0^2 \delta \int_0^1 \left(1 - \frac{u}{U_0}\right)\frac{u}{U_0}\,d\!\left(\frac{y}{\delta}\right)\right] = \tau_0. \tag{8.18}$$

Multiplying and dividing by δ inside the brackets and taking δ outside the integral is permissible because δ is a function of x only and is not a variable in the integration with respect to y. The limit of integration is changed to account for the change in variable (from y to y/δ). The integral in Eq. (8.18) now becomes a constant, say a.

$$\int_0^1 \left(1 - \frac{u}{U_0}\right)\frac{u}{U_0}\,d\!\left(\frac{y}{\delta}\right) \equiv a.$$

Then Eq. (8.18) becomes simply

$$\rho U_0^2 a\,\frac{d\delta}{dx} = \tau_0.$$

† See Schlichting (Reference 3) or the *Collected Works of Theodore von Kármán* (London: Butterworth, 1956).

The shear stress τ_0 can also be expressed in terms of the velocity profile

$$\tau_0 = \mu \left. \frac{du}{dy} \right|_{y=0} = \mu \frac{U_0}{\delta} \left(\frac{d(u/U_0)}{d(y/\delta)} \right)_{y=0}.$$

For simplicity, let

$$\beta = \left(\frac{d(u/U_0)}{d(y/\delta)} \right)_{y=0};$$

then Eq. (8.18) becomes

$$\rho U_0^2 a \frac{d\delta}{dx} = \mu \frac{U_0}{\delta} \beta, \tag{8.19}$$

which is a simple differential equation for δ. At the leading edge of the plate ($x = 0$), we assume that $\delta = 0$. Then we obtain

$$\delta = \sqrt{\frac{2\mu x \beta}{\rho U_0 a}}$$

or

$$\frac{\delta}{x} = \sqrt{\frac{2\beta}{a}} \frac{1}{\sqrt{\mathrm{Re}_x}},$$

where Re_x is the Reynolds number based on distance from the leading edge. The shear stress coefficient is

$$c_f = \frac{\sqrt{2\beta a}}{\sqrt{\mathrm{Re}_x}}, \tag{8.20a}$$

and the drag coefficient for one side of plate of length L is

$$C_D = \frac{2\sqrt{2\beta a}}{\sqrt{\mathrm{Re}_L}}. \tag{8.20b}$$

Suitable velocity profiles may now be selected and the values of a and β then determined. The results from shear stress and boundary layer thickness are fairly insensitive to the exact functional relationship of u/U_0 on y/δ. A good approximation can be obtained by *assuming* a "reasonable" shape for this relationship. By "reasonable" is meant that $u(y)$ should satisfy the boundary conditions $u(0) = 0$, $u/U_0 = 1$ at $y = \delta$. Furthermore, u/U_0 should blend smoothly into the constant value at the edge of the boundary layer. Simple functions such as

$$\frac{u}{U_0} = \sin \left(\frac{\pi}{2} \frac{y}{\delta} \right)$$

or

$$\frac{u}{U_0} = 2 \left(\frac{y}{\delta} \right) - \left(\frac{y}{\delta} \right)^2$$

satisfy these conditions. More complicated polynomial or transcendental functions have been used to satisfy more conditions of the original boundary layer equations

(8.2), and these approach more closely the exact Blasius solution. Fortunately, the selection of the profile is not crucial, as the following example will illustrate.

EXAMPLE 8.2

The simplest possible approximation for the velocity profile is the straight line $u/U_0 = y/\delta$. Then $\beta = d(u/U_0)/d(y/\delta) = 1$ and

$$a = \int_0^1 \left(1 - \frac{u}{U_0}\right) \frac{u}{U_0}\, d\left(\frac{y}{\delta}\right) = \int_0^1 \left(\frac{y}{\delta}\right)\left(1 - \frac{y}{\delta}\right)d\left(\frac{y}{\delta}\right) = \frac{1}{2} - \frac{1}{3} = \frac{1}{6}.$$

Then

$$\frac{\delta}{x} = \frac{\sqrt{12}}{\sqrt{\mathrm{Re}_x}} = \frac{3.46}{\sqrt{\mathrm{Re}_x}}$$

and

$$c_f = \frac{\sqrt{2/6}}{\sqrt{\mathrm{Re}_x}} = \frac{0.577}{\sqrt{\mathrm{Re}_x}},$$

which is surprisingly close to the exact value, 0.6641.

†8.8 Alternative Derivation of the Boundary Layer Integral

The boundary layer equations themselves [Eqs. (8.1)] contain the basic mechanics of the flow, including the conservation of momentum. The boundary layer integral can be obtained directly from these equations by integration from the wall to the edge of the boundary layer, $y = \delta$. This process is facilitated by multiplying Eq. (8.1c) by u and adding this to Eq. (8.1b) to obtain the result,

$$\frac{\partial}{\partial x}(u^2) + \frac{\partial}{\partial y}(uv) = -\frac{1}{\rho}\frac{dp}{dx} + v\frac{\partial^2 u}{\partial y^2}. \tag{8.21}$$

Now multiply each term by dy and integrate from $y = 0$ to $y = \delta$ with the conditions

$$u(x, 0) = 0,$$
$$v(x, 0) = 0,$$
$$u(x, \delta) = U_0,$$

to get

$$\int_0^\delta \frac{\partial u^2}{\partial x}\, dy + U_0 v(x, \delta) = -\frac{\delta}{\rho}\frac{dp}{dx} - v\frac{\partial u}{\partial y}(x, 0). \tag{8.22}$$

The continuity equation [Eq. (8.1c)] may be similarly integrated to get

$$v(x, \delta) = v(x, 0) - \int_0^\delta \frac{\partial u}{\partial x}\, dy.$$

---

† This section may be omitted without loss of continuity.

The x derivatives inside the integrals may be replaced by ones outside the integral by the relationships

$$\frac{\partial}{\partial x} \int_0^\delta u^2 \, dy = \int_0^\delta \frac{\partial u^2}{\partial x} \, dy + U_0^2 \frac{d\delta}{dx}$$

and

$$\frac{\partial}{\partial x} \int_0^\delta u \, dy = \int_0^\delta \frac{\partial u}{\partial x} \, dy + U_0 \frac{d\delta}{dx},$$

obtained with Leibnitz's rule for differentiating integrals. Assembling these results together and substituting into Eq. (8.22) term by term, we find

$$\frac{\partial}{\partial x} \int_0^\delta u^2 \, dy - U_0^2 \frac{d\delta}{dx} + U_0 \left[v(x, 0) - \frac{\partial}{\partial x} \int_0^\delta u \, dy + U_0 \frac{d\delta}{dx} \right] = -\frac{\delta}{\rho} \frac{dp}{dx} - v \frac{\partial u}{\partial y} (x, 0)$$

But $v(x, 0) = 0$ by assumption and

$$\tau_0 = \mu \frac{\partial u}{\partial y} (x, 0),$$

so that we get, after cancellation of terms,

$$\frac{\partial}{\partial x} \int_0^\delta u^2 \, dy - U_0 \frac{\partial}{\partial x} \int_0^\delta u \, dy = -\frac{\delta}{\rho} \frac{dp}{dx} - \frac{\tau_0}{\rho}, \tag{8.23}$$

which is precisely the same relation (except for division by ρ) derived from the momentum equation leading to Eq. (8.15).

PROBLEMS

8.6   Find the shear, the thickness of the boundary layer δ, and the total drag on one side of a flat plate of length l, assuming the boundary layer to be laminar over the whole length. Compare the results obtained by assuming (a) a sinusoidal, (b) a parabolic, and (c) a straight-line profile. Using the results of part (a), what is the numerical value of the drag force on a plate 0.3 m wide and 0.3 m long in water at 21°C if $U_0 = 0.7$ m/s? Also find the boundary layer thickness at the end of the plate.

8.7   **(a)** Show that the cubic approximation for the velocity profile

$$f(\eta) = \tfrac{3}{2}\eta - \tfrac{1}{2}\eta^3$$

results in the expression for boundary layer thickness,

$$\frac{\delta}{x} = 4.64(\mathrm{Re}_x)^{1/2}.$$

(b) Calculate the shear stress coefficient for this assumed profile.

8.8   A laminar boundary layer forms on a *porous* flat surface that removes fluid from the main flow at a constant velocity v_0, as shown. Using the approximate integral method and assuming that $u/U_0 = f(y/\delta)$ only, show that

$$c_f = \frac{2\tau_w}{\rho U_0^2} = \frac{2d\delta}{dx} a + \frac{2v_0}{U_0},$$

where

$$a = \int_0^1 (f - f^2) \, d\left(\frac{y}{\delta}\right).$$

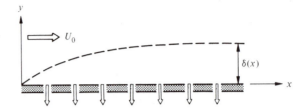

Problem 8.8

8.9 Vapor condenses on a vertical surface to form a liquid film. The film moves under gravity and forms a laminar liquid boundary layer.

   (a) Derive an expression for the mass flow rate, $\dot{m}$ of the film, as a function of the local film thickness, δ. Neglect any velocity components in the y-direction.

   (b) The local rate of condensation is related to the parameters of the problem by the relation

$$\frac{k\Delta T}{\delta} = \lambda \frac{d\dot{m}}{dx},$$

where k is the conductivity, ΔT the constant temperature difference across the film, λ the heat of evaporation, and δ the local film thickness. Determine δ as a function of x. (Neglect the viscosity of the vapor.)

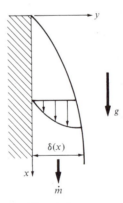

Problem 8.9

8.10 The slowing down of a stream of liquid of oncoming speed U_0 is to be investigated. The liquid density is ρ, the viscosity is μ, and the initial depth is d, as shown in the sketch. The pressure in the liquid is to be assumed constant along the plate. The boundary layer grows and eventually reaches the surface at a length x_0. Note that the boundary layer thickness there is *not* the same as the initial depth d. (Why?) Using the cubic velocity profile described in Problem

8.7, calculate the ratios δ_0/d and x_0/d in terms of the Reynolds number based on initial depth and velocity.

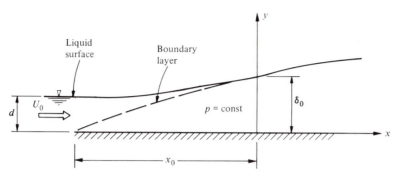

Problem 8.10

8.11 Consider again the flow described in Problem 8.10. When $x > x_0$, the velocity at the top surface is $U_t(x) < U_0$ because of friction on the plate.

    (a) Show that the momentum integral [Eq. (8.23)] in this instance becomes

$$\frac{d}{dx}\int_0^\delta u^2\,dy - U_t(x)\frac{d}{dx}\int_0^\delta u\,dy = -v\left.\frac{\partial u}{\partial y}\right|_0.$$

    (b) Show that a simple differential equation for the velocity at the top surface $U_t(x)$ can be obtained if the similar profile assumption is made *and* conservation of mass is imposed.

8.9 The Turbulent Boundary Layer: Approximate Solution

It may be recalled from Fig. 8.1 and the discussion in Section 8.1 that the flow in a turbulent boundary layer is much more complicated than that in the laminar case. The turbulent boundary layer mixes with a surrounding nonturbulent free-stream flow at the edge of the boundary layer. This "interface" between the nonturbulent and the turbulent flow is an unsteady, corrugated surface, as shown in Fig. 8.1. The flow there is characterized by being intermittently turbulent. This intermittency dies away in the nonturbulent flow of the surrounding free stream with increasing distance from the edge of the boundary layer. The turbulent boundary layer can be described in terms similar to that of the turbulent pipe flow of Chapter 7; there is an inner region, an intermediate turbulent portion (satisfying the law of the wall), and an outer region. The inner region holds right down to the wall, and part of this inner region has the logarithmic form of the law of the wall in Chapter 7. The outer region contains a new formulation of the velocity profile called the "law of the wake" (see Reference 9), which accounts for the velocity profile near the edge of the boundary layer. Both the logarithmic part of the profile and the modified outer parts of the velocity profile depend on the local shear stress coefficient, so that the velocity profiles do not retain a similar form as the flow proceeds down the plate. This considerably complicates the computation of drag. In any event the structure of the turbulent boundary layer is only known empirically.

    Despite the complications described above, useful results for turbulent boundary layer flow can still be obtained in cases such as that of a flow over a flat plate with no imposed pressure gradient. The basic momentum relation for flow past a flat plate [e.g., Eq. (8.17)] is still applicable for a turbulent flow, and in principle one can proceed in the same manner as for a laminar boundary layer. In that case, it may be recalled that the shear was evaluated from the slope of the assumed velocity profile at the wall. For turbulent flow it is still true that the shear right at the wall is proportional to the velocity gradient at that point. But as the discussion above indicates, the turbulent boundary layer is much more complicated, having transitions from the viscous-controlled region to the turbulent core with rapid changes in velocity profile. As a consequence, we can no longer expect that the slope of an assumed velocity profile will be adequate to give an acceptable value for the shear stress. We therefore have to take a different approach and refer directly to empirical results for turbulent shear stress near a wall. One of the simplest of these empirical derivations follows from the power law approximation for the velocity profile in turbulent pipe flow [Eq. (7.11)]; that is,

$$\frac{u}{u_\tau} = 8.7\left(\frac{yu_\tau}{\nu}\right)^{1/7}. \tag{7.15}$$

It then follows that

$$u_\tau = \frac{u^{7/8}(\nu/y)^{1/8}}{8.7^{7/8}} = 0.1506u\left(\frac{\nu}{uy}\right)^{1/8},$$

and since $u_\tau = \sqrt{\tau_0/\rho}$, we have $u = U_0$ at $y = \delta$, or

$$\tau_0 = 0.0227\rho U_0^2\left(\frac{\nu}{U_0\delta}\right)^{1/4}. \tag{8.24}$$

This expression for τ_0 is then used in Eq. (8.18), and upon integration we obtain

$$\delta^{5/4} - \delta_0^{5/4} = 0.0284\left(\frac{\xi}{a}\right)\left(\frac{\nu}{U_0}\right)^{1/4}, \tag{8.25}$$

where δ_0 is the boundary layer thickness at the point of transition to turbulent flow, and ξ is the distance along the plate measured from that point. Frequently, we are interested in the properties of the boundary layer relatively far downstream of the plate edge, in which case δ_0 may be neglected and ξ becomes equal to x, the distance from the edge of the plate. With this understanding the expressions for the boundary layer thickness and the shear on the plate become

$$\delta = 0.058\left(\frac{\nu}{U_0}\right)^{1/5}\left(\frac{x}{a}\right)^{4/5}, \tag{8.26}$$

$$\tau_0 = 0.0463\left(\frac{a}{\mathrm{Re}_x}\right)^{1/5}\rho U_0^2, \tag{8.27}$$

and the drag coefficient for one side is

$$C_D = 0.116\left(\frac{a}{\mathrm{Re}_L}\right)^{1/5}. \tag{8.28}$$

(The omission of δ_0 is particularly justified if we induce turbulence close to the leading edge of the plate by a trip wire or other transition devices.) By comparison of the expressions for δ in laminar and turbulent flow, it is seen that the turbulent layer grows much faster. The former grows at the rate of $x^{1/2}$, whereas the turbulent boundary layer increases as $x^{4/5}$. Equations (8.26 to 8.28) are limited to a Reynolds number based on an L of about 10^7 as a result of the inherent limitation in the empirical shear stress formula. This equation and the foregoing results are also limited to smooth plates.

A comment may now be added on the computation of a, where a was defined as

$$a = \int_0^1 \left(1 - \frac{u}{U_0}\right) \frac{u}{U_0} \, d\left(\frac{y}{\delta}\right).$$

As stated previously, acceptable results for the boundary layer characteristics will be obtained as long as a reasonable function

$$\frac{u}{U_0} = \frac{u}{U_0}\left(\frac{y}{\delta}\right)$$

is selected, but the results will, of course, be the better, the closer the selected function approaches the actual one. Again, from the empirical results on the velocity profiles in turbulent pipe flows, it was seen [see Eq. (7.15)] that this profile was adequately approximated by the equation

$$\frac{u}{u_\tau} = 8.7\left(\frac{y u_\tau}{v}\right)^{1/7} \tag{7.15}$$

in the fully turbulent portion of the flow. Evaluating this expression at $y = \delta$ where $u = U_0$ and taking the ratios of these velocities, one obtains

$$\frac{u}{U_0} = \left(\frac{y}{\delta}\right)^{1/7}.$$

If this velocity profile is used in computing a instead of the ones used for laminar flow, reasonable results may be expected. One may now recall that the logarithmic profile

$$\frac{u}{u_\tau} = 5.75 \log_{10}\left(\frac{y u_\tau}{v}\right) + 5.5$$

gave an even better fit to the experimental results than did Eq. (7.15). This logarithmic equation, however, cannot be brought into the form

$$\frac{u}{U_0} = \frac{u}{U_0}\left(\frac{y}{\delta}\right).$$

This observation emphasizes the fact that even the general assumption that the velocity profiles are similar and of the form

$$\frac{u}{U_0} = \frac{u}{U_0}\left(\frac{y}{\delta}\right)$$

is not strictly fulfilled and that it involves a certain approximation to the experimentally determined facts. Nevertheless, the similarity assumption is certainly sufficiently accurate for use in the approximate boundary layer calculations presented in this chapter.

The results of the turbulent boundary layer theory have been compared extensively with measurements and have been found to match experimental data within a few percent if δ is computed on the basis of Eq. (7.15) and if the imposed limitations on Reynolds number are observed. On the basis of direct measurements, other empirical formulas have been developed. One of these, proposed by Schlichting (Reference 3), gives the drag coefficient for one side of a smooth flat plate that is turbulent from the leading edge as

$$C_D = \frac{0.455}{[\log_{10} (U_0 l/v)]^{2.58}} . \tag{8.29}$$

This expression gives accurate results up to a Reynolds number of 10^9.

More recently, Granville (Reference 6) has proposed a drag formula which takes the "law of the wake" function into account that is valid for Reynolds numbers from about 3×10^6 to 10^9; namely,

$$C_D = \frac{0.0776}{[\log_{10} (U_0 L/v) - 1.88]^2} + \frac{60}{(U_0 L/v)} . \tag{8.30}$$

This equation gives a result somewhat higher than Eq. (8.29) at low Reynolds numbers and is below it at the higher Reynolds numbers.

One might then be tempted to ask why in the presence of such experimental information the theory above is needed at all. In reply, it may be stated that the methods discussed in the foregoing lend themselves to being extended to take into account pressure gradients, unsteady flow conditions, as well as compressibility effects. It then becomes very useful in predicting the drag over more complicated shapes than flat plates and for more difficult flow conditions for which experimental data are lacking.

A particularly important extension of the ideas above includes the effect of surface roughness on the turbulent boundary layer. The same sort of reasoning used in Section 7.9 for the effect of roughness in pipe flows can be used to estimate the local skin friction and overall drag coefficient on a flat plate. The procedure is more elaborate and is not repeated here; a useful result (given in Reference 5) for the drag coefficient on one side of a fully rough plate is

$$C_D = 0.024 \left(\frac{\varepsilon}{L} \right)^{0.167} .$$

The theory as presented here was originally developed by von Kármán in 1922 (Reference 3). The line of reasoning and the type of assumptions made were representative of engineering work at that time.

PROBLEMS

8.12 The velocity profile in a turbulent boundary layer on a flat plate ($U_\infty =$ constant) is approximated by the form

$$\frac{\bar{u}}{U_\infty} = \left(\frac{y}{\delta} \right)^{1/7} .$$

(a) Find the profile parameter a for this profile.
(b) If the wall shear stress τ_0 for this *turbulent* boundary layer is assumed to be given by the empirical formula in Eq. (8.24), solve the resulting momentum integral to find the thickness $\delta(x)$ of the turbulent boundary layer as a function of the coordinate, x, measured along the plate. Assume that the boundary layer first becomes turbulent at a position $x = x_0$, and that the thickness at this point is δ_0. The answer is in terms of x_0, δ_0, v, and U_∞ as well as x.

8.13 Use the seventh-root profile and compute the drag force and boundary layer thickness on a plate 6 m long and 3 m wide (for one side) if it is immersed in a flow of water of 6 m/s velocity. Assume turbulent flow to exist over the entire length of the plate. What would the drag be if laminar flow could be maintained over the entire surface?

8.14 A flat duct, as shown in the figure, is used to provide coolant air for the brakes of an automobile. The breadth of the duct is much larger than its height, so that the flow over the top and bottom *interior* surfaces can be considered as two-dimensional. The entrance to the duct is such that a *turbulent* boundary layer is assumed to grow right from the intake ($x = 0$). The height of the duct is constant at 1 in., and the *lateral* walls are slightly divergent, so that the free-stream flow velocity U_0 remains constant and no pressure drop occurs along the inlet. For a free-stream velocity U_0 of 100 ft/sec, compute the distance from the inlet x at which the upper and lower boundary layers meet.

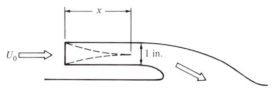

Problem 8.14

8.15 (a) A typical 12-m yacht has a length along the waterline of approximately 55 ft. Assuming a speed of 12 ft/sec, and that the short laminar portion of the boundary layer is negligible, find the total skin drag on a 1-ft strip at the waterline.

(b) What would the maximum admissible roughness of the surface be at mid-length to ensure a hydraulically smooth condition?

8.16 Consider a turbulent boundary layer on a flat plate (constant and uniform velocity and pressure in the flow outside the boundary layer). The plate is very rough, the size of the roughnesses, e, being very much greater than the laminar sublayer thickness that would occur in the absence of the roughness. It is anticipated that the velocity distribution within the turbulent part of the boundary layer can be approximated by

$$u^* = K\left(\frac{y}{e}\right)^{1/7},$$

where K is a constant, y the distance from the wall, $\bar{u}$ the mean velocity, and $u^* = \bar{u}/u_\tau$, where the friction velocity, $u_\tau = (\tau_0/\rho)^{1/2}$, τ_0 being the wall shear stress and ρ the fluid density. Using approximate boundary layer theory, find an expression for the boundary layer thickness, δ, as a function of x, the distance along the plate from the leading edge. Assume initial conditions $\delta = 0$ at $x = 0$; the result contains e, K, and the profile parameter a (equal to 0.0972).

8.10 Drag of Bodies

A body immersed in a flow experiences a force called the *drag* force in the direction of the flow. The drag force is of enormous practical importance in propulsion and in analysis of objects moving within a fluid, and it is important to understand the origin of this force.

We may recall from Section 2.6 that the force acting on an element of a surface is composed of a normal and a tangential stress vector. The drag force is then the sum of the components of these forces in the main flow direction. The tangential force is due entirely to the viscous shear stress and it is termed the "skin friction" drag. A *slender* body such as a two-dimensional wing section or aircraft fuselage has drag due primarily to skin friction. The normal stress is composed of the pressure (inwardly directed onto the surface) as well as a viscous normal shear stress. For Reynolds numbers (based on typical body dimensions) larger than about 100, the normal viscous stress may be neglected compared to the pressure. The component of drag due to the distribution of pressure is usually called "form" drag. Drag of *bluff* bodies such as a disk perpendicular to the flow or sphere at a high Reynolds number is primarily form drag due to the pressure distribution. The presence of form drag is due primarily to separation of the flow in regions of adverse pressure gradient.† Drag due to pressure may also arise in *compressible* flow (Chapter 10), or from the formation of surface waves caused by motion of a ship, for example, on the surface of the ocean.

The drag of a slender shape, then, such as a thin airfoil or even a ship hull, may be estimated by the boundary layer skin friction. The graph in Fig. 8.5 shows the drag coefficient for a flat plate (including *both* sides of the plate) predicted on the

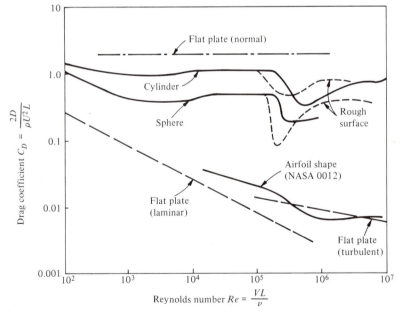

FIG. 8.5 Drag coefficient of slender and bluff bodies, showing the great differences that occur. The airfoil is the same as that described in Fig. 5.8, and the drag coefficient is based on *length*. The complete data for the cylinder and sphere are also shown in Fig. 5.7; note that the drag coefficient is based on *diameter*. The flat plate (normal) is similar to that shown in Fig. 8.6, while the lines for the flat plate marked (laminar) and (turbulent) are from Eqs. (8.6) and (8.29) for *both* sides of the plate.

---

† It may be recalled from Chapter 6 that the ideal flow of an unbounded uniform stream past a finite body *without* separation has no drag force, but it may have a lift force [i.e., there is no form drag (skin friction is zero by definition in ideal flow)].

basis of Eq. (8.6) for laminar flow and Eq. (8.29) for turbulent flow. The actual drag data for the slender airfoil described in Fig. 5.8 are also shown for purposes of comparison. The actual drag follows the laminar boundary layer result up to a Reynolds number of about 10^5, after which part of the boundary layer becomes turbulent. At Reynolds numbers approaching 10^7, the remaining laminar portion of the boundary layer becomes less important, and the total drag is nearly that predicted by the turbulent flat-plate formula.

Slender bodies such as the symmetrical airfoil or strut shown in Fig. 5.8 have much lower drag coefficients than "bluff" ones. The drag on a bluff body is due almost entirely to the pressure distribution, and the flow field around a bluff body typically is characterized by a large region of separated flow. The fluid pressure within the separated or "wake" region is typically at or somewhat below that of the approaching flow. An extreme case of such a flow is that past a sharp-edged plate held perpendicular or normal to the oncoming flow. A photograph of the flow past a normal flat plate in a water tunnel is shown in Fig. 8.6. Here the edge of the separated flow is made visible by *cavitation* or vaporization of the liquid water to form small vapor bubbles by reduction of the fluid pressure. The flow separates smoothly from the sharp edge of the plate. The region between the separated flow and the outside nonseparated flow is called a *shear flow*. This region is extremely unsteady and is a zone of intense mixing; in many ways the shear flow region resembles a boundary layer. The flow over the face of the normal plate is nearly that of an inviscid fluid forming a thin laminar boundary layer in a favorable pressure gradient. This boundary layer skin friction, however, does not contribute directly to drag. Drag coefficients for normal bluff body flows such as that shown in Fig. 8.6 are typically around 2 (Reference 7).

FIG. 8.6 Flow past a flat plate *normal* to the oncoming flow in a water tunnel. The flow is made visible by the formation of bubbles (cavitation) due to the sharp reduction in pressure behind the plate. The flow is from the right. [From T. J. O'Hern, "Cavitation Scale Effects," Ph.D thesis, California Institute of Technology, Pasadena, Calif., 1987.]

An interesting example of both skin friction and form drag is that of a sphere or very long cylinder held normal to an oncoming flow. The drag coefficients for these bodies were given in Fig 5.7 and are also shown in Fig. 8.5 for easy reference. Considering the cylinder, for example, at low Reynolds numbers, we find that the drag is composed of both skin friction and form drag, but at Reynolds numbers above approximately 10^4, the form drag is predominant. The drag coefficient shown in Fig. 8.5 is nearly constant (similar to the normal flat plate), indicating a well-developed broad wake. The laminar boundary layer formed on the forward side of the cylinder grows as it proceeds around the cylinder, separating at a position somewhere past $90°$ from the stagnation point. As the Reynolds number increases to about 2×10^5, the boundary layer becomes turbulent and is able to travel farther around the cylinder before separating. The wake becomes somewhat smaller, and as a result, the drag coefficient drops suddenly. Surface roughness becomes important, as it influences the formation of the turbulent boundary layer; making the cylinder rough will cause the boundary layer to become turbulent at a lower Reynolds number. Thus a smooth cylinder has a higher drag than one that is rough (in this range of Reynolds numbers). A sphere behaves in a similar way, as can be seen in Fig. 8.5. The idea of making a sphere rough (the "dimples" on a golf ball) in order to decrease its drag in a certain speed range was first pointed out by Prandtl, and the reference at the end of the chapter contains many interesting photographs of the flow in such cases. In particular at lower Reynolds numbers (60 up to 5000), the flow in the wake of a cylinder is unsteady, with vortices being shed in a periodic way from alternate sides of the cylinder. The vortex row or "street" so formed is shown in a qualitative way in Fig. 6.1d.

Further data on fluid-dynamic drag on both smooth and bluff bodies is tabulated extensively in monographs devoted to that subject. Particularly to be recommended is Reference 7 and the more recent summary in Reference 8.

REFERENCES

1. L. Prandtl, *Essentials of Fluid Mechanics*, Hafner Press, New York, 1952, especially Chap. 3.
2. H. Blasius, "The Boundary Layers in Fluids with Little Friction," *Tech. Mem. 1256*, National Advisory Committee for Aeronautics, 1950.
3. H. Schlichting, *Boundary Layer Theory*, 7th ed., McGraw-Hill, New York, 1979.
4. B. Carnahan and J. O. Wilkes, *Applied Numerical Methods*, Wiley, New York, 1969.
5. F. M. White, *Viscous Fluid Flow*, McGraw-Hill, New York, 1974.
6. P. S. Granville, "Drag and Turbulent Boundary Layer of Flat Plates at Low Reynolds Numbers," *J. Fluid Eng. Res.*, Vol. **21**, No. 1, 1977, pp. 30–39.
7. S. F. Hoerner, *Fluid-Dynamic Drag*, Dr-Ing. S. F. Hoerner (publisher), Midland Park, N.J., 1965.
8. R. D. Blevins, *Applied Fluid Dynamics Handbook*, Van Nostrand Reinhold, New York, 1984, especially Chap. 10.
9. D. E. Coles, "The Law of the Wake in the Turbulent Boundary Layer," *J. Fluid Mech.*, Vol. **1**, 1956, pp. 191–226.

CHAPTER 9

Compressible Fluids— One-Dimensional Flow

9.1 Introduction

When a fluid moves at speeds much less than the speed of sound, its density remains nearly constant, as was pointed out in Chapter 3. However, in many flows of engineering interest fluid speeds exceed the speed of sound, and density changes can be quite large. Examples are the high-speed flight of aircraft, internal flows in rocket nozzles, diffusers, wind tunnels, and shock tubes. These flows are termed *compressible flows*, and will be elaborated upon in both this and the following chapter.

The most striking feature of compressible flows is that the *basic nature* of the flow field changes as the fluid speed increases from less than to greater than sonic velocity. The occurrence of shock waves in supersonic flows, for example, profoundly influences the procedure for calculating pressure fields around a moving body. The remaining portion of this chapter is devoted to a discussion of some primary features of one-dimensional compressible flows.

9.2 Thermodynamic Preliminaries

Before taking up a discussion of compressible flows, a brief review of important thermodynamic results will be given. No attempt is made to derive these results here, and the following comments are intended to serve only as a short summary. A more complete presentation of thermodynamics is given in Appendix 4.

Compressible fluids are described by their *properties*, for example, pressure and temperature. Since temperature changes may result in heat flow across the boundaries of a fixed mass of fluid, additional thermodynamic properties such as internal energy must be considered. The properties of a substance define the *state* of the substance. For our purposes, not all properties of a fluid are relevant; for example, color or odor are not. It has been found by experience that thermodynamic states are sufficiently well specified by any *two* independent properties. The relationship between properties for a specific fluid is called the *equation of state* for that fluid.

If a substance undergoes a transformation from one state to another, the successive incremental changes describe the transformation *process*. A useful idealization is that of a reversible process, where the process can return the fluid to its initial state with no effect on the surroundings. In reality, effects of friction and heat flow cause all processes to be *irreversible*.

The interplay of thermodynamic effects is governed by thermodynamic laws which are universally valid. Our primary interest is in the *first law*, which declares the equivalence of heat and work as energy forms, and the *second law*, which limits the direction of natural processes.

For a system of *fixed mass* undergoing an incremental process, the first law can be stated as

$$dQ + dW = dE, \tag{9.1}$$

where dQ represents an increment of heat added to the system, dW an increment of work done on the system, and dE an increment in the total energy E of the system. The total energy E usually consists of internal energy U, as well as kinetic and potential energies. In some cases it may also include other forms of energy, such as elastic, chemical, electrical, or magnetic. However, these other forms are not considered further in the present discussion.

When kinetic and potential energies are negligible or identically zero, the law states that an increment in internal energy dU of a substance is determined by the incremental amounts of heat dQ and work dW added to a system. In the special case of a reversible process, the incremental work done per unit mass by surface forces acting on the mass can be expressed as

$$dw = -pd\left(\frac{1}{\rho}\right),$$

so that

$$du = dq - pd\left(\frac{1}{\rho}\right).$$

The first law of thermodynamics as stated in Eq. (9.1) applies strictly to the same collection of fluid particles, that is, to a fixed mass system. In many instances it is much more convenient to reformulate the law for a fixed control volume, much as the momentum theorem was derived for a control volume in Chapter 4. This transformation is detailed in a following section, and the resulting form of the first law for one-dimensional steady flow can be written as

$$q + w_s = \Delta\left(u + \frac{p}{\rho} + \frac{V^2}{2} + gy\right), \tag{9.2}$$

where w_s is work done per unit mass of flowing substance by a shaft, turbine, or pump, and Δ represents the difference in the quantity in parentheses between the outlet and inlet of a control volume. The combination of fluid properties $u + p/\rho = h$ is called the *enthalpy*, and is in itself a property of the fluid. Because very little use of head h is made in compressible flows, no confusion should arise from the use of the symbol h for enthalpy. The statement of the first law as given before is completely general and independent of the process. Thus frictional effects are accounted for in this form of the equation. The units of the equation are energy per unit mass; this is kJ/kg for both heat and work in SI units. However, in the British (or engineering) system, care must be exercised to relate heat units (Btu) to work units (ft-lb) [1 Btu $=$ 778 ft-lbf].

In the same way that changes in internal energy are defined by the first law, the second law of thermodynamics defines another property, called *entropy*. The entropy S is a measure of the internal molecular disorder of a substance, and the increment in entropy dS of a fixed mass of substance is defined as $dS = dQ/T$. Heat transfer processes must be imagined to be reversible for this definition of entropy to be valid, that is, $dS = dQ/T$ only for *reversible* processes. Irreversible processes result in larger degrees of molecular disorder and therefore changes in entropy, so that $dS > dQ/T$ for *irreversible* processes. The second law thus states that in general the entropy of a fixed mass system must increase during a process in which no heat transfer occurs (termed an *adiabatic* process). In the limit of a reversible adiabatic process, the entropy remains constant (an *isentropic* process).

The first and second laws of thermodynamics can be written in combined form for the incremental reversible processes of a fixed mass as

$$T\ ds = du + pd\left(\frac{1}{\rho}\right)$$ (9.3a)

or

$$T\ ds = dh - \frac{dp}{\rho}.$$ (9.3b)

Although the combined statement of the first and second laws above was derived on the basis of an incremental reversible process, we see that the result involves only changes in the properties of the system and so must be independent of the process. Thus Eqs. (9.3a) and (9.3b) are equally valid for incremental irreversible processes of a fixed mass.

The laws of thermodynamics are independent of the substance considered. However, many common gases, such as air, carbon dioxide, helium, and so on, obey the *perfect gas* equation of state

$$p = \rho R T$$ (9.4)

for a wide range of pressures and temperatures. It has also been established experimentally that for many of these gases changes in internal energy or enthalpy are simply related to the heat capacities of the gas by

$$du = c_v\ dT$$ (9.5a)

$$dh = c_p\ dT$$ (9.5b)

or, assuming that the specific heats are constant,

$$h = c_p(T - T_r) + h_r$$

$$u = c_v(T - T_r) + u_r,$$

where the subscript r denotes some reference state.

Thus for perfect gases, by combining Eqs. (9.3a), (9.3b), (9.5a), and (9.5b), we obtain

$$c_p - c_v = R.$$ (9.6)

A particularly useful expression relating pressures, temperatures, and densities during isentropic compression of a fixed mass of perfect gas can be obtained from the first law as

$$\frac{T_2}{T_1} = \left(\frac{p_2}{p_1}\right)^{(\gamma-1)/\gamma} = \left(\frac{\rho_2}{\rho_1}\right)^{\gamma-1},$$ (9.7)

where $\gamma = c_p/c_v$ is the ratio of specific heats of the gas. Further useful perfect gas relationships are given in Appendix 4.

9.3 The Speed of Sound

The physiological effects of speech and hearing are related to the transmission and detection of tiny pressure disturbances, and the speed at which these disturbances

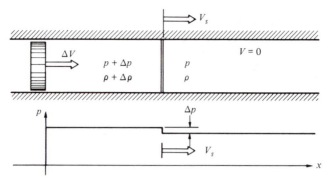

FIG. 9.1   Propagation of a small pressure pulse along a tube filled with a compressible fluid.

travel is commonly known as the speed of sound. We wish to examine how infinitesimal pressure or density disturbances propagate through a compressible fluid. One might expect that the speed of sound would depend on the specific properties of the transmitting fluid, and perhaps on the magnitude of the pressure disturbances themselves.

In order to relate these effects in a qualitative way, consider the piston-tube arrangement shown in Fig. 9.1. A long, thermally insulated tube is filled with a compressible fluid, and has a movable, close-fitting but frictionless piston in one end. If the piston is made to move very slowly toward the right at a constant speed ΔV, the fluid is compressed in front of the piston and the pressure and density increase slightly to values $p + \Delta p$, $\rho + \Delta \rho$. Experience shows that this change will move along the tube at some steady speed V_s. Behind this advancing front, the fluid properties remain constant at their increased value as long as the piston continues to move steadily. Because of continuity conditions, the fluid behind the advancing front moves at the same speed as the piston, while the fluid in front of the advancing pressure pulse is still stationary. Under the restriction of small pressure disturbances, frictionless movement, and no heat flow, the process can be considered reversible and isentropic. The qualitative discussion of the propagation of pressure pulses given before also applies to the mechanism of sound propagation, and the speed of the advancing front V_s is the speed of sound for the medium. It should be noted that the flow situation described is *unsteady* in a fixed reference frame, and calculation of how the speed of sound is related to fluid properties would be needlessly complicated in this reference frame. To obtain a situation where the advancing front is stationary and the flow field is steady, we consider a reference frame moving steadily at the speed V_s. The flow field on either side of the front is shown in Fig. 9.2.

We can now proceed to calculate the speed V_s by applying the equations governing conservation of mass and momentum, neglecting any viscous effects or elevation changes. From the continuity equation we have

$$\rho V_s = (\rho + \Delta \rho)(V_s - \Delta V) \tag{9.8}$$

and by applying the momentum theorem to a control volume as shown, we obtain

$$\Delta p = -(\rho + \Delta \rho)(V_s - \Delta V)^2 + \rho V_s^2$$

or, using Eq. (9.8) and rearranging, we have

$$\Delta p = \rho V_s \, \Delta V. \tag{9.9}$$

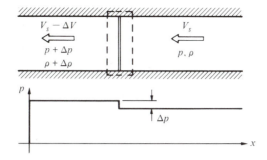

FIG. 9.2   An infinitesimal pressure rise in the steady flow of a compressible fluid.

If we now solve Eq. (9.8) for ΔV and substitute into Eq. (9.9), with some rearrangement we obtain

$$V_s^2 = \frac{\Delta p}{\Delta \rho}\left(1 + \frac{\Delta \rho}{\rho}\right).$$

Because we are considering only infinitesimal disturbances, in the limit as the disturbance tends to zero, the ratio $\Delta p / \Delta \rho$ can be interpreted as the derivative of p with respect to ρ. However, $dp/d\rho$ is still undetermined. To form this derivative, we have to ascertain first just how the pressure, p, depends on the density, ρ, or in other words we have to specify the thermodynamic process that brings about the changes dp and $d\rho$. In this case we remember that the compression was assumed to occur reversibly and adiabatically, so that the entropy of the fluid remains constant. Thus the limit of the ratio $\Delta p / \Delta \rho$ must be interpreted as the *partial* derivative of p with respect to ρ at constant entropy S, or $\Delta p / \Delta \rho \to (\partial p / \partial \rho)_s$.

Thus the expression for the speed of sound in a medium becomes (since $\Delta \rho / \rho \ll 1$)

$$V_s^2 = a^2 = \left(\frac{\partial p}{\partial \rho}\right)_s. \tag{9.10}$$

Experiments have shown that for many gases, over most audible frequencies (30 to 18,000 Hz), propagation of sound occurs with very little heat transfer and Eq. (9.10) is closely followed. At higher frequencies, however, some heat transfer does occur and the propagation of sound is more nearly isothermal. In other instances, such as the propagation of sound inside small-diameter tubes or at very low frequencies, isothermal conditions may also prevail.

Equations of state of gases are often given in terms of density and temperature rather than density and entropy, so that Eq. (9.10) cannot be used directly. However, a general thermodynamic result (derived in Appendix 4) states that $(\partial p / \partial \rho)_s = \gamma(\partial p / \partial \rho)_T$, so that for perfect gases Eq. (9.10) becomes

$$a^2 = \gamma RT. \tag{9.11}$$

Thus the speed of sound increases with temperature in a gas, or, comparing gases at the same temperature, more dense gases (smaller R) have a smaller speed of sound. For example, air at atmospheric pressure and $T = 293$ K has $a = 343$ m/s, while at $T = 1000$ K, $a = 634$ m/s. The rather simple form of Eq. (9.11) will be useful in following discussions.

For a compressible fluid flowing at some velocity V, the entire phenomenon of sound propagation is simply convected along with the moving fluid. For example, with a pressure pulse traveling in the same direction as the bulk fluid, the velocity of a signal measured in a stationary frame is $V + a$. The relative magnitude of fluid speed and the speed of sound is of great importance and is usually measured by the *Mach number*, $M = V/a$. Further mention of the Mach number and its importance will be made throughout this chapter. However, one interesting feature that may be pointed out here is that the Mach number has a physical significance much like the Reynolds, Froude, and other dimensionless groups discussed in Section 5.3. It was pointed out that the inertial force on a small fluid cube could be estimated by $F_{in} \simeq \rho d^2 V^2$. On the other hand, for a perfect gas the compressive, or elastic force on the fluid cube must be proportional to the density, temperature, and area (i.e., $F_{el} \propto pd^2$, or $F_{el} \propto \rho T d^2$). Thus the ratio of inertial to elastic or compressive forces on a fluid cube becomes

$$\frac{F_{in}}{F_{el}} \simeq \frac{\rho d^2 V^2}{\rho d^2 T} \simeq M^2;$$

that is, the Mach number is proportional to the square root of the ratio of inertial to elastic forces. The Mach number can also be interpreted as the ratio of the time necessary for a pressure signal to travel a fixed distance to the time necessary for a fluid particle to cover the same distance.

PROBLEMS

9.1   For a perfect gas subject to no friction or heat transfer, the relation between pressure and density may be expressed by the equation $p/\rho^\gamma = $ const, where γ is a constant and is equal to the ratio of the specific heats at constant pressure and volume, respectively. In addition, such gases obey the equation of state $p = \rho RT$, where R is the gas constant for the particular gas and T is the absolute temperature. Show that $a = (\gamma RT)^{1/2}$ for such a gas and evaluate a for air at $T = 520°R$, for which $\gamma = 1.4$ and $R = 53.3$ lbf–ft/lbm · °R. Repeat the calculation for $T = 290$ K and $R = 287$ J/kg · K.

9.2   The bulk modulus of a substance is defined by the equation $E = \rho(\partial p/\partial \rho)$. Given that $E = 2070$ MPa for water, determine the velocity of sound, letting $\rho = 1000$ kg/m³.

9.3   The temperature of the atmosphere decreases with altitude according to the relationship $T = T_0 - \beta h$, where h is the altitude in feet above sea level. If the sea-level temperature is $60°F$, and $\beta = 0.0036°F/ft$, how much faster in miles per hour would an aircraft fly at sea level than at 35,000 ft if, in both cases, the Mach number $M = 0.8$?

9.4   A small pressure pulse of magnitude Δp advances into a stationary fluid with pressure and temperature p_0 and T_0, respectively. Show that changes in the gas velocity and temperature after the pulse has passed are

$$\Delta V = \frac{(\gamma R T_0)^{1/2}}{\gamma p_0} \Delta p$$

and

$$\Delta T = \frac{(\gamma - 1)T_0}{(\gamma R T_0)^{1/2}} \Delta V.$$

9.5 Normal conversation results in sound waves with pressure variations of approximately 0.04 Pa. Using the results of Problem 9.4, find the corresponding velocity and temperature fluctuations at atmospheric conditions of 100 kPa and 27°C.

9.6 Deviations from perfect gas behavior can be calculated from the van der Waals equation of state

$$p = \frac{\rho RT}{1 - \rho \bar{b}} - \rho^2 \bar{a},$$

where $\bar{a}$ and $\bar{b}$ are constants. For a gas obeying the van der Waals equation, show that the speed of sound is given by

$$a^2 = \frac{\gamma RT}{(1 - \rho \bar{b})^2} - 2\gamma \rho \bar{a}.$$

What is the value of a for air at 100 kPa and 21°C? Compare this answer with the result obtained using the perfect gas model. For air

$$\bar{a} = 162 \text{ N} \cdot \text{m}^4/\text{kg}^2, \qquad \bar{b} = 1.26 \times 10^{-3} \text{ m}^3/\text{kg}.$$

9.7 A small pressure disturbance traveling through a stationary fluid strikes and is reflected from a solid wall. The reflected pulse travels back into the fluid with the same speed at which it approached the wall. Find the pressure rise at the wall resulting from the reflection (to first-order differential quantities only).

9.8 Consider an incompressible liquid (like water) with air bubbles uniformly dispersed throughout. The volume of air occupies a fraction α of the total volume of air and water mixture. Imagine that the mixture is compressed a slight amount from an original pressure p to a new pressure $p + dp$, thereby changing the overall density of the mixture from ρ to $\rho + d\rho$. Assume that the fraction of air in the mixture is compressed isentropically, and that the air behaves like a perfect gas while the liquid remains at constant density ρ_L.
(a) Show that the pressure and density changes for such a mixture are related by

$$\frac{dp}{p} = \frac{\gamma}{\alpha} \frac{d\rho}{\rho}.$$

(b) Show that the speed of sound in the mixture is approximately given by

$$a^2 \simeq \frac{\gamma p}{\rho_L \alpha(1 - \alpha)}.$$

(Note that for *no* air present the speed of sound is infinite; this is a result of our assumption that the liquid is incompressible.)
(c) Estimate the speed of sound in water at 20°C, which has 2 percent air (by volume) dispersed throughout. Compare this with the result of Problem 9.2.

9.9 Imagine a small pressure pulse propagating along a tube as shown in Fig. 9.1; however, the tube wall is made of an *elastic* material so that the cross-sectional area A depends on the internal pressure $A = A_0 + A_1 p$, where p is the pressure in the tube and A_0 and A_1 are constants. In addition, assume that the fluid in the tube is an *incompressible* liquid of density ρ. Find the speed at which such a small pressure pulse moves along the tube.

9.4 The Energy Equation

(a) Energy Equation for a Control Volume

The first law of thermodynamics, usually referred to as the energy equation, must be carefully applied to the same collection of fluid particles as they move about. In many cases it is simpler to reformulate the law in terms of a control volume, in much the same manner as the laws of motion were related to the momentum theorem in Chapter 4.

Consider an arbitrary body of fluid at two instances of time separated by a small time increment, dt, as shown in Figs. 4.3 and 9.3. During this increment in time, the fluid body or system may receive heat from the surrounding fluid and will have work done on it by forces acting on its surface as indicated in the figure. In order to apply Eq. (9.1) to this incremental process, we first consider the change in the energy dE of the system. The total energy of the system can be written as

$$E = \int_{\mathscr{V}'} \rho e \, d\mathscr{V},$$

where e is the energy per unit mass and the volume symbol $\mathscr{V}'$ has the same meaning as in Chapter 4. Thus the change in energy dE can be represented as

$$dE = \left(\frac{dE}{dt}\right) dt = \left(\frac{D}{Dt} \int_{\mathscr{V}'} \rho e \, d\mathscr{V}\right) dt.$$

The time derivative of the integral of energy over the instantaneous volume of the system can be evaluated in a manner similar to that used for the momentum theorem in Chapter 4. In fact, the final result for momentum change as given in Eq. (4.4a) can be taken over directly by substituting energy ρe for momentum $\rho \mathbf{v}$, so that

$$\frac{D}{Dt} \int_{\mathscr{V}'} \rho e \, d\mathscr{V} = \left(\frac{\partial E}{\partial t}\right)_{\mathscr{V}_0} + \int_{S_0} \rho e \mathbf{v} \cdot \mathbf{n} \, dS.$$

Thus the rate of change of energy for the system equals the rate of change within the original volume (now interpreted as a control volume), plus the rate at which energy flows out over the surface of the control volume.

Next we consider the incremental work done on the system. All of the forces that act on a surface element of the system can be represented by a normal force $\sigma_n \, dS$ and a surface shear force $\tau \, dS$, as pointed out in Section 2.6. In a time dt, the surface element dS will have moved a distance $V \, dt$ in the direction of the fluid motion.

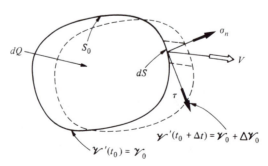

FIG. 9.3 An arbitrary body of fluid used in the derivation of energy equation. The fixed volume $\mathscr{V}_0$ represents the control volume.

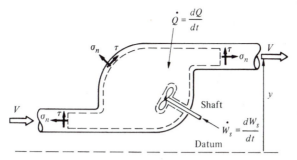

FIG. 9.4 Steady one-dimensional flow of a compressible fluid with heat addition and shaft work.

Thus the work done on the surface element is $(\sigma_n + \tau)\,dS \cdot \mathbf{v}\,dt$, or for the entire system

$$dW = dt \int_{S_0} (\sigma_n\,dS + \tau\,dS) \cdot \mathbf{v}.$$

Finally, we note that while heat is added to the actual system during the incremental process, we are at liberty to make the movement of the surface of the system arbitrarily small. Therefore, one can simply think of heat as being added through the original surface or control volume. If we collect the expressions for heat, work, and energy, the first law as applied to a control volume can be written in a very general manner as

$$\frac{dQ}{dt} + \int_{S_0} (\sigma_n + \tau) \cdot \mathbf{v}\,dS = \left(\frac{\partial E}{\partial t}\right)_{V_0} + \int_{S_0} \rho e \mathbf{v} \cdot \mathbf{n}\,dS. \qquad (9.12)$$

In particular, let us focus our attention on the *steady* flow of a fluid through a duct as shown in Fig. 9.4. The flow is one-dimensional, and additional work can be done on the fluid by a shaft representing a pump or turbine. For most problems, the energy of the fluid per unit mass e comprises internal, kinetic, and potential energies

$$e = u + \frac{V^2}{2} + gy,$$

where u is the *internal energy* per unit mass of the fluid. Evaluating the integrals over the surface of the control volume, we obtain

$$\dot{Q} + \dot{W}_s = [(\rho e - \sigma_n)VA]_{\text{out}} - [(\rho e - \sigma_n)VA]_{\text{in}},$$

where $\dot{W}_s$ represents the integral on the left-hand side of Eq. (9.12) over all of the surface except the surfaces through which flow takes place. That latter portion of the integral is represented by the terms that are evaluated at the inflow and outflow and contain σ_n. Note that no contribution is made by shear forces τ acting on the wall of the duct, since the fluid velocity $\mathbf{v}$ must be zero at the walls. As a consequence, $\dot{W}_s$ is just the rate at which work is added through the shaft and transmitted to the fluid by shear and normal forces at the surfaces surrounding the shaft. Further, we recall that normal stresses at the inlet and outlet comprise a pressure plus viscous stress as indicated in Eq. (2.20). However, we imagine the inlet and outlet to the control volume to be in a portion of the duct where the flow is not changing substantially, that is, $\partial u/\partial x = 0$. For this case, the normal force is simply the pressure, $\sigma_n = -p$.

Thus the steady-flow energy equation between the inlet and outlet of the control volume can be expressed as

$$\dot{Q} + \dot{W_s} = \dot{m}\Delta\left[\left(u + \frac{p}{\rho}\right) + \frac{V^2}{2} + gy\right],$$
(9.13)

where $\dot{m} = \rho A V$ is the mass flow rate (kg/s), and the symbol Δ represents outlet minus inlet conditions.

(b) Simple One-Dimensional Flows

For steady, one-dimensional flows between two fixed points 1 and 2 in the fluid, the energy equation can be stated as

$$q + w_s = h_2 + \frac{V_2^2}{2} + gy_2 - \left(h_1 + \frac{V_1^2}{2} + gy_1\right),$$

where q and w_s refer to heat added to and work done on the fluid per unit mass, respectively, between points 1 and 2. For a great many situations of interest in high-speed flow of gases, the potential energy term gy is much smaller than remaining terms and will be omitted from the following discussion.

Let us examine the simple situation of adiabatic flow along a duct or external streamline with no shaft work being done on the fluid. The energy equation for this case becomes

$$h_1 + \frac{V_1^2}{2} = h_2 + \frac{V_2^2}{2}.$$

It should be emphasized that this equation is valid regardless of frictional effects. If at some point 0 in the flow the fluid velocity is reduced to zero, the enthalpy h_0 at that point must be given by

$$h_0 = h_1 + \frac{V_1^2}{2} = h_2 + \frac{V_2^2}{2}.$$
(9.14)

The enthalpy h_0 is called the *total enthalpy* (as in the case of flow from a large reservoir) or *stagnation enthalpy* (as in the case of a stagnation point on a body). Equation (9.14) can be used to calculate fluid velocity if the total and local enthalpies are known.

EXAMPLE 9.1 The Subsonic Pitot Tube

A Pitot tube similar to the one shown in Fig. 3.8 is used to measure speed for a subsonic airplane. The difference between static and dynamic pressures is 14 kPa and the pressure far in front of the airplane is 70 kPa. The free-stream temperature is estimated to be $-50°C$. One would like to calculate the speed of the airplane.

Solution: From Eq. (9.14), $V^2 = 2(h_0 - h_1)$. Assuming that air is a perfect gas, the change in enthalpy $h_0 - h_1 = c_p(T_0 - T_1)$. If we neglect frictional effects the gas can be assumed to be compressed isentropically, and the temperatures are related by Eq. (9.7). Thus we have

$$V^2 = 2c_p\,T_1\left[\left(\frac{p_0}{p_1}\right)^{\gamma - 1/\gamma} - 1\right]$$

or

$$V = \left\{ 2(10^3)(223) \left[\left(\frac{84}{70} \right)^{0.4/1.4} - 1 \right] \right\}^{1/2}$$

$$= 154 \text{ m/s}.$$

If the flowing fluid is a perfect gas, some interesting features of the gas temperature can be deduced from Eq. (9.14). Because the enthalpy of a perfect gas can be expressed as $h = c_p T + \text{const}$, Eq. (9.14) may be written as

$$c_p T_0 = c_p T_1 + \frac{V_1^2}{2} = c_p T_2 + \frac{V_2^2}{2}. \tag{9.15}$$

In a manner similar to that for enthalpy, the temperature T_0 is called the *total*, or *stagnation temperature* of the fluid. If the fluid is moving with a velocity V_1, then its temperature T_1 is called the *free-stream temperature*. The free-stream temperature is necessarily less than the total temperature, and can be directly measured only by a measuring probe moving at the same speed as the fluid. If a stationary probe such as a thermometer or thermocouple is inserted in the flow, the fluid is brought to rest on the probe and a portion of the probe near the stagnation point will be at the total temperature. Because of heat conduction through the probe itself, the average probe temperature, and therefore the indicated probe temperature, will differ from the total temperature. This effect is usually expressed by a *recovery coefficient*, defined as

$$\lambda = \frac{T_{\text{indicated}} - T_1}{T_0 - T_1}.$$

For example, a bare thermocouple has a recovery coefficient of about 0.9.

In the following sections it will be useful to express the energy equation for a perfect gas in terms of the Mach number. If we write Eq. (9.15) once again,

$$c_p T_0 = c_p T_1 + \frac{V_1^2}{2} = c_p T_2 + \frac{V_2^2}{2},$$

by rearranging we obtain

$$T_0 = T_1 \left(1 + \frac{V_1^2}{2c_p T_1} \right) = T_2 \left(1 + \frac{V_2^2}{2c_p T_2} \right).$$

However, from Appendix 4 and Eq. (9.11),

$$c_p T_1 = \frac{\gamma R T_1}{\gamma - 1} = \frac{a_1^2}{\gamma - 1},$$

so that the energy equation becomes

$$T_0 = T_1 \left(1 + \frac{\gamma - 1}{2} M_1^2 \right) = T_2 \left(1 + \frac{\gamma - 1}{2} M_2^2 \right). \tag{9.16}$$

Equation (9.16) is valid between any two points in a one-dimensional, adiabatic, steady flow of a perfect gas, and includes any frictional or dissipative process that may occur in between.

If at some point the fluid velocity and temperature change so that the local velocity equals the speed of sound based on local conditions, the local Mach number is unity and Eq. (9.16) becomes

$$T_0 = T_1\left(1 + \frac{\gamma - 1}{2}\right).$$

The fluid temperature at this point is given the designation T^* and is called the *critical temperature*, that is,

$$T^* = \frac{2T_0}{\gamma + 1}. \tag{9.17}$$

In a similar way one can define a *critical velocity* as

$$V^* = \left(\frac{2\gamma R T_0}{\gamma + 1}\right)^{1/2}. \tag{9.18}$$

The physical significance of critical velocity is that when the fluid velocity exceeds critical velocity the flow is supersonic, while for velocities less than critical the flow is subsonic.

In dealing with compressible flows, it is often convenient to establish a fixed reference for flow quantities. Either the stagnation or critical condition can be used in this way. For example, it is possible to refer the Mach number to stagnation rather than to local conditions as follows:

$$M' = \frac{V}{\sqrt{\gamma R T_0}} = \frac{V}{\sqrt{\gamma R T}}\left(\frac{T}{T_0}\right)^{1/2},$$

or, using Eq. (9.16), we obtain

$$M' = M\left(1 + \frac{\gamma - 1}{2} M^2\right)^{-1/2}. \tag{9.19a}$$

In a similar way, the Mach number based on critical conditions is defined as

$$M^* = \frac{V}{V^*} = M\left(\frac{T}{T_0}\right)^{1/2}\left(\frac{T_0}{T^*}\right)^{1/2}$$

or

$$M^* = M\left(\frac{\gamma + 1}{2}\right)^{1/2}\left(1 + \frac{\gamma - 1}{2} M^2\right)^{-1/2}. \tag{9.19b}$$

Note that when M varies from zero to infinity, M^{*2} varies from zero to $(\gamma + 1)/(\gamma - 1)$.

PROBLEMS

9.10 Steady, laminar incompressible flow through a circular tube is called *Poiseuille flow*, as noted in Chapter 7. For such a flow, show that the rise in bulk temperature T_b of the fluid for a length L of insulated tube is given by

$$\Delta T_b = \frac{1}{c}\frac{fL}{d}\frac{V^2}{2},$$

where c is the specific heat of the fluid and

$$T_b = \frac{\displaystyle\int_0^{d/2} u(r)T(r)\pi r \; dr}{\displaystyle\int_0^{d/2} u(r)\pi r \; dr}.$$

9.11 A journal bearing may be simulated as a cylindrical shaft rotating in a concentric housing. The gap is very small and filled with a Newtonian fluid, so the flow can be considered a Couette flow as described in Chapter 1. A particular bearing houses a 5-cm-diameter shaft, the gap is 0.01 cm and it is filled with an oil having a viscosity similar to SAE 30. If the shaft turns at 360 rad/sec, how much heat must be removed if the outer part of the bearing is to be maintained at 40°C?

9.12 Consider the flow of a fluid between two flat plates as shown below and also discussed in Example 1.1. If we now allow the fluid to be compressible, temperature across the fluid will vary and heat will be conducted according to the expression

$$q = -k \frac{dT}{dy},$$

where q is the rate of heat flow per unit area and k is the thermal conductivity of the fluid. The moving surface is maintained at T_0 and the stationary surface at T_w. Thermal conductivity k and viscosity μ of the fluid both vary with temperature, but for many fluids their ratio $\mu/k = A$ is nearly constant. By applying the complete energy equation to the elemental control volume, show that

$$\mu \frac{d}{dy}\left(\frac{T}{A} + \frac{u^2}{2}\right) = \text{const} = -q_w.$$

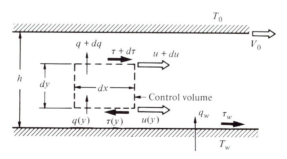

Problem 9.12

Integrating further, find a relationship between the temperature T and the velocity u at any point in the fluid.

Note: μ is not constant, but can be determined from the expression for shear stress τ.

9.13 The Concorde supersonic transport cruises at a Mach number $M = 2.2$, while the proposed HOTOL (horizontal takeoff and landing) plane would have a

Mach number of 5 at the same altitude. For each of these aircraft find the temperature at the stagnation point if the ambient temperature is 220 K.

9.14 Consider the steady flow of a perfect gas along a duct, as shown. The duct is insulated so that no heat flow occurs, and frictional effects can be neglected. By applying the momentum theorem and energy equation directly to the differential control volume, show that $p = \text{const } (\rho^\gamma)$ for such a flow.

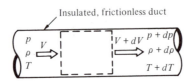

Problem 9.14

9.15 A scale model of an airplane is to be tested in a wind tunnel at $M = 2$. The tunnel is supplied from an air reservoir at 21°C, and the working section is held at atmospheric pressure 100 kPa.
(a) Assuming isentropic flow, what must the reservoir pressure be?
(b) What is the airspeed at the test section?
(c) If the cross-sectional area at the test section is 0.2 m², what is the mass rate of flow?

9.16 A rocket motor can be analyzed as a supersonic nozzle discharging a perfect gas with $R = 53.3$ ft lbf/lbm°R, $\gamma = 1.2$ from a reservoir. From test stand data, the reservoir temperature is 5000°R, and the gas leaves the nozzle at atmospheric pressure, 14.7 psia, and $M = 2.5$. Calculate the mass flow rate from the nozzle in lbm/sec per square foot of discharge area.

9.17 Air flows steadily through a Venturi tube having an upstream cross-sectional area of 2.4×10^{-3} m² and a throat area of 0.6×10^{-3} m². The flow can be considered isentropic, and the upstream and throat pressures are 2100 kPa and 1680 kPa, respectively. The upstream temperature is 590 K. Find the airflow rate in kg.

9.18 For steady, isentropic, one-dimensional flow show that

$$\frac{p_0}{p} = \left(1 + \frac{\gamma - 1}{2} M^2\right)^{\gamma/(\gamma - 1)}$$

and

$$\frac{\rho_0}{\rho} = \left(1 + \frac{\gamma - 1}{2} M^2\right)^{1/(\gamma - 1)}.$$

Note that the pressure ratio p_0/p and density ratio ρ_0/ρ are functions of the local Mach number only. For convenience these ratios are tabulated in Appendix 5.

9.19 (a) Show that for the steady, isentropic flow of a compressible fluid at a low Mach number the local static pressure p can be expressed as

$$p = p_0 - \frac{\rho V^2}{2}\left(1 + \frac{M^2}{4} + \frac{2 - \gamma}{24}M^4 + \cdots\right).$$

(b) Show that the percentage error in velocity, as obtained from a Pitot tube by neglecting fluid compressibility, is given by

$$\text{error} = \left\{\left[\frac{[(\gamma - 1)/\gamma](p_0/p_\infty - 1)}{(p_0/p_\infty)^{(\gamma - 1)/\gamma} - 1}\right]^{1/2} - 1\right\} \times 100,$$

where p_∞ is the free-stream static pressure.

9.20 Air expands irreversibly and adiabatically through a nozzle. The pressure in the tank (where the velocity is ≈ 0) is 3.5 MPa, and the temperature is 420 K. The exhaust pressure is 103 kPa. Successive states of the irreversible process are given by the relation $p/\rho^{1.3} = \text{const}$. (This does not describe the path of the expansion; an irreversible process cannot be described in this manner.) Find the actual exhaust velocity and the velocity coefficient of the nozzle. Assume air to be a perfect gas.

9.21 Air at 20°C is drawn through a convergent nozzle and accelerated to form a jet with a speed of 150 m/s. There is no friction or heat transfer.
(a) Calculate the temperature in the jet.
(b) Calculate the Mach number.
(c) Calculate the density and pressure of the jet.

9.22 Air is sucked from the atmosphere ($p_0 = 1$ atmosphere, $T_0 = 293$ K) *isentropically* into a pipe where the static pressure is $0.9p_0$.
(a) Calculate the velocity in the pipe.
(b) What would have been the velocity had the flow been incompressible?
(c) What is the Mach number at this point?

9.23 In a jet pump the expansion of a high-pressure fluid is used to pump a fluid originally at a low pressure. A particular pump is designed for air. The high-pressure "driver" fluid is at 3.3×10^5 N/m² and it expands to atmospheric pressure of about 10^5 N/m². The flow rate of the driver fluid is 10^{-3} kg/s and the inlet temperature is $T = 295$ K. The fluid to be pumped is also air. It is originally at a pressure of 0.5×10^5 and is elevated to 10^5 N/m². Its flow rate is 0.7×10^{-3} kg/s. Both airflows enter the jet pump at 295 K. Determine the efficiency of the pumping process by forming the ratio of the power needed to compress the low-pressure stream isentropically to the power derived by isentropically expanding the high-pressure stream. Assume air to be a perfect gas.

9.5 Flow in a Duct of Gradually Varying Area

Many devices of engineering interest, such as nozzles and Venturi tubes, are ducts whose cross-sectional area changes gradually along the axis of the tube. In this section we discuss the features of compressible flow in such ducts in a qualitative manner, with more detailed presentations in following sections.

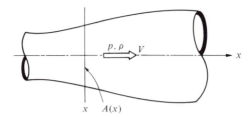

FIG. 9.5   Compressible fluid flow through a duct of gradually varying cross-sectional area.

Consider the flow through a duct as shown in Fig. 9.5. If the area changes very gradually along the duct, the flow can be considered one-dimensional, and in what follows we omit any frictional effects. We wish to relate fluid velocity, pressure, and temperature changes to duct area changes.

First let us restrict the flow to be adiabatic, so that with no friction the flow is also isentropic. The continuity equation requires that the total mass flow remain constant

$$\rho V A = \text{const,}$$

or in differential form,

$$\frac{d\rho}{\rho} + \frac{dV}{V} + \frac{dA}{A} = 0. \tag{9.20}$$

The equation of motion [Eq. (2.12)] for one-dimensional steady flow is simply

$$V \, dV = -\frac{dp}{\rho}. \tag{9.21}$$

Because our flow is isentropic, we can relate the pressure and density differences in Eq. (9.21) through the definition of the sonic velocity in Eq. (9.10):

$$a^2 = \frac{dp}{d\rho} = \left(\frac{\partial p}{\partial \rho}\right)_s.$$

By combining Eqs. (9.20) and (9.21), we obtain

$$\frac{dA}{A} = -\frac{dV}{V}\left(1 - \frac{V^2}{a^2}\right) = -\frac{dV}{V}(1 - M^2). \tag{9.22}$$

Several interesting features of compressible flow are evident from Eq. (9.22). First, consider the case of a constant-area duct (i.e., $dA = 0$). One way that Eq. (9.22) can be satisfied for this case is if $dV = 0$, the trivial case where the flow simply continues unchanged. However, small changes in V are permitted if the approaching flow exactly equals the sonic velocity ($M = 1$). Reexamining Fig. 9.2 for the derivation of sonic velocity, we see that infinitesimal pressure disturbances or sound waves comply exactly with Eq. (9.22).

For a *converging* duct as in Fig. 9.6a, dA/A is negative, and if the flow is subsonic ($M < 1$), dV/V must be positive. This corresponds to the familiar case of increasing velocity with decreasing area in incompressible flow. However, if the flow is supersonic ($M > 1$), dV/V becomes negative and the flow must decelerate in a converging duct. Although this seems puzzling at first, one must remember that the density also increases, so that total mass flow remains the same.

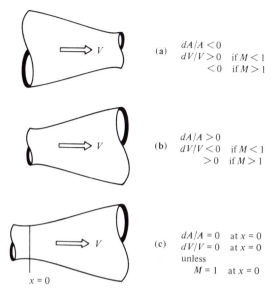

$$
\begin{array}{ll}
\text{(a)} & \begin{array}{l} dA/A < 0 \\ dV/V > 0 \quad \text{if } M < 1 \\ < 0 \quad \text{if } M > 1 \end{array}
\end{array}
$$

$$
\begin{array}{ll}
\text{(b)} & \begin{array}{l} dA/A > 0 \\ dV/V < 0 \quad \text{if } M < 1 \\ > 0 \quad \text{if } M > 1 \end{array}
\end{array}
$$

$$
\begin{array}{ll}
\text{(c)} & \begin{array}{l} dA/A = 0 \quad \text{at } x = 0 \\ dV/V = 0 \quad \text{at } x = 0 \\ \text{unless} \\ M = 1 \quad \text{at } x = 0 \end{array}
\end{array}
$$

FIG. 9.6 Summary of flow behavior in different ducts: (a) converging; (b) diverging; (c) throat.

In *diverging* ducts (Fig. 9.6b), dA/A is positive and the situation is exactly reversed from that of a converging duct: For subsonic flow the fluid slows down, whereas in supersonic flow the fluid accelerates and the density drops.

The limiting case of $M = 1$ can occur if $dA = 0$, as discussed before. In a duct of varying area, this implies that $M = 1$ can occur only at a location of minimum area in the duct as shown in Fig. 9.6c. Such a minimum is called a *throat*. It may be remarked that while the flow can equal the sonic velocity only at a throat, the existence of a throat does not mean that the flow is necessarily sonic at that point. A more detailed discussion of this question is given in the following sections.

We now consider some features of flow in a constant-area duct with *heat addition* or *withdrawal* as indicated in Fig. 9.7. As before, we restrict the discussion to frictionless flows only, and in addition assume that the fluid obeys the perfect gas law. We shall see that although the duct has constant area, heat addition can accelerate the flow to sonic conditions much as in a throat.

The energy equation for the differential element shown can be expressed as

$$
dq = dh + d\!\left(\frac{V^2}{2}\right)
$$

or equivalently,

$$
dq = c_p \, dT_0 = c_p \, d\!\left(T + \frac{V^2}{2c_p}\right). \tag{9.23}
$$

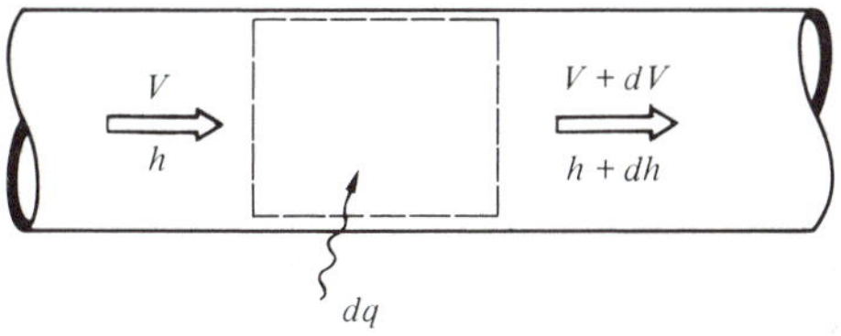

FIG. 9.7 Control volume for frictionless flow in a duct with heat addition.

The term in parentheses in Eq. (9.23) is the total temperature of the fluid, so heat addition can be viewed simply as increasing the total temperature by an amount $dT_0 = dq/c_p$. However, there are other effects of heat addition, as can be seen by substituting Eqs. (9.20) and (9.21) and the perfect gas law into Eq. (9.23). After some manipulation we obtain

$$\frac{dT_0}{T} = \frac{dV}{V}(1 - M^2).$$

Thus we see that in subsonic flow heat addition ($dT_0 > 0$) increases fluid velocity, while for supersonic flow heat addition reduces fluid velocity.

As a subsonic fluid is heated, it steadily increases velocity until it reaches $M = 1$. At this point it cannot be accelerated further. In a similar manner, a supersonic flow can be decelerated by heating, but will not be reduced to subsonic conditions.

PROBLEMS

9.24 One of the problems of wind tunnel testing at airspeeds near sonic speed is the large effect on flow conditions of a small change in the test configuration. Suppose that sonic velocity in a wind tunnel with 1-m^2 cross section is 300 m/s. Estimate the percentage change in test section airspeed when a model of 0.01-m^2 cross section is placed in the test section if the airspeed in the empty tunnel is 30, 270, and 330 m/s.

9.25 Show that for a frictionless flow of a perfect gas through a duct with heat addition

$$\frac{dT}{T} = \frac{dT_0}{T}\left(\frac{1 - \gamma M^2}{1 - M^2}\right).$$

Note: From this result heat addition for Mach numbers M in the range $1/\sqrt{\gamma} < M < 1$ will cause the gas temperature to *decrease*.

9.26 A duct has a converging–diverging shape like that shown in Fig. 9.6c. A perfect gas flows steadily and isentropically through the duct. Sketch the pressure variation along the duct for both subsonic and supersonic inlet conditions. The flow is not necessarily sonic at the throat.

9.27 Consider the one-dimensional frictionless flow of a perfect gas *with* heat addition in a duct of constant area. Because of the heat addition the total (or stagnation) temperature T_0 will increase as discussed in the preceding section. Using the differential forms of the basic equations, verify that the expression for the fractional gain in velocity is

$$\frac{dV}{V} = \frac{dT_0}{T}(1 - M^2)^{-1},$$

where M is the Mach number.

9.28 Air (assumed to be a perfect gas) flows without friction in a pipe of constant area. The flow rate is 1436 kg/m$^2 \cdot$s. The conditions are such that the Mach number is equal to 0.3 at a temperature of 250 K. What is the maximum amount of heat that could be added to this flow? This maximum occurs when $M = 1$ at the exit.

9.29 A long tube of unit cross-sectional area has a steady, frictionless, adiabatic flow of a perfect gas with a speed u, density ρ, and temperature T. At some location an infinitesimal amount of mass flow $dm \equiv \rho u \, di$ having the same initial temperature is injected into the tube with *no* streamwise momentum and mixes with the surrounding fluid. As a result, the velocity, pressure, and so on, all change by small amounts. In terms of the variables ρ, u, M, and di, the fractional mass injection rate:

(a) Obtain the appropriate form of the continuity equation.

(b) With the aid of the momentum theorem, develop an expression for the pressure change dp in terms of ρ, u, di, and so on.

(c) Similarly apply the energy equation to obtain the change dh of enthalpy.

(d) Using perfect gas relations, obtain an equation for du/u in terms of di and the Mach number.

9.6 Normal Shock Waves

In a preceding section we examined the conditions under which a steadily flowing fluid in a constant-area duct could have a sudden small change in pressure and velocity. It was seen that if the flow velocity was exactly sonic, such a small jump could occur. It is interesting to see what conditions are necessary to have *finite* changes in the flow, and if they also occur in an abrupt manner.

First, we wish to see in a qualitative manner how a *finite* discontinuity can be formed by reexamining the piston-tube arrangement described in Section 9.3. A slight, steady movement of the piston produces a pressure wave that propagates through the stationary gas. If the piston velocity is now increased slightly, a second pressure pulse forms and also moves along the tube. However, this second pulse is superimposed upon the slight bulk movement produced by the first pulse. In addition, the pressure, density, and temperature of the gas compressed by the first pulse are slightly higher than the static fluid, so that the sonic velocity will be higher. The combination of these effects results in a higher absolute propagation velocity for the second pulse, so that in time it can overtake the first pulse at some point down the tube. If we now imagine that the piston continues to accelerate up to some finite velocity, a further series of pulses are formed and travel down the tube. The pressure profile along the tube at successive times will continue to steepen as the pulses overtake one another. Finally, as indicated in Fig. 9.8a, very steep gradients in pressure and velocity can occur. Behind this steep compression front the gas moves at the terminal piston speed, while ahead it is still stationary. The compression front itself moves at a speed which can be much greater than the speed of sound in the undisturbed region. This compression front is termed a *normal shock wave*. It remains as a thin, moving region because of the very large internal viscous stresses and heat transfer rates that develop to prevent an actual discontinuity from forming.

By way of comparison, if the piston is drawn away from rather than into the fluid, a reverse process to that just described occurs. The resulting pressure profile, termed a *rarefaction wave*, continues to broaden along the tube as shown in Fig. 9.8b.

It is often convenient to refer to formation of compression and rarefaction waves on a time–distance plot as on the lower portions of Fig. 9.8a and b. The light lines represent loci of small pressure pulse locations. In the case of the compression front these lines grow less steep as the piston advances, and eventually begin to overlap, indicating formation of a shock wave. For the rarefaction front, however, these lines fan out in an *expansion fan*. The simple ideas of superimposed pressure pulses are

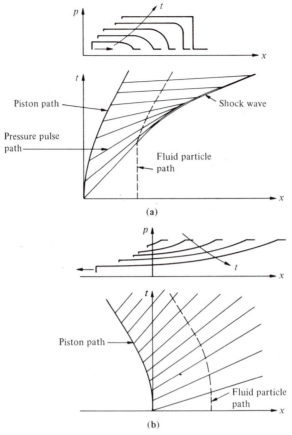

(a)

(b)

FIG. 9.8 (a) Development of a shock wave by a piston accelerating into a gas-filled tube; (b) development of a rarefaction wave by a piston accelerating out of a gas-filled tube.

not suitable for detailed analysis of the shock structure, since they neglect viscous and thermal effects precisely in the region where they are dominant.

To simplify the analysis of a shock wave, we use a moving reference frame in which the shock is stationary and the resulting flow steady. We consider the flow to be in a constant-area duct, with no heat addition or frictional effects. To simplify the presentation further, we take the fluid to be a perfect gas. Let us assume that an abrupt, finite change in the flow takes place at a fixed location in the duct as indicated in Fig. 9.9. Conditions in front and behind the discontinuity are denoted by the subscripts 1 and 2, respectively. Note that V_1 is the velocity at which the shock wave moves through a still fluid. Our aim is to relate pressures, temperatures, and velocities on either side of the discontinuity. From the continuity equation

$$\rho_1 V_1 = \rho_2 V_2, \tag{9.24}$$

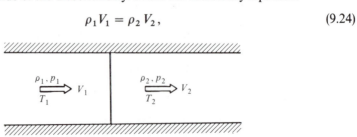

FIG. 9.9 Steady flow of a compressible fluid through a shock wave.

while the momentum equation (Section 4.2) leads to

$$p_1 - p_2 = \rho_1 V_1 (V_2 - V_1).$$ (9.25)

Equation (9.15), the energy equation, remains valid:

$$T_0 = T_1 + \frac{V_1^2}{2c_p} = T_2 + \frac{V_2^2}{2c_p}.$$ (9.15)

We now eliminate the pressure terms from the momentum equation using the perfect gas law and continuity equation, so that

$$V_2 - V_1 = \frac{p_1}{\rho_1 V_1} - \frac{p_2}{\rho_2 V_2} = \frac{RT_1}{V_1} - \frac{RT_2}{V_2}.$$ (9.26)

If we now substitute the temperature T from Eq. (9.15), we obtain

$$V_2 - V_1 = \frac{R}{V_1}\left(T_0 - \frac{V_1^2}{2c_p}\right) - \frac{R}{V_2}\left(T_0 - \frac{V_2^2}{2c_p}\right).$$ (9.27)

Finally, after some manipulation we obtain *Prandtl's equation*,

$$V_1 V_2 = \frac{2\gamma R T_0}{\gamma + 1} = V^{*2}.$$ (9.28)

The term on the right-hand side of this equation is the square of the critical velocity V^* defined in Eq. (9.18). Thus Eq. (9.28) states that if the approaching flow is supersonic ($V_1 > V^*$), the downstream flow is subsonic ($V_2 < V^*$). Actually, Eq. (9.28) also permits changes from subsonic to supersonic velocities. However, only the transition from supersonic to subsonic is observed to occur in reality because of limitations resulting from the second law of thermodynamics. We deal with this point later in this section.

Although the preceding results were obtained for a constant-area duct, they are also valid in a gradually varying duct. This stems from the result of detailed calculation and observation that the shock region is very thin; in fact, it is only on the order of a few molecular mean free path lengths. The area change across the shock is then usually negligible for practical purposes. In the preceding results the effects of viscosity near the wall have also been ignored. Since the velocity must still be zero at the wall, a supersonic flow must revert to subsonic in the near-wall region. The simple normal shock wave may become two-dimensional in the wall or boundary layer; the description of this type of flow is complex and beyond the simple introduction intended here. It can be said, nevertheless, that since the viscous region is small (particularly in pipe flow), the preceding derivations are still useful for the main portion of the flow.

Shock waves may occur in a variety of compressible flow situations. The flow around a cylinder held in a supersonic stream is shown in Fig. 9.10, and a curved shock may be seen to form. Near the stagnation point the shock is approximately normal; however, it begins to turn as it moves from the front of the cylinder. The exact curve of the shock depends on the body shape and free-stream Mach number. Some further discussion of two-dimensional compressible flows will be given in Chapter 10.

EXAMPLE 9.2

A blunt Pitot tube as shown in Fig. 3.8 is used to find the velocity in a supersonic airstream. In this case a shock wave similar to the type shown in Fig. 9.10 is formed in front of the tube. The flow through the center portion can be considered to be a

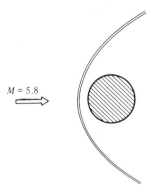

FIG. 9.10 The formation of a shock wave around a cylinder held in a uniform supersonic flow. The corresponding free-stream Mach number is $M = 5.8$. (Approximately to scale.)

normal shock. The pressure reading at the head of the Pitot tube is 120 kPa, and the total temperature is 480 K. The pressure immediately behind the shock is 106 kPa. We would like to find the velocity and pressure of the supersonic airstream.

Solution: The compression taking place between the downstream face of the shock and the head of the Pitot tube can be considered isentropic. Therefore, the temperature T_2 behind the shock is

$$T_2 = T_0\left(\frac{p_2}{p_{ST}}\right)^{(\gamma - 1)/\gamma} = 480\left(\frac{106}{120}\right)^{1/3.5} = 463.3 \text{ K}.$$

The velocity behind the shock can be obtained as in Example 9.1 for subsonic flow:

$$V_2 = [2c_p(T_0 - T_2)]^{1/2}$$
$$= [2(10^3)(480 - 463.3)]^{1/2}$$
$$= 182.8 \text{ m/s}.$$

The velocities across the shock are related by Eq. (9.28), so that

$$V_1 = \frac{2\gamma R T_0}{V_2(\gamma + 1)}$$
$$= \frac{2(1.4)(287)(480)}{182.8(2.4)}$$
$$= 879.2 \text{ m/s}.$$

The pressure upstream of the shock p_1 can be determined using the momentum equation, Eq. (9.25),

$$p_1 = p_2 + \rho_2 V_2(V_2 - V_1) = p_2 + \frac{p_2 V_2}{R T_2}(V_2 - V_1)$$

or

$$p_1 = 106 - \frac{106(182.8)(879.2 - 182.8)}{287(463.3)}$$
$$= 4.52 \text{ kPa}.$$

Actually, the calculation just outlined can be shortened considerably by the use of charts and tables of flow properties that are described later in this section.

The relationship between upstream and downstream Mach number for a shock wave can be obtained by rewriting Eq. (9.28) as

$$M_1 M_2 = \frac{2}{\gamma + 1} \frac{T_0}{\sqrt{T_1 T_2}}.$$

By relating the temperature ratios T_0/T_1 and T_0/T_2 to M_1/M_2 through Eq. (9.16), we obtain, after reduction,

$$M_2^2 = \frac{M_1^2(\gamma - 1) + 2}{2\gamma M_1^2 - \gamma + 1}. \tag{9.29}$$

The pressure ratio across the shock can be computed from the momentum equation, Eq. (9.25), and the continuity equation, Eq. (9.24), by using the perfect gas law. We obtain

$$p_1 - p_2 = \frac{p_2 V_2^2}{RT_2} - \frac{p_1 V_1^2}{RT_1}$$

or

$$p_1 - p_2 = \gamma(p_2 M_2^2 - p_1 M_1^2),$$

so that

$$\frac{p_1}{p_2} = \frac{1 + \gamma M_2^2}{1 + \gamma M_1^2}. \tag{9.30}$$

Since we have stated that for real shocks $M_1 > M_2$, we see that $p_1 < p_2$; that is, the gas is *compressed* by the shock wave as indicated in our previous discussion.

To find the density ratio, we write the perfect gas law as

$$\frac{\rho_1}{\rho_2} = \frac{p_1 T_2}{p_2 T_1}.$$

However, the temperature ratio has already been given in Eq. (9.16) as

$$\frac{T_1}{T_2} = \frac{1 + [(\gamma - 1)/2]M_2^2}{1 + [(\gamma - 1)/2]M_1^2}. \tag{9.16}$$

Substituting Eqs. (9.16) and (9.30) into the perfect gas law, we obtain

$$\frac{\rho_1}{\rho_2} = \frac{(1 + \gamma M_2^2)(1 + [(\gamma - 1)/2]M_1^2)}{(1 + \gamma M_1^2)(1 + [(\gamma - 1)/2]M_2^2)}. \tag{9.31}$$

We note that pressure, temperature, and density ratios across a shock are functions of the Mach numbers M_1 and M_2. However, since M_2 can be related to M_1 by Eq. (9.29), these ratios can be expressed solely as functions of M_1. The calculation of these quantities is ideally suited to a spreadsheet program on a personal computer. A column containing increments of upstream Mach number is first formed; a second column of corresponding downstream Mach numbers is calculated with Eq. (9.29). Columns for pressure, temperature, and density ratios follow directly. By placing the ratio of specific heats in a remote or protected cell for use in the calculations above, all the ratios can be quickly recalculated for any value of γ. For convenience, tables of pressure, temperature, and density ratios for air at various Mach numbers are presented in Appendix 5 and are shown graphically in Fig. 9.11.

We now return to a discussion of why only compression shocks are observed in practice. Recall that while no assumptions were made regarding frictional effects, the

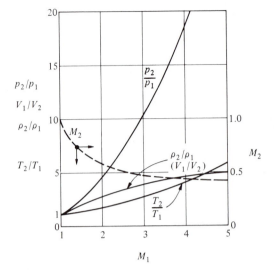

FIG. 9.11 Change of flow properties through a normal shock wave in air ($\gamma = 1.4$).

shock process was still considered adiabatic. The second law of thermodynamics states that an adiabatic process must increase the entropy of the fluid, or in the limit of a reversible adiabatic process, the entropy remains constant. Shock waves are highly irreversible, since very large velocity and temperature gradients occur through the shock itself and hence frictional, or dissipative effects, must be present. Only shock waves across which entropy increases can thus be allowed. Entropy change across a shock can be computed from the expression for entropy changes in a perfect gas, as taken from Appendix 4.

$$s_2 - s_1 = c_v \ln \frac{T_2}{T_1} + R \ln \frac{\rho_1}{\rho_2}.$$

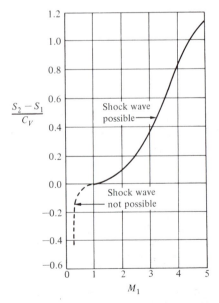

FIG. 9.12 Entropy rise across a normal shock wave in air ($\gamma = 1.4$).

Substituting Eqs. (9.29), (9.16), and (9.31) into the preceding equation, we obtain

$$s_2 - s_1 = \frac{R\gamma}{\gamma - 1} \ln \left[\frac{2}{(\gamma + 1)M_1^2} + \frac{\gamma - 1}{\gamma + 1} \right] + \frac{R}{\gamma - 1} \ln \left(\frac{2\gamma M_1^2}{\gamma + 1} - \frac{\gamma - 1}{\gamma + 1} \right).$$

The expression for entropy rise is shown as a function of Mach number in Fig. 9.12. It can be seen that shock waves which change the gas from supersonic to subsonic conditions produce a rise in entropy, and thus are the only possibility allowed.

PROBLEMS

9.30 A Pitot tube that senses the stagnation pressure, p_s, at its forward opening is often used to measure the speed of an airplane. Such a device is incorporated into the nose of a supersonic airplane for the purpose of finding the Mach number, M, at which the airplane is traveling ($M > 1$). Assume that the ambient pressure of the air through which the airplane is traveling, p_A, is known. If a bow shock (like a normal shock for present purposes) forms ahead of the Pitot tube, find the relation between the measured quantity, p_s/p_A, and the required quantity, M.

Note: The relation includes γ for air; it cannot be written explicitly in the form $M = $ function $(\gamma, p_s/p_A)$. Therefore, the answer should be stated in the form $p_s/p_A = $ function (M, γ).

9.31 For a stationary shock wave in a perfect gas, derive the following, where M_1 is the upstream Mach number:

$$\frac{p_2}{p_1} = \frac{2\gamma}{\gamma + 1} M_1^2 - \frac{\gamma - 1}{\gamma + 1}$$

$$\frac{T_2}{T_1} = \frac{\left(1 + \frac{\gamma - 1}{2} M_1^2 \right) \left(\frac{2\gamma}{\gamma - 1} M_1^2 - 1 \right) 2(\gamma - 1)}{(\gamma + 1)^2 M_1^2}$$

$$\frac{p_{02}}{p_{01}} = \left(\frac{\frac{\gamma + 1}{2} M_1^2}{1 + \frac{\gamma - 1}{2} M_1^2} \right)^{\gamma/(\gamma - 1)} \left(\frac{2\gamma}{\gamma + 1} M_1^2 - \frac{\gamma - 1}{\gamma + 1} \right)^{-1/(\gamma - 1)}.$$

9.32 The pressure and density ratios across a shock wave in a perfect gas are related by the *Rankine–Hugoniot relationship*.

(a) Show that this relationship is

$$\frac{\rho_2}{\rho_1} = \frac{\frac{\gamma + 1}{\gamma - 1} \frac{p_2}{p_1} + 1}{\frac{p_2}{p_1} + \frac{\gamma + 1}{\gamma - 1}}.$$

(b) What is the maximum density ratio for air ($\gamma = 1.4$) as the pressure ratio p_2/p_1 becomes very large?

(c) For very small pressure differences (a weak shock) show that this expression becomes equal to the isentropic relationship $dp/p = \gamma \, d\rho/\rho$.

9.33 Air, considered to be a perfect gas with $\gamma = 1.4$, flows supersonically in a nozzle. A normal shock reduces the velocity to 300 m/s, and increases the pressure from 170 to 680 kPa.

(a) Use the results of Problem 9.32 and the continuity equation to find the velocity before the shock wave.

(b) Use the momentum and speed of sound relationships to find the Mach numbers before and after the shock wave.

9.34 A shock wave moves into a stationary fluid having pressure p_1, sonic speed a_1, and specific heat ratio γ. The pressure behind the shock is p_2. Show that the speed of propagation of the shock V_s is given by

$$V_s = a_1 \left(\frac{\gamma - 1}{2\gamma} + \frac{p_2}{p_1} \frac{\gamma + 1}{2\gamma} \right)^{1/2}.$$

9.35 A ground-level explosion causes a blast wave to propagate into still air at 14.7 psia and 70°F. A recording instrument on the ground registers a maximum gage pressure of 100 psi as the wave passes. Considering the wave as a traveling shock, calculate its speed in ft/sec, and the wind speed following the shock in ft/sec.

9.36 Air at a pressure of 105 kPa, a temperature of 20°C, and a Mach number of 3 passes through a normal shock. Assume air to be a perfect gas.

(a) What are the velocities upstream and downstream of the shock wave?

(b) How much does the pressure rise through the shock wave?

(c) What would be the pressure if the velocity had been reduced isentropically from the initial velocity to that behind the shock?

9.37 Consider a normal shock in a general compressible fluid that is not a perfect gas. Using the approach of Section 9.6, show that

$$h_2 - h_1 = \frac{1}{2} \frac{\rho_1 + \rho_2}{\rho_1 \rho_2} (p_2 - p_1),$$

where station 1 is upstream and station 2 is downstream of the shock. This equation is useful when calculating the characteristics of a shock wave in a gas whose properties are given in tabular form only.

9.38 A very weak shock wave has a pressure rise $\Delta p = p_2 - p_1$, where $\Delta p / p_1 \ll 1$. Using the relationship for entropy rise across a shock in a perfect gas given in Appendix 5, as well as the Rankine–Hugoniot relationship of Problem 9.32, show that for a weak shock wave

$$s_2 - s_1 \approx \left(\frac{\Delta p}{p_1} \right)^3 + \text{higher-order terms}.$$

From this result, we see that the isentropic assumption for sound waves is justified.

9.39 The blast wave described in Problem 9.35 strikes and is reflected from a rigid wall. Find the static pressure on the wall and the speed of the reflected shock.

9.40 A shock tube is a device suitable for producing high temperatures and velocities with very simple equipment. It consists of a long tube divided into two sections by a membrane. The pressure in one side of the tube (the "driver" gas) is raised considerably above the other (the "test" gas), but the temperature of both sections is the same. When the membrane is broken, the driver gas compresses a slug of the test gas, much as the moving piston described earlier in this section, and generates a shock wave that moves into the stationary test gas. For most purposes, the "contact surface" between the driver and test gas moves at a constant speed. The shock itself then also moves at constant speed. In the driver section, however, the gas is at rest until affected by the rarefaction wave produced by the rupturing of the membrane. The flow in the driver section can be considered isentropic.

(a) In a particular shock tube the test gas is air at 20 kPa and 15°C. The pressure behind the shock in the compressed slug of test gas is 120 kPa. Find the gas temperature and speed behind the shock, as well as the shock propagation velocity.

(b) Discuss qualitatively how one might proceed to calculate the required initial pressure in the driver section.

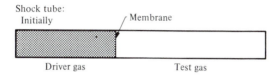

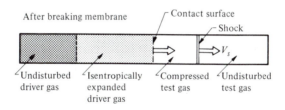

Problem 9.40

9.7 Isentropic Flow Through a Duct of Variable Area

In Section 9.5 we discussed the qualitative aspects of isentropic flow through a duct of variable area. For actual engineering applications such as nozzles or diffusers, however, we shall require specific relationships between pressures, velocities, and so on. For purposes of discussion, consider a flow issuing from a large reservoir at p_0, T_0, through a duct of variable area as shown in Fig. 9.13. The flow is steady and the total mass flow rate $\dot{m}$ is held constant. In addition, we assume that the fluid acts as a perfect gas. Several important features of the flow can be demonstrated by first expressing the mass flow in terms of local fluid properties along the duct. The mass flow at any point can be given as

$$\dot{m} = \rho V A.$$

However, from Eq. (9.15) this can be written as

$$\dot{m} = \rho A [2c_p(T_0 - T)]^{1/2}. \tag{9.32}$$

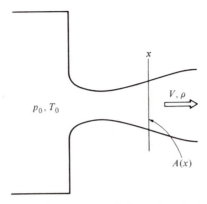

FIG. 9.13 Isentropic flow from a large reservoir through a duct of gradually varying area.

Since the flow is isentropic, the pressures, temperatures, and densities are related by Eq. (9.7), that is,

$$\frac{T}{T_0} = \left(\frac{p}{p_0}\right)^{(\gamma-1)/\gamma}$$

and

$$\frac{\rho}{\rho_0} = \left(\frac{p}{p_0}\right)^{1/\gamma}.$$

Thus the expression for mass flow [Eq. (9.32)] becomes

$$\dot{m} = \rho_0 (2c_p T_0)^{1/2} A \left(\frac{p}{p_0}\right)^{1/\gamma} \left[1 - \left(\frac{p}{p_0}\right)^{(\gamma-1)/\gamma}\right]^{1/2}. \qquad (9.33)$$

Equation (9.33) is shown plotted for several values of the constant $\dot{m}/\rho_0 \sqrt{2c_p T_0}$ in Fig. 9.14. For each constant value of mass flow, the curve of A versus p/p_0 has a minimum, and the duct area cannot be reduced below this value, which we designate as A^*. The pressure ratio at this point can easily be calculated from Eq. (9.33), and solving for p/p_0, we obtain

$$\left(\frac{p}{p_0}\right)_{\min} = \frac{p^*}{p_0} = \left(\frac{2}{\gamma+1}\right)^{\gamma/(\gamma-1)}. \qquad (9.34)$$

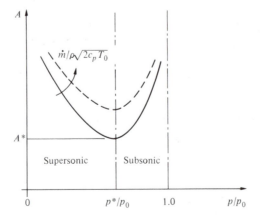

FIG. 9.14 Duct area required to pass a given mass flow as a function of pressure.

The pressure at this minimum, p^*, is the *critical* pressure, and the ratio p^*/p_0 depends only on $c_p/c_v = \gamma$. For air ($\gamma = 1.4$) the critical pressure ratio $p^*/p_0 = 0.528$. By comparing Eqs. (9.34), (9.17), and (9.7), we see that the temperature of the fluid at the minimum area point is the critical temperature, and the flow must be exactly sonic there. The minimum area point is evidently a *throat*, as discussed previously. The flow is subsonic for pressures greater than the critical, and supersonic for pressures less than the critical.

The critical area A^* can be expressed as

$$A^* = \frac{\dot{m}}{\rho_0(\gamma R T_0)^{1/2}} \left(\frac{p^*}{p_0}\right)^{-(\gamma + 1)/2\gamma} \tag{9.35}$$

and is a function of mass flow, reservoir conditions, and specific heat ratio for the gas.

It may be observed from Eq. (9.16) that the temperature ratio T/T_0 at any point in the duct depends only on the local Mach number. Further, the ratios p/p_0, A/A^*, and V/V^* can be shown to be functions of only the Mach number. Once again, a simple spreadsheet program on a personal computer can be used to calculate these ratios easily; for convenience they are tabulated for air in Appendix 5 and are also shown graphically in Fig. 9.15.

Using the information in Fig. 9.15 or in Appendix 5, we can proceed with the design of a duct for any specified mass flow, reservoir, and exit condition. For example, we see that if the discharge pressure is *greater* than the critical pressure, the duct has a simple converging shape and the flow is subsonic along its entire length. If the discharge pressure is *less* than the critical pressure, the fluid must be supersonic at the exit. The duct must first be converging until the critical area is reached, then diverging as the flow accelerates and becomes supersonic. This is the reason for the converging–diverging shape of rocket engine nozzles and supersonic

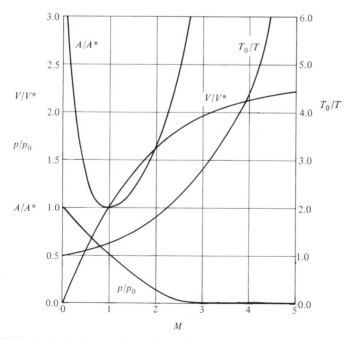

FIG. 9.15 Variation of flow properties with a Mach number for the isentropic one-dimensional flow of air ($\gamma = 1.4$). The subscript zero refers to stagnation conditions.

wind-tunnel inlets. The actual variation of area with length can be chosen somewhat arbitrarily, although certain limitations are imposed on the pressure gradient in the diverging section. If the divergence is quite rapid, the resulting pressure gradient will be very large and separation of the flow occurs as discussed briefly in Chapter 6. The resulting flow would depart seriously from the one-dimensional assumptions we have employed to this point and would have a much different pressure drop than predicted by isentropic flow theory. For conically shaped diverging sections, an included angle of less than 15° seems to produce good results, although much larger angles are sometimes used where the overall length of the nozzle is important. Details of the converging portion appear to be less critical.

A nozzle configuration frequently used in practice for discharging fluid from a reservoir is the simple convergent nozzle ending in a throat of area A_t, as shown in Fig. 9.16. The mass flow through the nozzle for given reservoir and exit conditions can be computed from Eq. (9.33). The discharge rate is plotted in Fig. 9.16 as a function of pressure ratio, and exhibits a maximum at a pressure ratio equal to the critical pressure ratio p^*/p_0. As the pressure outside the nozzle is reduced further, the discharge rate would *drop* according to Eq. (9.33). However, physical reasoning does not make this seem likely. Upon closer examination, we see that when critical conditions exist at the throat, the throat velocity is sonic and further reductions in pressure at that point are not possible since they would create supersonic flow at the throat. The discussion in Section 9.5 rules out such an occurrence. The pressure at the throat remains at its critical value and the pressure of the issuing jet is diffused through a complex structure of shock waves until it reaches outside conditions. The mass flow rate remains at the maximum value,

$$\dot{m}_{\max} = \rho_0 (\gamma R T_0)^{1/2} A_t \left(\frac{p^*}{p_0}\right)^{(\gamma+1)/2\gamma}, \qquad (9.36)$$

whenever the discharge pressure is less than the critical value p^* given in Eq. (9.34).

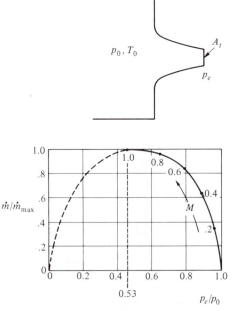

FIG. 9.16   Flow of air from a convergent nozzle.

9.8 The de Laval Nozzle

We now consider the flow through a converging–diverging nozzle having a specified cross-sectional area at any point along its axis. Such a nozzle is called a *de Laval nozzle*, in honor of the man who first sought to patent such a device. It is interesting to examine the flow in such a nozzle as the exit pressure p_e is reduced. We restrict our attention to steady flow of a perfect gas, with specified reservoir conditions.

As the exit pressure p_e is reduced below p_0, flow through the nozzle begins. If p_e is only slightly less than p_0, the flow throughout the nozzle is subsonic and the pressure profile along the axis would be like curve a in Fig. 9.17. Reducing p_e increases the mass flow rate $\dot{m}$, which can be computed from Eq. (9.33). As the flow rate increases, the pressure at the throat decreases until it reaches the critical pressure as indicated by curve b. The exit pressure p_e which exactly corresponds to sonic conditions at the throat can easily be determined from the isentropic flow tables in Appendix 5 by first computing A_e/A^*. The flow is subsonic everywhere in the nozzle except at the throat, and mass flow is the maximum possible for the given nozzle and reservoir conditions.

Suppose that the exit pressure is now reduced to a value corresponding to curve g in Fig. 9.17. The exit pressure at g is such that the expansion is still isentropic; however, the flow is supersonic in the diverging portion of the nozzle. This value of pressure is again simply obtained from the tables in Appendix 5. The pressure within the nozzle exit cannot be reduced further, and when the external pressure is lowered to h the fluid leaving the nozzle changes its pressure through a complicated flow pattern outside the nozzle. Thus the curves b and g represent the two limiting cases of exit pressure for isentropic flow in such a nozzle. For exit pressures below that at b, a shock wave forms within the diverging part of the nozzle, changing the flow from supersonic to subsonic and compressing the gas exactly enough to match exit conditions. Because of the entropy rise across the shock, the overall flow

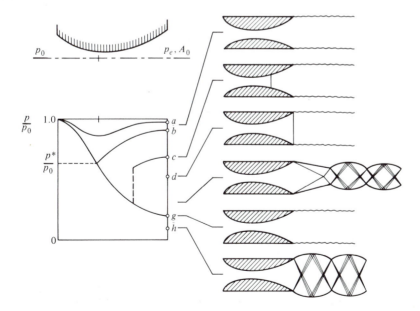

FIG. 9.17 Flow through a convergent–divergent nozzle, showing the pressure profiles and the exit jet configuration for various pressure ratios.

through the nozzle is not isentropic, although the flow on either side of the shock can still be considered isentropic. The lower limit for this kind of flow pattern is given by a shock occurring exactly at the exit as indicated by curve *d*. The flow conditions for exit pressures between *b* and *d* may be computed with the aid of both Tables 6-A and 7-A in Appendix 5. At still lower exit pressures the flow adjusts itself through a series of two- or three-dimensional shock waves, and the average exhaust velocity is generally still supersonic.

The designer must choose an appropriate condition from the foregoing possibilities for the particular application. When the flow leaves the nozzle at supersonic speeds and its pressure exactly equals that of the surroundings (curve *g*), the nozzle is called *correctly expanded*. If the exit area is less than the correctly expanded value for a given back pressure, the nozzle is *underexpanded* and the fluid leaving the nozzle has a pressure greater than that of the surroundings (curve *h*). On the other hand, if the exit area is too large, shock waves form within or just outside the nozzle and the flow is then called *overexpanded*. Sketches of the flow patterns expected for each of these conditions are also shown in Fig. 9.17. The particular mode of operation of any nozzle can be quickly checked by first establishing the limiting pressure curves *b* and *d* and comparing them with the specified exit pressure.

EXAMPLE 9.3

Air at 1000°R and 500 psia (considered to be a perfect gas) flows from a large tank through a convergent–divergent nozzle to another very large tank held at 495 psia. The exit area is 1.0 in.^{2} and the throat area is 0.1 in.^{2}. We wish to find the mass flow rate assuming isentropic conditions (except for possible normal shocks).

Solution: For the given area ratio $A_e/A^* = 10$, Table 7-A of Appendix 5 gives p_e for curve *b* as $p_{e_b} = 500(0.998) = 499.0$ psia. The exit pressure for curve *d* is obtained by first calculating the Mach number just upstream of the shock. From Table 7-A, $p_1/p_0 = 0.007$, $M_1 = 3.93$, so that from Table 6-A $p_{e_d} = 500(0.007)(16.0) = 57.6$ psia. Since the actual exit pressure lies within these limits, a normal shock is set up in the nozzle. The flow is given by Eq. (9.36), which for $\gamma = 1.4$ can be written as

$$\dot{m}_{\mathrm{max}} = 0.53 \, \frac{p_0 A_t}{\sqrt{T_0}},$$

where $\dot{m}_{\mathrm{max}}$ is in lbm/sec, p_0 in psia, T_0 in °R, and A_t in in.^{2}. Thus the flow rate is

$$\dot{m}_{\mathrm{max}} = 0.53 \left[\frac{500(0.1)}{\sqrt{1000}} \right] = 0.835 \text{ lbm/sec.}$$

Often it is necessary to find the shock location within a particular nozzle for a given reservoir and exit conditions. One method that can be used assumes a shock location and calculates the resulting pressure at the exit. The shock location is then changed until a proper exit pressure is obtained by trial and error. A more precise but involved method uses a fictitious converging–diverging nozzle attached to the end of the nozzle in question as shown in Fig. 9.18. Conditions at the second throat are denoted by a double asterisk, and stagnation pressures upstream and downstream of the shock are $p_0 = p_{01}$, and p_{02}, respectively. Since the same fluid passes

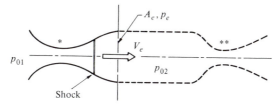

FIG. 9.18   Reference sketch for calculating shock position in a convergent–divergent nozzle.

through both throats, the critical pressure ratios are equal:

$$\frac{p^*}{p_{01}} = \frac{p^{**}}{p_{02}} = \left(\frac{2}{\gamma + 1}\right)^{\gamma/(\gamma - 1)}.$$

Thus, from Eq. (9.36), the mass flow can be expressed as

$$\dot{m} = \left[\frac{\gamma}{RT_0}\left(\frac{2}{\gamma + 1}\right)^{(\gamma + 1)/(\gamma - 1)}\right]^{1/2} p_{01}A^* = \left[\frac{\gamma}{RT_0}\left(\frac{2}{\gamma + 1}\right)^{(\gamma + 1)/(\gamma - 1)}\right]^{1/2} p_{02}A^{**},$$

so that we obtain

$$p_{01}A^* = p_{02}A^{**}. \tag{9.37}$$

We can use this last relationship to locate the shock position by first dividing both sides of Eq. (9.37) by the specified product $p_e A_e$ and rearranging,

$$\frac{p_e A_e}{p_{02} A^{**}} = \frac{p_e A_e}{p_{01}A^*}. \tag{9.38}$$

The right-hand side of Eq. (9.38) is a fixed quantity for given nozzle, reservoir, and exit conditions. Thus the ratio p_e/p_{02} can be obtained by forming the product of the pressure and area ratios for isentropic flow and equating them to Eq. (9.38) (only one value of p_e/p_{02} can satisfy this requirement). Hence the ratio of stagnation pressures can be calculated by rewriting Eq. (9.38) as

$$\frac{p_{02}}{p_{01}} = \frac{p_e}{p_{01}}\left(\frac{p_{02}}{p_e}\right).$$

However, the ratio of stagnation pressures across a shock can be shown to be a unique function of the upstream Mach number (see Problem 9.31), and is also tabulated in Table 6-A. Thus, knowing p_{02}/p_{01}, we determine M_1, and A_1/A^* from Table 7-A for the same Mach number M_1. Since the shape of the nozzle is given, knowledge of A_1/A^* thus locates the shock.

Finally, it should be remembered that the preceding discussion has been limited to one-dimensional flows with no frictional effects. Actual shock waves in ducts are spread out in complicated patterns near the duct walls due to viscous forces. Measurements of pressures in a de Laval nozzle at various discharge pressures are shown in Fig. 9.19. Although isentropic flow predictions give good agreement with experiments, the actual shock patterns produce compression that appears over a finite distance.

Our treatment of the limiting case of a shock just at the exit must also be modified to include two-dimensional effects in the leaving jet. The assumed plane shock at the exit forms into a regular pattern of oblique compression shocks as indicated

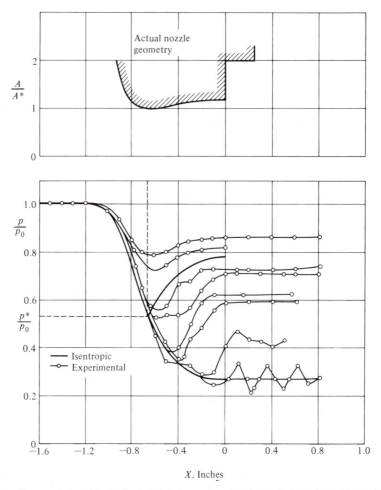

FIG. 9.19 Comparison of experimental pressure ratios for a de Laval nozzle with isentropic flow results. (Courtesy of G. V. Parkinson, University of British Columbia.)

in Fig. 9.17. Pressure traverses as shown in Fig. 9.19 for this region exhibit a sawtooth form of pressure profile through the compression and rarefaction areas. Further discussion of two-dimensional flows is left for Chapter 10.

PROBLEMS

9.41 A tank of air at 15 MPa and 20°C temperature has a valve knocked off to form a *convergent* nozzle with a throat diameter of 0.01 m. The ambient pressure is 100 kPa. Neglect friction and heat transfer.
  (a) Calculate the pressure and velocity at the end of the nozzle.
  (b) If the flow were to expand isentropically to 100 kPa, calculate the velocity, temperature, and flow area.
  (c) Calculate the mass flow in kg/s.
  (d) Calculate the thrust of this nozzle in N.

9.42 Air (assumed to be a perfect gas) expands from a reservoir at $p_0 = 4 \times 10^5$ N/m^2 to a pressure of 10^5 N/m^2. The expansion is assumed to be isentropic.

Find the cross-sectional areas of the flow passage for the points where the pressure is 3×10^5, 2×10^5, and 10^5 N/m^2. The flow rate is to be 10 kg/s.

9.43 Consider a frictionless convergent–divergent nozzle, as shown. The nozzle is supplied from a large reservoir at a pressure p_0 and a temperature T_0. The fluid is air, which may be treated as a perfect gas. There is no heat transfer from the nozzle and the flow may be assumed one-dimensional and frictionless. The atmospheric pressure at the discharge end is equal to p_2, where p_2 is only slightly lower than p_0, so that the flow issuing from the nozzle is subsonic. What is the minimum ratio of the throat area to the discharge area (A_t/A_e) that will not reduce the mass flow rate through the nozzle? Give the answer in terms of p_2/p_0 and γ only.

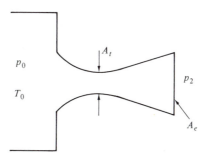

Problem 9.43

9.44 In a hypersonic wind-tunnel installation air is pumped into a large receiver tank. The air from the receiver tank expands through a convergent–divergent nozzle to a Mach number M in the test section. The temperature of the flowing gas in the test section has to be above $T_c = 80$ K to avoid condensation of any of the components of air. Assuming that air behaves like a perfect gas down to a temperature of 80 K, determine the minimum reservoir temperature T_0 that is required in the reservoir to obtain a Mach number M in the test section without any condensation. Neglect the gas velocity in the reservoir. Give the answer in terms of M, γ, and T_c. If the reservoir temperature is limited to 800 K for structural reasons, what is the maximum Mach number that can be reached in the tunnel test section without any condensation? (Let $\gamma = 1.4$.)

9.45 A supersonic wind tunnel has a rectangular throat 130 by 1.0 mm. The pressure in the tank (where the velocity is negligible) is 2 MPa, and the temperature is 530 K. Find the discharge rate in kg/s. Assume that air can be treated as a perfect gas and that the expansion process is isentropic.

9.46 A blow-down supersonic wind tunnel is supplied with air from a large reservoir, as shown. The Mach number in the test section is 2, and the pressure is below atmospheric, so that a shock wave is formed just at the exit. The pressure at point 3 immediately behind the shock is 14.7 psia.
(a) Find p_0, p_1, p_2, and M_3.
(b) A Pitot tube is placed in the exit jet as shown. What is the pressure p_4? Why is $p_4 < p_0$? How does T_4 compare with T_0?

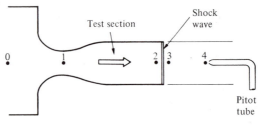

Problem 9.46

9.47 For the purposes of an experiment we wish to design a convergent–divergent nozzle (a de Laval nozzle), which will be supplied from a compressed air reservoir ($\gamma = 1.4$). It is required that there be a normal shock across the exit, and that the jet emerging downstream of the shock have a Mach number of 0.523.

(a) Find the ratio of the area of cross section at the exit to the area of the throat.

(b) Find the ratio of the ambient pressure downstream of the exit to the throat pressure.

(c) What is the ratio of the ambient pressure downstream of the exit to the reservoir pressure?

9.48 A converging–diverging nozzle may occasionally have sufficient length so that frictional effects in the diverging portion become important. In these instances, the maximum flow rate given in Eq. (9.36) may be reduced from the frictionless value. This is usually accounted for by a discharge coefficient c, where

$$\dot{m} = \rho A V = c\dot{m}_{max}.$$

Show that for supersonic operation, the Mach number M at some point in the diverging portion where the area and pressure are A and p, respectively, is obtained from

$$M\sqrt{1 + \frac{\gamma - 1}{2} M^2} = c\frac{p_0/p}{A/A_t} \sqrt{\left(\frac{2}{\gamma + 1}\right)^{(\gamma + 1)/(\gamma - 1)}}.$$

9.49 In a supersonic wind tunnel (tank pressure 2 MPa and temperature 530 K) the circuit is designed so that a normal shock occurs in the divergent part of the nozzle at a point where the airspeed is 950 m/s. Find the temperature, pressure, and Mach number just ahead of the shock, as well as the velocity, temperature, and pressure just after the shock. How much does the entropy rise across the shock?

9.50 Each of the five F-1 engines of the first stage of the Saturn booster rocket (used for the manned missions to the moon) gave a thrust of 1.5 million pounds. Assuming a chamber temperature of 6500°R, a chamber pressure of 500 psia, and an exhaust pressure of 4 psia, find the diameter of the *throat*. Assume the gas to be perfect and all processes to be isentropic. Let $R = 2100$ ft-lbf/°R-slug and $\gamma = 1.2$.

9.51 Rocket nozzles are usually designed as convergent–divergent nozzles for most operating conditions, and the flow is shock-free and nearly isentropic through-out.

(a) Referring to Example 4.1, and assuming a perfect gas, show that the thrust F is given by

$$F = p_0 A_t \gamma \left\{ \left[\left(\frac{1}{\gamma - 1} \right) \left(\frac{2}{\gamma + 1} \right)^{(\gamma + 1)/(\gamma - 1)} \right] \left[1 - \left(\frac{p_e}{p_0} \right)^{(\gamma - 1)/\gamma} \right] \right\}^{1/2}$$

$$+ A_e (p_e - p_a).$$

(b) Show that the thrust F is a maximum when the ratio A_e/A_t is adjusted so that $p_e = p_a$.

9.52 A one-dimensional diffuser, as shown, is designed for a supersonic approach velocity of M. To reduce the velocity at the diffuser discharge to a subsonic speed, the Mach number at the throat has to be equal to unity. Find an expression for the area ratio A_e/A_t, where A_e is the entrance area and A_t is the area of the throat. This expression should contain only the variable M and the ratio γ. As a check, show that the required area ratio is equal to 1 for $M = 1$.

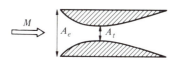

Problem 9.52

9.53 An airplane travels at a Mach number $M = 2.2$ and uses a diffuser as described in Problem 9.52 to provide engine air. Call the inlet area A_1.
(a) Calculate the throat area A_t for isentropic compression as a fraction of A_1.
(b) What Mach number results when the flow is further diffused back to area A_1?

9.54 Air flows through a simple convergent nozzle ending in a throat of 2.5×10^{-4} m^2. The entrance pressure and temperature are 1.0 MPa and 310 K, and the nozzle discharges into a large tank at 680 kPa pressure. The entrance velocity is negligible.
(a) Find the discharge rate in kg/s.
(b) What is the maximum discharge rate obtainable with the given supply pressure? What is the corresponding maximum receiver tank pressure?

9.55 The air in an automobile tire has a pressure and temperature of 50 psia and 530°R. A hole of 0.05-in. diameter is accidentally punched in the tire and assumes the shape of a convergent nozzle. Neglecting frictional effects, find the discharge rate from the hole (the atmospheric pressure is 14.7 psia). What would the jet velocity and Mach number be if the hole were correctly shaped for an isentropic expansion to atmospheric pressure?

9.56 A race car is to be driven by heated compressed hydrogen that is to be expanded (rearward) through a correctly designed nozzle. The hydrogen delivered to the nozzle is at a temperature of 1500°R and at a pressure of 300 psia. The exhaust pressure is 15 psia. The power delivered to the car is to be equivalent to 500 hp at 200 mph. Assume hydrogen to behave like a perfect gas with $R = 770$ (ft-lbf/lbm-°R), $c_p = 3.4$ Btu/lbm-°F, and $\gamma = 1.4$. Determine:
(a) The exhaust velocity of the hydrogen.
(b) The mass flow rate.
(c) The throat area of the nozzle.

9.57 A rocket motor is supposed to deliver a thrust of 10^6 N. The gas in the chamber is at a pressure of 15 MPa and the temperature in the combustion chamber is at 3400 K. The pressure at the exit of the nozzle is 10 kPa. Assume the gas to behave like a perfect gas with the following properties:

$$R = 319 \text{ J/kg} \cdot \text{°C},$$

$$c_p = 2000 \text{ J/kg} \cdot \text{°C} \quad \text{(from which } \gamma = 1.20\text{)}.$$

(a) Find the exhaust velocity.

(b) Using the momentum theorem, determine the flow rate in kg/s required for specified thrust.

9.58 A crude de Laval nozzle with a throat area A^* and a diffuser exit area 16 times larger ($A_E = 16A^*$) is made using a straight-sided conical diffuser, as indicated. The nozzle is fed from an air reservoir ($\gamma = 1.4$) of pressure p_0; the external pressure of the air downstream of the diffuser exit is p_E. Find the ratio of p_0/p_E at which a normal shock will form halfway down the diffuser (i.e., at $x/l = 0.5$).

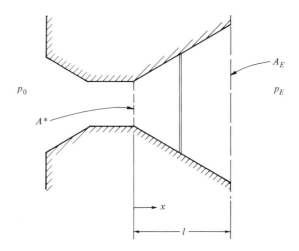

Problem 9.58

9.59 We have adiabatic, steady, frictionless supersonic flow of air ($\gamma = 1.4$) in a tube as shown. The diameter of the tube is 1 m.

(a) The tube is reduced in diameter to form a "throat" having diameter $D_t = 0.75$ m; then the tube expands back to its original diameter. Find the Mach number at the throat and *sketch* the velocity and pressure distributions through this nozzle.

(b) What is the throat diameter (in meters) for $M = 1$ there?

(c) What do you think would happen if D_t is made *less* than the value for your answer in part (b)?

(d) Suppose that at the appropriate location, a normal shock with $M_1 = 2.00$ is experienced.

    (1) What is the diameter where $M_1 = 2.0$?

    (2) What is the Mach number M_2 after the shock?

    (3) After the normal shock takes place, what would the diameter D_t have to be to make $M = 1$ there? Why is it different from your answer in part (b)?

    (4) Sketch the pressure distribution up to the throat in this case.

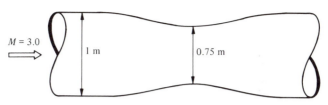

Problem 9.59

9.60 An airplane is in flight at a Mach number of 2.2, where the ambient temperature is $-56°C$ and the pressure is 7 kPa. Assume air to be a perfect gas with $\gamma = 1.4$. The air is drawn into the airplane through an intake duct.

(a) Calculate the stagnation temperature in °C.

(b) Calculate the stagnation pressure for isentropic flow.

(c) Suppose that the intake diameter is 1 m. What diameter would be needed to bring the stream to sonic conditions with isentropic flow? What Mach number would result if the flow were diffused back to the original diameter?

(d) Suppose that a normal shock were to occur at $M = 2.2$ at the intake. What then are the total pressure and Mach number after the shock? Suppose that this flow is now expanded to sonic conditions; what is the required throat diameter? Compare your answer in part (c). Explain the difference.

9.61 Consider the flow of a perfect gas through a convergent nozzle with given reservoir conditions (p_0, T_0). For the present problem, assume that the flow is *isothermal* (constant temperature) and that even the propagation of small pressure pulses takes place under isothermal conditions. The pressure of the surroundings is sufficiently low to assure maximum flow.

(a) Find the maximum velocity at the exit of the convergent nozzle.

(b) Starting with the differential form of the momentum equation (neglect friction), determine the pressure, p_e, right at the discharge section (but still inside the nozzle) in terms of p_0 and T_0. (The pressure of the surroundings must be equal to or lower than p_e.)

(c) Find the mass flow rate if the exit area of the nozzle is A_e.

9.9 Effect of Friction on Flow in Ducts

As a final topic in one-dimensional compressible flows, we now consider the effects of fluid friction on the flow in ducts. Two types of flow will be examined—adiabatic and isothermal. With adiabatic flow the gas particles pass through the duct before a significant amount of heat transfer occurs. It should be noted that although the flow may be adiabatic, frictional or dissipative processes will cause the entropy of the fluid to increase as it moves along the duct, and the isentropic-flow relations derived previously are not applicable.

Isothermal flow may occur in very long ducts such as pipelines, or at lower flow velocities. In this case, the fluid is in thermal equilibrium with its surroundings and the temperature remains constant. Heat must flow across the duct walls, and as in the previous case the fluid entropy changes steadily.

First we consider in a qualitative form the steady, adiabatic flow through a duct of constant area as shown in Fig. 9.20. From Section 9.4 we see that the energy

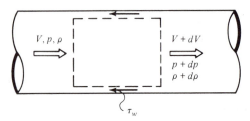

FIG. 9.20 Steady, adiabatic flow through a duct with friction.

equation is given in Eq. (9.14) as

$$h_0 = h + \frac{V^2}{2},$$

or, using the continuity equations $\dot{m} = \rho A V$,

$$h_0 = h + \frac{(\dot{m}/A)^2}{2\rho^2}. \tag{9.39}$$

Equation (9.39) has been plotted in Fig. 9.21a to show the variation of enthalpy with density for given stagnation conditions and several mass flow rates. Lines of constant mass flow are called *Fanno lines* for this flow. It is instructive to replot the Fanno lines as enthalpy versus entropy as in Fig. 9.21b. This is accomplished by first computing lines of constant s in the h versus $1/\rho$ diagram, and then transposing to form h versus s curves as shown. Typical lines of $p = $ const for a perfect gas are also shown in Fig. 9.21b.

Imagine that gas enters the duct with fixed mass flow and entropy. In Fig. 9.21b the Fanno lines have two branches, so two values of enthalpy can satisfy the given conditions. As gas moves along the duct the entropy must increase according to our previous discussion until the limiting case at point c, where the branches meet. It is interesting to calculate the velocity at point c. The differential form of the combined first and second laws of thermodynamics is $T\,ds = dh - dp/\rho$. However, at point c, $ds = 0$, so that $dh = dp/\rho$. The different form of the energy equation $dh + V\,dV = 0$, and continuity equation $dV/V + d\rho/\rho = 0$ can be combined to give, finally, $V^2 = dp/d\rho$. Since the flow is isentropic at point c, we conclude that the velocity must exactly equal the local speed of sound. Further, by checking the direction of variation of pressure with the Mach number for isentropic flow (valid in the imme-

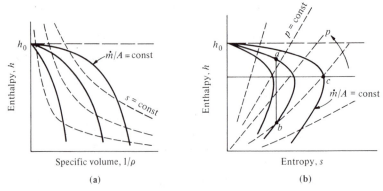

FIG. 9.21 Fanno curves for adiabatic flow in a duct with friction.

diate region of point c), the flow must be *subsonic* on the upper and *supersonic* on the lower branch of the Fanno line.

We can now summarize the expected behavior of various flow situations. If fluid enters a duct at subsonic conditions (corresponding to point a), pressure, density, and temperature *decrease*, while the velocity and Mach number *increase* as the fluid moves along the duct. At a particular duct length, the flow will become sonic at the exit and cannot be accelerated further. Conversely, if the fluid enters the duct at supersonic conditions (point b), pressure, temperature, and density *increase*, while the velocity and Mach number *decrease* along the duct. The flow cannot be reduced to subsonic conditions by a continuous transition because entropy must always increase along the duct.

So far we have considered only qualitative aspects of frictional flows. To relate pressure drops, Mach numbers, and so on, we apply the momentum theorem to the control volume shown in Fig. 9.20 and obtain

$$-dp - \frac{\tau_0 P \, dx}{A} = \rho V \, dV,$$

where P is the perimeter of the duct. The shear stress τ_0 can be defined in terms of a friction factor f, where, as before, $\tau_0 = (f/4)(\rho V^2/2)$. For a circular pipe the ratio A/P is equal to $D/4$, where D is the diameter of the duct. Further, since $\rho V^2 = \gamma p M^2$, the preceding equation can be written as

$$-\frac{dp}{p} - \left(\frac{\gamma M^2}{2}\right) \frac{f \, dx}{D} = \gamma M^2 \frac{dV}{V}. \tag{9.40a}$$

It is possible to arrive at an expression relating the dimensionless shear stress term $f \, dx/D$ to changes in the Mach number only. To do this, we assume a perfect gas $p = \rho RT$, or

$$\frac{dp}{p} = \frac{d\rho}{\rho} + \frac{dT}{T}. \tag{9.40b}$$

For a perfect gas, the Mach number can be expressed as $M^2 = V^2/\gamma RT$, and

$$\frac{dM^2}{M^2} = 2\frac{dV}{V} - \frac{dT}{T}. \tag{9.40c}$$

In addition, we can write the continuity equation as

$$\frac{d\rho}{\rho} + \frac{dV}{V} = 0 \tag{9.40d}$$

and the energy equation Eq. (9.15) as

$$\frac{dT}{T} = -(\gamma - 1)M^2 \frac{dV}{V}. \tag{9.40e}$$

By combining these equations, we obtain

$$\frac{f \, dx}{D} = \frac{1 - M^2}{\gamma M^2} \left(1 + \frac{\gamma - 1}{2} M^2\right)^{-1} \frac{dM^2}{M^2}. \tag{9.41}$$

Suppose that we integrate this expression from a duct inlet ($x = 0$), where the Mach number M is given, to the length L_{max}, where the Mach number becomes unity. The left-hand side of Eq. (9.41) can be written as $\bar{f} L_{max}/D$, where $\bar{f}$ is an average friction coefficient over the length as given by $\bar{f} = (1/L_{max}) \int_0^{L_{max}} f \, dx$. The term $\bar{f} L_{max}/D$ is a

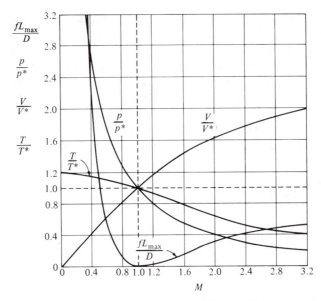

FIG. 9.22 Variation in flow properties for air ($\gamma = 1.4$) in an adiabatic duct with friction at various inlet Mach numbers.

unique function of the inlet Mach number M and is tabulated in Table 8-A in Appendix 5. To every set of inlet conditions, there corresponds an exit condition that makes the flow just sonic there. The properties for this sonic flow are designed by superscript asterisks. It is then possible to derive for each property the ratio of its value at the inlet to that at the corresponding exit (e.g., p/p^*, T/T^*, etc.). These ratios are given in Table 8-A of Appendix 5 and in graphical form in Fig. 9.22. The ratio of the values of any property between two arbitrary points is found simply by

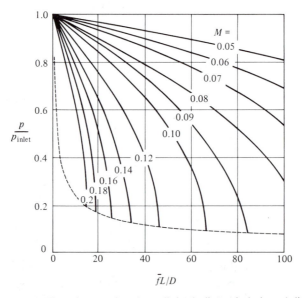

FIG. 9.23 Pressure profiles along a duct for adiabatic flow of air ($\gamma = 1.4$) for various inlet Mach numbers.

obtaining the tabulated values for the two conditions, or a simple spreadsheet calculation can be made as before. For low-speed duct flow it is also instructive to present the foregoing relationships in the form shown in Fig. 9.23. The presssure along a duct deviates appreciably from the incompressible flow value when the duct becomes sufficiently long.

If the duct length is increased beyond its value L_{max} as given before for subsonic inlet conditions, a reduction in mass flow occurs. The effect of friction is thus said to *choke* the flow. This behavior is very similar to that of a frictionless flow through a converging nozzle with an attached duct, as shown in Fig. 9.24a. As with the converging nozzle of Section 9.7, decreasing the back pressure increases mass flow until a critical condition is reached at the tube exit. At this point the flow is a maximum and remains constant with further decreases in back pressure. By using the frictional flow relationships given before, the mass flow can be computed for any given duct length and exit and reservoir conditions. Results of this computation are shown in Fig. 9.24b.

If duct lengths are increased beyond the value L_{max} for supersonic inlet conditions, choking does not occur. However, shock waves may be set up at different locations in the duct as illustrated in Fig. 9.25. With a long duct, a shock wave may

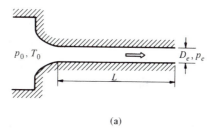

(a)

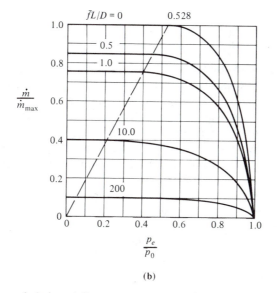

(b)

FIG. 9.24 Discharge of air ($\gamma = 1.4$) versus pressure ratio through adiabatic ducts with different values of $\bar{f}L/D$.

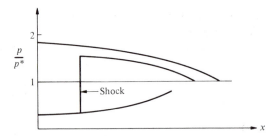

FIG. 9.25 Typical pressure profiles along a duct for both subsonic and supersonic inlet conditions.

even form right at the duct inlet. The interested reader may wish to consult the references at the end of this chapter for more complete discussion of these effects.

Finally, we consider the case of isothermal flows with friction. We can proceed in a manner similar to that for adiabatic flows and compute parallel relationships. There is, however, one very important difference between isothermal and adiabatic duct flows, in that isothermal choking occurs at a Mach number $M = 1/\sqrt{\gamma}$ rather than $M = 1$ as in adiabatic flows. The governing equations, Eqs. (9.40a) through (9.40d), are unchanged except that $dT = 0$ for isothermal flow. The energy equation, corresponding to Eq. (9.40e), is now obtained by differentiating Eq. (9.16):

$$\frac{dT_0}{T_0} = \frac{(\gamma - 1)M^2}{1 + \frac{\gamma - 1}{2} M^2} \frac{dM}{M}. \tag{9.42}$$

Combining Eqs. (9.40a) through (9.40d), we obtain with $dT = 0$,

$$\frac{dp}{p} = \frac{d\rho}{\rho} = -\frac{dV}{V}$$

or

$$\frac{dp}{p} = -\frac{dM}{M} = -\frac{\gamma M^2}{2(1 - \gamma M^2)} f \frac{dx}{D}. \tag{9.43}$$

From Eq. (9.43) we see that $M = 1/\sqrt{\gamma}$ represents a limiting value for the direction of change in flow quantities, much as $M = 1$ was the limit for adiabatic flow. For isothermal flows with $M < 1/\sqrt{\gamma}$, pressure and density *decrease* while the velocity and Mach number *increase* along the duct. The reverse occurs for $M > 1/\sqrt{\gamma}$. If we take $M = 1/\sqrt{\gamma}$ at a maximum duct length L_{max} for any inlet condition, Eq. (9.43) can be integrated from $x = 0$ to $x = L_{\text{max}}$, yielding pressure, density ratios, and so on, as unique functions of the inlet Mach number similar to adiabatic flows. These results are shown in Fig. 9.26.

For a great many problems of engineering interest, the results of isothermal computation do not differ greatly from adiabatic results. Because neither strictly adiabatic nor isothermal flows actually occur in practice, the two computations give upper and lower limits which can be used for estimating purposes.

EXAMPLE 9.4

Air enters a circular pipe at a pressure $p_1 = 30$ psia and a temperature $T_1 = 80°$F. The pipe is cast iron, 6 in. in diameter and 1310 ft long, and the airflow rate is 200

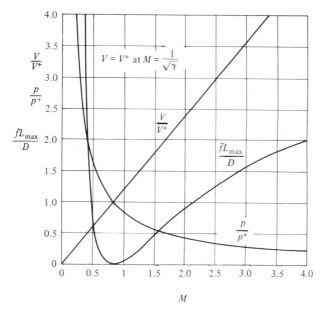

FIG. 9.26 Flow properties at $L = L_{max}$ for the flow of air ($\gamma = 1.4$) in an isothermal duct with friction at various inlet Mach numbers.

lbm/min. The flow can be assumed to be adiabatic, and air acts like a perfect gas with $\gamma = 1.4$. We wish to find the pressure, temperature, density, and Mach number at the pipe exit.

Solution: At the inlet, the air velocity is

$$V_1 = \frac{\dot{m}}{\rho A} = \frac{\dot{m}RT}{pA}$$

$$= \frac{200(53.3)(540)(144)(4)}{60(30)\pi(6)^2} = 113 \text{ ft/sec.}$$

The local speed of sound at the inlet is

$$a = (\gamma RT)^{1/2} = [1.4(32.2)(53.3)(540)]^{1/2} = 1140 \text{ ft/sec.}$$

Thus the inlet Mach number is approximately $M_1 = 0.1$. The Reynolds number at the inlet is

$$(\text{Re})_1 = \frac{V_1 D}{\nu} = \frac{113(6)}{12(0.75 \times 10^{-4})} = 7.5 \times 10^5.$$

From Fig. 5-4 for cast iron pipe $f = 0.02$. The friction factor is seen to be nearly independent of the Reynolds number in this region and can be taken as constant without a significant loss of accuracy.

The conditions at the pipe outlet can be related to inlet conditions by use of Table 8-A in Appendix 5. From Eq. (9.41)

$$\frac{\bar{f}L}{D} = \int_1^2 \frac{f\,dx}{D} = \left(\frac{\bar{f}L_{max}}{D}\right)_1 - \left(\frac{\bar{f}L_{max}}{D}\right)_2.$$

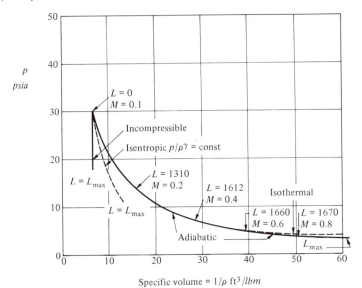

Specific volume = $1/\rho$ ft³/lbm

FIG. 9.27 The pressure–density relationship for the frictional flow of air in a pipe under adiabatic and isothermal conditions for the following conditions: $p_1 = 30$ psia, $T_1 = 80°F$, $\dot{m} = 200$ lbm/min. $D = 6$ in., $f = 0.02$.

Thus

$$\left(\frac{\bar{f}L_{max}}{D}\right)_2 = \left(\frac{\bar{f}L_{max}}{D}\right)_1 - \frac{fL}{D}.$$

From Table 8-A, Appendix 5, $\bar{f}L_{max}/D = 66.922$ at $M_1 = 0.1$. Also, we have

$$\frac{\bar{f}L}{D} = \frac{0.02(1310)12}{6} = 52.4,$$

so that

$$\left(\frac{\bar{f}L_{max}}{D}\right)_2 = 66.922 - 52.4 \approx 14.5$$

or $M_2 = 0.2$ at the pipe exit.

The pressure, temperature, and velocity at the pipe exit can be found by noting that

$$p_2 = p_1 \frac{(p/p^*)_2}{(p/p^*)_1},$$

so that at $M_2 = 0.2$,

$$p_2 = 30\left(\frac{5.455}{10.944}\right) = 14.92 \text{ psia},$$

$$T_2 = 540\left(\frac{1.190}{1.198}\right) \simeq 540°R,$$

and

$$V_2 = 113\left(\frac{0.218}{0.109}\right) = 225 \text{ ft/sec}.$$

We note that for the given inlet conditions, the maximum length of pipe that can pass the specified flow is

$$L_{max} = (66.922)\,\frac{D}{\bar{f}} = 1673 \text{ ft.}$$

Some features of this particular flow are summarized in Fig. 9.27. The pressure variation with density or velocity has been plotted from inlet conditions to sonic conditions (maximum duct length). Flow along an isothermal duct with the same inlet conditions is also shown for comparison. For most purposes, the discrepancy between the two cases is negligible as long as the inlet Mach number is low enough. The pressure–density variations for incompressible and isentropic flow through a duct of equivalent length are also shown.

PROBLEMS

9.62 Air, originally contained in a large reservoir at $T_0 = 700$ K and $p_0 = 350$ kPa, flows through a pipe of a constant cross-sectional area of 6×10^{-4} m^2. The pipe is insulated. Different lengths of pipe are to be considered. When the pipe is very short, frictional effects may be neglected, but those effects have to be taken into account for long pipes. The pressure at the discharge end of the pipe may be reduced at will.
  (a) What are the maximum values of the discharge velocity (measured right at the pipe exit) and the mass flow rate for a short pipe?
  (b) As the pipe is made longer and longer, how does the maximum discharge velocity change? Explain.
  (c) How does the mass flow rate change as the pipe increases in length? Explain.

9.63 A long adiabatic duct of diameter D is arranged so that conditions at the exit are exactly sonic (i.e., $M = 1$). The length of duct required to obtain this state is L_{max} for any given inlet Mach number M. By integrating Eq. (9.41) directly, show that

$$\bar{f}\,\frac{L_{max}}{D} = \frac{1 - M^2}{\gamma M^2} + \frac{\gamma + 1}{2\gamma}\,\ln\frac{(\gamma + 1)M^2}{2 + (\gamma - 1)M^2}.$$

9.64 At some point in an adiabatic duct the Mach number is 0.6, the pressure is 810 kPa, and the temperature is 560 K. Air flows through the duct at a rate of 1465 kg/s · m^2. For this flow, accurately plot a Fanno curve in terms of temperature versus entropy over the range $M = 0.2$ to $M = 4$. Use the expression for entropy change for a perfect gas given in Appendix 5, with the foregoing conditions taken as a reference for entropy. (Table 8-A, Appendix 5 can also be used.)

9.65 Consider the adiabatic flow of a perfect gas along a duct with friction. Show for this flow that the pressure–density relationship of the gas is

$$p = A\rho - \frac{B}{\rho},$$

where

$$A = \frac{p^*}{\rho^*} + \frac{\gamma - 1}{2\gamma} \left(\frac{\dot{m}}{A}\right)^2 \frac{1}{\rho^{*2}}$$

and

$$B = \frac{\gamma - 1}{2\gamma} \left(\frac{\dot{m}}{A}\right)^2.$$

The asterisk refers to the value of a quantity at a point in the duct where $M = 1$. Compare this p–ρ relationship with that for adiabatic and frictionless (isentropic) flow.

9.66 (a) For a frictional, adiabatic flow of a perfect gas in a duct, show that Eqs. (9.40c) through (9.40e) can be arranged as

$$\frac{d\rho}{\rho} = -\frac{dM^2}{M^2[2 + (\gamma - 1)M^2]}.$$

(b) By integrating this result of part (a), show that

$$\frac{\rho}{\rho^*} = \left[\frac{2 + (\gamma - 1)M^2}{(\gamma + 1)M^2}\right]^{1/2},$$

where $\rho = \rho^*$ at $M = 1$.

(c) For a given mass flow $\dot{m}$ and stagnation temperature T_0, show that the density ρ^*, where $M = 1$, is given by

$$\rho^* = \sqrt{\frac{\gamma + 1}{2\gamma R T_0}} \left(\frac{\dot{m}}{A}\right).$$

9.67 By rearranging Eqs. (9.40b) through (9.40e) and integrating, derive the results

$$\frac{pM}{p^*} = \frac{1}{M}\frac{V}{V^*} = \left(\frac{T}{T^*}\right)^{1/2} = \left[\frac{\gamma + 1}{2 + (\gamma - 1)M^2}\right]^{1/2},$$

where the asterisk denotes values at a point where $M = 1$.

9.68 When dealing with an adiabatic duct flow of a perfect gas, it is very useful to have a relationship linking the state of the gas at two points separated by a length L. Show that the required relationship is

$$\rho_2^2 = \rho_1^2 - \rho^{*2}\left(\frac{2\gamma}{\gamma + 1}\frac{\bar{f}L}{D} + 2\ln\frac{\rho_1}{\rho_2}\right).$$

Hint: First rearrange Eq. (9.40a) as an equation involving only $f\,dx/D$, $d\rho/\rho$, and M. Then use the result of Problem 9.66(b) to eliminate M in terms of ρ.

9.69 Show that for an isothermal, frictional flow of a perfect gas through a vertical duct, including the effect of gravity, the relationship between the Mach number and the distance is obtained from

$$\frac{dM}{M} = \frac{\gamma M^2}{2(1 - \gamma M^2)}\left(\frac{f}{D} \pm \frac{2g}{\gamma R T M^2}\right)dx,$$

where the plus sign refers to the upward flow. Sketch the possible variations in M versus x with downward flow for various values of the parameter $2Dg/\gamma R T f$.

9.70 For an isothermal, frictional flow in a duct show that

(a)
$$\frac{\bar{f}L_{\max}}{D} = \frac{1 - \gamma M^2}{\gamma M^2} + \ln \gamma M^2.$$

Note: $M = 1/\sqrt{\gamma}$ at $L = L_{\max}$.

(b) Show that

$$\frac{p}{p^+} = \frac{\rho}{\rho^+} = \frac{V^+}{V} = \frac{1}{\sqrt{\gamma}} \frac{1}{M},$$

where the superscript plus denotes the value of a quantity at a point where $M = 1/\sqrt{\gamma}$.

9.71 Prove that for an isothermal, frictional flow the pressure at two points in a duct separated by a length L is related by

$$p_2^2 = p_1^2 - p^{+2}\left(\frac{\bar{f}L}{D} + 2 \ln \frac{p_1}{p_2}\right),$$

where

$$p^+ = \sqrt{RT}\,\frac{\dot{m}}{A}.$$

9.72 A reservoir of air at 61.5 psia and 80°F discharges through a smooth converging nozzle into a long pipe as shown. The pipe has a diameter $D = 0.8$ in. and a friction factor $f = 0.025$. The pressure at the duct inlet is 60 psia, and the flow in the converging nozzle portion may be considered isentropic. Find the length of duct required to reduce the pressure to 15 psia for both adiabatic and isothermal flow.

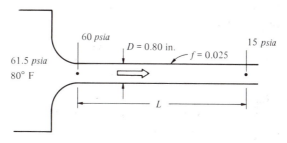

Problem 9.72

9.73 A tank of O_2 gas at 15 MPa pressure and 20°C temperature is attached to a pipe of 1 cm diameter. The flow into the pipe is isentropic. The friction factor may be assumed to be $f = 0.02$. The mass rate of flow is adjusted to give a Mach number of $M_1 = 0.5$ at the inlet.

(a) Calculate the pressure and velocity at the inlet.

(b) Calculate the length of pipe needed to bring the Mach number to a value of $M = 0.8$. Calculate the temperature and velocity there.

9.74 A straight pipe of length l and diameter d is attached to a large tank that contains air at a pressure p_0 and a temperature T_0. The entrance to the pipe is well rounded, so that the flow proceeds smoothly into the pipe. The friction coefficient for the pipe, f, is constant and given. The outside pressure, p_e, is sufficiently low, so that the flow rate through the pipe is at its maximum. This maximum flow rate (in terms of mass flow rate per unit cross-sectional area) is to be determined. The following approach is suggested: Find the Mach number, M_1, at the inlet to the pipe; then find the flow rate corresponding to the isentropic expansion from p_0 and T_0 to the conditions corresponding to M_1. The numerical values are

$$p_0 = 1 \text{ MPa}, \qquad f = 0.020,$$
$$T_0 = 310 \text{ K}, \qquad l = 1.34 \text{ m},$$
$$d = 0.025.$$

9.75 Flow takes place in a constant-diameter pipe without heat transfer but with friction. The flow rate per unit area is $\rho V = 1436 \text{ kg/s} \cdot \text{m}^2$. The stagnation temperature is 604 K. The substance is air, which is assumed to behave like a perfect gas. Find the densities, Mach numbers, and entropies at the two points where the temperature differs by $\pm 80°C$ from that at $M = 1$. Give the entropy in respect to that at $M = 1$ (i.e., let $s = 0$ at $M = 1$).

9.76 Consider the steady, *frictionless* flow of a perfect gas in a constant-area duct with either heat addition or cooling. At a particular point along the duct the Mach number is M, the temperature T, and the flow rate per unit area is ρV.

(a) Using only the conservation of mass, momentum, and the equation of state (and *not* conservation of energy), show that the change in entropy ds along the duct can be related to the temperature change dT by

$$ds = c_p \frac{dT}{T} \frac{M^2 - 1}{\gamma M^2 - 1}.$$

(b) Show that the change in Mach number can be given in terms of the change in temperature dT by

$$\frac{dM}{M} = \frac{1}{\gamma - 1} \frac{dT}{T} \left[\frac{M^2 - 1}{\gamma M^2 - 1} - (\gamma + 1) \right].$$

Note: Starting at any given point in such a duct flow, the results of parts (a) and (b) can be used to calculate the variation of temperature (or equivalently, enthalpy) with entropy. Such a curve is called the *Rayleigh line* for the flow.

9.77 (a) Using the results of Problem 9.76, plot a Rayleigh line for the same conditions as given for the Fanno line in Problem 9.64.

(b) The Fanno and Rayleigh lines intersect at a point on the *supersonic* branch of the Fanno line and at another point on the subsonic branch. Find these points.

(c) Explain how these two intersection points described in part (b) would be related to a shock wave if the flow were initially on the supersonic branch of the Fanno line.

9.78 Check if the normal shock indicated in the graph of the Rayleigh and Fanno lines does actually correspond to the relations we developed for such a shock. Specifically, check M_2, T_2 and Δs for the given initial conditions M_1, T_1, and ρV.

9.79 Consider once again the Fanno and Rayleigh curves corresponding to Problems 9.64 and 9.77. Assume now that the flow in a *frictionless* pipe meets the conditions above, and that this flow is being *heated* starting at a point where $M = 0.3$. Find the maximum amount of heat that can be added.

REFERENCES

1. H. Liepmann and A. Roshko, *Elements of Gasdynamics*, Wiley, New York, 1956, Chaps. 3 and 5.
2. A. H. Shapiro, *Compressible Fluid Flow*, Vol. I, Ronald Press, New York, 1953.
3. H. W. Emmons, ed., *Fundamentals of Gas Dynamics*, Princeton University Press, Princeton, N.J., 1956.
4. Y. B. Zel'dovich and Y. P. Razier, *Physics of Shock Waves and High-Temperature Hydrodynamic Phenomena*, Vol. I, Academic Press, New York, 1966.
5. Equations, Tables, and Charts for Compressible Flow, *NACA TR 1135*, 1953.

CHAPTER 10

Elements of Two-Dimensional Gas Dynamics

10.1 Introduction

In a great many compressible flow situations of practical interest, the flow fields are two- or three-dimensional. For example, the flows over wings or other bodies, or supersonic flows within variable-area channels, are of this type. Such two-dimensional flows can be considerably different from the incompressible two-dimensional flows such as we discussed in Chapter 6.

The simple one-dimensional results of Chapter 9 will prove helpful, although incomplete, for considering the details of two-dimensional flows. In analogy with the de Laval nozzle, where downstream changes are not felt upstream of the throat, entire regions of a two-dimensional flow can be uninfluenced by downstream effects. Several of these ideas will now be considered in further detail.

10.2 Infinitesimal Disturbances in Two-Dimensional Flows

The intuitive arguments about the propagation of a tiny pressure pulse in a long tube presented in Chapter 9 led to the conclusion that such small pressure pulses would travel at the speed of sound in the medium. The same kind of arguments can be used for two-dimensional flows. Consider a long cylinder of small diameter situated in a stationary compressible fluid. If the cylinder is made to pulsate slightly, pressure waves will travel radially outward from the cylinder at the speed of sound. As long as the fluid is isotropic and the cylinder pulsates only radially, the pressure fronts or sound waves will be circles which grow continuously with time, finally traversing all parts of the fluid. This situation is shown for two such pulses in Fig. 10.1a. The circular wave front labeled t_1 was generated by a pulse t_1 seconds earlier, and has traveled a distance at_1 during the interval. A second wave front emitted at a time t_2 seconds earlier (where $t_2 > t_1$) is also shown.

Some basic properties of two-dimensional flows can be demonstrated by assuming the point source to now move steadily through the fluid at a velocity V. First let us consider the case where V is less than the speed of sound. Choosing a reference frame moving with the source, the fluid appears to flow steadily with a velocity V past the source. Pressure pulses travel at the speed of sound with respect to the moving fluid, so that in the moving coordinate system circular wave fronts will remain circular, but their centers will appear displaced along the direction of motion as shown in Fig. 10.1b. Alternatively, this figure can be interpreted from a stationary observer's point of view as showing the source and wave front position after a time

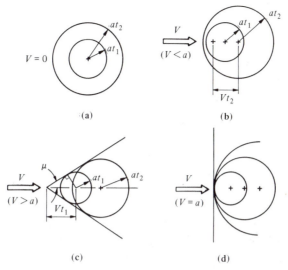

FIG. 10.1 Wave fronts from a pulsating line source moving through a compressible fluid at various velocities.

interval has elapsed (two different time intervals are shown together). It is important to note that the circular wave fronts will still reach all portions of the fluid, since the source moves at less than sonic velocity.

If the source moves faster than the speed of sound, the approaching flow appears supersonic in the moving coordinate system. The cylindrical wave fronts generated at two different times will be centered at distances Vt_1 and Vt_2 downstream of the source, but their radii at_1 and at_2 will be smaller than the displacement distance, as shown in Fig. 10.1c. Thus a limiting envelope to the circular wave fronts is formed, the fluid upstream of the envelope being completely unaware of the presence of the source. This situation is distinctly different from the subsonic case, where pressure waves would eventually reach all parts of the fluid, and illustrates the fundamental difference between sub- and supersonic two-dimensional flows.

The limiting envelope is called the *Mach wave*, and has a semi-apex angle μ, called the *Mach angle*, which is simply related to the approaching flow Mach number by

$$\mu = \sin^{-1} \frac{at_2}{Vt_2} = \sin^{-1} \frac{1}{M}. \tag{10.1}$$

The flow field and disturbance wave fronts for the limiting case of $M = 1$ are shown in Fig. 10.1d. It should be emphasized that the flow behavior described before, and in particular the relationship in Eq. (10.1), is valid only for a line source of very weak pulses. Finite bodies, which generate qualitatively similar results, will be discussed in a following section.

It is interesting to see how the foregoing results can be used in describing supersonic flow over a wall. Neglecting viscosity, supersonic flow over a wall with a small disturbance on its surface will produce results similar to the line source. If the wall is slightly curved, we can also calculate the effect that curvature will have on the fluid pressure, density, and velocity.

Consider the simple case shown in Fig. 10.2, where the wall is bent through a small angle $d\theta$. The flow upstream of the Mach wave will not be influenced by the wall change, but downstream the fluid must flow in the new direction parallel to the

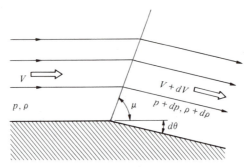

FIG. 10.2 Supersonic isentropic flow over a wall with a slight change in direction.

wall and will have its pressure, density, and velocity changed slightly. In the limit of very small deflection angles, all fluid properties can be considered constant throughout the flow, except for the small changes that take place across the Mach wave. In the absence of viscosity or heat transfer, the flow will evidently remain isentropic.

Denoting the x- and y-components of the vector increment in velocity $d\mathbf{V}$ as du and dv, respectively, flow quantities on either side of the Mach wave can be related by applying the principles of conservation of mass and momentum to the control volume in Fig. 10.3. The length of the diagonal of the control volume is arbitrarily chosen as unity. The continuity equation becomes

$$\rho V \sin \mu = (\rho + d\rho)[V \sin \mu + du \sin \mu + dv \cos \mu]$$

or to first order of differential quantities

$$-V \sin \mu \, d\rho = \rho(du \sin \mu + dv \cos \mu). \tag{10.2}$$

Applying the momentum theorem, we obtain for the x-direction,

$$-dp \sin \mu = -\rho V^2 \sin \mu + (\rho + d\rho)[(V + du)^2 \sin \mu + (V + du) \, dv \cos \mu],$$

which for small changes reduces to

$$-dp \sin \mu = (V^2 \, d\rho + 2\rho V \, du) \sin \mu + \rho V \, dv \cos \mu. \tag{10.3}$$

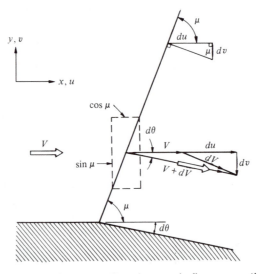

FIG. 10.3 The control volume for computing changes in flow properties during a slight turn.

In a similar way, for the y-direction the momentum theorem yields

$$dp \cos \mu = (\rho + d\rho)[-(V + du) \, dv \sin \mu + (dv)^2 \cos \mu],$$

or for small changes

$$dp \cos \mu = -\rho V \, dv \sin \mu. \tag{10.4}$$

The pressure and density changes in Eqs. (10.2) through (10.4) can be eliminated, and after some manipulation we obtain the simple result

$$\frac{du}{dv} = \tan \mu. \tag{10.5}$$

Referring to Fig. 10.3, we see that the geometrical interpretation of Eq. (10.5) is that the vector increment in velocity dV must be in a direction *normal* to the Mach wave. In particular, the component of velocity tangential to the Mach line remains unchanged across such a weak disturbance.

Further we note that for very small deflections the increment in flow speed dV is approximately du, while the angular deflection $d\theta$ very nearly equals $(-dv/V)$. We can now rewrite Eq. (10.5) as

$$\frac{dV}{V} = -\tan \mu \, d\theta. \tag{10.6}$$

Since a clockwise rotation $d\theta$ is negative, we see that turning the wall away from the flow causes the fluid speed to increase, while turning the wall into the flow will reduce the speed. Correspondingly, pressure changes are associated with small deflections of a supersonic flow. For very small deflections, the one-dimensional steady flow equation of motion, Eq. (9.21), is applicable and can be expressed as

$$\frac{dV}{V} = -\frac{dp}{\rho V^2} = -\frac{dp}{\gamma M^2 p}.$$

Upon combining this result with Eq. (10.6), and noting that $\tan \mu = (M^2 - 1)^{-1/2}$, the pressure–deflection relationship becomes

$$\frac{dp}{p} = \frac{\gamma M^2 \, d\theta}{\sqrt{M^2 - 1}}. \tag{10.7}$$

Since the pressure drops when the wall is turned away from the flow, the deflection is called an *expansion turn*, while an opposite deflection is called a *compression turn*.

A gradual, continuous turning of a flow through a finite angle θ can be viewed as a succession of the infinitesimal deflections described before. At each point on the wall, Mach waves representing the local flow conditions are formed as shown in Fig. 10.4. The final Mach number is a unique function of initial Mach number and total angular deflection. If we note that $V^2 = M^2(\gamma R T)$, or

$$\frac{dV}{V} = \frac{dM}{M} + \frac{1}{2} \frac{dT}{T},$$

and also that from the energy equation [Eq. (9.16)],

$$T_0 = T\left(1 + \frac{\gamma - 1}{2} M^2\right),$$

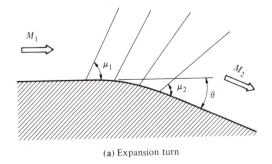

(a) Expansion turn

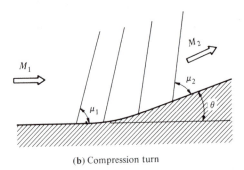

(b) Compression turn

FIG. 10.4 A supersonic, isentropic flow over a continuous gradual turn. An expansion turn reduces fluid pressure; a compression turn increases fluid pressure.

the resulting increment in Mach number for an incremental turn $d\theta$ can be deduced from Eq. (10.6) as

$$d\theta = -\frac{\sqrt{M^2 - 1}\; dM}{M\left(1 + \dfrac{\gamma - 1}{2} M^2\right)}.$$

By integrating the foregoing equation between two specified values of the Mach number, an expression for the total deflection θ_T is obtained (or conversely, an expression for the final Mach number in terms of deflection angle and the initial Mach number). Integration yields

$$\theta_T = -\int_{M_1}^{M_2} \frac{\sqrt{M^2 - 1}\; dM}{M\left(1 + \dfrac{\gamma - 1}{2} M^2\right)} = \theta_T(M_2, M_1).$$

Because of the algebraic complexity of the integral, it is more convenient to rewrite it as follows:

$$\theta_T = \int_{1}^{M_1} \frac{\sqrt{M^2 - 1}\; dM}{M\left(1 + \dfrac{\gamma - 1}{2} M^2\right)} - \int_{1}^{M_2} \frac{\sqrt{M^2 - 1}\; dM}{M\left(1 + \dfrac{\gamma - 1}{2} M^2\right)}$$

or

$$\theta_T = v(M_1) - v(M_2),$$

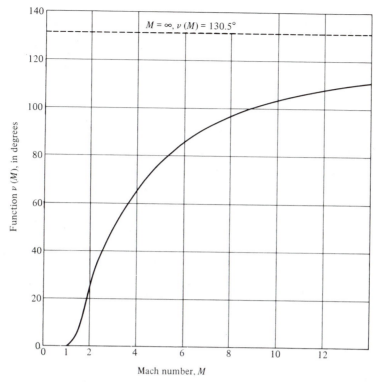

FIG. 10.5 The Prandtl-Meyer function as a function of the Mach number for air ($\gamma = 1.4$).

where

$$v(M) = \int_1^M \frac{\sqrt{M^2 - 1}\, dM}{M\left(1 + \dfrac{\gamma - 1}{2} M^2\right)}. \tag{10.8}$$

The function $v(M)$ is a unique function of the Mach number and is called the *Prandtl–Meyer function*. It is shown in Fig. 10.5, as well as tabulated in the isentropic flow tables in Appendix 5. The total deflection angle θ_T between two given Mach numbers is thus the difference in the Prandtl–Meyer function at those Mach numbers [i.e., $\theta_T = v(M_1) - v(M_2)$].

EXAMPLE 10.1

A duct as shown in Fig. 10.6 first expands a supersonic flow with $M_1 = 3.5$ by turning through an expansion turn of $20°$, then recompresses the flow in a $25°$ compression turn. We wish to find the final Mach number.

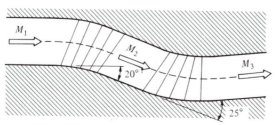

FIG. 10.6 Supersonic, isentropic flow in a curved duct as discussed in Example 10.1.

Solution: Referring to Table 7-A, we find that $v(M) = 58.53°$ at $M_1 = 3.5$. Thus $v(M_2) = v(M_1) - \theta_T = 58.53° - (-20°) = 78.53°$, and $M_2 = 5.175$. For the compression turn, $v(M_3) = 78.53° - 25° = 53.53°$, so that the final Mach number is $M_3 = 3.20$.

10.3 Supersonic Flow over Slender Bodies

The incompressible potential flows discussed in Chapter 6 led to the result that a finite body could not experience a drag force. The situation for supersonic flows is somewhat different, however, even if frictional effects are neglected. A drag force associated with forming the wave field, called *wave drag*, will be present no matter how low a viscosity the fluid may have.

The wave drag on a thin airfoil at small angles of attack can be calculated easily by using the results from the preceding section. If the airfoil is very thin compared to its length, and at a very small angle of attack, the isentropic pressure–deflection relationship given in Eq. (10.7) will be valid. First we consider the simplest airfoil, a flat plate at an angle of attack α as shown in Fig. 10.7.

Mach lines are formed above and below the leading edge and are inclined to the flow at the Mach angle μ. On the upper surface, the flow is deflected downward through the angle α and a small pressure drop occurs, while the pressure increases a corresponding amount on the lower surface. The flow is parallel to the upper and lower surfaces until it reaches the end of the plate, where it returns to its original direction.

The force per unit length acting normal to the upper surface is $(p - dp)l$, while the force on the lower surface is $(p + dp)l$. Thus the net force on the plate acting in the upward direction normal to the plate surface is simply $2\,ldp$. The drag force D is the x-component of this force,

$$D = 2\,ldp \sin \alpha$$

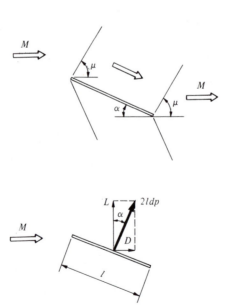

FIG. 10.7 Supersonic flow over a thin plate at small angles of attack.

while the lift L is

$$L = 2\,ldp\,\cos\alpha.$$

The lift coefficient C_L becomes

$$C_L = \frac{2L}{\rho V^2 l} = \frac{4\,dp\,\cos\alpha}{\rho V^2},$$

and similarly, we define the drag coefficient C_D as

$$C_D = \frac{2D}{\rho V^2 l} = \frac{4\,dp\,\sin\alpha}{\rho V^2}.$$

The pressure increment dp may be calculated from Eq. (10.7) with α replacing $d\theta$, and further expanding $\cos\alpha$ and $\sin\alpha$ for small values of α, we finally arrive at the results

$$C_L = \frac{4\alpha}{\sqrt{M^2 - 1}} \qquad (10.9a)$$

and

$$C_D = \frac{4\alpha^2}{\sqrt{M^2 - 1}}. \qquad (10.9b)$$

These formulas give a rough estimate of the lift and drag on a supersonic airfoil, although they neglect frictional effect, thickness, and camber. It should be noted that the drag coefficient C_D is very small for small angles of attack, although wave drag can become appreciable for moderate values of the angle α.

Lift and drag coefficients for more general types of airfoil profiles can be computed by approximating the smooth surfaces with a series of infinitesimal flat plates (Fig. 10.8). By computing the net force on the upper and lower surfaces as for flat plates, the expressions for lift and drag coefficients for small angles of attack can be shown to be

$$C_L = \frac{2L}{\rho V^2 l} \simeq -\frac{2}{l\sqrt{M^2 - 1}} \int_0^l \left(\frac{dy_l}{dx} + \frac{dy_u}{dx}\right) dx \qquad (10.10a)$$

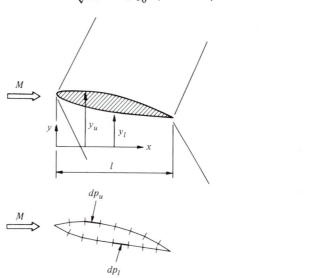

FIG. 10.8 Supersonic flow over a thin airfoil of a general shape at small angles of attack.

and

$$C_D = \frac{2D}{\rho V^2 l} \simeq \frac{2}{l\sqrt{M^2 - 1}} \int_0^l \left[\left(\frac{dy_l}{dx}\right)^2 + \left(\frac{dy_u}{dx}\right)^2\right] dx. \qquad (10.10b)$$

Once again, with sufficiently thin airfoils at small angles of attack, the foregoing equations give fairly accurate estimates of C_L and C_D for an airfoil having specified upper and lower surfaces.

From the expression for C_D just given, it may be seen that to minimize wave drag a supersonic airfoil must have a sharp leading edge. In practice, supersonic airfoil sections are very thin, but not sharp at their leading edges. The effects of aerodynamic heating due to friction are actually more severe for sharp edges, and some compromise in design is usually necessary.

PROBLEMS

10.1 A small source of disturbance moves at a uniform speed in a straight line through the air at 20°C. The position of two wave fronts generated at two different times is shown in the illustration. Find the speed of the source.

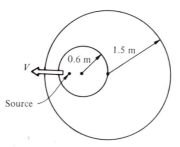

Problem 10.1

10.2 A supersonic aircraft flying at an altitude of 1 mile passes directly over an observer who hears nothing until the plane has flown 2 miles beyond him. Assuming that the aircraft acts as a point disturbance, and the speed of sound $a = 1080$ ft/sec, find:
  (a) The Mach number and Mach angle.
  (b) The position of the airplane when the sound the observer first hears was generated.

10.3 Consider the supersonic flow over a wall with a slight turn as described by Figs. 10.2 and 10.3.
  (a) Verify that the velocity components tangential to the Mach wave before and after the wave are the same.
  (b) Show that the velocity components normal to the Mach wave differ by an amount consistent with the simple one-dimensional result given in Eq. (9.9).

10.4 A supersonic flow with $M = 2$ is deflected gradually through an expansion turn of 30°. Find the final Mach number and the ratio of downstream to upstream pressure.

10.5 An airfoil with a wedgelike shape as shown is exposed to a supersonic flow of air at an angle of attack α. Determine the *lift* coefficient as well as the *drag* coefficient. All angles (and even the sum of two angles) are assumed to be sufficiently small so that the cosines may be set equal to unity and the sines may be replaced by the angles themselves (in radians). Find the numerical value for the lift for the following data: $M = 3$, wing area $= 10 \text{ m}^2$, $T = 290 \text{ K}$, $\rho = 0.5 \text{ kg/m}^3$, and $\alpha = \beta = 5°$.

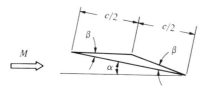

Problem 10.5

10.6 The illustration shows a supersonic airfoil with a flat underside mounted close to the floor of the test section of a supersonic wind tunnel. Calculate the maximum ratio c/h that will cause no aerodynamic interference from the floor with the airfoil. Take α to be a very small angle.

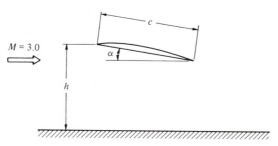

Problem 10.6

10.7 **(a)** Find the lift coefficient and wave drag coefficient of the airfoil shown for $M > 1$.
  (b) Estimate the lift coefficient for $M < 1$. Why might this estimate be inaccurate?

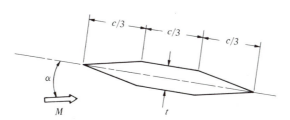

Problem 10.7

10.8 A symmetrical diamond-shaped airfoil is held at zero angle of attack in a uniform supersonic flow.
  (a) Calculate the wave drag coefficient C_D.
  (b) By comparison with a symmetrical airfoil having an arbitrarily curved surface as shown, deduce that the diamond shape produces the smallest value of C_D.

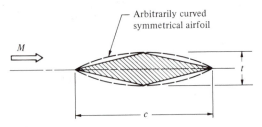

Problem 10.8

10.9 A high-speed aircraft uses a very thin symmetrical airfoil section with leading and trailing edge sections which can be deflected downward as shown. If the drag coefficient $C_D = 0.2$, $M = 2.0$, and the angle of attack is 15°, find δ in degrees. Use the thin airfoil theory.

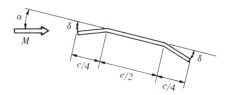

Problem 10.9

10.10 A simple airfoil is formed from two circular arcs as shown. The ratio of thickness to chord t/c for this section is 0.1, and the approaching flow has a Mach number $M = 2$. Find the drag coefficient C_D.

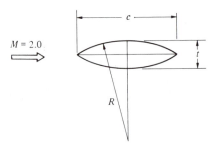

Problem 10.10

10.11 A thin, flat-plate airfoil is fitted with a trailing edge flap as shown. The airfoil has the dimensions shown in the figure and is held in a uniform supersonic flow with a Mach number M. Find the lift and drag coefficients as functions of the flap-deflection angle ϵ. Assume that the thin airfoil theory remains valid.

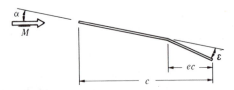

Problem 10.11

10.12 The *Busemann biplane* employs two thin airfoil sections as shown in the illustration. For a sufficiently high approach Mach number M, the Mach waves from one airfoil do not affect the other, but below some value M_0 interaction may occur. Find M_0 and indicate for this arrangement why $C_D = 0$ at $M = \sqrt{2}$.

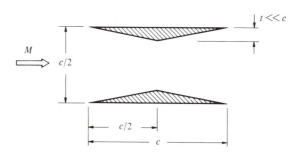

Problem 10.12

10.13 Consider the compressive supersonic turn shown below with $M_1 = 2.00$. The upper wall is turned through an angle of 14.475°.
(a) Find, using charts or tables for air, the downstream Mach number M_2; give reasons for each step you take.
(b) Determine the pressure rise coefficient $c_p = (p_2 - p_1)/(\rho_1 u_1^2/2)$ across the compression (use tables when applicable).
(c) Find the angles μ_1 and μ_2.
(d) Given the curve B–C, *describe* how curve D–E is to be found.

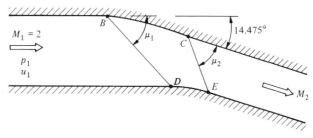

Problem 10.13

10.4 Prandtl–Meyer Expansion Fan

Not all regions of the flow field are adequately described by simple isentropic-turning concepts. Consider a compression turn as depicted in Fig. 10.9a. The Mach waves generated at the wall are not parallel, but inclined together because of the continuous turning of the flow. At some point in the flow the Mach waves will cross, and the resulting flow becomes more complicated. Qualitatively, a pressure rise inappropriate to the expected isentropic flow pressure rise must now occur at a point, suggesting the formation of a shock wave. In practice a shock wave inclined to the main flow direction, called an *oblique shock*, is observed to occur. Further

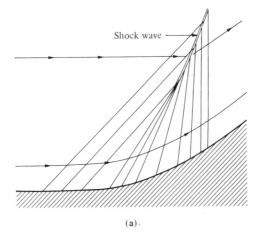

(a).

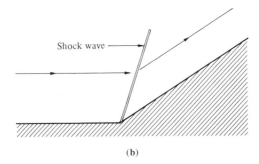

(b)

FIG. 10.9 Illustration of the formation of an oblique shock wave during a compression turn: (a) gradual turn; (b) sharp turn (qualitative only).

calculations are not possible using isentropic flow results. A more complete discussion of oblique shock flows is presented in the following section. For the limiting case of a sharp corner as in Fig. 10.9b, an oblique shock is formed right at the wall (neglecting boundary layer effects), and the entire flow turning is nonisentropic. It is interesting to note the qualitative similarity between the steady oblique shock discussed before and shock formation in the unsteady one-dimensional flow generated by a piston (Section 9.6; see Fig. 9.8).

In contrast to compression turns, abrupt discontinuities cannot be formed in an expansion turn. For supersonic flow over a continuous turn as depicted in Fig. 10.10a, the Mach waves diverge as they spread through the flow, so that no mechanism for oblique shock formation is present. The flow remains isentropic throughout and is called a *simple* expansion.

With a sharp turn as in Fig. 10.10b all the Mach waves for the turn emanate from a single point, and the wave pattern is called a *centered wave*. The resulting isentropic expansion occurs through this "fan" of waves, called the *Prandtl–Meyer expansion fan*. Values of the Mach numbers and Mach angles can be calculated for any given value of the turning angle θ_T using the results of Section 10.2.

Finally, it may be noted that for air a maximum value of the Prandtl–Meyer function v of $130.5°$ occurs as $M \to \infty$. Thus, for flow of air past a sharp, reentrant corner (Fig. 10.10c) having a turning angle greater than $130.5°$, a separated region is formed where the fluid pressure tends to zero. More elaborate calculations, beyond the purpose of the present discussion, are required to describe this flow.

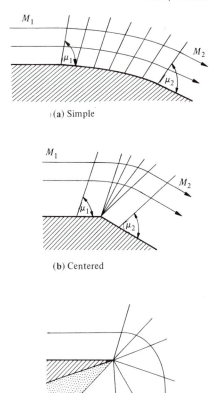

(a) Simple

(b) Centered

(c) Separated

FIG. 10.10 (a) Simple expansion turn; (b) centered expansion turn; (c) flow past a sharp edge with the formation of a void.

In many cases, such as the underexpanded de Laval nozzle discussed in Chapter 9, the fluid pressure is specified in the region after a sharp corner. Thus the Mach number is fixed by isentropic flow requirements, and the flow turns through an expansion fan to the proper pressure and Mach number.

We can now give a qualitative description of the flow leaving an underexpanded nozzle, such as at point h of Fig. 9.17. As shown in Fig. 10.11, the flow expands to the appropriate back pressure through expansion fans centered at O and O'.

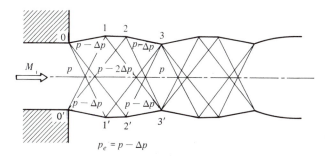

FIG. 10.11   Flow pattern leaving an underexpanded nozzle.

Examining the lower jet boundary, for example, we see that after the expansion fan O'–1–2 the flow proceeds until it first intersects with the expansion fan O–$1'$–$2'$ from the upper corner. The two expansion fans would tend to reduce the pressure at the jet boundary below the specified back pressure. Because the pressure must be constant across the jet boundary $1'$–$2'$, the expansion fan O–$1'$–$2'$ is reflected as an equal and opposite compression turn $1'$–$2'$–3. The flow is thus reverted to the ambient pressure level and is also turned toward the centerline of the jet. At point $3'$ the flow intersects the compression turn from the upper boundary 1–2–$3'$, and is compressed back to the original jet pressure. The flow also becomes parallel to the jet centerline once again. The entire pattern is repeated after point 3 in a series of compressions and expansions until viscous effects dissipate the jet entirely. Although somewhat oversimplified, the foregoing description serves to illustrate some of the features of such a jet.

10.5 Oblique Shock Waves

From our previous discussion it may be seen that a shock wave can be formed at some angle of inclination to the main flow. Such a shock wave is called an *oblique shock* wave, and occurs frequently in supersonic flows over aircraft wings. The gas is compressed across the discontinuity as in the normal shock wave, and also is deflected in some other flow direction. In addition, we shall see that the flow after the shock may be either subsonic or supersonic, depending on the flow deflection angle.

Oblique shock waves are most conveniently described by referring to velocity components in the normal and tangential directions to the shock as shown in Fig. 10.12. We first note that for infinitesimal flow deflections as discussed in Section 10.2, the velocity component tangential to the Mach line remained unchanged. Similarly, for larger flow deflections no physical mechanism is present which could accelerate the flow tangentially as it crosses the shock. Thus only the normal velocity component plays a role in governing the shock characteristics. Since the normal velocity component is reduced across a normal shock, the addition of a tangential velocity serves only to turn the flow into a new direction after the shock.

The foregoing conclusions can alternately be deduced directly from the equations of continuity and conservation of momentum. Neither of these equations is altered by transforming a reference frame moving at constant velocity, so that we need consider only normal velocity components as before. Thus all *static* quantities such as pressure, temperature, and density ratios are unaffected by a tangential velocity

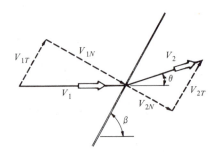

FIG. 10.12 Oblique shock wave geometry.

and can be obtained from the normal shock relationships of the preceding chapter by replacing V_1 and V_2 with $V_1 \sin \beta$ and $V_2 \sin (\beta - \theta)$, respectively. For example, the pressure ratio across an oblique shock, from Eq. (9.30), becomes

$$\frac{p_1}{p_2} = \frac{1 + \gamma M_2^2 \sin^2 (\beta - \theta)}{1 + \gamma M_1^2 \sin^2 \beta},$$

while the relationship between M_2 and M_1 can be obtained from Eq. (9.29) as

$$M_2^2 \sin^2 (\beta - \theta) = \frac{(M_1^2 \sin^2 \beta)(\gamma - 1) + 2}{2\gamma M_1^2 \sin^2 \beta - \gamma + 1}. \tag{10.11}$$

Alternatively, the tables of normal shock relations in Appendix 5 can be used for obtaining static quantity ratios by replacing M_1 with $M_1 \sin \beta$. Note that the normal component of velocity must be supersonic for an oblique shock, that is, $M_1 \sin \beta > 1$.

The specific relationship between deflection angle θ and shock wave angle β can be obtained from the continuity equation. From Fig. 10.12 we first note that

$$\frac{V_{2N}}{V_{1N}} = \frac{\rho_1}{\rho_2} = \frac{\tan (\beta - \theta)}{\tan \beta}.$$

However, the density ratio is obtained from Eq. (9.31) with the substitutions of $M_1 \sin \beta$ for M_1, $M_2 \sin (\beta - \theta)$ for M_2, and by expressing M_2 in terms of M_1 from Eq. (10.11). The result is the following expression:

$$\tan \theta = 2 \cot \beta \, \frac{M_1^2 \sin^2 \beta - 1}{M_1^2(\gamma + \cos 2\beta) + 2}.$$

Curves of β versus θ for a constant upstream Mach number are included in Appendix 5 and can be used for rapid determination of oblique shock properties. Appendix 5 also contains a curve depicting the pressure rise across an oblique shock for various upstream Mach numbers M_1 and deflection angles θ. Further interpretation of these curves will be given in the following pages.

The energy equation, unlike the continuity and momentum equations, must be modified for oblique shocks since the tangential velocity contributes to the kinetic energy of the flow. The relationship between total velocities V_1 and V_2 for normal shock waves given in Eq. (9.28) becomes somewhat changed. It is useful to obtain such an expression for oblique shocks. The continuity and momentum equations are unchanged if we simply replace V_1 and V_2 with V_{1N} and V_{2N}. However, Eq. (9.27) becomes

$$V_{2N} - V_{1N} = \frac{R}{V_{1N}} \left[T_0 - \frac{(V_{1N} + V_{1T})^2}{2c_p} \right] - \frac{R}{V_{2N}} \left[T_0 - \frac{(V_{2N} + V_{2T})^2}{2c_p} \right].$$

Recognizing that $V_{2T} = V_{1T}$, we obtain, after reduction, the extended form of Eq. (9.28):

$$V_{1N} V_{2N} = \frac{2\gamma R T_0}{\gamma + 1} - \frac{\gamma - 1}{\gamma + 1} V_T^2. \tag{10.12}$$

The normal shock relationship is recovered by setting $V_T = 0$.

The description of oblique shock waves given in the preceding paragraphs can be used as a basis for all practical calculations. However, a further useful and illustrative presentation is given by the *shock-polar diagram*, which relates x and y velocity components after the shock to the upstream velocity. To construct the shock polar, consider the velocity triangle for the oblique shock as shown in Fig. 10.13a. Here the upstream and downstream velocity triangles have been superimposed, and for convenience the x-direction is taken coincident with the upstream flow. From the diagram we obtain

$$V_{1N} = V_1 \sin \beta = \frac{V_1(V_1 - u_2)}{\sqrt{(V_1 - u_2)^2 + v_2^2}}$$

and

$$V_{2N} = V_{1N} - \sqrt{(V_1 - u_2)^2 + v_2^2}$$

and also

$$V_T = V_1 \cos \beta = \frac{V_1 v_2}{\sqrt{(V_1 - u_2)^2 + v_2^2}}.$$

If these equations are substituted into Eq. (10.12), we obtain a relationship between v_2 and u_2 with V_1 as a parameter. The final result becomes

$$v_2^2 = (V_1 - u_2)^2 \frac{V_1 u_2 - V^{*2}}{[2/(\gamma + 1)]V_1^2 - V_1 u_2 + V^{*2}}. \tag{10.13}$$

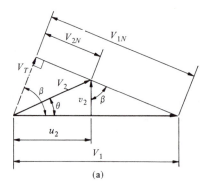

(a)

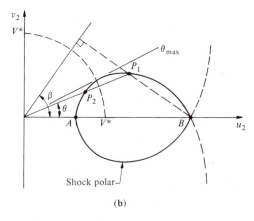

(b)

FIG. 10.13 (a) Velocity triangles for an oblique shock wave; (b) construction of a shock polar diagram.

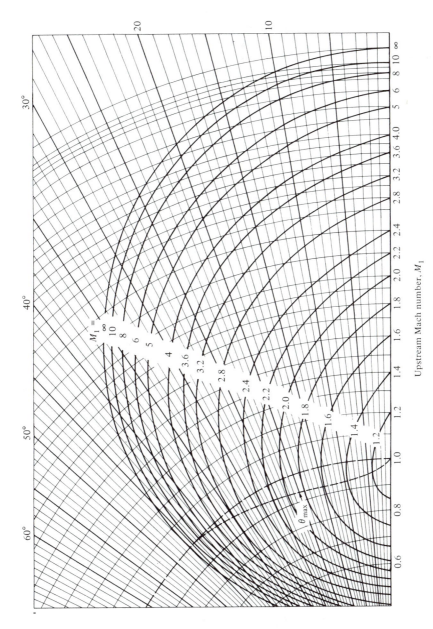

Upstream Mach number, M_1

FIG. 10.14 Shock polar diagram for air ($\gamma = 1.4$). Example 10.2 illustrates the use of this diagram.

Thus, for a supersonic flow with V_1 and V^* [defined in Eq. (9.28)] specified, we can plot values of v_2 for varying values of u_2, as shown in Fig. 10.13b. The plane formed by the velocity components u, v is called the *hodograph plane* and is sometimes more useful than an x, y plane in discussing two-dimensional supersonic flows. The curve in Fig. 10.13b actually continues to the right of point B. However, this region has no physical interest and will not be considered further.

It can be seen that for a given value of V_1 and deflection angle θ, two possible downstream flows corresponding to points P_1 and P_2 are admissible. To see the distinction between these, consider the case of the normal shock with $v_2 = 0$. From Eq. (10.13), this can occur for $V_1 = u_2$, that is, no shock at all, and so would correspond to point B. The second possibility is $V_1 u_2 = V^{*2}$, which is the usual normal shock relationship. This case occurs at point A, and it can be seen that the circle of radius V^* will fall between points A and B. Thus the solution on the P_1 branch of the shock polar (a weak shock) corresponds to supersonic flow after the shock, while subsonic flow occurs on the P_2 branch (a strong shock). From the properties of the construction it can further be seen that the shock inclination angle β is determined by finding the normal to a line connecting the appropriate point P to the point B on the u_2 axis. From Fig. 10.13b it is also evident that a given flow has a maximum deflection angle θ_{max} through a shock. The downstream velocity at θ_{max} is slightly less than sonic for all upstream Mach numbers (except the limiting cases $M_1 = 1$ and $M_1 = \infty$).

Rapid estimates of oblique shock characteristics can be obtained from Fig. 10.14, which combines several shock polars for different approaching flow Mach numbers.

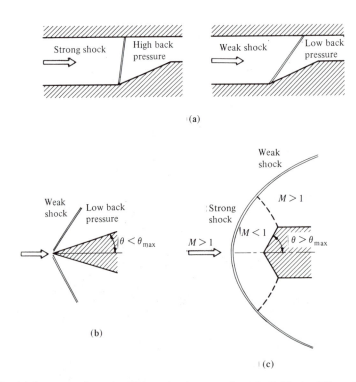

(a)

(b)

(c)

FIG. 10.15 (a) Strong and weak oblique shock waves in a duct due to different back pressures; (b) weak oblique shock formation on a wedge in a uniform free stream flow; (c) detached oblique shock on a blunt body. Far away from the body the shock reverts to a weak oblique shock.

The u and v axes have been made dimensionless by dividing by the critical velocity V^*, and they are actually scaled with Mach number using the relationship between M and M^* developed in Chapter 9. The angles θ and β, and the Mach numbers M_1 and M_2 can be read directly from the plot by using the superimposed polar coordinate grid.

The practical situation determines whether the strong shock or weak shock occurs. Consider, for example, an oblique shock with v_2 nearly zero. The weak shock solution very nearly approaches the Mach line and only a small pressure rise occurs, while the strong shock is very nearly a normal shock with a large pressure rise. Thus, for flow in a duct with one wall bent through an angle θ as shown in Fig. 10.15a, the back pressure in the duct determines the flow geometry. For a wedge-shaped body moving through a stationary fluid as in Fig. 10.15b, only the weak shock is observed since the fluid pressure must return to its free-stream value far away from the body.

If the wall or wedge deflection angle is increased beyond the value θ_{max}, the shock cannot be maintained at its previous location, and a detached shock as in Fig. 10.15c is observed in practice (also see Fig. 9.10). The shock front becomes curved in a way that depends on detailed body shape. Over some portion of the flow, the shock structure corresponds to a strong shock and the downstream flow is subsonic. Farther from the body the shock curves back and corresponds to a weak shock with supersonic flow downstream. Although predictions of the flow patterns can be made for some simple cases, detached shock flows are extremely complicated and results are usually obtained from experiments.

EXAMPLE 10.2

An oblique shock inclined at $30°$ to an approaching flow of $M_1 = 3$ strikes and is reflected from a solid wall as shown in Fig. 10.16a. We wish to calculate the final Mach number and the reflected shock position.

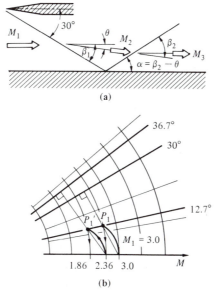

FIG. 10.16 (a) Reflection of an oblique shock from a plane wall; (b) Illustration for Example 10.2.

Solution: For $\beta_1 = 30°$ and $M_1 = 3$, the figures in Appendix 5 give $\theta = 12.7°$. Further, $M_1 \sin \beta_1 = 1.5$, so that from Eq. (10.11) or Table 6-A, Appendix 5, $M_2 \sin (\beta_1 - \theta) = 0.701$, or $M_2 = 2.36$. The reflected shock must return the flow parallel to the wall; that is, $\theta = 12.7°$ for the reflected shock. Thus, once again from Appendix 5, for $M_2 = 2.36$ we obtain $\beta_2 = 36.7°$, and in a way similar to the foregoing $M_3 = 1.86$. The final inclination angle $\alpha = \beta_2 - \theta = 24°$.

Alternatively, the problem can be solved by reference to Fig. 10.14 as shown in Fig. 10.16b. A line normal to the 30° ray and passing through $M_1 = 3$ on the horizontal axis locates point P_1, and so θ can be read directly. The downstream Mach number M_2 is found moving along a circular arc from P_1 and reading $M_2 = 2.36$ from the scale. Next, the flow must turn through $\theta = 12.7°$, so that by following the $M_2 = 2.36$ shock polar to the 12.7° ray we find point P'_1. Once again following a circular arc to the axis, we obtain $M_3 = 1.86$. The angle β_2 can be read by finding the ray normal to a line passing through P'_1 and $M_2 = 2.36$ on the horizontal axis.

As a final comment regarding oblique shocks we note that operation of an over-expanded de Laval nozzle between points d and g on Fig. 9.17 will result in formation of oblique shocks at the nozzle exit. Referring to Fig. 10.17, we see that the fluid in the nozzle is at a lower pressure than the exhaust region, so oblique shocks are formed from points O and O'. The jet is deflected inward as well as being compressed, and further compression takes place across the reflection shocks 1–2 and 1–2' which form from the intersection of the first oblique shock pair. The jet is thus raised to a pressure above the exhaust region, and the rest of the jet expands through a series of expansion and compression fans as with the underexpanded nozzle discussed in the preceding section.

For some angles of shock interaction, the reflection process described here and in Example 10.2 is not possible, and a much more complicated interaction occurs. A discussion of these topics is beyond the scope of this book, and the reader may wish to consult the references listed at the end of the chapter.

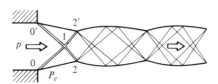

FIG. 10.17 Flow patterns showing reflections in a jet from an overexpanded nozzle.

PROBLEMS

10.14 A flow of $M_1 = 2.0$ is compressed by a circular arc turn to a final Mach number of 1.4, after which the wall is straight. Make a *neat scale drawing* of the turn and the Mach lines at $M = 2.0$, 1.8, 1.6, and 1.4, as well as two or three streamlines. Indicate the region where an oblique shock might form.

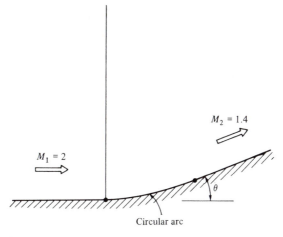

Problem 10.14

10.15 A parallel stream with a pressure of 70 kPa and a Mach number of 2 approaches a sharp 10° corner as shown.
  (a) Assuming that an oblique shock of the weak type is formed, calculate the final Mach number, pressure, entropy change, and ratio of final flow per unit area to the initial flow per unit area.
  (b) Make a sketch to scale showing streamlines, the shock line, and the Mach waves upstream and downstream from the shock.
  (c) Compare the results of part (a) for the final Mach number and pressure with the analogous result for a 10° turn based on a Prandtl–Meyer corner-type flow.

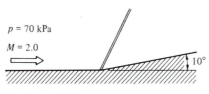

Problem 10.15

10.16 A supersonic flow of $M_1 = 2.0$ is turned through an angle of 10° by a centered expansion fan.
  (a) Using the charts, determine the new Mach number.
  (b) The wall (see sketch) then abruptly changes back to the original direction. Describe what happens.
  (c) Determine the final Mach number M_3.

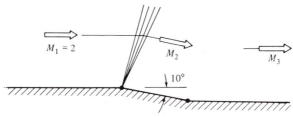

Problem 10.16

10.17 A supersonic flow of Mach number M_0, pressure p_0, and density ρ_0 is turned through a centered expansion turn of θ_0 degrees. Derive an expression for the stream function $\psi(r, \theta)$ for such a turn.

10.18 Two intake designs for a supersonic aircraft are to be investigated to determine the final pressure obtained. The first design simply compresses the air through a normal shock, whereas the second involves an oblique shock plus a normal shock as shown in the figure. Find the final pressure for the two configurations if the approaching stream has a Mach number of 2.5 and a pressure of 20 kPa. Comment on any differences that result.

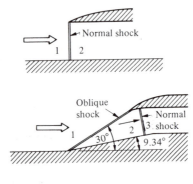

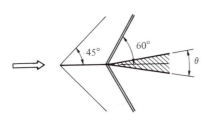

Problem 10.18

10.19 Observations of the flow over a two-dimensional wedge show that an oblique shock is formed at an angle of 60° to the approaching flow. A very slender probe attached to the front of the wedge as shown produces a Mach wave inclined at 45° to the flow. Find the wedge angle θ.

Problem 10.19

10.20 The airfoil described in Problem 10.6 is to be tested at an angle of attack $\alpha = 20°$. Using oblique-shock results, find the minimum ratio of h/c for no interference with the oblique shock reflected from the wall.

10.21 A thin wedge with a blunt base is placed in a uniform supersonic flow with $M = 2$. When the sharp end of the wedge is forward, the pressure at point A is measured to be 12 psia. With the wedge turned blunt end forward in the same flow, a detached bow shock is formed, and the pressure recorded at point B is 410 kPa. Find the free-stream pressure p_∞ and the wedge angle θ.

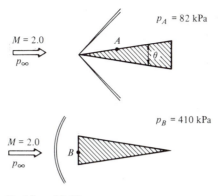

Problem 10.21

10.22 The exhaust jet of a large rocket motor under test at sea-level conditions is shown in the illustration. The Mach number in zone 1 is determined to be $M_1 = 2$.

(a) Find the altitude at which the jet would be free of shock or expansion waves for the same flow conditions in zone 1. A table of atmospheric properties is in Appendix 2.

(b) Calculate the Mach number and the direction of the flow in zone 2 for the conditions shown.

(c) Describe how you would calculate the oblique shock wave angle between zones 2 and 3, and the pressure and Mach number in zone 3.

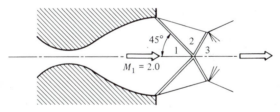

Problem 10.22

10.6 Compressible Potential Flow

We now turn to the problem of calculating stream functions and velocity potentials for a compressible flow. We recall from Chapter 6 that in many cases the flow of an incompressible fluid can be considered irrotational, allowing the formulation of a velocity potential from which the velocity field can be calculated. It was mentioned in Section 6.1 that when heating takes place and the fluid density is a function of both pressure and temperature, the flow is usually rotational. Thus a velocity potential cannot be formed for the most general type of compressible flow. However, in the special case of frictionless adiabatic (or isentropic) flow of a perfect gas the fluid acts as a *barotropic fluid* because pressure can be related to density alone as in Eq. (9.7). Barotropic fluid flow is often irrotational, and the potential flow methods of Chapter 6 are directly applicable. Thus velocity fields for the isentropic flow of a perfect gas can also be calculated from $\mathbf{v} = \nabla\varphi$, where φ is the appropriate velocity potential function.

The equation satisfied by φ for a compressible flow is considerably more complex than that for an incompressible flow. Recalling that the condition of irrotationality requires that p be a function of ρ, we may write $p = p(\rho)$, so that in the absence of body forces the Eulerian equations can be written

$$\frac{\partial \mathbf{v}}{\partial t} + (\mathbf{v} \cdot \nabla)\mathbf{v} = -\frac{1}{\rho}\left(\frac{\partial p}{\partial \rho}\right)_s \nabla \rho = -a^2 \frac{\nabla \rho}{\rho}. \tag{10.14}$$

For the sake of simplicity we shall now limit ourselves to two dimensions and select a Cartesian coordinate system. The equation of motion may then be written as

$$\frac{\partial u}{\partial t} + u\frac{\partial u}{\partial x} + v\frac{\partial u}{\partial y} = -\frac{a^2}{\rho}\frac{\partial \rho}{\partial x} \tag{10.14a}$$

$$\frac{\partial v}{\partial t} + u\frac{\partial v}{\partial x} + v\frac{\partial v}{\partial y} = -\frac{a^2}{\rho}\frac{\partial \rho}{\partial y} \tag{10.14b}$$

and the continuity equation becomes

$$\frac{\partial \rho}{\partial t} + \rho\left(\frac{\partial u}{\partial x} + \frac{\partial v}{\partial y}\right) + u\frac{\partial \rho}{\partial x} + v\frac{\partial \rho}{\partial y} = 0. \tag{10.14c}$$

Now if we differentiate the Bernoulli equation [Eq. (6.2b)] with respect to time, we obtain

$$\frac{\partial^2 \varphi}{\partial t^2} + \frac{1}{2}\frac{\partial}{\partial t}(u^2 + v^2) + \frac{a^2}{\rho}\frac{\partial \rho}{\partial t} = 0 \tag{10.14d}$$

by making use of the relationship

$$\frac{\partial}{\partial t}\int_{\rho_0}^{\rho}\frac{dp}{\rho} = \frac{\partial}{\partial t}\int_{\rho_0}^{\rho}\frac{dp}{d\rho}\frac{d\rho}{\rho} = \frac{\partial}{\partial t}\int_{\rho_0}^{\rho}\frac{a^2\,d\rho}{\rho}.$$

Let us next multiply Eq. (10.14a) by u, Eq. (10.14b) by v, and add the two. Next we eliminate the terms $u\,\partial\rho/\partial x + v\,\partial\rho/\partial y$ by means of Eq. (10.14c) and the term $\partial\rho/\partial t$ by means of Eq. (10.14d). The resulting equation becomes

$$\frac{\partial^2 \varphi}{\partial t^2} + \frac{\partial}{\partial t}(u^2 + v^2) + \frac{\partial^2 \varphi}{\partial x^2}(u^2 - a^2) + \frac{\partial^2 \varphi}{\partial y^2}(v^2 - a^2) + 2uv\frac{\partial^2 \varphi}{\partial x\,\partial y} = 0. \tag{10.15}$$

This equation governs the behavior of a compressible, two-dimensional, potential flow of a barotropic fluid.

Because the velocity components still appear in the coefficients multiplying derivatives of φ, the equation is nonlinear and very difficult to solve in general. A few special cases satisfying the complete equation have been worked out, for example, the Prandtl–Meyer flow described in Section 10.2. However, no simple general solutions to Eq. (10.15) are available.

Before considering two-dimensional flows in detail, it is very instructive to return to the unsteady one-dimensional problem of small pressure pulses being propagated along a tube as discussed at the beginning of Chapter 9. For small pressure changes the sonic velocity a is nearly constant and may be evaluated with respect to the undisturbed flow. Further, the fluid velocity u is much smaller than a, so that terms involving u^2 can be neglected, and Eq. (10.15) reduces to

$$\frac{\partial^2 \varphi}{\partial t^2} = a^2\frac{\partial^2 \varphi}{\partial x^2}. \tag{10.16}$$

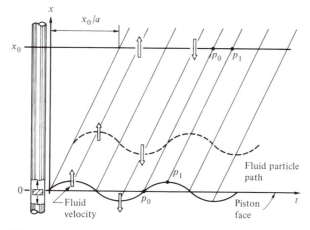

FIG. 10.18   Propagation of pressure waves along a tube.

Equation (10.16) is a hyperbolic partial differential equation and is often simply called the *wave equation*. Possible solutions to this equation are of the form $\varphi = f(x \pm at)$, as may be verified by direct substitution. This result shows that the velocity potential (and therefore all other relevant quantities) will be constant for constant values of $x \pm at$. Thus, relative to some fixed point, small disturbances travel or propagate at the speed of sound a, which is the same conclusion as that reached earlier. To illustrate this effect, let us reconsider the piston-tube arrangement used for the discussion of the speed of sound in Chapter 9. The lines on the t–x plot in Fig. 9.8 are lines of constant $(x - at)$, as discussed previously. If the piston were to oscillate with a small amplitude instead of moving steadily, the fluid velocity and pressure at the piston face are propagated through the fluid unchanged along lines of constant $(x - at)$. As indicated in Fig. 10.18, at some downstream point x_0 in the tube the fluid pressure and velocity are equal to those on the piston face, but are delayed in time by an amount $t = x_0/a$. It may be noted that Eq. (10.16) and its three-dimensional counterpart form the basis of studies of *acoustics*. The analysis of acoustic phenomena such as transmission, reflection, and diffraction of sound waves is represented in the design of loudspeakers, sonar systems, and other acoustic equipment.

Returning to our discussion of two-dimensional flows, we first consider the steady flow over a very thin body held in a uniform *supersonic* stream of speed V_∞. If the body is very thin, its presence can be considered to perturb the free-stream flow only slightly. Thus, to a good degree of approximation, the velocity potential φ is composed of the uniform flow potential φ_∞ plus an additional small quantity φ_1 due to the thin body

$$\varphi = \varphi_\infty + \varphi_1 = V_\infty x + \varphi_1(x, y).$$

In the same way, fluid velocities will have additional small components. Taking the x-axis in the main flow direction, as shown in Fig. 10.19a, we find that the fluid velocity field can be represented as

$$u = V_\infty + u_1$$

$$v = v_1,$$

where the ratios u_1/V_∞ and v_1/V_∞ are much less than unity. The speed of sound a varies throughout the flow, but for small disturbances the flow can be approximated

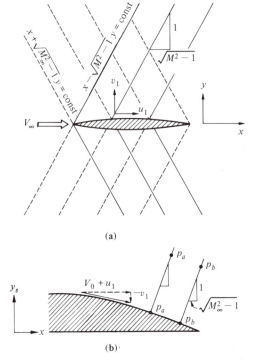

FIG. 10.19 (a) Characteristics for supersonic flow over a thin body; (b) boundary conditions on a body surface.

by its constant free-stream value a_∞. Substituting these expressions into Eq. (10.15), and neglecting such terms as u_1/V_∞, v_1/V_∞, we arrive at the expression

$$\frac{\partial^2 \varphi_1}{\partial y^2} = (M_\infty^2 - 1) \frac{\partial^2 \varphi_1}{\partial x^2}. \tag{10.17}$$

Thus, for a *supersonic* two-dimensional steady compressible flow, the velocity potential is also governed by the wave equation, and the solution can be expressed as $\varphi_1 = f(x \pm \sqrt{M^2 - 1}\, y)$. All flow quantities of interest are constant along lines of constant $x \pm \sqrt{M^2 - 1}\, y$. By starting at some point in the flow where these quantities are specified, for example on the thin body, the entire flow field can be constructed. Note that the slope of lines of constant $x - \sqrt{M_\infty^2 - 1}\, y$ is just $(\sqrt{M_\infty^2 - 1})^{-1}$, so that these lines are the Mach waves described in Section 10.2. For this particular problem the lines of constant $x + \sqrt{M_\infty^2 - 1}\, y$ are lines directed *into* the approaching flow and need not be considered further.

The technique just described for calculating φ is called the *method of characteristics*, the *characteristics* being the lines of constant $x \pm at$ or $x \pm \sqrt{M_\infty^2 - 1}\, y$. This method can be used to solve more general forms of partial differential equations than the simple wave type in Eqs. (10.16) and (10.17); it is a very powerful tool in analysis. The simple wave equation can also be solved in principle by methods such as *separation of variables*. This method will be described further in discussing subsonic potential flows. However, we shall use the method of characteristics for supersonic flow because of its simplicity and the insight it offers into the problem.

To compute the pressure field around a thin body in supersonic flow, we return to the equations of motion [Eq. (10.14)]. If the pressure is written in terms of the free-stream value p_∞ and a small perturbation value p_1 (i.e., $p = p_\infty + p_1$), by substituting the potential function $\varphi = \varphi_\infty + \varphi_1$ into the x-direction equation and keeping only terms of first-order magnitude, we obtain

$$\rho_\infty V_\infty \frac{\partial^2 \varphi_1}{\partial x^2} = -\frac{\partial p_1}{\partial x}$$

or

$$p_1 = -\rho_\infty V_\infty \frac{\partial \varphi_1}{\partial x}. \tag{10.18}$$

An identical result is obtained from the y-direction equation.

Examining the flow on the body itself as in Fig. 10.19b, we find that the condition that the flow be tangent to the surface can be expressed as

$$\frac{dy_s}{dx} = \frac{v_1}{V_\infty + u_1} \simeq \frac{v_1}{V_\infty} = \frac{1}{V_\infty}\left(\frac{\partial \varphi_1}{\partial y}\right)_{y=y_s}. \tag{10.19}$$

Since $\varphi_1 = \varphi(x - \sqrt{M_\infty^2 - 1}\, y)$, we have

$$\left(\frac{\partial \varphi_1}{\partial y}\right)_{y=y_s} = -\sqrt{M_\infty^2 - 1}\, \varphi', \qquad \frac{\partial \varphi_1}{\partial x} = \varphi',$$

where the prime denotes differentiation with respect to the argument $(x - \sqrt{M_\infty^2 - 1}\, y)$. Substituting the expressions for $\partial\varphi_1/\partial x$ and $\partial\varphi_1/\partial y$ into Eqs. (10.18) and (10.19), we obtain

$$p_1 = p - p_\infty = \frac{\rho_\infty V_\infty^2}{\sqrt{M_\infty^2 - 1}}\left(\frac{dy_s}{dx}\right). \tag{10.20}$$

Thus, for any given body shape, the value of p_1 can be computed at some point on its surface (such as a or b in Fig. 10.19b), and this pressure will be constant on a line of constant $x - \sqrt{M_\infty^2 - 1}\, y$ through that point. In particular, for a flat plate at an angle of attack α to the main flow, the pressure coefficient C_p on the upper surface becomes

$$C_p = \frac{2p_1}{\rho_\infty V_\infty^2} = -\frac{2\alpha}{\sqrt{M_\infty^2 - 1}},$$

which is the same result as that derived earlier by flow-turning considerations. Lift and drag coefficients for a slender body in supersonic flow can also be computed, the results being the same as those in Eq. (10.10).

As a final point regarding supersonic flow, we note that the velocity potential itself is constant along the characteristics. To solve for φ_1, we need to compare its value on the body surface and relate it to the entire flow field. From Eq. (10.19) we must have $(\partial\varphi_1/\partial y)$ evaluated at the body surface equal to the slope of the surface $y_s(x)$. If the body is very thin, however, derivatives can be evaluated at $y = 0$ with sufficient accuracy. Thus the boundary condition on φ_1 can be written

$$\left(\frac{\partial \varphi_1}{\partial y}\right)_{y=0} \simeq V_\infty y_s'(x). \tag{10.21}$$

The solution for φ_1 now becomes

$$\varphi_1(x, y) = -\frac{V_\infty}{\sqrt{M_\infty^2 - 1}}\, y_s(x - \sqrt{M_\infty^2 - 1}\, y), \tag{10.22}$$

as can be verified by direct substitution.

For *subsonic* flows over thin bodies or other small disturbances, the procedure used to derive the equation governing the potential function φ is still valid and yields the same equation, Eq. (10.17). However, since $M_\infty < 1$, this equation is written as

$$\frac{\partial^2 \varphi_1}{\partial y^2} + (1 - M_\infty^2) \frac{\partial^2 \varphi_1}{\partial x^2} = 0. \tag{10.23}$$

Equations such as (10.23) and Laplace's equation [Eq. (6.3)] for incompressible potential flows are *elliptic* partial differential equations. Solutions to elliptic equations are generally more difficult to find than for the wave equation, since they require imposition of boundary conditions over an entire curve bounding the region of interest. For example, in a uniform subsonic flow over bodies the conditions on both the body itself and those in the free stream must be considered.

As mentioned earlier, one possible method for solving equations like (10.23) is *separation of variables*. In this method the unknown function $\varphi_1(x, y)$ is assumed to be a product of functions of x and y alone, $\varphi_1(x, y) = X(x)Y(y)$. Substituting into Eq. (10.23), we obtain

$$\frac{1}{X} \frac{d^2 X}{dx^2} = \frac{1}{M_\infty^2 - 1} \frac{1}{Y} \frac{d^2 Y}{dy^2} = -k^2,$$

where k is a constant. The functions $X(x)$ and $Y(y)$ thus must be of the form

$$X(x) = A \cos kx + B \sin kx$$

$$Y(y) = Ce^{\sqrt{1 - M_\infty^2} \, ky} + De^{-\sqrt{1 - M_\infty^2} \, ky}, \tag{10.24}$$

where A, B, C, and D are further constants which must be selected to match the boundary conditions. The actual form of the boundary conditions to be applied is the same as in supersonic flow; for example, at the surface of a *thin* body

$$\left(\frac{\partial \varphi_1}{\partial y} \right)_{y=0} = V_\infty \, y_s'(x), \tag{10.25a}$$

while far away in an undisturbed flow $u \to V_\infty$ and $u_1 \to 0$, so that

$$\left(\frac{\partial \varphi_1}{\partial y} \right)_{y \to \infty} \to 0. \tag{10.25b}$$

EXAMPLE 10.3

In Example 6.5 we calculated the solution for incompressible potential flow over a sinusoidally corrugated wall, as shown in Fig. 6.12. We would like to find the corresponding situation for both subsonic and supersonic flow.

Solution
Supersonic Flow: For this case the results of the method of characteristics presented earlier can be taken over directly. For example, if $y_s = -y_0 \sin 2\pi x/\lambda$, the supersonic perturbation potential $\varphi_1(x, y)$ is obtained from Eq. (10.22):

$$\varphi_1(x, y) = \frac{V_\infty}{\sqrt{M_\infty^2 - 1}} y_0 \sin \left[\frac{2\pi}{\lambda} \left(x - \sqrt{M_\infty^2 - 1} \, y \right) \right].$$

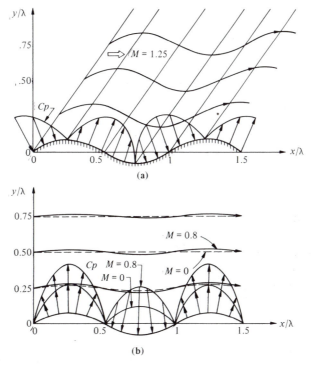

FIG. 10.20   (a) Supersonic flow over a wavy wall; (b) subsonic flow over a wavy wall.

The pressure distribution along the wall can be expressed in terms of the pressure coefficient C_p from Eq. (10.20):

$$C_p = -\frac{2}{\sqrt{M_\infty^2 - 1}} \frac{2\pi y_0}{\lambda} \cos \frac{2\pi x}{\lambda}.$$

Figure 10.20a shows the streamlines and characteristics for $M = 1.25$. Note that the pressure coefficient peak is shifted to the location $y_s = 0$. Also note the resemblance to Fig. (10.18).

Subsonic Flow: The solution for the perturbation potential can be obtained from Eq. (10.24) as

$$\varphi_1(x, y) = (A \cos kx + B \sin kx)(Ce^{\sqrt{1 - M_\infty^2} ky} + De^{-\sqrt{1 - M_\infty^2} ky}).$$

First, imposing the boundary condition far from the wall Eq. (10.25b), we see that $C = 0$, so that

$$\varphi_1(x, y) = e^{-\sqrt{1 - M_\infty^2} ky}(AD \cos kx + BD \sin kx).$$

Next, imposing the boundary condition at the wall (taken as $y = 0$) from Eq. (10.25a) yields

$$\frac{\partial \varphi_1}{\partial y} = -\frac{V_\infty 2\pi y_0}{\lambda} \cos \frac{2\pi x}{\lambda} = -\sqrt{1 - M_\infty^2}\, k(AD \cos kx + BD \sin kx).$$

Thus we must have

$$k = \frac{2\pi}{\lambda}, \qquad BD = 0, \qquad AD = \frac{V_\infty y_0}{\sqrt{1 - M_\infty^2}}$$

and finally,

$$\varphi_1(x, y) = \frac{V_\infty y_0}{\sqrt{1 - M_\infty^2}} e^{-\sqrt{1 - M_\infty^2}\,(2\pi y/\lambda)} \cos \frac{2\pi x}{\lambda}.$$

The pressure coefficient for the subsonic case is obtained from Eq. (10.18):

$$C_p = -\frac{2}{V_\infty}\left(\frac{\partial \varphi_1}{\partial x}\right)_{y=0} = \frac{4\pi y_0}{\lambda\sqrt{1 - M_\infty^2}} \sin \frac{2\pi x}{\lambda}.$$

The streamlines are shown in Fig. 10.20b for both $M = 0.8$ and the incompressible case $M = 0$. For subsonic flow the peak in pressure coefficient occurs where the wall has maximum amplitude.

For subsonic flow, the compressible potential flow equation can also be solved by first transforming it to Laplace's equation and then using the potential flow methods outlined in Chapter 6. If we define a new potential function $\Phi = \sqrt{1 - M_\infty^2}\,\varphi_1$, and new coordinates $x = \xi$ and $\eta = \sqrt{1 - M_\infty^2}\,y$, Eq. (10.23) becomes Laplace's equation,

$$\frac{\partial^2 \Phi}{\partial \xi^2} + \frac{\partial^2 \Phi}{\partial \eta^2} = 0.$$

This transformation of coordinates is called the *Prandtl–Glauert transformation*. It is easily seen that the boundary condition for the function Φ on the body is [from Eq. (10.25a)]

$$\frac{\partial \Phi}{\partial \eta} = \frac{\partial \varphi}{\partial y} = V_\infty\,y_s'(\xi),$$

and the pressure coefficient C_p is given by

$$C_p = -\frac{2u_1}{V_\infty} = -\frac{2}{V_\infty}\frac{\partial \varphi_1}{\partial x} = -\frac{2}{V_\infty\sqrt{1 - M_\infty^2}}\frac{\partial \Phi}{\partial \xi}.$$

From these results it may be concluded that the function Φ is the solution for the *incompressible* flow past a thin body having the same shape as the one in the subsonic compressible flow. From this we are able to deduce a simple and useful result, the *Prandtl–Glauert rule*, namely, that

$$C_p = \frac{C_p \text{ (incompressible)}}{\sqrt{1 - M_\infty^2}}.$$

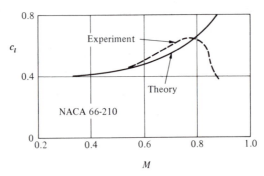

FIG. 10.21 Comparison of the Prandtl–Glauert rule with an experiment for a particular airfoil. (Data from *NACA TN* 1396, 1947.)

Correspondingly, values of lift and drag coefficients for subsonic flows can be estimated by dividing their incompressible values by $\sqrt{1 - M_\infty^2}$. Some results of experimental lift coefficient C_L on a particular airfoil section are shown in Fig. 10.21, along with the values predicted by the Prandtl–Glauert rule. Good agreement may be seen for moderate Mach numbers, but deviations begin to appear as M_∞ approaches unity. The approximations of small perturbation flow just given are not valid near Mach numbers of unity, and a more careful evaluation of orders of magnitude in the potential flow equations is then required, but this is beyond our present scope.

PROBLEMS

10.23 Consider a tube of length L, one end fitted with a movable piston, and the other end closed. The tube is filled with air, and the piston is made to oscillate harmonically with a small amplitude δ and frequency ω.
   (a) Using the small-disturbance approach, find the velocity and pressure perturbation at any point along the tube.
   (b) Discuss the phase relationship between the pressure and velocity fluctuations at the piston face.
      Hint: Use the separation of variables technique first to solve for the perturbation velocity potential.

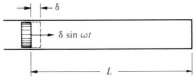

Problem 10.23

10.24 Consider the same piston-tube arrangement as in Problem 10.23, but now the end opposite the piston is open to the atmosphere, so that the pressure at the end of the tube must remain constant. Find the velocity and pressure along the tube in this case.

10.25 Consider the piston-tube arrangement of Problem 10.23 once again! Now a second tube of length L is connected at right angles to the first at the end opposite the piston. The second tube is closed at its end away from the joint. Discuss how this arrangement will affect the pressure at the place where the two tubes are joined and also at the piston face.

10.26 Consider the supersonic flow of a gas in a duct as shown in the illustration. All disturbances are assumed small so that the small-disturbance theory is valid. Also assume that $(M_0^2 - 1)^{-1/2} > 2h/L$. (What is the physical significance of this requirement?)

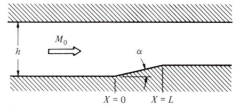

Problem 10.26

(a) Sketch the wave pattern, indicating the compression and expansion waves.

(b) A wave leaves the corner $x = 0$ and is reflected from the upper plane wall. Find the pressure downstream of the reflected wave.

10.27 We have the steady linearized supersonic flow of air (a perfect inviscid gas) in the two-dimensional channel shown. The approaching Mach number is $M_0 = \sqrt{2}$. There is a shallow wedge O–O' of unit length whose slope is $0 = 0.1$ rad (5.73°).

(a) Calculate the Mach angle μ.

(b) Calculate the pressure along O–O', relative to the atmosphere.

(c) What is the pressure at a point just downstream of the wedge?

(d) The Mach waves leaving O and O' strike the upper wall at R and R'. The upper wall is a *solid* surface. This wall adds another disturbance to the flow, which is necessary to cause the wall to be a streamline. First (this is linearized flow), what would be the pressure and slope of the streamline at a point just downstream of R in the *absence* of the wall? Second, what is the pressure along R–R' with the wall present?

(e) Sketch a streamline of the flow halfway between the upper and lower surfaces.

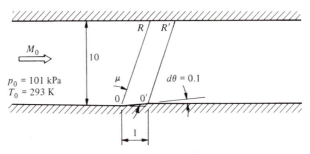

Problem 10.27

10.28 Consider a subsonic flow past a wavy wall in the presence of a plane upper wall as shown. The equation of the lower wall is $y_s = -y_0 \sin (2\pi x/\lambda)$, and the disturbance is small enough so that small-disturbance theory may be used. Find the perturbation velocity potential throughout the channel and the pressure distribution along each wall.

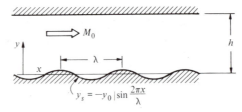

Problem 10.28

10.29 Show how the Prandtl–Glauert transformation may be used in Problem 10.28 to transform an incompressible flow with a certain free-stream velocity and wall wavelength into a compressible flow with $M < 1$ for the same free-stream velocity and wall wavelength. Assume that all disturbances are very small and write down all relevant mathematical steps.

10.30 A helicopter is supported by a three-bladed rotor. Each blade has constant chord c and constant angle of attack α along the span. The lift L on any radial increment dr of blade can be calculated from the subsonic, flat-plate result [see Eq. (6.34)], $C_L = 2\pi\alpha$.

(a) Assuming that wakes from upstream blades do not interfere with an oncoming blade, and taking account of the fact that the blade tip Mach number may be as large as 1.0, show that the local spanwise lift at any radial position r may be approximated as

$$\frac{dL}{dr} \simeq \frac{2\pi\alpha\rho c a^2}{2} \frac{M^2}{\sqrt{1 - M^2}},$$

where

$$M = \frac{\omega r}{a}.$$

(b) Prove that the net lift per blade is given by

$$L = \frac{2\pi\alpha\rho c a^3}{4\omega}\left(\theta_t - \frac{\sin 2\theta_t}{2}\right),$$

where

$$\sin \theta_t = M_{\text{tip}}.$$

(c) Find the required angle of attack for a three-bladed rotor turning at 600 rpm and supporting an 18,000-lb payload if $M_{\text{tip}} = 1$ and $c = 1$ ft.

REFERENCES

1. H. Liepmann and A. Roshko, *Elements of Gasdynamics*, Wiley, New York, 1956.
2. L. D. Landau and E. M. Lifshitz, *Mechanics of Continuous Media*, Addison-Wesley, Reading, Mass., 1959.
3. A. H. Shapiro, *Compressible Fluid Flow*, Vol. II, Ronald Press, New York, 1953.
4. M. J. Zucrow, *Aircraft and Missile Propulsion*, Vol. I, Wiley, New York, 1958.
5. Equations, Tables, and Charts for Compressible Flow, *NACA TR* 1135, 1953.
6. Notes and Tables for Use in the Analysis of Supersonic Flow, *NACA TN* 1428, 1947.

CHAPTER 11

Flow in Open Channels

11.1 Introduction

Engineers have long been concerned with problems of the flow of water in rivers, canals, and other hydraulic structures. Such flows may generally be termed flows in "open channels." Unlike the flow in a filled conduit, the depth of the flow in an open channel with a free surface becomes a variable of the problem, and for this reason open channel flow is in general more difficult to analyze than flows in full-flowing channels. A wide variety of flows—both steady and nonsteady—may be encountered. An example of a typical steady flow consists of the flow of water in an irrigation canal or flood drainage channel. One may anticipate that for such a flow one must in general take into account the effect of friction just as in the case of pipe flow. When the flow is nonsteady, surface waves will be a feature of importance. The variation of depth with time will then render the problem more complex; on the other hand, it is sometimes possible to neglect the effects of friction as, for example, in treating ocean waves. In this chapter we discuss successively some aspects of each of these two types of flow.

In the first portion of this chapter we consider open channel flows which have very gradual changes of liquid depth, cross-sectional area, and slope of the free surface. These approximations permit us to apply the momentum and the continuity equations directly to the flow process. The simplified equations that result are very similar in form to those developed in Chapter 9 for the flow of *compressible* fluids in ducts of gradually varying area. In fact, the similarity is sufficiently close that a "hydraulic analogy" to certain compressible flows can be established. From this remark we may anticipate that many of the phenomena observed in compressible flows have their counterparts in open channel flows. We shall see, for example, that sound waves are analogous to the propagation of long water waves; normal shocks have their counterpart in hydraulic jumps, and the phenomena of sub- and super-sonic gas flows in ducts have their analogs in open channel flows also.

The simplified equations referred to before require that the velocity of the flow be essentially constant over the cross section of the flow. This restriction is removed in the second portion of this chapter for the special case of *infinitesimal* water waves.

11.2 Simplified Equations of Steady Open Channel Flow

Let us consider a steady flow in an open channel, as shown in Fig. 11.1. The cross section of the channel will be taken to be rectangular for simplicity, but the subsequent method of analysis is equally applicable to other types of cross sections (see,

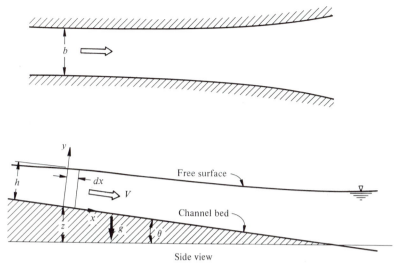

FIG. 11.1   Flow in a rectangular open channel.

e.g., Reference 1). The breadth of the channel is denoted by b and the elevation of the channel bed above an arbitrary datum plane is denoted by z. The slope of the bed is given by the angle θ. The depth of the water measured normal to the bed will be called h and the distance along the channel bed x. In general, the velocity V will be a function of the depth as well as of the lateral position. In the equations to follow, however, we shall take V to be constant over the cross section. We shall further assume the direction of V to deviate so slightly from the x-direction that accelerations normal to x are small compared to the gravitational constant. This second assumption is justifiable whenever the changes of slope and depth of channel are very gradual, a condition that is usually met in practice. The pressure will then be distributed hydrostatically with depth. The first assumption implies that V now represents a certain average of the true velocity distribution. To be exact, the proper average for the continuity equation would differ from that applicable to the momentum equation. Fortunately, the two averages do not materially differ for the type of problem in question, and the concept of an average is quite adequate for the derivation of results that are of engineering use.

We now apply the momentum theorem to a control volume bounded by the bed, the side walls, the water surface, and two surfaces perpendicular to the bed separated by the distance dx in Fig. 11.1. With x being measured positive as shown, we obtain

$$d(bhV^2) = ghb\,dx\,\sin\theta - d\left(\frac{gbh^2}{2}\right)\cos\theta - \frac{\tau_0 P}{\rho}\,dx + \frac{gh^2}{2}\,db\,\cos\theta.$$

The term on the left-hand side is the momentum flux through the faces of the control volume. The first term on the right-hand side is the component of weight in the x-direction. The second is the contribution of the hydrostatic pressure force on the two faces of the volume element. The quantity τ_0 is the wall shear stress (assumed constant around the periphery), and P is the wetted perimeter, $P = b + 2h$. The next-to-last term on the right-hand side is the resisting viscous stress. Finally, the last term represents the component of hydrostatic force on the lateral sides of the control volume in the direction of motion. After simplification this equation becomes

$$hV\,dV + gh\,dh\cos\theta - gh\sin\theta\,dx + \frac{\tau_0 P\,dx}{\rho b} = 0, \tag{11.1}$$

in which the continuity equation

$$Q = bhV = \text{const} \tag{11.2}$$

has been used. From the geometry of Fig. 11.1 we see that

$$\sin\theta\,dx = -dz,$$

and furthermore for most applications the channel slope θ is sufficiently small, so that $\cos\theta$ can be replaced by unity. Equation (11.1) may then be written

$$hV\,dV + gh\,dh + gh\,dz + \frac{\tau_0 P\,dx}{\rho b} = 0. \tag{11.3}$$

The velocity derivative in Eq. (11.1) may be eliminated with the continuity relation Eq. (11.3) in differential form, that is,

$$\frac{db}{b} + \frac{dh}{h} + \frac{dV}{V} = 0,$$

and the momentum equation, Eq. (11.3), finally becomes

$$dh\left(1 - \frac{Q^2}{h^3 b^2 g}\right) - \frac{V^2}{gb}\,db = -dx\left(\frac{\tau_0 P}{\rho gbh} + \frac{dz}{dx}\right). \tag{11.4}$$

Equation (11.4) holds the key to a large number of problems concerning the flow in open channels. Although relatively simple in form, there are several interesting implications, and we shall spend much of the remainder of this chapter in interpreting this equation. For the purpose of simplicity we shall take b to be a constant.

One aspect of Eq. (11.4), which may be seen immediately, is that the quantity in parentheses on the left-hand side may be positive or negative depending on the dimensionless parameter $Q^2/h^3 b^2 g$. The effect of the slope or friction on the water depth h will depend very much on the sign of the term in parentheses. In fact, if the term in parentheses $(1 - Q^2/h^3 b^2 g)$ is positive, a negative slope dz/dx will tend to increase the depth h, whereas the opposite effect will occur if the term in parentheses $(1 - Q^2/h^3 b^2 g)$ has a negative value. This feature of the equation is one of the indications that shows the importance of the parameter $Q^2/h^3 b^2 g$ in determining the character of the flow. Names have been given to the two types of flow, the one corresponding to $Q^2/h^3 b^2 g < 1$ and the other for $Q^2/h^3 b^2 g > 1$. The former is called *subcritical flow* and the latter *supercritical flow*. When this parameter is examined, it is seen to be the square of the Froude number (Fr) based on the average channel velocity and the channel depth. The fact that the Froude number appears as an important parameter in open channel flow is consistent with our previous discussion on dimensionless parameters in Chapter 5. The Froude number in the foregoing flows plays much the same role as the Mach number for the compressible flows discussed in Section 9.5.

11.3 The Shear Term

Before discussing any particular examples, methods of evaluating the shear stress at the wall τ_0 will have to be discussed. The origin of this stress is exactly the same as that for the shear in a pipe. Therefore, one would expect again that it should be possible to express the wall shear stress in the form

$$\tau_0 = c_f \frac{\rho V^2}{2},$$

where the coefficient c_f would be a function of Reynolds number and the relative roughness (for pipe flow $c_f = f/4$). In studying pipe flow, we have seen, however, that the friction factor becomes independent of the Reynolds number if the Reynolds number is sufficiently large and if the surface is not perfectly smooth. Since in most open channel flows these conditions prevail, the effect of Reynolds number on the shear is generally neglected and the effect of roughness is expressed empirically. Many empirical equations have been proposed. One of the most popular of these is that due to Manning (1890), which may be written in the form[†]

$$\tau_0 = 29.14 \frac{n^2}{R^{1/3}} \frac{\rho V^2}{2}. \tag{11.5}$$

The quantity R is the ratio of the cross-sectional area to the wetted perimeter P of the channel cross section. It is taken here as the characteristic dimension of the channel, so that n^6/R is equivalent to the pipe roughness ratio ε/d, which was used for different types of surfaces. Values of n for several kinds of surfaces are given in Table 11.1. However, it is usually preferable to express the shear stress in terms of the friction factor f. In lieu of data on the actual channel in question, a useful approximation to f may be obtained from pipe data or other data by employing the "hydraulic radius," $m = 4R$, mentioned in Section 5.4. It is seen, for a fully wetted

TABLE 11.1 Table of Roughness Coefficient n[a]

Type of Channel	n
Artificial Channels of Uniform Cross section	
Sides and bottom lines with well-planed timber evenly laid	0.009
Neat cement plaster, smoothest pipes	0.010
Cement plaster (3 cement to 1 sand), smooth iron pipes	0.011
Unplaned timber evenly laid, ordinary iron pipes	0.012
Ashlar masonry, best brickwork, well-laid new sewer pipe	0.013
(This last value should be used for the previous categories if in doubt as to the excellence of construction and maintenance; free from slime, rust, or other growths and deposits.)	
Average brickwork, foul planks, foul iron pipes, ordinary sewer pipes after average uneven settlement and average fouling	0.015
Good rubble masonry, concrete laid in rough forms, poor brickwork, heavily encrusted iron pipes	0.017
Channels Subject to Nonuniformity of Cross Section	
Excellent clean canals in firm gravel, of fairly uniform section; rough rubble, "dry paving"	0.020
Ordinary earth canals and rivers in good order, free from large stones and heavy weeds	0.025
Canals and rivers with many stones and weeds	0.03–0.04

[a] Data from L. Marks and T. Baumeister, eds., *Marks Mechanical Engineers Handbook*, 8th ed., McGraw-Hill Book Company, New York, 1979.

[†] The term R, when used in this equation along with the values of n in Table 11.1, must be evaluated in British units, that is, $R \equiv$ ft.

circular pipe of diameter D, that $m = D$. Values of f for noncircular cross sections are then determined with m playing the role of D. A convenient compilation of n values for natural streams with stable beds is given by Barnes (see Reference 4).

The values given in Table 11.1 do not reflect the fact that friction factors of rivers and canals with sand beds are variable and depend strongly on the size and spacing of dunes that develop naturally in the bed. The tendency is for these dunes to diminish in size as velocity and discharge increase, thus also diminishing the friction factor. For example, the Manning roughness coefficient n for the Mississippi River at St. Louis (Reference 5) had a value of 0.025 when the discharge was 291,000 ft^3/sec (velocity = 5.11 ft/sec, depth = 34 ft) and a value of 0.038, when the discharge was 59,800 ft^3/sec (velocity = 2.33 ft/sec, depth = 17 ft). (The bed sand had a mean size of 0.3 mm.) Additional information on streams in sand beds may be found in an article by Brownlie (Reference 6), who has presented relations for predicting the velocity and depth of a sand-bed stream with a known discharge. These relations also indicate the velocity and depth for which dunes will be obliterated and the bed will become flat and devoid of dunes.†

11.4 Uniform Flow

When the flow in an open channel is of constant depth and shape, it is said to be *uniform* flow. It can be seen from Eq. (11.4) that the requirement for uniform flow is

$$-\frac{dz}{dx} = \frac{\tau_0 P}{\rho g A}, \tag{11.6}$$

where the cross-sectional area bh of a rectangular channel has been replaced by A, the area of a channel of general shape. [Equation (11.6) is still valid for arbitrary cross sections, whereas Eq. (11.4) is not.] When there is no shear, the slope for uniform flow is of course zero, just as the pressure drop is zero in a frictionless closed conduit. Hydraulic engineers often use the slope for uniform flow as a means of expressing the shear. Letting this slope be S_u, we find that the shear is

$$\tau_0 = \rho g S_u \frac{A}{P} = \rho g S_u R. \tag{11.7}$$

If Eq. (11.7) is substituted for τ_0 into Eq. (11.4), the basic momentum equation for channels of constant breadth becomes

$$dh\left(1 - \frac{Q^2}{b^2 g h^3}\right) = -\left(\frac{dz}{dx} + S_u\right) dx, \tag{11.8}$$

a form that frequently occurs in the literature. The quantity $-dz/dx$ is again the slope of the channel bed, for which the symbol S_0 is often used.

11.5 Two Applications

(a) The Backwater Curve

Before proceeding with the interpretation of the momentum equation, let us discuss two examples illustrating the use of this equation. One important problem that can be solved with the use of Eq. (11.4) is the determination of the water depth in a channel upstream of a dam. The curve determined by these water levels is called the

---

† The data and reference were provided through the courtesy of V. A. Vanoni, Professor of Hydraulic Engineering, California Institute of Technology.

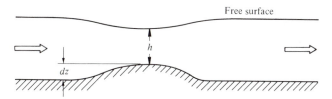

FIG. 11.2   Flow over a bump in an open channel.

backwater curve. The flow rate and water depth at the dam are usually known, and the flow upstream of the dam is generally subcritical. The channel width b and the slope are also known.

The first step in the solution of this problem is to substitute for τ_0 in terms of Manning's formula, Eq. (11.5), or an equivalent expression for the shear stress. The velocity V may next be eliminated by means of the continuity equation, Eq. (11.2). The resulting differential equation then has only one remaining unknown, the desired depth h. This equation may, therefore, be solved for the depth h. Since the equation is usually quite nonlinear, graphical or numerical methods of integration have to be used to obtain the solution.

(b) Flow over a Bump

As a second illustration let us consider the flow over a bump in a frictionless channel, as indicated in Fig. 11.2. The characteristics of the approaching flow are given, and it is desired to determine whether the water surface will hump up or dip down when passing over the bump in the channel.

To answer this question, let us simplify Eq. (11.8) by assuming frictionless flow and solve for the change in depth dh, so that

$$dh = \frac{-dz}{1 - (Q^2/b^2gh^3)}. \tag{11.9}$$

Let us now consider first the case of a subcritical approaching flow. The denominator of Eq. (11.9) will then be positive but smaller than unity. The depth change dh is consequently opposite in sign from the change in channel floor elevation dz, and the absolute value of dh will exceed that of dz. As a result the water surface will dip as indicated in Fig. 11.2.

If, on the other hand, the approaching flow is supercritical, dh and dz will have the same sign. As a consequence, the water surface will show a hump whenever the channel floor does. The hump in the water surface will be more pronounced than the hump on the channel floor as the change dh is additive to dz.

11.6 The Infinitesimal Surge

In this and the following sections several special phenomena occurring in the flow of open channels will be discussed. Let us consider first the flow in a horizontal frictionless channel. The differential equation for this problem is again derived from Eq. (11.4) and is given by

$$dh\left(1 - \frac{Q^2}{b^2gh^3}\right) = 0. \tag{11.10}$$

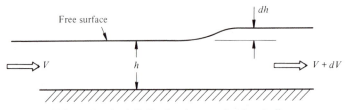

FIG. 11.3 An infinitesimal surge in steady flow.

This equation has two solutions: first, dh may be equal to zero. Physically, this means that the flow continues along the frictionless, horizontal channel without any change in depth. Inspection of Eq. (11.10) shows, however, that there is a second solution possible. In this solution dh is not equal to zero but the term in parentheses is equal to zero. The physical interpretation of this second solution is illustrated in Fig. 11.3. The flow approaches with a depth h. At some position the depth may change from h to $h + dh$, and the flow continues at this new depth. Such an occurrence is, however, possible only when the approaching velocity has a value such that the term in parentheses in Eq. (11.10) is zero. That is, the velocity of approach must be equal to

$$V_{cr} = \sqrt{gh}. \tag{11.11}$$

The subscript cr has been attached to the velocity symbol to indicate that there is only one particular value for which the infinitesimal change is possible. This special velocity is called the *critical velocity*, hence the subscript cr. From (11.11) we see that the Froude number for this flow (Fr $= V/\sqrt{gh}$) is unity. It follows that an infinitesimal stationary discontinuity in h may occur if the Froude number of the approaching stream is unity.

So far it has only been demonstrated that an infinitesimal elevation change or jump can exist, but it has not been shown how the location of the jump along the channel may be determined. Let it suffice for the moment to say that the location is determined by the downstream conditions in the channel, and that infinitesimal jumps do actually occur.

A phenomenon closely related to the infinitesimal jump can be derived from Eq. (11.10) with little additional work. Equation (11.10) was obtained from the momentum equation and the continuity equation. Both of these equations are valid with respect to either a coordinate system at rest or with respect to one that is moving at constant velocity. We may imagine, therefore, to have solved the problem of the infinitesimal jump in a system that moves at a constant velocity (say, c) with respect to the stationary system. The result of the solution is, of course, that an infinitesimal jump is possible if the approach velocity (with respect to the moving system) is equal to

$$V'_{cr} = \sqrt{gh},$$

where the prime denotes velocities measured in the moving system. To a stationary observer the approach velocity now appears to be $V'_{cr} + c$, and the jump location is no longer stationary but moves with a velocity c. A particularly interesting case results if the velocity c of the moving system is chosen equal to V'_{cr} in magnitude but opposite in direction. To a stationary observer the flow then appears as indicated in Fig. 11.4. The velocity of the "approaching" stream (to the left of the infinitesimal jump) is equal to zero. The location of the discontinuity moves to the left with the

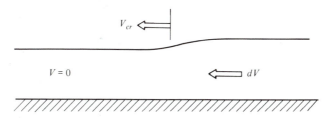

FIG. 11.4   An infinitesimal surge advancing into fluid at rest.

velocity V_{cr}. The water to the right of the continuity has a slight velocity dV. The magnitude and direction of this velocity may be determined from the continuity equation.

The resulting flow picture is that of an infinitesimal surge or wave advancing into an undisturbed fluid. An actual case in which such a surge occurs can be imagined rather easily. Consider, for example, a long horizontal channel subdivided into two portions by a partition. Let the water level on the right-hand side exceed that on the left-hand side by an increment dh. Let us further imagine that the level on the right can be maintained at the given value, even after the partition is removed. When the partition is then removed, a small surge will travel into the fluid of lower level. The velocity of the crest will be $V_{cr} = \sqrt{gh}$. The increment dh, incidentally, can be either positive or negative. The resulting surge will, correspondingly, either increase or decrease the elevation of the undisturbed fluid.

It is interesting to note that by means of the artifice of using two coordinate systems that move with respect to each other, it was possible to solve a nonsteady-state problem without having to resort to any of the nonsteady equations. It turned out that the solution in one system could be obtained from that in the other, simply by superimposing the relative velocity of the two systems in respect to each other. The arguments regarding the validity of the momentum and continuity equations in any coordinate system moving with a constant velocity had to be introduced, however, in order to show that this superposition was permissible.

11.7 The Hydraulic Jump

In Section 11.6 the propagation of an infinitesimal surge was discussed. It is rather challenging to investigate the possibility of a surge of finite magnitude. If such a surge exists, it would mean that water could approach a certain section in the channel at a depth h_1, then suddenly jump to a depth h_2 and continue flowing at this new depth. This flow picture is illustrated in Fig. 11.5.

In investigating such a surge, we proceed in a way exactly analogous to that followed in the study of the infinitesimal surge. We again assume a horizontal frictionless channel. The flow rate is steady, and the data for the approaching stream are all given. We now apply the momentum theorem to a control volume as indicated in Fig. 11.5. The control volume is so located that it crosses the stream at points where the velocity may be assumed to be horizontal and uniform. Using the momentum equation, we then obtain the relation

$$\rho V_1^2 h_1 + \tfrac{1}{2} h_1^2 \rho g = \rho V_2^2 h_2 + \tfrac{1}{2} h_2^2 \rho g. \tag{11.12}$$

Using the continuity equation, that is,

$$Q = b h_1 V_1 = b h_2 V_2,$$

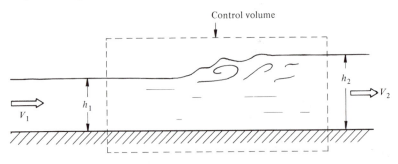

FIG. 11.5 The hydraulic jump.

Eq. (11.12) may be transformed into the expression

$$(h_1 - h_2)\left[h_1 + h_2 - \frac{2Q^2}{gb^2h_1h_2}\right] = 0. \tag{11.13}$$

An inspection of this equation shows that it has two solutions. The equation may be satisfied by setting either the term in brackets or the term in parentheses equal to zero. The solution corresponding to the term in parentheses being equal to zero leads to the rather obvious conclusion that the flow may simply proceed without any disturbance. The second possible solution, however, shows that a change in depth is possible. Solving for the new surface height, we obtain

$$h_2 = -\frac{h_1}{2} + \sqrt{\frac{h_1^2}{4} + \frac{Q^2}{b^2gh_1^3}\,2h_1^2}. \tag{11.14}$$

The flow picture corresponding to Eq. (11.14) is that shown in Fig. 11.5. A stream with constant flow rate Q approaches with a depth h_1. A region of elevation change then occurs, and beyond that the flow continues with a depth h_2. This flow phenomenon is called a *hydraulic jump*. It should be noted from Eq. (11.14) that for a given flow rate Q, only one particular depth h_2 corresponds to each initial depth h_1.

Some additional information may be gained by inspection of Eq. (11.14). The quantity $Q^2/b^2gh_1^3$ is the square of the Froude number for the approaching flow. If this number happens to be unity, Eq. (11.14) yields $h_1 = h_2$, which means that the jump has become infinitesimally weak. This conclusion agrees with that reached in our discussion on the infinitesimal jump, where it was shown that a vanishingly weak jump could occur if the initial Froude number was equal to unity. If the initial Froude number is not unity, the jump will be finite, and the Froude number of the flow leaving the jump will differ from that of the approaching stream. The relation between the two Froude numbers can be derived from Eq. (11.14). Again using the symbol Fr for the Froude number, it is seen that

$$Fr_2 = Fr_1\left(\frac{h_1}{h_2}\right)^{3/2} = \frac{1}{8Fr_1^2}(1 + \sqrt{1 + 8Fr_1^2})^{3/2}. \tag{11.15}$$

Inspection of this expression shows that when $Fr_1 > 1$, then $Fr_2 < 1$, and when $Fr_1 < 1$, then $Fr_2 > 1$.

Since $Fr > 1$ corresponds to supercritical flow and $Fr < 1$ to subcritical flow, we may conclude that in a hydraulic jump the flow changes from supercritical to subcritical, or vice versa. In other words, in a hydraulic jump the flow changes from one flow regime to the other. The conclusions drawn so far are correct, but they are not yet complete. The basis of our analysis so far has been the momentum theorem and the continuity equation. We have not as yet taken into account any limitations that may be imposed by energy considerations. Writing the Bernoulli equation for any streamline between points 1 and 2 of the hydraulic jump, we obtain (using the channel floor as the baseline for elevation)

$$\frac{V_1^2}{2g} + h_1 - h_f = \frac{V_2^2}{2g} + h_2, \tag{11.16}$$

where h_f represents any friction losses that occur between the two sections. These friction losses are of a special kind. They are *not* caused by friction against the channel walls—in fact, we have assumed these walls to be perfectly frictionless in our analysis. The friction losses are caused by shear forces set up internally within the fluid in the transition region of the jump where the flow is extremely complicated. Without attempting to analyze these friction forces in any more detail, we may state with conviction that these friction forces will lead to losses and that h_f in Eq. (11.16) is positive. The quantity $(h_2 + V_2^2/2g)$ downstream of the jump, therefore, must be less than $(h_1 + V_1^2/2g)$ for the approaching stream. In the light of this conclusion we must now reexamine our previous statements regarding the possible occurrence of hydraulic jumps. Upon computing the change in the quantity $(h + V^2/2g)$ for the cases discussed before, it can be shown that this quantity shows an *increase* for the type of jump in which the flow changes from subcritical to supercritical. For the jump in which the flow changes from supercritical to subcritical, the quantity $(h + V^2/2g)$ decreases. We must conclude, therefore, that only those jumps are possible in which the flow changes from supercritical to subcritical. This conclusion agrees with experimental evidence, and only jumps of the latter type have been observed.

11.8 The Positive Surge, or Elevation Wave

The advance of a high-water front into an undisturbed region of lower water level we shall call a *positive surge*, or *elevation wave*. Such an elevation wave is illustrated in Fig. 11.6. The solution for this surge can be obtained from that for the hydraulic jump by simple superposition. The method is quite analogous to that used to calculate the advance of the infinitesimal surge (Section 11.6). As in that case, it can be shown that superposition of velocities is permissible. We may, therefore, take the solution for a hydraulic jump of depths h_1 and h_2 and superimpose a velocity equal in magnitude and opposite in direction to that of the approaching flow. The result will be an advancing surge, as illustrated in Fig. 11.6. The velocity of advance will be V_1, and the velocity of the deep water will be $(V_1 - V_2)$, where V_1 and V_2 are the velocities for the corresponding jump.

Experimentally, one can produce a positive surge by dividing a long channel into two portions by a partition. Let the water to the left be at a lower level than that on the right, and let us assume that the higher level can be maintained constant at all

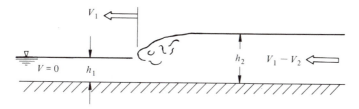

FIG. 11.6   An elevation wave advancing into fluid at rest.

times. Upon removal of the partition a positive surge will occur. If the side of high water level cannot be maintained at constant elevation, its depth will start to decrease as soon as the partition is removed. The advancing front then cannot be maintained at its original strength, and a more complicated surge will result.

11.9 The Negative Surge, or Depression Wave

In Section 11.8 the propagation of a high-water level into a region of undisturbed water at a lower level was discussed. We shall now examine the reverse problem, the encroachment of a low-level region into one of higher level. Experimentally, this condition can again be created by means of a partitioned channel with different fluid levels on the two sides of the partition. In this example, however, provisions have to be made to maintain the level of the lower side. Upon removal of the partition the lower level will propagate into the high-level region.

In the case of the elevation jump we were able to solve the problem simply by superimposing a velocity on the solution for the hydraulic jump. If now the kind of jump would exist in which a sudden decrease in depth occurs, then a corresponding negative surge should also exist, derivable from this type of jump by simple superposition. We have argued before, however, on the basis of energy considerations, that a jump in which the depth decreases should not exist. Similarly, the corresponding depression wave should also not exist in nature, and this conclusion agrees with observations.

How, then, does a lower level propagate into a body of fluid that is at a higher level? The answer is that the propagation takes place by means of a series of infinitesimal steps instead of by one abrupt step. The infinitesimal step was discussed in Section 11.6, and it was seen that the velocity of propagation of such a step is $V_{cr} = \sqrt{gh}$. To illustrate the method of propagation, let us consider now a body of water in a channel as shown in Fig. 11.7. To the right the water is contained by a baffle. In this illustration we shall assume that the negative surge impulses are created by moving the baffle to the right. As soon as the baffle is set in motion, the first infinitesimal surge will propagate to the left with a velocity $V = \sqrt{gh}$. The depth in the region of water traversed by the surge has therefore decreased somewhat. When the next infinitesimal pulse is sent out from the baffle, it will also travel to the left with a velocity $\sqrt{gh}$. The depth h, however, is now below that of the initial channel, and in addition the fluid in which the pulse is traveling is moving to the right at a small velocity. As a result, the velocity of the second pulse is slower than that of the first. The second pulse will never catch up with the first; on the contrary, the distance between them will increase with time. Similarly, additional pulses will be sent out by the baffle until the summation of these equals the full decrease in depth. The relation between consecutive pulses will be similar to that between the first and the second. At a particular time the water level may look as

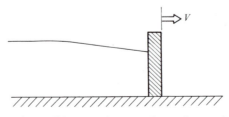

FIG. 11.7 A possible way of generating a depression wave.

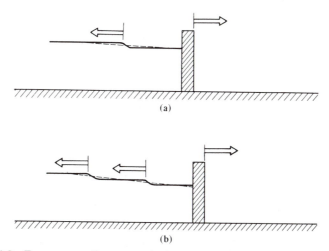

FIG. 11.8 Two consecutive stages in the propagation of a depression wave.

shown in Fig. 11.8a. At a later time the steps will be more separated (see Fig. 11.8b), and the region over which the decrease in depth takes place will have increased. The surface shape in this region will therefore change with time. In the actual case the steps mentioned in the discussion are infinitesimal and not distinct. The resulting surface shape will, therefore, be more like the smooth curve indicated by the dashed line rather than consisting of individual steps.

The discussion above is only qualitative in nature. Nevertheless, it indicates how an elevation decrease is propagated, and it leads to one important quantitative fact. The foremost edge of the depression wave advances into the undisturbed fluid with a velocity of $\sqrt{gh}$, where h is the elevation of the undisturbed fluid. All other parts of the wave have *slower* propagation velocities. This is quite different from the positive jump, or *bore*, in which the propagation velocity exceeds the value $\sqrt{gh}$, where h is again the depth of the undisturbed fluid.

It is still necessary to point out that we have neglected effects of surface curvature. The slopes, as illustrated in Fig. 11.8, were assumed to be sufficiently gradual to allow this simplification. Otherwise, the maximum velocity of the wave edge differs from the value given.

11.10 Flow over a Crest

We may now discuss a somewhat more general problem. Let us consider a very large reservoir filled with water. Water flows from this reservoir over a crest into a channel as shown in Fig. 11.9. A second crest is located downstream in the channel. This second crest is to be adjustable and can be used to regulate the water level in the channel.

At first let the second crest be higher than the water level in the reservoir. In that case there is, of course, no flow at all. Next, the second crest is lowered somewhat to allow a small rate of flow of water over the dam. Applying Eq. (11.4), which was derived from the momentum theorem, we may analyze the changes in velocity, depth, and water level as the water flows over the dam. Since the flow rate is small, the Froude number will also be small, and we shall assume here that the rate is small enough so that the Froude number is smaller than unity throughout the channel.

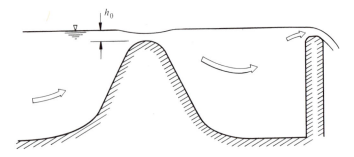

FIG. 11.9 Flow over the crest of a reservoir.

The factor multiplying dh on the left-hand side will then be positive. For the approach to the dam the derivative dz/dx is positive, and the friction term is, of course, also positive. It is seen, therefore, that the water depth decreases as the crest of the dam is approached. The velocity will correspondingly increase. The slope of the water surface can also be computed by forming the quantity $(z + h)$. It will be seen that this surface drops as the crest is approached. Beyond the crest we shall now assume that the channel slope is sufficiently steep so that the quantity $(dz/dx + \tau_0 P/\rho ghb)$ is negative. In that case the water depth will again increase and the velocity will correspondingly decrease.

As the second crest is lowered further, the downstream water level will fall correspondingly and the flow rate will increase. The flow behavior will remain qualitatively the same until the velocity at the crest becomes equal to the critical velocity. Before proceeding further with our example, we have to realize that the effect of a falling downstream water level can be communicated to the fluid upstream only by means of a depression wave. This depression wave has to advance upstream against the flow direction. We have seen in a preceding section that the maximum velocity of such a depression wave is just equal to the critical velocity of the stream into which the wave is advancing. If the velocity of the stream itself is now equal to or greater than the critical velocity, no depression wave can travel upstream. We must conclude, therefore, that decreasing the downstream water level below that at which the velocity at the first crest becomes critical can have no effect on the flow conditions at the crest. The velocity, the depth, and particularly the flow rate will remain unchanged as the downstream water level is lowered beyond this critical point.

Let us imagine now that the downstream water level is fixed at a level below that at which the flow over the crest of the dam first becomes critical. At the crest the Froude number of the flow is then just unity. Although the flow rate is fixed, the velocity may increase further as the fluid passes over the dam. The flow will then enter the supercritical region, and a falling channel slope will lead to an increase in velocity, as can be seen from Eq. (11.4). (It was again assumed here that the term dz/dx outweighs the friction term.) One may note that this flow over a crest is quite analogous to the flow through a throat of a nozzle as discussed in Section 9.7.

By integrating Eq. (11.4), it now seems possible in principle to determine the velocity and water depth at any point downstream of the crest of the dam. Such a calculation can actually be carried out. However, since the flow is supercritical, the possibility of a hydraulic jump occurring must be kept in mind. In our case, for example, the water level far downstream is specified. If this level does not happen to coincide with the value computed from the integration mentioned above, then a hydraulic jump will occur. The location at which the jump will establish itself will be such that the specified level will be reached.

A brief summary of the flow history in this last example is as follows: The water accelerates from rest as it approaches the crest of the dam. Critical flow conditions are reached at the crest. Beyond the crest the flow becomes supercritical, and the water further accelerates up to the point at which the jump occurs. The jump occurs at such a point that the changes in water level during the subsequent subcritical flow will lead to the initially specified downstream level.

It must be mentioned that there is a certain range of water levels that cannot be met by any location of the hydraulic jump. If the specified downstream level is within this range, the water will assume a certain two-dimensional flow pattern which cannot be described by our one-dimensional methods. The analysis of this two-dimensional pattern is beyond the scope of the present discussion.

11.11 The Critical Flow Rate

From the previous discussion it was seen that in certain cases—such as the flow over the crest of a dam—the flow rate is limited to a maximum value. It was seen that the velocity at this flow rate is equal to the critical velocity. Let us now derive a quantitative relation for this flow and consider again the flow over the crest of a dam, as illustrated in Fig. 11.9. Writing the Bernoulli equation between a point in the reservoir and the top of the crest and neglecting friction between these two points, we obtain

$$h_0 = \frac{V^2}{2g} + h, \tag{11.17}$$

where h is the depth at the crest, V the velocity at the crest, and h_0 the water level in the reservoir measured above the location of the crest. For critical flow conditions to be obtained at the crest,

$$V_{cr}^2 = gh.$$

If we insert this relation in the Bernoulli equation [Eq. (11.17)], it follows that

$$h_0 = \tfrac{3}{2} h_{cr}. \tag{11.18}$$

In other words, Eq. (11.18) states that for critical flow conditions the depth at the crest is equal to two-thirds of the water level in the reservoir, the water level being measured with respect to the crest. From Eq. (11.18) we can derive the flow rate itself,

$$Q_{cr} = b V_{cr} h_{cr}$$

or

$$Q_{cr} = b \sqrt{g} \, (\tfrac{2}{3} h_0)^{3/2}. \tag{11.19}$$

As a special application the concept of the maximum flow rate may be applied for estimating flow rates in certain circumstances. In the case of flow over the crest of a dam (Fig. 11.10), for example, the downstream slope may be very steep and the downstream water level may be very low. It may then be possible to predict the existence of critical flow conditions at the crest without any detailed calculations. Once it has been determined that the flow is critical, the flow rate can be determined from a single depth measurement, as seen from Eq. (11.19). Two typical examples of circumstances under which critical flow conditions are likely to occur are shown in Fig. 11.10. The configuration of Fig. 11.10b is usually called a *broad-crested* weir.

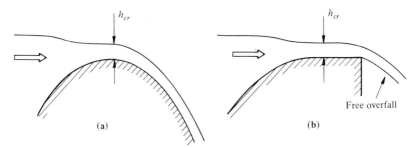

FIG. 11.10 Two flows in which the water depth assumes the critical value.

The principal limitations of the foregoing results are due to the assumptions that the velocity is uniform and one-dimensional, that the curvature of the water surface is negligible, and that the pressure distribution is hydrostatic. These assumptions are justified when analyzing flow profiles in channels in which the change of surface slope is very small. In locations of rapidly varying slope, however, the previously derived equations will be inaccurate. If, for example, the crest of the dam which was discussed in Section 11.11 has a small radius of curvature, the slope of the surface will also vary rapidly, and the velocities and flow rates computed by the one-dimensional method will be somewhat in error. This applies also to the value of the critical flow rate. Similarly, in the examples shown in Fig. 11.10 the critical flow rate occurs at a place at which the surface curvature may be significant. Flow rates determined on the basis of the assumption of flow critically should, therefore, be regarded as estimates in these instances.

11.12 Compressible Flow Analogy

In preceding sections of this chapter we have pointed out certain similarities between the flow of an incompressible fluid in a channel and the flow of a compressible fluid through a duct. The basis for these similarities may now be developed more specifically by comparing the continuity and momentum equations for simple one-dimensional flows in both instances.

For the one-dimensional flow of a compressible fluid in a frictionless duct of constant area the continuity equation [Eq. (9.20)] reduces to

$$\frac{dV}{V} + \frac{d\rho}{\rho} = 0$$

and the momentum equation [Eq. (9.21)] to

$$V \, dV + \frac{dp}{\rho} = 0,$$

which may also be written

$$V \, dV + \frac{dp}{d\rho} \frac{d\rho}{\rho} = 0.$$

Assuming an isentropic relationship between p and ρ, we find that $dp/d\rho$ may be set equal to a^2, so that

$$V \, dV + a^2 \frac{d\rho}{\rho} = 0,$$

where a is the velocity of sound.

For a horizontal open channel of constant width, the continuity equation is given by

$$\frac{dV}{V} + \frac{dh}{h} = 0.$$

Upon comparing this equation to the corresponding equation for the compressible fluid, the similarity in the form of the equations is evident. There is a direct correspondence between the velocity terms, and the liquid depth h seems to be analogous to the density ρ. Next let us examine the momentum equation for the channel flow [Eq. (11.3)]. For a horizontal channel in the absence of friction this equation is given by

$$V \, dV + g \, dh = 0,$$

but which we shall write for purposes of comparison as

$$V \, dV + gh \frac{dh}{h} = 0.$$

The analogy suggested by the continuity equation would apply to the momentum equation provided that the quantity gh may be identified with the velocity of sound. Now, we have already shown in Section 11.6 that $\sqrt{gh}$ is the velocity of propagation, c, of an infinitesimal discontinuity in elevation h. Since h corresponds to ρ and the velocity of sound may be interpreted as the propagation velocity of an infinitesimal discontinuity in density, one may conclude that there is at least a qualitative correspondence between the two propagation velocities. To obtain a quantitative relationship, let us consider the expression for the acoustic velocity

$$a = \sqrt{\gamma R T}.$$

By means of the equation of state and the isentropic relationship for a perfect gas ($p/\rho^\gamma = \kappa = $ const), this equation becomes

$$a = \sqrt{\kappa \gamma \rho^{\gamma - 1}}.$$

Recalling that ρ corresponds to h, we find that γ has to be set equal to 2 and κ equal to $g/2$ in order to make a numerically equal to c. If we specify γ and κ in this way, the differential equations for V and ρ for the flow of a perfect gas are identical to those determining V and h for the flow on an incompressible fluid in a channel. The solution of one set of equations (be it accomplished analytically or experimentally) serves as an immediate solution to the analogous set of equations.

The analogy is restricted by the fact that γ had to be set equal to a particular numerical value. Nevertheless, it is sometimes convenient to perform experiments in open channels or water tables in order to study wave phenomena, flow through convergent–divergent nozzles, behavior of inlet diffusers, and so on.

PROBLEMS

When not otherwise specified, the fluid is water and the properties are to be evaluated at 20°C and 1 atmosphere pressure.

11.1 A uniform flow of water takes place in a concrete channel ($n = 0.013$). The slope of the bed is 10 percent, the channel is 8 ft wide, and the water is 2 ft deep. Compute the velocity, flow rate, and Froude number of this flow.

11.2   A wide stream (i.e., $b \gg h$) has a friction factor $f = 0.02$, and the slope is 0.001.
(a) Find the flow velocity if the flow has a *uniform* depth of 0.5, 1, and 2 m.
(b) Calculate the Froude number for the same conditions as in part (a). Is the flow sub- or supercritical?

11.3   Consider the flow in a very wide channel at a uniform depth of 3.00 ft. The flow is fully turbulent and the velocity profile can be considered to satisfy the law of the wall [Eq. (7.14)]. The velocity is measured at two heights above the bottom and found to be: at $y = 0.30$ ft, $u = 2.57$ ft/sec; at $y = 2.4$ ft, $u = 3.60$ ft/sec. Find:
(a) The friction velocity u_τ.
(b) The average channel velocity V.
(c) The friction factor f.
(d) The equivalent sand grain roughness. (Is the channel surface rough or smooth?)
(e) The slope of the channel.

11.4   Water flows down a right-angled gutter (see the diagram). The gutter is made of smooth concrete and is on a slope of 0.09. If the height of the fluid is 1 ft, find the velocity and flow rate.

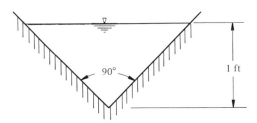

Problem 11.4

11.5   Fluid flows without friction over a horizontal bed with velocity V and depth h, $V > \sqrt{gh}$. To what height must the channel bottom be raised to make the flow just critical? Give numerical values for the case of $V = 6$ m/s and $h = 0.3$ m.

11.6   In a very wide channel, water flows on a frictionless, horizontal bed at a speed of 8 m/s. The water is 1.0 m deep.
(a) The bed elevation is raised smoothly to a height h_m above the approaching bed. Find the maximum value of h_m to which the bed can be raised and still have steady flow.
(b) What is the maximum possible height of the water surface?

11.7   A stream of 5 ft depth and great breadth flows down a surface with a constant slope of 1/1000 ($n = 0.035$) and discharges into a reservoir whose surface is at 1000 ft elevation. At this point the depth of the stream is 15 ft. Calculate and plot on a suitable scale the profile of the flow from a point where the stream is 15 ft deep to a point where the stream is 5.5 ft deep. Also determine the distance between these points, and the elevation where the depth is 5.5 ft.
   Hint: This problem is best done numerically using a spreadsheet program on a personal computer. Compute the lengths required for depth changes of 5, 2, 1, 0.5 ft. Arrange the computations in tabular form.

11.8 A wide stream of depth h_1 which flows with a supercritical velocity V_1 is directed onto a horizontal bed for which $n = 0.033$. For $V_1 = 15$ ft/sec and $h_1 = 1$ ft, calculate and plot to a suitable scale the stream height as a function of distance along the bed until the depth becomes 0.8 of the critical depth.

11.9 A rectangular channel is 3.0 m wide and carries 3 m³/s. The slope and roughness are such that the flow is uniform at a depth of 0.3 m. Suppose now that an obstruction such as a submerged weir is placed across the channel, rising to a height 0.1 m above the channel bottom. Will this obstruction cause a hydraulic jump to form upstream? Show the reason for your answer. (The discharge continues at 3 m³/s.)

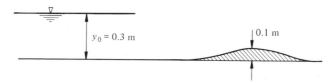

Problem 11.9

11.10 A wide, supercritical stream of velocity $V_1 = 15$ ft/sec and with an initial depth $h_1 = 1$ ft is directed onto a horizontal bed for which $n = 0.033$ as in Problem 11.8. However, an obstruction is placed far downstream so that the downstream depth of the channel is maintained at 3 ft. Under these conditions a hydraulic jump will take place. Find the location of this jump in terms of the distance l from the start of the channel and determine the elevations h_1 and h_2 before and after the jump, respectively. Neglect friction on the bed downstream of the jump.

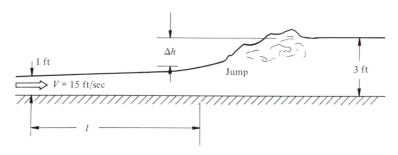

Problem 11.10

11.11 A horizontal stream of water has a depth of 0.1 m and a speed of 3 m/s, and is made to undergo a hydraulic jump. Calculate the downstream depth and the total head loss.

11.12 The flow in Problem 11.11 takes place over a hydraulically smooth surface.
   (a) Calculate the distance for the stream to rise 0.2 m. (This must be done numerically; friction must be taken into account.)
   (b) How would you find the distance required for the stream to rise to the critical depth?

11.13 A hydraulic jump that moves at a steady speed into a *still* water channel, which we termed a positive surge wave in Section 11.8, is sometimes called a hydraulic "bore." If the speed of the bore is 4 m/s and the still water has a depth of 0.5 m, find the depth of the water and the flow velocity behind the bore.

11.14 A supercritical flow takes place in a frictionless, horizontal channel at a speed V_1 and depth h_1. The flow is dammed suddenly, stopping the stream and causing a hydraulic jump to move upstream at a speed V_J, while the height of the stream rises to h_2. Neither the speed V_J nor the height h_2 are known. Using the continuity equation and conservation of momentum, derive an equation from which h_2/h_1 can be found.

11.15 A level rectangular channel is filled with still water to a depth d. The gate at the left end of the channel is raised suddenly, allowing water to flow into the channel from the reservoir. A surge wave of height h travels down the channel at a velocity V. Friction along the channel sides and bottom is to be neglected.

(a) Determine V in terms of h and d.

    Hint: Use a reference system moving to the right with the speed of the surge wave V, and consider a control volume bounded by sections 1 and 2, which move with this reference system.

(b) What is the limiting value of V as $h \to 0$?

(c) What must be the rate of flow from the reservoir into the channel (per unit width) required to maintain a uniform depth $d + h$ behind the surge wave?

(d) By writing the Bernoulli equation with a loss term and expressing velocities in terms of d and h, show that for h small compared to d the losses vary as h^3.

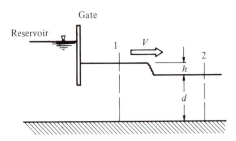

Problem 11.15

11.16 A frictionless liquid flows along a flat, open channel of variable breadth b subject only to the influence of gravity g.

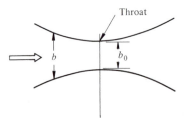

Problem 11.16

(a) Obtain the expression

$$\frac{dh}{h} = -\frac{1}{1 - gh/V^2}\frac{db}{b},$$

where h is the depth of the liquid at any point and V is the fluid velocity.

(b) Deduce from this equation that V^2/gh can equal unity only where $db = 0$ (i.e., at a "throat").

(c) For a given Bernoulli constant $[(V^2/2g) + h = h_0]$ show that the maximum rate of flow Q_{max} through a channel with a throat of breadth b_0 occurs when $V/\sqrt{gh} = 1$ at the throat.

(d) Sketch the elevation of the water surface for the flow from a reservoir through a converging–diverging channel (see the figure) for two cases: (1) $Q < Q_{max}$ and (2) $Q = Q_{max}$.

11.17 A horizontal, frictionless channel of width 1.0 m has a flow of 1-m depth and a velocity of 1 m/s.

(a) The width of the channel is decreased at one point to 0.8 m. Calculate the water depth there.

(b) What is the smallest width that can still allow this flow?

11.18 Consider the Venturi flume shown in the accompanying figure. The flow is horizontal and may be considered to be frictionless. The upstream depth is 0.3 m and the depth at the throat is 0.25 m, whereas the upstream breadth is 0.6 m and the breadth of the throat is 0.3 m. Find the flow rate through the flume.

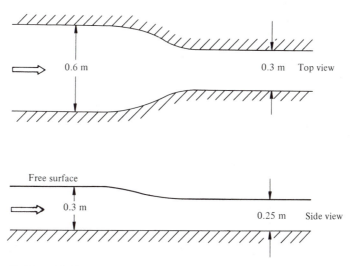

Problem 11.18

11.19 A rectangular channel 1.2 m wide carries a discharge of 1.0 m³/s. The slope and roughness of the channel are such that uniform flow occurs at a depth of 0.6 m. At point B, as shown in the sketch, the cross section is contracted laterally to a width of b_0 ft. This contraction may be assumed to be frictionless, and note that there is no bump or hollow in the bed of the channel.

(a) Find the minimum allowable value of b_0 for which the flow condition and depth upstream will remain undisturbed.

(b) If the channel is contracted further than this, what will happen to the water surface upstream?

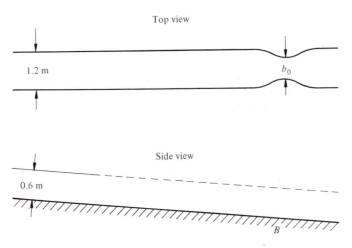

Problem 11.19

11.20 The crest of a broad-crested weir is 0.5 m below the level of an upstream reservoir. What is the maximum flow rate per unit width that could pass over the weir?

11.21 A channel has a rectangular cross section and is 20 ft wide. It is designed for a depth of flow of 8 ft and the corresponding velocity is to be one-third the critical value. Find the required channel slope for uniform flow assuming that the channel surfaces are equivalent to cement plaster.

11.13 Waves of Finite Wavelength

In previous sections we have discussed problems in which it was not necessary to determine the detailed flow pattern throughout the channel. We assumed that the velocity was constant over any particular cross section. Even in the case of discontinuities (such as a hydraulic jump) it was possible to assume a constant-velocity profile in front of and behind the discontinuous portion of the flow. If, however, we now desire to study the motion of water waves of finite wavelength, the assumption of constant velocities is no longer appropriate. We shall have to compute the motion of the liquid starting from the basic equations of fluid mechanics.

As an example of the type of wave motion that occurs in channels, let us consider the following situation: A quantity of water is contained in an infinitely long rectangular channel of mean depth h. There are waves on the surface as illustrated in Fig. 11.11, but except for them, the fluid is at rest (i.e., there is no average horizontal velocity). It is desired to determine the motion of the waves and of the fluid beneath the waves.

In Fig. 11.11 an arbitrary wave form of wavelength λ has been shown. The height of the waves is measured from the mean water position (shown as a light line), and it is designated by the symbol η. In general η is a function of time and position [i.e., $\eta(x, t)$]. The coordinate system is oriented so that the x-axis coincides with the mean surface.

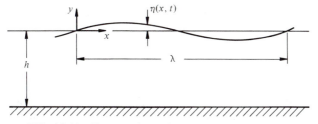

FIG. 11.11 Nonsteady surface wave in a channel.

In treating this problem, we restrict ourselves to waves of amplitudes small compared to the channel depth and to the wavelength. We assume, furthermore, that the viscosity is negligible and that, as before, the density is constant; that is, if we add the supposition that the fluid was at rest (or at least in irrotational motion) prior to the generation of waves, we may conclude that the motion of the water will be irrotational. Consequently, it is possible to describe the motion by means of a velocity potential φ, where φ must satisfy Laplace's equation,

$$\nabla^2 \varphi = 0. \tag{11.20}$$

(a) Waves Without Surface Tension

Before attempting to find appropriate solutions Eq. (11.20), we must first determine the necessary boundary conditions. One of these can be determined rather easily. The velocity component in the y-direction must go to zero at the channel bottom. Consequently, we may write

$$v = \frac{\partial \varphi}{\partial y} = 0 \qquad \text{at} \quad y = -h. \tag{11.21}$$

The second boundary condition is determined by the fact that—neglecting surface tension—the pressure at the free surface must be equal to the atmospheric pressure above the liquid. To express this condition in terms of the velocity potential, let us first write the Bernoulli equation. For nonsteady potential flow the Bernoulli equation can be given in the form

$$\frac{p}{\rho} + \frac{V^2}{2} + gy + \frac{\partial \varphi}{\partial t} = \text{const}, \tag{11.22}$$

as was shown in Chapter 6 [see Eq. (6.2)]. At the surface where $y = \eta(x, t)$, the pressure $p = p_0 = \text{const}$, and Eq. (11.22) therefore can be simplified to read

$$\frac{V^2}{2} + g\eta + \frac{\partial \varphi}{\partial t} = \text{const}. \tag{11.23}$$

Furthermore, for waves of small amplitudes it is now permissible to neglect the velocity term $V^2/2$ compared to the other two terms, although this must be justified after a solution has been obtained. Thus we can put

$$g\eta + \frac{\partial \varphi}{\partial t} = \text{const}, \tag{11.24}$$

and upon differentiating this equation with respect to time, we have

$$g \frac{\partial \eta}{\partial t} + \frac{\partial^2 \varphi}{\partial t^2} = 0. \tag{11.25}$$

Since the velocity components of the fluid u, v are small, as we have already assumed in deriving Eq. (11.24), and since the amplitude η is small, it follows that the vertical velocity component at the surface is approximately given by

$$v = \frac{\partial \eta}{\partial t}. \tag{11.26}$$

The component v, however, is also equal to $\partial \varphi / \partial y$, so that the boundary condition at the free surface [Eq. (11.25)] finally becomes

$$g \frac{\partial \varphi}{\partial y} + \frac{\partial^2 \varphi}{\partial t^2} = 0. \tag{11.27}$$

We may now proceed with the solution of the governing differential equation, Eq. (11.20). Many solutions to this equation are known, as discussed in Chapter 6. A solution that turns out to be most convenient for the present problem is

$$\varphi = (Ae^{my} + Be^{-my}) \cos m(x - ct). \tag{11.28}$$

That this expression for φ satisfies the Laplace equation can be verified by substitution. The quantities A, B, m, and c are constants, whose meaning we can interpret. The constants A and B determine the amplitude of the wave motion. The significance of m can be determined by analyzing the cosine term which can be written as $\cos (mx - mct)$. The cosine term will take on identical values whenever the argument changes by an amount 2π. Let us now increase x by an increment λ such that $m\lambda = 2\pi$. Examining Eq. (11.28), it is seen that the dependence of the velocity potential on y and t at the position x is identical to that at $x + \lambda$. Similarly, it can be shown that the velocities and velocity potentials are repeated at the same interval. The interval λ is, therefore, the wavelength of the motion. The constant m can then be expressed as a function of the wavelength as follows:

$$m = \frac{2\pi}{\lambda}.$$

Next let us interpret the constant c. The argument of the cosine in Eq. (11.28) can be maintained constant by simultaneously varying x and t in such a way that Δx is equal to $c \, \Delta t$. The flow picture that at time t existed at point x will, therefore, also occur at a point $(x + c \, \Delta t)$ at a time $t + \Delta t$. We may think of the flow picture as having been shifted by the distance $c \, \Delta t$ during the time interval Δt, and we may say that the flow picture has been translated with a velocity c. This velocity is called the *wave velocity*. It is not to be confused with the actual velocity of the fluid particles themselves.

Let us return now to Eq. (11.28). This expression for φ represents a possible solution of our problem. We have to see, however, if the boundary conditions of Eqs. (11.21) and (11.27) can be properly satisfied. Let us try then to satisfy the first boundary condition Eq. (11.21). At $y = -h$ we obtain by differentiation that

$$\left(\frac{\partial \varphi}{\partial y} \right)_{y = -h} = m(Ae^{-mh} - Be^{mh}) \cos m(x - ct) = 0.$$

This equation must be satisfied at all times. To comply with this requirement, it is necessary for the first term in parentheses to be equal to zero; that is,

$$Ae^{-mh} - Be^{mh} = 0$$

or

$$B = Ae^{-2mh}.$$

Inserting this value into Eq. (11.28) gives the velocity potential

$$\varphi = \frac{2A}{e^{mh}} \cosh m(y + h) \cos m(x - ct).$$

No special physical meaning has so far been attached to the constant A in the expression above, and it is convenient to define a new constant

$$A' = 2A \frac{\cosh mh}{e^{mh}},$$

so that the velocity potential appears as

$$\varphi = A' \frac{\cosh m(y + h)}{\cosh mh} \cos m(x - ct).$$

The constant A' is now the amplitude of the potential at the mean height $y = 0$. The second boundary condition now requires that at the surface

$$\frac{\partial^2 \varphi}{\partial t^2} + g \frac{\partial \varphi}{\partial y} = 0.$$

Since we are limiting ourselves to waves of small amplitudes, or to be more precise to $\eta/h \ll 1$, we may make the approximation that the surface is located at $y = 0$ and evaluate the differentials in Eq. (11.27) at $y = 0$ instead of at the exact location of the surface. These steps lead directly to the expression

$$c^2 = \frac{g}{m} \tanh mh,$$

or, substituting for m,

$$c^2 = g \frac{\lambda}{2\pi} \tanh \frac{2\pi h}{\lambda}. \tag{11.29}$$

We already have identified c with the wave velocity. Equation (11.29) is consequently the expression for the velocity of a wave of small amplitude. It is of interest to note that the wave velocity depends on the wavelength λ. The relationship is sketched in Fig. 11.12. When λ/h is small compared to unity, the wave velocity is proportional to $\sqrt{\lambda}$. When $\lambda/h \gg 1$, the wave velocity becomes, however, practically independent of the wavelength and approaches the value $\sqrt{gh}$. We may point out that this is the same expression as that obtained previously for the velocity of propagation of a small jump. The fact that these two expressions are the same is not a coincidence. One may actually regard the infinitesimal jump as a wave of infinite wavelength, in which case the two problems actually become identical. The hydraulic jump of finite magnitude, on the other hand, cannot be compared to the foregoing wave solution. The two phenomena differ basically, because friction losses are of major importance in the transition zone of the hydraulic jump, whereas our wave solution was based on frictionless flow.

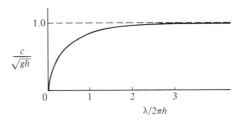

FIG. 11.12   Dimensionless wave velocity as a function of the ratio of wavelength to channel depth. The effect of surface tension is neglected.

Let us now return briefly to Eq. (11.29). Names have been given to the waves encompassed by this equation depending on the magnitude of the ratio λ/h. Waves for which $\lambda/h \gg 1$ are called *tidal waves*, or *shallow-water waves*. It was seen that the velocity of these waves is nearly independent of the wavelength. As the first name indicates, tidal waves in the ocean—produced, for example, by an underwater earthquake—are examples of this type of wave. Assuming a depth of 2 miles as typical, the theory of tidal waves yields a velocity of roughly 165 m/s. Therefore, tidal waves originating at the coast of Japan would be expected to reach the Pacific coast in about 20 hours or so. Actual observations are in close agreement with the predicted values. A word may be added here as to the name *shallow-water waves*, which also applies to the wave just discussed. A depth of 2 miles is not generally considered shallow. We should remind ourselves, therefore, that the word "shallow" has to be interpreted in respect to the wavelength.

Waves for which $\lambda/h \ll 1$ also have been given special names. They are called *deep-water waves*, or *surface waves*. These waves are generally more familiar to us. The ordinary wind-generated ocean waves, for example, are of this kind. The velocity of these waves depends on the wavelength. If waves of different wavelength are generated, they will be sorted out according to wavelength, because of their different speeds.

Waves of the type discussed above have the property that the velocity, the velocity potential φ, and so on, are constant for a point or observer that moves to the right in such a way that $(x - ct)$ is a constant. That is, the wave appears to be translating to the right at the velocity c. For this reason such waves are called *progressive*, or *traveling*, *waves*. A particular characteristic of traveling waves is that they transmit energy in the direction of the wave propagation.

The particular wave type we have studied above is by no means the only type possible. For example, the velocity potential

$$\varphi = \cosh m(y + h) \cos m(x + ct)$$

also satisfies the boundary conditions that are met by Eq. (11.28). This function is seen to represent a wave traveling to the left at velocity c. Another class of wave motion is given by velocity potentials of the form

$$\varphi = \cosh \frac{2\pi}{\lambda} (y + h) \cos \frac{2\pi}{\lambda} x \cos \frac{2\pi}{\lambda} ct.$$

The particular characteristic of these waves is that at the points $x = \lambda/4, 3\lambda/4$ and so on, the velocity potential and hence the vertical velocity is always zero, whereas when $x = 0, \lambda/2, \lambda$ and so on, the horizontal velocity is always zero. This type of wave is known as a *standing wave*, and the points $x = \lambda/4, 3\lambda/4$, and so on, where the vertical motion of the wave is zero, are called *nodes*. A sketch of a standing wave

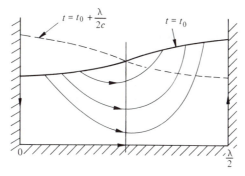

FIG. 11.13   A standing wave in a container.

at the extremities of its motion is shown in Fig. 11.13 together with some representative streamlines.

Note that since there is no flow through the planes $x = 0$, $\lambda/2$, we can regard them as being solid surfaces. There is, of course, a close connection between standing waves and traveling waves. In fact, standing waves can be obtained by superposing two trains of traveling waves, one moving to the right and the other to the left. The velocity potential of two such trains may be expressed as

$$\varphi = \cosh \frac{2\pi}{\lambda}(y + h)\left[\cos \frac{2\pi}{\lambda}(x + ct) + \cos \frac{2\pi}{\lambda}(x - ct)\right],$$

or, by rearranging,

$$\varphi = 2 \cosh \frac{2\pi}{\lambda}(y + h) \cos \frac{2\pi}{\lambda} x \cos \frac{2\pi}{\lambda} ct,$$

which is exactly the potential function for a standing wave as quoted before.

The methods outlined above can, of course, be extended to waves traveling in more than one direction. Either standing waves or progressing waves can be worked out for many problems of interest, although the mathematical complexity will increase. Numerical methods such as the boundary element method mentioned briefly in Chapter 6 become an important and practical way to solve such problems (see Reference 7).

(b) Effect of Surface Tension

There is an additional effect, that due to surface tension, which must be taken into account for small wavelengths. For water this effect becomes important for wavelengths of the order of a few inches or less. Surface tension can be taken into account rather easily as follows: The basic differential equation for the motion [Eq. (11.20)] remains unchanged. Similarly, the boundary condition that there should be no flow through the channel floor remains unchanged. The effect of surface tension enters only into the second boundary condition on the free surface. If the surface tension is of importance, it can no longer be said that the pressure at the surface of the wave is equal to the atmospheric pressure. The pressure at the surface in the liquid in that case is given by the expression

$$p_s = p_a - \sigma \frac{\partial^2 \eta}{\partial x^2},\tag{11.30}$$

where σ is the surface tension, p_a the atmospheric pressure, p_s the pressure at the surface of the water, and η the vertical coordinate of the water surface. Since the amplitude of the wave motion is small, $\partial^2\eta/\partial x^2$ is a sufficiently accurate approximation to the curvature of the surface. Inserting this expression for p_s into Eq. (11.23) and evaluating it at the surface, where $y = \eta$ leads to the equation, we obtain

$$\frac{V^2}{2} + g\eta - \frac{\sigma}{\rho}\left(\frac{\partial^2\eta}{\partial x^2}\right) + \frac{\partial\varphi}{\partial t} = \text{const.}$$

This relation corresponds to Eq. (11.23). Neglecting again the term in V^2 leads to

$$g\eta - \frac{\sigma}{\rho}\left(\frac{\partial^2\eta}{\partial x^2}\right) + \frac{\partial\varphi}{\partial t} = 0.$$

Upon differentiating and making the same approximations as before, the new boundary condition becomes

$$g\frac{\partial\varphi}{\partial y} - \frac{\sigma}{\rho}\left(\frac{\partial^3\varphi}{\partial x^2\,\partial y}\right) + \frac{\partial^2\varphi}{\partial t^2} = 0. \tag{11.31}$$

We may now solve the problem, following the same method as before, obtaining as the general solution [see Eq. (11.28)],

$$\varphi = (Ae^{2\pi y/\lambda} + Be^{-2\pi y/\lambda})\cos\frac{2\pi}{\lambda}(x - ct). \tag{11.32}$$

The boundary condition on the channel bottom leads again to the simplification that

$$\varphi = A'\frac{\cosh(2\pi/\lambda)(y + h)}{\cosh(2\pi h/\lambda)}\cos\frac{2\pi}{\lambda}(x - ct). \tag{11.33}$$

We now substitute this expression for φ into the new boundary condition, Eq. (11.31), and obtain the following expression for the wave velocity, including the effect of surface tension:

$$c^2 = \left(\frac{g\lambda}{2\pi} + \frac{\sigma 2\pi}{\rho\lambda}\right)\tanh\frac{2\pi}{\lambda}h. \tag{11.34}$$

Equation (11.34) reduces to the result obtained previously when the term in σ is negligible. It is seen that surface tension effects become more important as λ decreases. For the usual magnitudes of σ and h the term in σ has to be considered only for wavelengths so small that $\tanh(2\pi h/\lambda)$ can be taken as unity. The relationship between the wave velocity c and the wavelength λ in that case is illustrated in Fig. 11.14. The curve is seen to have a minimum value. Waves of lengths shorter than that corresponding to this minimum are often called *ripples*. This kind of wave is produced, for example, when a small object is thrown into the water.

It is also of interest to divide Eq. (11.34) by c^2. The equation then becomes dimensionless and the first term corresponds to the Froude number (or the square of its reciprocal) and the second to the Weber number, which was introduced in Problem 5.2.

PROBLEMS

When not otherwise specified, the fluid is water at 20°C.

11.22 A ship 140 m long steams at velocity V_0 into a series of approaching waves. It is noticed that successive wave crests pass under the bow at intervals of 6 s

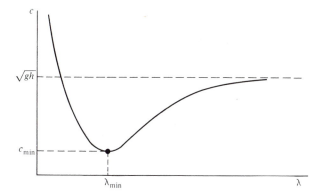

FIG. 11.14   Wave velocity as a function of wavelength, including the effects of surface tension.

and that the wavelength is just equal to the length of the ship. What is the velocity V_0 in m/s? (Assume deep-water waves.)

11.23   A deep-water wave of height $2a$ from crest to trough has a wavelength λ such that $a/\lambda \ll 1$.
   (a) What is the equation of the surface profile?
   (b) With the stipulation that the amplitude a of the wave motion is much smaller than the wavelength, show that the trajectories of the fluid particles are circles ($r = ae^{2\pi y/\lambda}$). Also find the depth in terms of the wavelength for which the amplitude of the motion is only 5 percent of its surface value. (Neglect surface tension.)

11.24   A sinusoidal wave of wavelength λ and half amplitude a travels on the surface of a body of water of mean depth h. A pressure-measuring gage is mounted on the bottom. The gage will indicate an average pressure (equal to the mean depth) and a fluctuating pressure due to the wave passing by.
   (a) Compute the fluctuating part of the pressure on the bottom.
   (b) Show that for $\lambda \gg h$ the variation in bottom pressure head is equal to the change in elevation of the fluid as the wave passes overhead. (Neglect surface tension.)

11.25   A certain harbor exhibited wave behavior that damaged ships, docks, and mooring lines. Such behavior suggests standing waves at resonant frequencies. The harbor is about 10 m deep and 1400 m long in a direction parallel to wave trains entering from the ocean.
   (a) Neglecting surface tension, compute the fundamental period of the basin.
   (b) For a wave of this period, compute the maximum horizontal velocity of the water surface for a resonant wave of 0.15 m total amplitude (crest to trough).

11.26   A deep-water wave of 0.15 m amplitude (the height of a crest over the mean water level) travels to the right with a wavelength of 30 m. Neglect surface tension.
   (a) Compute the wave speed.
   (b) Determine the maximum speed of a water particle on the surface.
   (c) In what direction is the fluid motion in the trough?

11.27 In Problem 11.26, calculate the velocity head at a wave crest and compare its value with that of the elevation. Is it reasonable to neglect the velocity head in computing the pressure at the surface?

11.28 Verify Eq. (11.34).

11.29 Assume that the wavelength is much less than the depth and determine the wavelength and wave velocity for the minimum wave velocity when surface tension is taken into account. Calculate these quantities for water.

11.30 Show that when $h/\lambda \to 0$ (i.e., for shallow-water waves), Eq. (11.27) can be written in the form

$$gh \frac{\partial^2 \varphi}{\partial x^2} = \frac{\partial^2 \varphi}{\partial t^2}$$

and compare this result with Eq. (10.16).

11.31 Standing waves are observed in a basin of mean depth h and width L. (The waves are two-dimensional.) The wave profile may be taken as

$$\eta = a \sin \sigma t \cos kx,$$

where $k = 2\pi/\lambda$, $\sigma = kc$.
(a) Show that the potential

$$\varphi = ac \cosh k(y + h)/\sinh kh \cos kx \cos \sigma t$$

satisfies the boundary conditions at the bottom and at the surface (it will be necessary to use the Bernoulli equation here).
(b) Determine the equations of the orbits for the individual water particles.

11.32 A very deep layer of dense fluid ($\rho = \rho_1$) underlies a very thick lighter one of density $\rho_2(<\rho_1)$. The effects of surface tension and friction at the interface are to be neglected. Small-amplitude progressive waves of length λ are excited on the interface and it is desired to calculate their velocity.
(a) Defend the statement that the pressure and velocity normal to the interface must be continuous across the interface but that the tangential velocity may be discontinuous across the interface.
(b) If φ_1 and φ_2 are velocity potentials in the lower and upper layers, respectively, show that the boundary condition at the interface (to be evaluated at the undisturbed level) is

$$\rho_1\left(\frac{\partial^2 \varphi_1}{\partial t^2} + g \frac{\partial \varphi_1}{\partial y}\right) = \rho_2\left(\frac{\partial^2 \varphi_2}{\partial t} + g \frac{\partial \varphi_2}{\partial y}\right).$$

What further assumptions are necessary to obtain this form, if any?
(c) Finally, by considering functions of the form

$$\varphi = Ae^{\pm 2\pi y/\lambda} \sin \frac{2\pi}{\lambda}(x - ct)$$

for the upper and lower layers, show that

$$c^2 = gk \frac{\rho_1 - \rho_2}{\rho_1 + \rho_2}.$$

REFERENCES

1. V. T. Chow, *Open-Channel Hydraulics*, McGraw-Hill, New York, 1959.
2. R. L. Daugherty and J. R. Franzini, *Fluid Mechanics*, 7th ed., McGraw-Hill, New York, 1977.
3. L. M. Milne-Thompson, *Hydrodynamics*, 4th ed., Macmillan, New York, 1960.
4. Harry H. Barnes, Jr., "Roughness Characteristics of Natural Channels," U.S. Geological Survey, Water Supply Paper 1849, U.S. Government Printing Office, 1967.
5. Paul R. Jordan, "Fluvial Sediment of the Mississippi River at St. Louis, Missouri," U.S. Geological Survey, Water Supply Paper 1802, U.S. Government Printing Office, 1965.
6. W. R. Brownlie, "Flow Depth in Sand-Bed Channels," *J. Hydraulic Eng.*, Vol. 109, 7, 1983.
7. J. L. K. Chan and S. M. Calisal, "A Numerical Procedure for Time Domain Non-Linear Surface Wave Calculations," *Computer Techniques in Environmental Studies*, Springer-Verlag, New York, 1988, pp. 309–329.

12

Turbomachines

12.1 General Description

Pumps, turbines, and fans are machines whose function is to change the energy level of the flowing medium. A pump or compressor increases the total head or pressure of the fluid and a turbine decreases it and extracts energy from the flow. The study of some of their elementary characteristics and properties is the subject of the present chapter and the one to follow. We shall, however, be concerned only with those machines in which the kinetic energy of the fluid plays a major role. The active elements in these machines are usually rotating, and this general type has therefore been termed a *turbomachine*. Positive-displacement machines such as gear or piston pumps, as well as electromagnetic or viscous pumps, will be excluded from the present discussion. Turbomachines are an important part of technology, and a knowledge of their general characteristics is essential for many branches of engineering. Furthermore, a study of their operation provides an excellent example of the application of fluid mechanics to engineering problems.

The geometry of turbomachines varies appreciably for the differing types that have been developed. Broadly, there are two classes. In the first class there is a pronounced change in radius from the inlet to the discharge; these may be said to be *centrifugal* turbomachines. This is an important type of turbomachine, and a great number of pumps, turbines, and compressors are of this type. The flow is usually toward the larger radius for a pump and radially inward for a turbine. It is, however, possible to design centrifugal pumps and turbines with flow directions opposite to those described, and examples of each kind are known. The other class consists of *axial* machines, in which the flow is largely parallel to the axis of rotation. Here again one finds a wide variety of turbines, fans, pumps, and compressors. Between these extremes are examples in which the flow may proceed along conical surfaces of revolution, and these are sometimes called *mixed-flow* turbomachines. In all these varying types, however, there must be a rotating member, usually called a rotor, or impeller, to do work on the fluid and one or more stationary members (called stator, volute, and guide vanes) to guide the flow before and/or after the impeller.

Figure 12.1 shows a typical centrifugal pump. The inlet is axial, although the leading edges of the blades may not be, and the discharge is primarily radial. The flow leaving the impeller is collected and the kinetic energy of the discharge flow subsequently converted into pressure in the pump housing. The collection and diffusion processes may be done in different ways. For example, in the sketch (Fig. 12.1) the fluid is first slowed down somewhat in an annular diffuser, and then it is collected in a spiral volute and subsequently diffused somewhat more in the circular

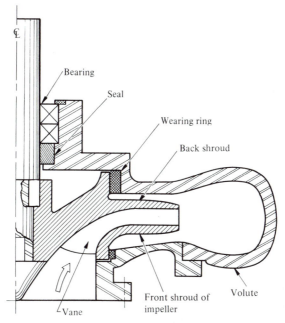

FIG. 12.1   Cross section of a typical centrifugal pump.

duct of the discharge pipe. The extent of the annular diffusion may vary from one type to another. Alternatively, the diffusion may be accomplished by a circular array of guide vanes downstream of the impeller before it is collected by the volute. The details of a centrifugal turbine are not greatly different except that the flow is usually

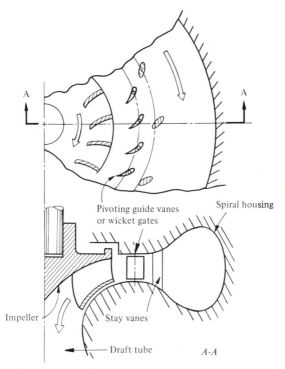

FIG. 12.2   Cross section of a typical centrifugal turbine. The flow is inward and the flow direction at the inlet to the rotor is determined by special guide vanes.

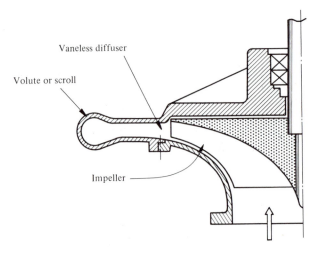

FIG. 12.3 Cross section of a typical centrifugal air compressor.

radially inwards and a support structure and set of movable guide vanes, or "wicket gates," is generally provided to regulate the flow through the turbine (Fig. 12.2). The cross section of a centrifugal compressor is shown in Fig. 12.3. In all of these machines the flow at the larger radius is essentially radial, which enables some simplifications to be made in the analysis. Finally, a cross-sectional view of a multistage axial-flow compressor is shown in Fig. 12.4.

The geometry is complex in all of these devices, and it is not usually possible to analyze the flow through them in detail. For this reason there has been a great amount of empiricism and experimental development in this field of engineering. There are, nevertheless, certain fundamental similarities of the flow through all turbomachines. In the remainder of this chapter these basic relations will be emphasized, and some important engineering characteristics of these machines will be discussed.

12.2 Momentum and Bernoulli Relations
(a) General Relationships

Figure 12.5 shows two views of an arbitrary rotor or stator consisting of a definite number of vanes, and the torque exerted by this device is to be determined. The

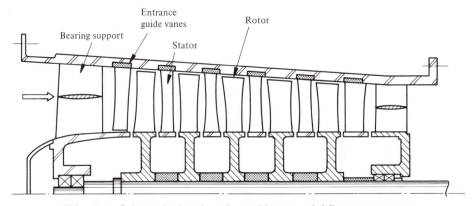

FIG. 12.4 Schematic drawing of a multistage axial-flow compressor.

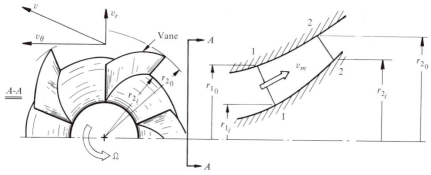

FIG. 12.5 Diagram of an arbitrary rotor, indicating pertinent velocity vectors. The radial cross section is called the meridional plane.

notation is generally that of a cylindrical coordinate system with the z-axis coinciding with the centerline. The reference frame is stationary (or nonrotating), and velocities measured in this frame will be called *absolute* velocities. The component of velocity parallel to the streamlines in the meridional plane is called the meridional velocity and denoted by the subscript m. For machines in which the axial velocity component is zero, v_m is identical to the radial velocity component v_r, and for machines in which $v_r = 0$, the meridional velocity v_m coincides with the axial component v_a. The circumferential velocity at any point on the rotor is called u (equal to Ωr). The absolute velocity will be denoted by V and its component in the circumferential direction by v_θ. The velocity relative to the rotating impeller will be denoted by w. The components u, V, w form a vector diagram (Fig. 12.6) called the velocity triangle. The leading edges of the blade system are given by the line 1–1 in Fig. 12.5 and the discharge or trailing edges by 2–2. The momentum theorem of Chapter 4 will now be applied to the control volume formed by the outer stream surface of revolution containing the blade system, the surfaces 1–1 and 2–2, and the inner surface of revolution. It is assumed that the mass flow rate is steady and that the rate of rotation is constant or zero. The latter case would apply to a stator. There is no change of angular momentum within the chosen volume with respect to time. Application of Eq. (4.10) then gives

$$T = \int_{2-2} \rho r v_\theta v_m \, dS - \int_{1-1} \rho r v_\theta v_m \, dS. \tag{12.1}$$

If there are no viscous shear forces exerted over the boundaries of the control volume, Eq. (12.1) is also the magnitude of the torque exerted by the vanes on the

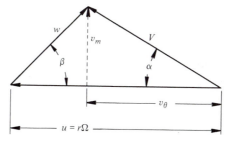

FIG. 12.6 Velocity vector diagram showing the relation between the absolute velocity V, the relative velocity w, and the blade velocity u. The tangential component of the absolute velocity v_θ and the meridional component v_m are also shown.

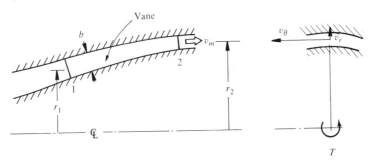

FIG. 12.7 Diagram of a possible turbomachine for which the "one-dimensional" assumption may be satisfactory.

fluid. The vane system may be rotating; in that event the work done by the impeller is $T\Omega$ per unit time, where Ω is the angular speed measured in rad/sec. If $T\Omega$ is positive (i.e., T and Ω of like sign), work is being done on the fluid, and we would expect to have a pumping action. However, if T and Ω are of unlike sign, power is being extracted from the fluid, and we would expect to have a turbine. Equation (12.1) cannot be evaluated unless detailed knowledge about the quantities ρ, r, v_θ, and v_m is available. We shall now consider a simple case—that of *one-dimensional* flow in which v_m, v_θ are constant at a given value of r. A possible geometry for which this might be true is shown in Fig. 12.7. For this approximation to be realistic, the number of vanes guiding the fluid must be relatively large, so that v_θ, v_m, and so on, do not vary significantly around the periphery of Sections 1–1 and 2–2. For the present we merely assume such a flow field and leave for a later section the problem of finding the number of blades required to make the assumption above acceptable for engineering purposes. We further assume the thickness b of the stream tube shown in Fig. 12.7 to be small compared to the radius. Then from Eq. (12.1) we have

$$T = 2\pi r_2 b_2 v_{m_2} r_2 v_{\theta_2} \rho_2 - 2\pi r_1 b_1 v_{m_1} r_1 v_{\theta_1} \rho_1$$

or

$$T = \dot{m}(r_2 v_{\theta_2} - r_1 v_{\theta_1}), \tag{12.2a}$$

since the mass-flow rate $\dot{m}$ is given by

$$\dot{m} = 2\pi r_2 b_2 v_{m_2} \rho_2 = 2\pi r_1 b_1 v_{m_1} \rho_1. \tag{12.2b}$$

In the notation of Fig. 12.7, T is positive if $r_2 v_{\theta_2} > r_1 v_{\theta_1}$, where the counterclockwise direction has been taken as positive. Whether the stream surface is axial or radial at the exit is immaterial. Even though Eq. (12.2a) is subject to the approximation of uniform flow around the vane system, it is an extremely useful result because this approximation is valid for many practical applications. Equation (12.2a) can be written in a form involving the concept of circulation. We then obtain

$$T = \frac{\dot{m}}{2\pi} (\Gamma_2 - \Gamma_1), \tag{12.2c}$$

where Γ is the value of the circulation. Although Eq. (12.2c) was obtained by a one-dimensional argument, it is also valid for two-dimensional potential flow in which v_θ, v_m vary around the periphery, as can be shown from Eq. (12.1).

Let us suppose now that the vane system is rotating and that Ω is positive (i.e., in the same direction as T). The power required by the impeller is then $T\Omega$ or

$$P = T\Omega = \dot{m}(u_2 v_{\theta_2} - u_1 v_{\theta_1}), \qquad (12.3)$$

where $u = \Omega r$ is the impeller speed at radius r.

(b) Relationships for Incompressible Fluids

The equations above are valid for compressible fluids as well as incompressible ones. In the subsequent analysis we shall restrict ourselves, however, to incompressible fluids only. In Eq. (12.3) the power transmitted to the impeller was given. If we now assume complete *absence of friction*, this power should equal the power increase (or decrease for a turbine) of the fluid, that is,

$$P = T\Omega = \rho g Q H,$$

where H is the change in total head.† From this equality it follows that

$$H = \frac{1}{g}(u_2 v_{\theta_2} - u_1 v_{\theta_1}). \qquad (12.4)$$

A sign convention is implicit in Eq. (12.4); namely that a positive H results when the total head *increases* in the direction of the flow. The corresponding increase in fluid power, $\rho g H Q$, is then also positive. Similarly, the power *input* to the impeller $T\Omega$ is considered positive when T and Ω are of like sign and negative when they are of opposite sign, as was mentioned earlier. With this convention it is not necessary to refer the signs of H, Ω, T, and so on, to any particular coordinate system.

The total head rise across the impeller, using the definition of total head, is

$$H = \frac{p_2 - p_1}{\rho g} + \frac{V_2^2 - V_1^2}{2g}.$$

We now substitute for H the expression from Eq. (12.4) and express V in terms of the components of the velocity triangle; that is,

$$V^2 = w^2 + u^2 - 2wu \cos \beta = w^2 + u^2 - 2u(u - v_\theta).$$

We then obtain the expression

$$\frac{p}{\rho g} + \frac{1}{2g}(w^2 - u^2) = \text{const.} \qquad (12.5)$$

This equation is sometimes called the *Bernoulli equation in rotating coordinates*.

The derivative of Eq. (12.5) as given above is restricted to axisymmetric flow. The equation applies, however, more generally to any two points on a "relative" streamline as seen from the rotating coordinate system, provided that the flow is steady in respect to this rotating system and as long as there is no friction. Because of the importance of Eq. (12.5) we shall present a second, less restrictive derivation in which the integration is carried out along a streamline in the rotating coordinate system. Let us begin by writing the equations of motion in an absolute coordinate system. For convenience we select cylindrical polar coordinates with the axis of rotation at the center, and we restrict ourselves to two dimensions (r, θ). The extension to the dimension in the axial direction can be accomplished without difficulty.

---

† This quantity is called M in Eq. (3.11). The notation has been changed here to be in agreement with the standard engineering notation in this subject.

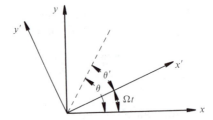

FIG. 12.8 Relation between stationary and rotating coordinate system.

Neglecting viscous effects and body forces, we then obtain (see Problem 2.21)

$$\frac{\partial v_r}{\partial t} + v_r \frac{\partial v_r}{\partial r} + v_\theta \frac{\partial v_r}{r\, \partial \theta} - \frac{v_\theta^2}{r} = -\frac{1}{\rho}\frac{\partial p}{\partial r} \tag{12.6a}$$

and

$$\frac{\partial v_\theta}{\partial t} + v_r \frac{\partial v_\theta}{\partial r} + v_\theta \frac{\partial v_\theta}{r\, \partial \theta} + \frac{v_r v_\theta}{r} = -\frac{1}{\rho r}\frac{\partial p}{\partial \theta}. \tag{12.6b}$$

The angular position of a point fixed relative to the rotating coordinate system, say θ', is now related to the fixed or absolute coordinate system by the relationship

$$\theta = \theta' + \Omega t,$$

as shown in Fig. 12.8, where the primes denote coordinates in respect to the rotating axes.

In order to rewrite Eqs. (12.6a) and (12.6b) in terms of the coordinates of the rotating system, we replace the independent variables θ, t, and r by θ', t', and r'. At the same time let us replace the absolute velocity components by the relative ones as measured in respect to the rotating system. Thus we have the following transformations of variables:

$$\theta' = \theta - \Omega t,$$
$$t' = t,$$
$$r' = r,$$
$$w_{\theta'} = v_\theta - r\Omega,$$
$$w_{r'} = v_r.$$

Furthermore, the partial differential in respect to t and θ may be expressed as

$$\frac{\partial}{\partial t} = \frac{\partial}{\partial t'}\frac{\partial t'}{\partial t} + \frac{\partial}{\partial \theta'}\frac{\partial \theta'}{\partial t} = \frac{\partial}{\partial t'} - \Omega \frac{\partial}{\partial \theta'},$$

$$\frac{\partial}{\partial \theta} = \frac{\partial}{\partial t'}\frac{\partial t'}{\partial \theta} + \frac{\partial}{\partial \theta'}\frac{\partial \theta'}{\partial \theta} = \frac{\partial}{\partial \theta'},$$

and

$$\frac{\partial}{\partial r} = \frac{\partial}{\partial r'}.$$

When the indicated substitutions are made, the equations of motion take the form

$$\frac{\partial w_r}{\partial t} + w_r \frac{\partial w_r}{\partial r} + \frac{w_\theta}{r}\frac{\partial w_r}{\partial \theta} - \frac{w_\theta^2}{r} - r\Omega^2 - 2w_\theta \Omega = -\frac{1}{\rho}\frac{\partial p}{\partial r} \tag{12.6c}$$

for the radial direction and

$$\frac{\partial w_\theta}{\partial t} + w_r \frac{\partial w_\theta}{\partial r} + \frac{w_\theta}{r} \frac{\partial w_\theta}{\partial \theta} + 2w_r \Omega + \frac{w_\theta w_r}{r} = -\frac{1}{\rho}\frac{1}{r}\frac{\partial p}{\partial \theta} \tag{12.6d}$$

for the tangential direction. In these expressions the primes have been dropped. Comparing Eqs. (12.6c) and (12.6d) to (12.6a) and (12.6b), we see that the effect of rotation is to introduce the additional terms $-r\Omega^2$, $-2w_\theta\Omega$, and $2w_r\Omega$. The first of these is the centrifugal acceleration, and the last two are the Coriolis forces brought about because of the noninertial coordinate system.

Let us now postulate that the flow is steady in respect to the rotating coordinate system so that $\partial w_r/\partial t$ and $\partial w_\theta/\partial t$ are equal to zero. We then proceed to integrate Eqs. (12.6c) and (12.6d) along a streamline in the rotating coordinate system in a manner analogous to that followed in the derivation of the Bernoulli equation for a stationary coordinate system. The streamline in our present system is defined by

$$\frac{w_r}{w_\theta} = \frac{dr}{r\,d\theta}. \tag{12.6e}$$

By multiplying Eqs. (12.6c) and (12.6d) by dr and $r\,d\theta$, respectively, and adding the resulting expressions, we obtain

$$\frac{\partial}{\partial r}\left(\frac{w_r^2}{2}\right) dr + \left\{\frac{w_\theta}{r}\frac{\partial w_r}{\partial \theta} - \frac{w_\theta^2}{r} - 2w_\theta\Omega\right\} dr - r\Omega^2\,dr + \frac{\partial}{\partial \theta}\left(\frac{w_\theta^2}{2}\right) d\theta$$

$$+ w_r\frac{\partial w_\theta}{\partial r} r\,d\theta + \left\{2w_r\Omega + \frac{w_\theta w_r}{r}\right\} r\,d\theta = -\frac{1}{\rho}\left\{\frac{\partial p}{\partial r} dr + \frac{\partial p}{r\,\partial \theta} d\theta\right\}.$$

If we substitute for the differentials that multiply the two terms in parentheses by means of Eq. (12.6e), the equation above may be simplified and, for constant density, written in the form of a total differential. For a given streamline in the rotating coordinate system this differential becomes

$$d\left(\frac{w^2}{2} + \frac{p}{\rho} - \frac{r^2\Omega^2}{2}\right) = 0. \tag{12.6f}$$

Integration of Eq. (12.6f) leads again to Eq. (12.5). The same equation is also valid for three-dimensional flows, as mentioned earlier.

PROBLEMS

12.1 Prove that the torque exerted on a steadily rotating two-dimensional impeller of constant breadth b in an incompressible, irrotational, frictionless flow is

$$\rho 2\pi b(r_2^2\,\bar{v}_{\theta_2}\,\bar{v}_{r_2} - r_1^2\bar{v}_{\theta_1}\bar{v}_{r_1}),$$

where the bar denotes average values around the periphery.

Hint: Use Eq. (12.1) and calculate the variable portions of v_r, v_θ from the general solution of potential flow in polar coordinates.

12.2 Assume that the absolute flow around a steadily rotating impeller is a potential flow. Evaluate the term $\partial\varphi/\partial t$ in the nonsteady Bernoulli equation, assuming that the relative flow is steady and thus prove Eq. (12.5).

Hint: Show first that for the conditions stated, $\partial/\partial t = -\Omega(\partial/\partial\theta)$.

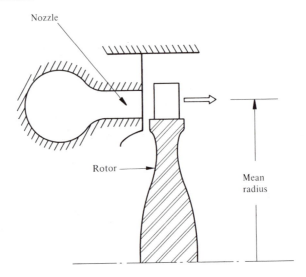

FIG. 12.9 Sketch of a single-stage impulse turbine.

12.3 The Impulse Turbine

Before continuing with the more general development of the theory of turbo-machines, we shall turn our attention to a particular machine, the *impulse turbine*. This turbine is an important and simple type of turbomachine, and its character-istics are discussed here to afford the reader a more concrete example of the types of machines under consideration in this chapter.

An impulse turbine consists of one or more stationary nozzles that direct a stream of high-velocity gas or liquid against a moving row of vanes, or buckets, as they are frequently called. The flow from the nozzles passes through the rotor row without a substantial change in static pressure. The force on the buckets therefore

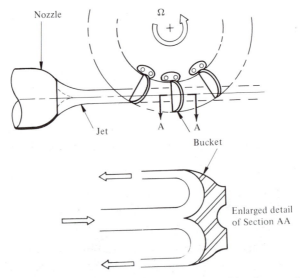

FIG. 12.10 Sketch of a Pelton turbine, a type of impulse turbine frequently used when the working fluid is water.

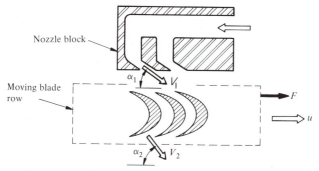

FIG. 12.11   "Two-dimensional" view of a simple impulse turbine of a type shown in Fig. 12.9, showing nozzles and a section of the rotor.

arises purely from momentum change, whence the name "impulse" turbine is derived. Figure 12.9 shows a sketch of such a turbine in which the meridional flow is axial through the rotor. This form of turbine is commonly used for gases. A slightly different arrangement is shown in Fig. 12.10, in which the flow is directed tangentially against the buckets. This form is known as a *Pelton turbine* and is used principally when the working fluid is a liquid. The height of the buckets of either arrangement is usually much smaller than the radius of the impeller. For most applications it is therefore permissible to assume that the fluid interacts with the rotor at a fixed radial distance, which is usually taken as the mean radius of flow passing through the rotor. We shall adopt this simplification in the following discussion. To facilitate analysis, we then consider a developed, or "unrolled," section of an impeller of the type shown in Fig. 12.9, including the jet as shown in Fig. 12.11. By application of the momentum theorem to the control volume indicated, the force F necessary to maintain the system of blades in equilibrium is seen to be

$$F = \dot{m}(V_2 \cos \alpha_2 - V_1 \cos \alpha_1), \qquad (12.7)$$

where $\dot{m}$ is the mass rate of flow through the nozzle, and the angles are measured as shown in Figs. 12.11 and 12.12.

The flow relative to the moving blade row is shown by the velocity diagram in Fig. 12.12. The special feature of the impulse turbine is that the pressure on each side of the impeller is the same. In the absence of friction this requires that $w_2 = w_1$. To account, however, for the effect of friction on the flow, we can introduce an empirical coefficient $c_b = w_2/w_1$, where c_b is always less than unity. (Typical values

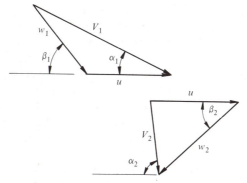

FIG. 12.12   Velocity diagrams at the inlet and discharge from the rotor of an impulse turbine.

might range from 0.8 to 0.95.) With reference to Fig. 12.12, it can be seen that

$$V_2 \cos \alpha_2 = u - w_2 \cos \beta_2,$$

where β_2 is the angle of the relative flow leaving the impeller. But

$$w_2 = c_b w_1$$

and

$$w_1 = (V_1^2 + u^2 - 2V_1 u \cos \alpha_1)^{1/2}.$$

The reaction on the vanes now becomes

$$F = \dot{m}[u - c_b(V_1^2 + u^2 - 2V_1 u \cos \alpha_1)^{1/2} \cos \beta_2 - V_1 \cos \alpha_1]. \qquad (12.8)$$

The power delivered by the blade row is

$$P = Fu,$$

and the power made available by the jet is

$$P_{\text{avail}} = \dot{m}\frac{V_1^2}{2}.$$

We note that if power is to be extracted from the fluid, F must be negative. Thus for a turbine the ratio of the power output to the available power of the jet is

$$\eta_b = \frac{-Fu}{\frac{1}{2}\dot{m}V_1^2} = 2\frac{u}{V_1}\left[c_b \cos \beta_2\left(1 + \frac{u^2}{V_1^2} - 2\frac{u}{V_1}\cos \alpha_1\right)^{1/2} + \cos \alpha_1 - \frac{u}{V_1}\right], \qquad (12.9)$$

where η_b is called the *blade efficiency*. This relation is somewhat cumbersome. It can be seen, however, that aside from the parameters c_b, α_1, β_2, the blade efficiency is dependent only on the ratio u/V_1. In usual turbine practice the power output for a given size and the blade efficiency is desired to be as high as possible. This requires β_2 and α_1 to be small. In a practical machine they can, of source, not be set equal to zero, because an axial velocity component always has to exist for any finite mass flow through the machine. For actual turbines, values of $\alpha_1 = 17°–25°$ and $\beta_2 = 30°$ are typical. To obtain an idea of the approximate behavior of Eq. (12.9), we may therefore set $\alpha_1 = 0$ without too great an error. The blade efficiency then becomes

$$\eta_b = 2(1 + c_b \cos \beta_2)\frac{u}{V_1}\left(1 - \frac{u}{V_1}\right) \qquad (12.10)$$

and has its maximum value of $(1 + c_b \cos \beta_2)/2$ at $u/V_1 = \frac{1}{2}$. Equation (12.10) is plotted in Fig. 12.13 with $c_b \cos \beta_2 = 0.80$, and it can be seen that when u/V_1 becomes less than about 0.25, the blade efficiency becomes low. This is due to the amount of kinetic energy remaining in the jet leaving the impeller. Operation at low velocity ratios (u/V_1) is, however, sometimes unavoidable, and modified impulse turbines, designed to recover some of the leaving energy at low values of u/V_1, have been devised. Some of these will be discussed presently. It is worth emphasizing at this point that the relations for the impulse turbine developed so far are not restricted to incompressible flow and are valid for compressible fluids.

From the foregoing discussion it can be seen that the blade efficiency is merely the fraction of jet energy converted into useful work and is not a ratio of input power to output power, as are, for example, the pump or turbine efficiencies defined in Chapter 3. To determine such a term here, it is convenient to consider an *incom-*

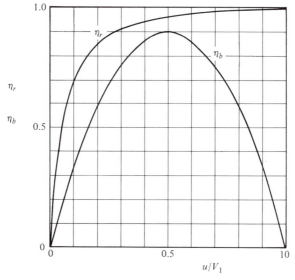

FIG. 12.13 Blade efficiency [Eq. (12.10)] and rotor efficiency [Eq. (12.12)] for a single-stage impulse turbine. The blade factor was taken as $c_b = 0.92$, and $\cos \beta_2$ was assumed to be 0.87.

pressible fluid, although a similar discussion for compressible fluids can be made. The change in total pressure across the moving blade row of an impulse turbine for an incompressible fluid is

$$p_{t_2} - p_{t_1} = \frac{\rho}{2} (V_2^2 - V_1^2),$$

which can be written (for $\alpha_1 = 0$)

$$p_{t_2} - p_{t_1} = -\frac{\rho}{2} u^2 \left(\frac{V_1}{u} - 1\right)\left[1 + c_b^2 + 2c_b \cos \beta_2 + \frac{V_1}{u} (1 - c_b^2)\right]. \quad (12.11)$$

The efficiency of the blade row, or *rotor efficiency*, defined (as in Chapter 3) as the power output divided by the power given up by the fluid, then becomes

$$\eta_r = \frac{P}{\dot{m}(p_{t_2} - p_{t_1})} = \frac{2(c_b \cos \beta_2 + 1)}{1 + c_b^2 + 2c_b \cos \beta_2 + (V_1/u)(1 - c_b^2)} \quad (12.12)$$

for $\alpha_1 = 0$. Equation (12.12) is also plotted in Fig. 12.13, and it can be seen that $\eta_r = 0$ at $u/V_1 = 0$ and increases to unity at $u/V_1 = 1$. The latter value is not of real interest, however, since there is no force on the blades and no power is produced by the turbine or given up by the fluid in that case.

The total pressure change across the rotor [Eq. (12.11)] is negative for all values of $0 \leq u/V_1 \leq 1$, confirming the turbining action of the blade row. On the other hand, when u/V_1 is greater than unity, the total pressure change across the rotor is positive (i.e., energy is being added to the fluid). Also, the blade reaction force becomes positive, so that positive work is required to move the row, and the machine now operates as a pump even though there is no increase in static pressure across the rotor row. The work done on the fluid results in an increase in the kinetic energy of the fluid. This kinetic energy could be transformed into an increase in pressure by means of a diffuser, and the pressure increase expected from a pump

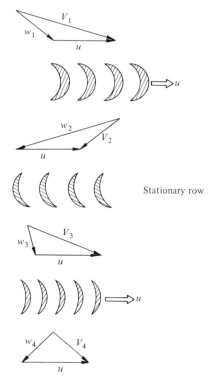

FIG. 12.14 Velocity diagrams for a two-stage impulse turbine. This type is also called a Curtis turbine.

could be obtained in this way. The transformation of kinetic energy into pressure in an unconfined jet is, however, not generally an efficient process, and for this reason impulse pumps are very rarely encountered.

It was mentioned previously that the blade efficiency was low for low values of u/V_1 because of the kinetic energy left in the flow. This energy could, however, be utilized in another set of moving vanes as shown in Fig. 12.14 in a configuration known as a two-stage, or "Curtis"-stage, turbine. This type of turbine consists of two rotors on the same shaft with a stationary row, or stator, between the two rotor rows. The pressure is constant throughout, and in the absence of friction $w_1 = w_2$, $w_3 = w_4$, and $V_2 = V_3$. To exhibit the essential points of the two-stage impulse turbine, let us assume that the absolute and relative flow angles are all zero and that the blade factor c_b is the same for all rows. Then

$$w_2 = c_b w_1 = c_b(V_1 - u),$$

$$V_2 = w_2 - u,$$

$$V_3 = c_b V_2,$$

$$w_3 = c_b V_2 - u,$$

$$w_4 = c_b w_3,$$

and the total reaction force on both rows is

$$F = \dot{m}(-w_2 - w_1 - w_3 - w_4) = -\dot{m}(1 + c_b)\{V_1(1 + c_b^2) - u(2 + c_b + c_b^2)\}.$$

The blade efficiency for this case is

$$\eta_b = 2(1 + c_b)\left[1 + c_b^2 - \frac{u}{V_1}(2 + c_b + c_b^2)\right]\frac{u}{V_1}. \tag{12.13}$$

The maximum blade efficiency for the two-stage "zero angle" turbine may be shown to be

$$\eta_{b_{max}} = \frac{1}{2}\frac{(1 + c_b)(1 + c_b^2)^2}{2 + c_b + c_b^2}. \tag{12.14}$$

This maximum occurs when

$$\frac{u}{V_1} = \frac{1 + c_b^2}{2(2 + c_b + c_b^2)}. \tag{12.15}$$

The two-stage and single-stage impulse turbines are compared in Fig. 12.15 for $\alpha_1 = \beta_2 = 0$ and $c_b = 0.9$. These curves are only illustrative and are not intended to represent accurately values achieved in practice. They do, however, bring out the principal features—namely, that the maximum blade efficiency occurs at a lower value of u/V_1 for the two-stage machine, but that this maximum value is below that for the single-stage turbine. If there were no friction, the maximum would occur for $u/V_1 = 0.25$. Due to friction, it occurs at a somewhat lower value.

Although many of the essential features of impulse turbines have been described, our discussion has been necessarily idealized, and a number of important considerations for real machines have been omitted. For example, the effect of friction cannot be fully described by a blade factor, since leakage, windage, and churning losses can be extremely important. Furthermore, the blade factor c_b is dependent on the angle between the incoming relative flow and the blade angle—the angle of attack—and the Reynolds number as well as the details of the blade design. If the fluid is a gas, the effect of partial admission (when the nozzle extends only over a fraction of the periphery) is extremely important, because certain irreversible losses occur during the filling and emptying of the passages between the blades. To

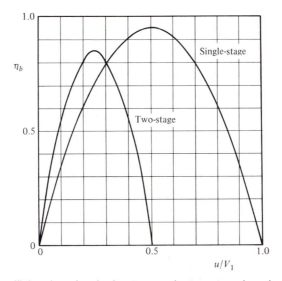

FIG. 12.15 Blade efficiencies of a single-stage and a two-stage impulse turbine. For these curves it was assumed that $\alpha_1 = \beta_2 = 0$ and that the blade factor $c_b = 0.9$ throughout. The jet velocity is equal to V_1, and no nozzle losses are taken into account.

account for all these effects is beyond the present scope of this book, and it is easy to see that some of them can only be determined by careful experimentation.

In the calculations above, the leaving angle of the flow was assumed to be known and was furthermore assumed to be independent of the velocity ratio. We have left unanswered the problem of how to design the blades to achieve the desired flow angle; and whether these angles are, in fact, dependent on the entering flow angle or the angle of attack. It should be appreciated that this is an involved problem and that no simple result is available, as the details of blade geometry, effects of friction, and compressibility are all important.

Nevertheless, for many impulse turbines the blades are sufficiently closely spaced so that the outlet angles can be accurately guessed and the friction effects can well be approximated by the factor c_b, as was suggested. The foregoing simplified analyses can then be expected to give estimates of power and efficiency within 10 to 20 percent of the true values.

Finally, it may be mentioned that the amount of energy available at the nozzle may be so great that if the expansion were carried out in one step, the resulting value of u/V_1 and therefore efficiency would be unacceptably low even if several impulse stages were to be used. This situation usually occurs only for a gas when a high-pressure ratio for the expansion is available. The difficulty is avoided by making the reduction in pressure in several steps, or stages. Each stage, then, consists of a nozzle and an impulse rotor, and since the pressure is different in each of the stages, a system of seals to prevent leakage between the stages is necessary. This arrangement is called *pressure staging*, or sometimes *Rateau staging*, after a well-known French engineer. The analysis of each of the stages does not differ from the single impulse stage and therefore we will not pursue it further.

PROBLEMS

When not otherwise specified, the properties of the fluids are to be evaluated at 20°C and 1 atmosphere pressure.

12.3 Consider an impulse stage with $\alpha_1 = 20°$, $\beta_2 = 30°$, and $c_b = 0.92$. Determine the value of u/V_1 that results in the best value of blade efficiency η_b, and for that condition, obtain the value of the leaving velocity V_2 in terms of V_1.

   Note: It is probably best to calculate a few points and determine the maximum graphically to avoid tedious algebra.

12.4 An impulse stage consists of a nozzle with a velocity coefficient c_v and a rotor with blade coefficient c_b. Obtain an expression for the stage efficiency (i.e., power output divided by the power input based on the pressure drop across the nozzle) in terms of α_1, β_2, c_v, c_b, and u/V_1.

12.5 A stream of gas with a velocity of 1100 m/s issues from a nozzle with an angle $\alpha_1 = 17°$ and enters a two-stage impulse turbine for which $\beta_2 = 30°$. The blade factor is $c_b = 0.92$ for each of the rotor and stator rows. Draw to a suitable scale the velocity diagrams for each of the rotors and stators for $u/V_1 = 0.2, 0.25, 0.3$. Assume that the angle of the absolute velocity leaving the intermediate stationary row is also at an angle of 17°.

12.6 Consider a three-stage impulse turbine with $\alpha_1 = \beta_2 = 0$ for all stages and a value of c_b equal to unity for all rows—rotor and stator.
   (a) What is the ratio of u/V_1 for optimum blade efficiency?

(b) At this optimum speed ratio what would be the overall blade efficiency η_b if all rotors and stators had a blade factor of 0.9?

12.7 A hydroelectric power plant station having a head of 2000 ft is to be designed to have a mechanical power output of 20,000 hp. A single-stage impulse turbine is to be used to drive a generator. By making reasonable assumptions about flow angles, blade factors, and nozzle coefficients determine the jet diameter, and *estimate* the rotor diameter, rotative speed, and overall efficiency of the turbine.

12.4 Axial-Flow Machines

A special type of axial-flow machine, the axial-flow impulse turbine, was discussed in Section 12.3. A schematic illustration of a more general type is shown in Fig. 12.16. Suppose that the machine in Fig. 12.16 is one in which energy is added to the fluid. It may then be called a compressor, or a pump. In such a device the first row of blades is a stationary set of *guide vanes*, or *entrance blades*. They direct the flow properly onto the following row of rotating blades. The function of this rotating element, or *rotor*, is to add energy to the fluid. Following the rotor row is another set of stationary blades, called a *stator*. The function of the stator is to remove all or part of the angular momentum from the fluid and thereby to increase the static pressure. Subsequent stages—each stage consisting of a rotor and stator—may be used to increase further the pressure rise. Axial-flow compressors for gas turbines or similar applications may employ 5, 10, or even more stages. Axial-flow pumps usually dispense with the entrance row for economy. Similarly, ventilating fans do not usually have entrance blades, and in some cases of small pressure rise the stator row may also be omitted.

Propellers are also examples of axial-flow machines, but except for a few rare cases, they do not have confining outer cases. Indeed, the outer streamlines are left free to adjust themselves to yield the amount of thrust commensurate with the power input and forward speed. Here again entrance and stator blades are rarely used.

If the machine in Fig. 12.16 is one in which work is abstracted from the fluid, it is called a turbine. The schematic diagram of Fig. 12.16 is still applicable. The first row of blades is now called a set of *entrance nozzles*. This set may be uniformly distributed over the entire circumference, or it may consist of a few individual nozzles. The first rotor abstracts energy from the fluid and imparts mechanical work to the shaft. The rotor is followed by a stator which generally increases the flow velocity and

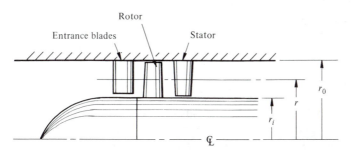

FIG. 12.16 Diagrammatic sketch of an axial-flow compressor.

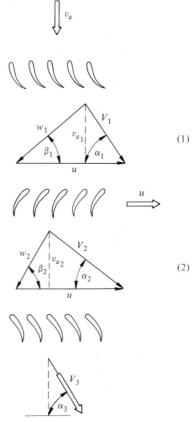

FIG. 12.17 Developed cylindrical surface section through an axial-flow compressor. The diagram shows the entrance vanes, the first rotor, and the subsequent stator.

redirects the fluid to produce the desired interaction with the following rotor. Again the machine may contain many stages, each stage consisting of a rotor and stator.

We shall first consider the simplest type of analysis applicable to an axial turbomachine. The principal assumptions involved in this analysis are that the flow takes place along cylindrical surfaces about the axis of the machine and that there exists complete axial symmetry.

The first of these assumptions is generally well satisfied when the blade height h is small compared to the radius r, that is, when the *hub ratio*, r_i/r_o, is close to unity. For blades longer than about 30 percent of the radius, however, this assumption is no longer tenable, particularly when the angular momentum change through the blade row is large. A more elaborate analysis, taking into account radial variations, will then have to be carried out. The second assumption regarding axial symmetry is made despite the fact that the actual blades are finite in number and thickness. This makes some deviation from axial symmetry unavoidable. Even for relatively wide blade spacing, however, the changes in a tangential direction are often small and the flow may be treated as uniform in this direction.

In addition, we shall assume the fluid to be incompressible. This is certainly permissible for liquids. The assumption can also frequently be used if the fluid is a gas, however, because the changes in density *per stage* are often small. Different densities may be assumed for successive stages.

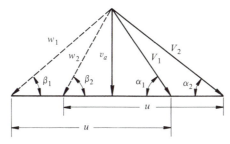

FIG. 12.18 Vector diagram for the flow through the rotor of an axial-flow compressor. Stations 1 and 2 correspond to those in Fig. 12.17.

As a first step in the simplified analysis we shall now develop a typical cylindrical flow surface onto a plane. The system of blades will then appear as in Fig. 12.17. Let us then consider the flow through a typical rotor as shown, designating conditions at the entrance to the rotor by subscript 1 and those at the downstream side by 2. The relative and absolute velocities are shown by vectors in between the blade rows. Frequently, the two vector diagrams are combined so that their vectors coincide. The diagram then appears as in Fig. 12.18. It offers a convenient way of surveying the vector angles and velocity changes which have to be produced by the rotor and stator. The two diagrams (Figs. 12.17 and 12.18) also serve to define the flow angles α and β. These angles have been chosen somewhat differently than in the discussion on the impulse turbine, but in a way which is more suitable for the subsequent analysis.

For the stream surface in question the total head rise across the rotor *in the absence of friction* can be determined by application of the Bernoulli equation across the rotor or directly from (12.4). We then obtain for the total head rise

$$H = \frac{u}{g}(v_{\theta_2} - v_{\theta_1}), \tag{12.16}$$

where u is the rotor velocity at the given radius. In Fig. 12.17, α designates the angle between the absolute velocity and the direction of motion, and β refers to the included angle between relative velocity and the local rotor velocity u. From the principle of conservation of mass and the previous assumptions, it follows that

$$v_{a_1} = v_{a_2} = v_a,$$

where v_a is the axial velocity component, which for the present flow is the same as the meridional velocity v_m. The peripheral components v_θ can be eliminated by the geometric relationships

$$v_\theta = u - v_a \cot \beta = v_a \cot \alpha.$$

Thus the total head rise across the rotor may be written as

$$H = \frac{u^2}{g}\left[1 - \frac{v_a}{u}(\cot \beta_2 + \cot \alpha_1)\right]. \tag{12.17}$$

This represents also the total head rise across the entire stage, since there is no change in total head across the stator, provided, of course, that friction may again be neglected. The angles α_1 and β_2 in Eq. (12.17) refer to the actual flow directions and not necessarily to the geometric angles of the trailing edges of the blades, which we shall designate by α_{1v}, β_{2v}, and so on. If the vanes are thin and very closed spaced, flow angles may be taken equal to the blade angles ($\beta_2 = \beta_{2v}, \alpha_{1v} = \alpha_1$, etc.).

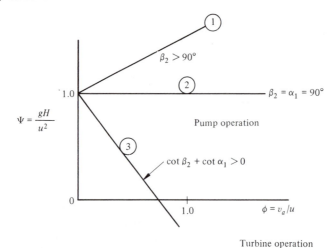

FIG. 12.19 Idealized relationship between the head coefficient Ψ and the flow coefficient ϕ for a purely axial-flow machine. Curves computed from Eq. (12.18).

Equation (12.17) may also be written in dimensionless form by introducing the parameters Ψ and ϕ, which are defined† by

$$\Psi = \frac{gH}{u^2},$$

$$\phi = \frac{v_m}{u}.$$

The head rise in dimensionless form then becomes

$$\Psi = 1 - \phi(\cot \beta_2 + \cot \alpha_1). \tag{12.18}$$

This expression is still applicable only to a single stream surface. For blades that are very short compared to the radius, however, Eqs. (12.17) and (12.18) would give a good approximation to the head increase for the entire flow. The values at the mean blade radius would be used to evaluate the angles and velocities. To obtain an understanding of the implications of Eq. (12.18), let us first consider such a machine for which the blade height is very small compared to the radius ($h/r \ll 1$). In Fig. 12.19 the total head coefficient is plotted as a function of the flow coefficient ϕ by means of Eq. (12.18), which now characterizes the flow through a stage of the short-bladed machine in question. Curves are shown for various flow angles β_2 and α_1, which are determined principally by the trailing edges of the rotor and stator blades, respectively. These angles are here considered constant and independent of the flow velocities.

Pumps and compressors are characterized by positive values of Ψ, whereas turbines have negative total head coefficients. Most pumps, fans, or compressors have values of β_2 and α_1 such that Ψ decreases with increasing ϕ; that is, the head decreases with increasing flow rate as indicated, for example, by curve (3) in Fig. 12.19. In the turbine regime the pressure coefficient always becomes increasingly negative with increasing flow rate. The equations derived previously for the impulse turbine can also be cast into the form of Eq. (12.18). The ideal characteristics of this turbine would correspond to the negative portion of a curve such as (3).

---

† These should not be confused with the stream function and velocity potential of potential flow.

The form of Eqs. (12.17) and (12.18) are particularly convenient when the rotative speed is constant. Sometimes, however, it is more desirable to consider the characteristics of a turbomachine when other variables, such as the flow rate, are kept constant. For example, in the application of turbines a fixed flow rate frequently is available, and it is desired to determine the torque (or total head change) as a function of blade speed. We shall here define a different set of parameters, which is useful in analyzing this type of problem. From our fundamental relations (12.2) we have for the torque

$$T = \rho Q r (v_{\theta_2} - v_{\theta_1}), \tag{12.19}$$

where we still consider here an axial-flow machine with sufficiently short blades so that the radius may be assumed constant. Defining a torque parameter as

$$\chi = \frac{T}{\rho Q r v_{\theta_1}},$$

Eq. (12.19) may be written in the form

$$\chi = \frac{v_{\theta_2}}{v_{\theta_1}} - 1 = \frac{u}{v_{\theta_1}} - (1 + \tan \alpha_1 \cot \beta_2). \tag{12.20}$$

If we continue to consider α_1 and β_2 to be constant, Eq. (12.20) represents the equation of a straight line. For small u the torque or force on the rotor is negative—that is, opposite to the direction of positive u. It is easy to show that when χ is negative, H is also negative, so that the machine operates as a turbine.

The power of the machine is equal to

$$P = T\Omega,$$

and we may define a power coefficient

$$\Pi = \frac{P}{\rho Q v_{\theta_1}^2} = \frac{u}{v_{\theta_1}} \chi. \tag{12.21}$$

Curves of χ and Π versus u/v_{θ_1} are shown in Fig. 12.20. The highest turbine power occurs when the torque coefficient is equal to one-half of its value at zero speed. This fact is illustrated in Fig. 12.20 and can be shown from Eq. (12.21).

Up to now we have totally neglected any effects of friction, and as a consequence the mechanical power ($T\Omega$) was always set equal to the hydraulic power ($HQ\rho g$). To recognize the existence of friction and similar losses, we introduce pump and turbine efficiencies, which are defined, respectively, by the equations

$$T\Omega \eta_p = HQ\rho g \tag{12.22}$$

and

$$T\Omega = \eta_t HQ\rho g. \tag{12.23}$$

These expressions may again be presented in dimensionless form, and we shall again use the blade speed u and the discharge area A_2 for this purpose as we did in defining Ψ and ϕ.

Dividing both sides by the same quantities and substituting Q by $v_m A_2$, we obtain for a pump

$$\frac{T\Omega \eta_p}{u^3 \rho A_2} = \frac{Hg}{u^2} \frac{v_m}{u} \frac{A_2}{A_2} = \Psi \phi$$

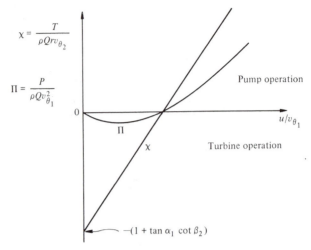

FIG. 12.20 Idealized relationship between the power coefficient, the torque coefficient, and the velocity ratio u/v_{θ_1}. Curves computed from Eqs. (12.20) and (12.21).

or

$$\tau\eta_p = \Psi\phi, \tag{12.24a}$$

where

$$\tau \equiv \frac{T\Omega}{u^3\rho A_2}.$$

The dimensionless parameter τ is called a *power coefficient*, or also a *torque coefficient*, although it differs, of course, from the definition of χ. Equation (12.24a) is particularly useful when analyzing experimental data. The parameter τ may then be determined from the measured power input and Ψ and ϕ from hydraulic measurements. The efficiency can then be derived from these dimensionless coefficients by Eq. (12.24a).

In a way analogous to that leading to Eq. (12.24a), we obtain for a turbine

$$\tau = \eta_t \phi\Psi. \tag{12.24b}$$

In the foregoing discussion we have limited ourselves to axial turbomachines with low hub ratios $r_i/r_o \sim 1$. The total head rise for the stream surface at midblade height could therefore be taken as representative for the entire flow. Examining the flow conditions along the blade in more detail, it is seen from Eq. (12.16) that the total head change through a rotor is proportional to $u\,\Delta v_\theta$. If it is desired to obtain the same total head change at all radii, Δv_θ will have to be proportional to $1/r$ since $u = \Omega r$. The blade shapes could be made to vary along the radius so as to yield velocity components such that the product $u\,\Delta v_\theta$ remains constant. It should be remembered, however, that in the present simplified analysis the assumption of flow along strictly cylindrical surfaces was based on the existence of relatively short blades. If the blades are so long as to require shape changes along the radius, it usually is also necessary to consider possible radial velocity components, and a more extensive analysis is required.

A remark should also be made in regard to the relation between blade angles (or more generally blade shapes) and flow angles. In this section we have concentrated mainly on the flow angles and have assumed the existence of the proper blades to bring about the angles. We shall discuss a few aspects of the flow around the blades themselves in Chapter 13.

PROBLEMS

12.8 By applying the Bernoulli equation to the relative flow in an axial impeller, show that the static pressure rise across the rotor is $p_2 - p_1 = \frac{1}{2}\rho[v_{\theta_1}^2 - v_{\theta_2}^2 + 2u(v_{\theta_2} - v_{\theta_1})]$ if there is no friction.

12.9 The reaction ratio R is defined as the ratio of the static pressure rise across the rotor row to the total pressure rise across the rotor. Show that in the absence of friction R can be expressed as $R = \frac{1}{2}[1 + \phi(\cot \alpha_1 - \cot \beta_2)]$.

12.10 Make neat sketches of possible velocity diagrams for the flow on either side of a rotor for values of the reaction ratio $R = 0$, $\frac{1}{2}$, 1, assuming frictionless, incompressible flow.

12.11 Axial-flow fans usually have no guide vanes, so that $\alpha_1 = 90°$, and frequently they have no stator blades, so that any rotational kinetic energy of the air stream leaving the rotor is lost. Express this loss as a fraction of the total head rise and give the result in terms of Ψ, ϕ. Neglect all other losses.

12.12 An axial blower 12 in. in diameter with a hub 8 in. in diameter turns at 6000 rpm and discharges 1500 cfm (cubic feet per minute) of air at standard atmospheric conditions with a total pressure increase of 0.25 ft of water. There are no inlet vanes. Assume that there is no friction (which is not a good assumption), and that the flow is strictly along cylindrical surfaces. For this blower:
  (a) Calculate the values of ϕ and Ψ at the tip, hub, and at a point midway between, based on the blade speed at these radii. Sketch the velocity diagrams at the midway point and determine the angles α_1 and β_2.
  (b) Determine the power required in hp.
  (c) Estimate the expected power and pressure rise at 1200 cfm by applying Eq. (12.18) at the mean diameter of 10 in. and by assuming α_1 and β_2 to remain unchanged from the values computed in part (a).

12.13 Water is maintained in a reservoir supplying a turbine at a height z above the level of the rotor blades. The flow leaving these blades is discharged to atmosphere as shown, so that the discharge pressure of the flow is always atmospheric. The relative flow angle leaving the turbine rotor may be assumed to be the same as the turbine discharge blade angle β_{2v}. Guide vanes installed upstream cause the flow approaching the rotor to have an angle α_1 in the direction of the rotor motion. The elevation of the turbine discharge may be assumed to be the same as that of the rotor itself. Effects of friction may be neglected.
  (a) By employing the Bernoulli equation obtain the relationship between the flow coefficient v_a/u, the angles α_1, β_{2v}, and the quantity $\sqrt{2gz}/u$ (where u is the rotor tip speed), required for the turbine discharge to be at the specified discharge pressure.
  (b) Obtain an expression for the decrease in total head across the turbine in terms of the tip speed u, α_1, β_{2v}, and v_a/u.
  (c) Plot the ratio of the total head decrease to z as a function of $u/\sqrt{2gz}$ for values of $\alpha_1 = 25° = \beta_{2v}$.

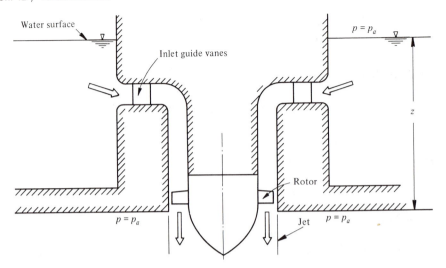

Problem 12.13

12.14 For a turbine arrangement and flow conditions such as those shown in Problem 12.13, explain how by varying the angle of the guide vanes the turbine can be made to vary its power output when the height of the reservoir is maintained at a constant level.

12.5 Centrifugal Machines

In the preceding section we have developed a simplified theory for an axial turbomachine. We now develop a similar theory for a centrifugal machine, again assuming frictionless flow and complete axial symmetry. The general notation is the same as that used previously. The width of the flow passage will be denoted by b, where b may vary from inlet to discharge. Subscript 2 denotes quantities at the outer radius and subscript 1 those at the inner radius.

Let us suppose first that the flow is radially outward. The torque to be supplied by the rotor is then equal to

$$T = \rho Q(r_2 v_{\theta_2} - r_1 v_{\theta_1}), \tag{12.25}$$

where the subscript θ denotes velocity components in the tangential direction. We eliminate v_{θ_2}, v_{θ_1} as before and, assuming no friction, obtain the dimensionless expression for the head coefficient

$$\Psi = \frac{gH}{u_2^2} = 1 - \phi_2\!\left(\cot \beta_2 + \frac{b_2}{b_1}\cot \alpha_1\right), \tag{12.26}$$

where β_2 and α_1 are measured in a radial plane. The flow coefficient is defined on the basis of quantities at the outer radius, that is,

$$\phi = \frac{Q}{A_2 u_2} = \frac{v_{r_2}}{u_2},$$

A_2 being the discharge area of the impeller normal to v_{r_2}. Equation (12.26) is quite similar to that for the head coefficient of an axial machine, and the torque or power coefficient may also again be defined as

$$\tau = \frac{T}{\rho r_2 A_2 u_2^2}.$$ (12.27)

As shown previously, when losses are neglected,

$$\tau = \phi \Psi.$$

In order to take losses into account, however, efficiencies may be introduced in the same way as for the axial machines, and one obtains, as before,

$$\tau \eta_p = \Psi \phi$$ (12.24a)

for a pump or compressor and

$$\tau = \eta_t \Psi \phi$$ (12.24b)

for a turbine. The characteristics of a centrifugal machine are seen to be similar to those of an axial machine as illustrated in Fig. 12.19. Thus when Ψ and τ are positive, the machine acts as a pump, and when Ψ and τ are negative, it acts as a turbine. In the present example the latter machine would represent a radial outflow turbine. This is not, however, a usual type of turbine, for it is more common for turbines to have radial *inflow*. Then also the angles α_2 and β_1 should be regarded as the independent angles rather than α_1 and β_2, and it would be preferable to express equations such as (12.26) in terms of α_2 and β_1.

This completes our preliminary survey of turbomachines based on the simplified theory. The idealized equations that have been obtained can be adjusted by introducing certain empirical coefficients. Some of these coefficients are to remove the inaccuracies of the theory that arise due to the assumed simplified flow patterns. Other coefficients are intended to account for friction losses and similar viscosity effects. Some of the coefficients of the first type can actually be computed analytically. The effect of the existence of a finite number of blades can be approximated by analyzing the potential flow of an inviscid fluid about a row of blades, a problem which involves the solution of Laplace's equation for the given boundary conditions. The second type of coefficient is much more difficult to predict, because it involves prediction of real fluid effects. As we have already seen, even the boundary layer on a flat plate presents some difficulty of solution. It can be fairly stated that except in certain rather simple situations the real fluid flow through a turbomachine cannot be calculated a priori, and for this reason hydraulic engineers have to depend on extensive experimentation to develop turbomachines to fit specific applications. In the next section we review briefly some characteristics of real machines and discuss some of the design variables now employed. Following this we discuss some two-dimensional flow problems that enable one to make more realistic estimates of performance and which provide some basis for understanding the flows involved.

PROBLEMS

12.15 Consider a radial turbomachine with inwardly directed flow. The flow impinging upon the outer radius of the impeller has an angle α_2 determined by the guide vanes and the relative flow leaving the impeller is at the angle β_1 to the peripheral direction. Obtain the head coefficient $\Psi_2 = gH/u_2^2$ as a function of $\phi_2 = v_{m_2}/u_2$, α_2, and β_1 and sketch the result for $\beta_1 = 30°$, $\alpha_2 = 10°$, $20°$, $30°$, and $r_1/r_2 = 0.5$, and $b_1 = b_2$. Label the portions of the diagram corresponding to pumping and turbining action. Neglect losses.

12.16 A radial-flow pump is constructed with the proportions $\alpha_1 = 90°$, $r_1/r_2 = 0.5$, and $b_2/b_1 = 0.75$. The blade angles $\beta_{1v} = \beta_{2v} = 25°$ and the flow angle β_2 is assumed to be equal to β_{2v}. Draw accurately the curve of Ψ versus ϕ_2. Calculate and mark on this diagram with the designation ϕ_d the value of ϕ_2 for which $\beta_1 = \beta_{1v}$ (i.e., no incidence angle at the inlet for $\phi_2 = \phi_d$). Neglect losses.

12.17 A water pump with the characteristics of that in Problem 12.16 and operating at the flow rate coefficient $\phi_2 = \phi_d$ is to pump 8 ft³/sec at a total head of 250 ft of water at a speed of 1800 rpm. Determine (a) the outside diameter d_2 and breadth b_2 of the pump impeller, and (b) the maximum power requirement this pump would have assuming that there is no friction.

    Note: This will not necessarily be equal to the power requirement at the design flow rate coefficient ϕ_d.

12.18 The combined water pump-storage units at Castaic, California, produce a flow of 2000 ft³/sec and turn at 257 rpm. The impeller outer diameter is 19.25 ft, the breadth there is 16 in., and the vane angle is 20°. The total head increase is to be 1040 ft, and there is no inlet swirl. For preliminary estimation purposes the flow can be assumed frictionless.

(a) Calculate the flow coefficient.

(b) Determine the head coefficient needed.

(c) Find the relative flow angle β_2 required to produce the head.

(d) Assume that the flow angle of part (c) is constant. Determine the total head for *no* flow, the flow rate (in ft³/sec) for *no* head, the fluid power for the operating point stated above, and the flow rate for maximum fluid power.

12.6 Examples of Performance Curves

We now examine a few typical performance curves of the various types of turbomachines previously mentioned. Figure 12.21 shows the characteristics of a centrifugal pump typical of modern practice. It is seen that the head coefficient Ψ is a continually decreasing function of the flow rate coefficient ϕ. This characteristic is in agreement with the results of the simplified theory [Eq. (12.26)] assuming a value of β_2 less than 90°. For comparison this equation is also shown in Fig. 12.21, it is assumed that the flow angles α_1 and β_2 are equal to the vane angles α_{1v} and β_{2v}. It may also be noted that the torque coefficient has a maximum value near the point of best efficiency where the efficiency is defined as the fluid power output to the shaft power input, as in Eq. (12.22). It is apparent, however, that a serious discrepancy exists between the results of the elementary theory and the actual measurements.

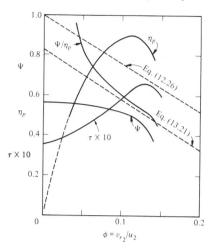

FIG. 12.21 Performance characteristics of a centrifugal pump with seven blades, $\beta_{2v} = 23°$, $\beta_{1v} = 19°$, $b_2/d_2 = 0.16$, and $r_1/r_2 = 0.61$. The graph shows measured values for the torque coefficient τ, the head coefficient Ψ, and the efficiency η_p. For comparison, predictions based on idealized flow conditions are shown by the dashed lines. (Data computed from B. L. Vanderboegh, "Model Tests of the Grandby Pumps," *Trans. ASME, vol.* **69**, 1947, p. 535.)

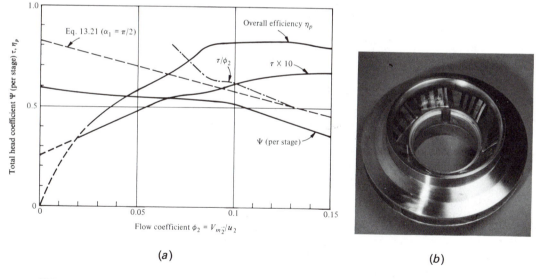

(a)

(b)

FIG. 12.22 (a) Performance of a four-stage boiler feedwater pump operating with water at a temperature of 173°C at 5483 rpm. At a flow of 0.357 m^3/s the total head rise was 1940 m and the power required was 7400 kW. The efficiency shown includes all losses, mechanical plus hydraulic. Design of such pumps is based upon reliability and long life rather than high efficiency. (b) Photograph of test impeller for the pump of part (a). The impeller diameter of each stage is 0.35 m; the discharge blade angle is 22.5° and there are seven impeller blades. The lines painted on the vanes are for reference during flow observation in a cavitation test. (Based upon data provided by D. Florjancic, Sulzer Brothers Ltd., used with permission.)

Part of the discrepancy is due to friction. But a greater portion of the difference must be ascribed to the lack of perfect guidance of the relative flow, which leads to a difference between the flow angles and the vane angles. The more correct flow angles may be determined from potential flow theories to be described later. Using the angles computed from such a theory, a new relation for Ψ versus ϕ is obtained [Eq. (13.21)], which is also shown in Fig. 12.21. It is seen that this new result approaches the measured performance more closely. We should recall that all estimates of the head coefficient based on perfect fluid theories neglect frictional effect. We should expect then that the measured head coefficient *divided* by the efficiency would compare more favorably with a potential theory. This is indeed the case, as Fig. 12.21 demonstrates. In fact, near the point of best efficiency the two functions are in good agreement. At much higher and lower flow values than the point of best efficiency, however, values of Ψ/η_p depart appreciably from the ideal calculations indicated.

Centrifugal *multistage* pumps are frequently used to supply boiler feedwater for steam power plants. An example of a performance curve for such a pump is given in Fig. 12.22a and a photograph of one of the impellers is shown in Fig. 12.22b. Here it may be seen that the flow through the impeller is primarily radial or "centrifugal." A somewhat less centrifugal type of machine is the example of a recirculating coolant pump shown in Fig. 12.23, where it may be seen that the meridional flow is not strictly radial. Sometimes turbomachines of this configuration are termed *mixed-flow* turbomachines, as distinct from centrifugal or axial ones.

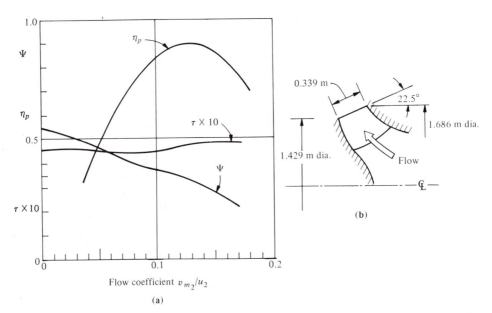

FIG. 12.23 Performance chart for a 3000 kW circulating water pump. At a rotative speed of 395 rpm, 3360 kW is required to pump 7.57 m³/s at a head of 40.2 m: (a) head coefficient based on the outer tip speed, the overall efficiency, and the torque coefficient versus flow coefficient; (b) meridional projection of the impeller blades with the overall dimensions. There are four vanes; the discharge vane angle at the larger tip diameter is 20° and that at the inner is 24°. The specific speed at the point of maximum efficiency is 1.28. In normal operation at 395 rpm this pump requires an absolute inlet head higher than the liquid vapor pressure of 7.93 m. (Based on data supplied by the Byron-Jackson Pump Company, used with permission.)

An example of a turbomachine for compressible fluids is the centrifugal air compressor rotor shown in Fig. 12.24a, with the corresponding performance curve given in Fig. 12.24b. The coefficient Ψ for a compressible fluid is defined as $(h_{2is} - h_1)/u_2^2$, where h_{2is} is the enthalpy at the discharge pressure p_2 for an isentropic compression. This definition of Ψ closely matches that used for incompressible fluids. Similarly, the efficiency is defined as the ratio of the total enthalpy rise for an isentropic

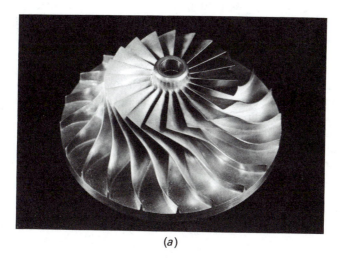

(a)

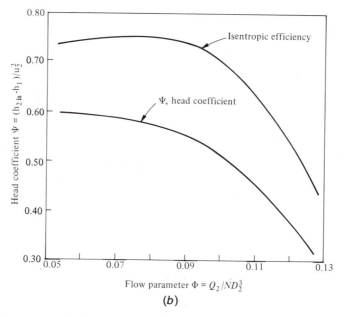

(b)

FIG. 12.24 (a) Photograph of a prototype centrifugal air compressor rotor; the discharge diameter is 0.0181 m and the exit blade angle is 50° from the tangential direction. The rotative speed varies from 48,000 to 76,000 rpm; at the higher speed the pressure ratio is 3.6 at a mass flow rate of 0.5 kg/s. (Courtesy D. Japikse, Concepts E.T.I., used with permission.) (b) Performance data for the rotor shown in part (a) at a rotative speed of 48,000 rpm; the coefficient for a compressible fluid is defined in terms of the rise in enthalpy for an isentropic process. (See the explanation in the text.)

process to the actual total enthalpy rise. The flow parameter ϕ is based on the discharge flow rate and the term ND^3 (where N is the rotative speed in revolutions per second) rather than the exit tip speed and meridional area. The behavior of compressible flow turbomachines is generally quite similar to incompressible ones, except, of course, Mach number effects may become important at sufficiently high rotative speeds or near the outer tip of a large-diameter rotor.

The last example of this section is the high-pressure oxidizer pump of the Space Shuttle main engine; a schematic diagram of a cross section through this combined turbine-pump assembly is shown in Fig. 12.25. It may be seen that the flow through the turbine is axial and that the flow through the pump impeller is divided into two equal portions, which approach the impeller *axially*. This arrangement is typical of many centrifugal pumps, and such pumps are called *double-suction* pumps. The impeller thus is symmetric back to back. Each inlet portion of the present pump also contains special axial blading in addition to the centrifugal blading. The purpose of the axial blading is to give a slight boost to the inlet pressure, which is necessary to prevent the liquid from *cavitating* in the centrifugal portion of the impeller (see Section 13.8). A photograph of this portion of the impeller is shown in Fig. 12.26 and a pressure performance curve is given in Fig. 12.27. (See page 445.)

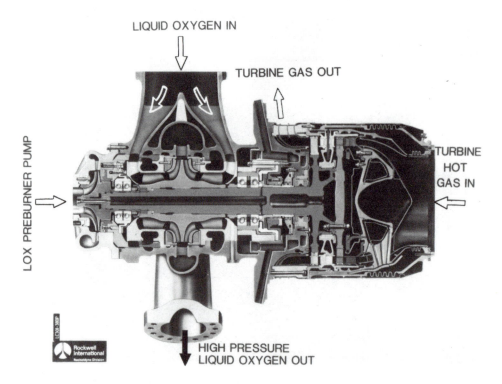

FIG. 12.25 Schematic diagram of a cross section through the high-pressure oxidizer turbopump assembly of the Space Shuttle main engine. The pump is driven by a two-stage axial-flow turbine (at the right of the diagram) at speeds up to 30,000 rpm. The fluid being pumped is liquid oxygen. The inflow to the pump impeller is symmetric from both sides; the discharge is collected in a central diffuser-volute. The nominal flow rate is 486 kg/s and the total pressure rise is nominally 27 MPa. The centrifugal impeller has a diameter of 0.173 m and the breadth of the discharge (both sides) is 0.029 m. (Courtesy of E. Jackson, Rockwell International, Rocketdyne Division, used by permission).

FIG. 12.26 Photograph of the impeller for the high-pressure oxidizer pump in the Space Shuttle main engine. The impeller has been cut in half in order to be used in a single-suction test apparatus. The axial inlet portion of the impeller is specially designed for operation with cavitation; this portion of the impeller is called the *inducer*. The tip diameter has been slightly trimmed for test purposes to a diameter of 0.169 m. There are eight main blades with a discharge blade angle of about 36°.

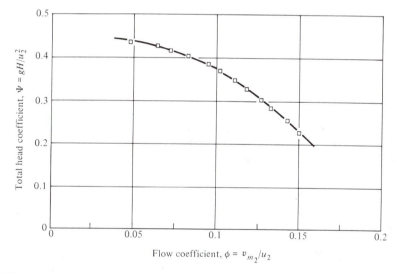

FIG. 12.27 Performance curve for the test impeller of 12.26. (Data provided by N. Arndt and R. Franz, California Institute of Technology, Pasadena, Calif., used with permission.)

The smaller pump on the left end of the shaft is a "preburner" liquid oxygen pump used to provide fluid for the turbine at startup, which further increases the pressure rise somewhat.

12.7 The Affinity Laws

In the preceding examples the performance of turbomachines was presented in terms of the dimensionless coefficients ϕ, Ψ, and τ. These relationships depend, of course, on the geometry of the machine. Even for a machine with fixed geometrical dimensions, however, these relationships may depend upon the absolute speed and size or, more properly, the Reynolds number. In general, as the Reynolds number increases, the efficiency increases slowly, and the head coefficient Ψ and torque coefficient may change slightly as well. The Reynolds number may be defined here as $u_2 r_2/v$, where u_2 is the rotor tip speed, r_2 the rotor radius, and v the kinematic viscosity of the fluid. However, if the Reynolds number is not too low, the relationships between Ψ, η_p, τ, and ϕ become essentially independent of Reynolds number. This occurs at a value of Reynolds number $(\text{Re} = u_2 r_2/v)$ of about 5×10^5.

The foregoing conclusion may also be explained by pointing out that for sufficiently high Reynolds numbers the flow angles become independent of Reynolds number, and losses, like the developed head, are proportional to the square of the velocities, as in fully rough turbulent pipe flow.

For the above conditions then it may be assumed that a given value of ϕ determines definite values of Ψ, η_p, and τ. It follows that

$$H = \frac{\Psi u_2^2}{g} \sim r_2^2 \Omega^2,$$

$$Q = \phi_2 u_2 A_2 \sim r_2^3 \Omega, \tag{12.28}$$

$$T = \tau \rho r_2 A_2 u_2^2 \sim \rho r_2^5 \Omega^2,$$

and

$$P = T\Omega \sim \rho r_2^5 \Omega^3.$$

The set of relations in Eqs. (12.28) are sometimes called the *affinity laws*. It may be well to repeat that the affinity laws in this form neglect any effects of Reynolds number and that they are restricted to incompressible flow.

We shall now study several typical problems encountered in hydraulic engineering practice that involve the use of Eqs. (12.28).

(a) Effect of Impeller Speed

In Fig. 12.28 the head–discharge $(H–Q)$ characteristic is shown for a typical pump at a given speed Ω_1. Suppose that it is now required to predict the $H–Q$ curve and corresponding efficiency for the same pump running at a speed Ω_2. Let us consider then a particular point, say O on the first $H–Q$ curve, and let us designate the head and discharge at this point by H_1 and Q_1. According to Eqs. (12.28) the corresponding point at the new speed for the same ϕ and Ψ would yield

$$H_2 = H_1 \left(\frac{\Omega_2}{\Omega_1}\right)^2$$

and

$$Q_2 = Q_1 \frac{\Omega_2}{\Omega_1}.$$

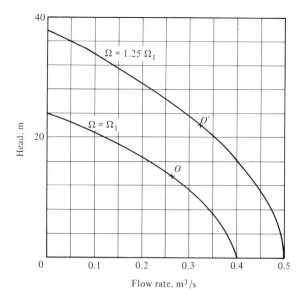

FIG. 12.28   Graph illustrating the computation of performance characteristics for a given pump at various speeds.

With ϕ and Ψ constant the efficiency at this new point O' is expected to be the same as that at O within the assumptions on which Eqs. (12.28) are based. Point by point the entire new $Q–H$ curve may be plotted in this way. Basically, of course, there is nothing more involved here than the acceptance of the relationship between the dimensionless quantities ϕ, Ψ, and η_p.

(b) Effect of Impeller Diameter

In a similar way we now examine the effect of the increase in size of a geometrically similar impeller. Again let a given impeller have the characteristics as shown in Fig. 12.29. A new, geometrically similar impeller is then constructed with all dimensions scaled by the ratio r_2/r_1, where r_1 and r_2 are the radii of the first and second

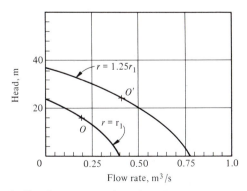

FIG. 12.29   Graph illustrating the computation of performance characteristics for a pump design at a given speed but for different sizes of the impeller.

impeller. If we select a point O of the H—Q curve of the first impeller, the corresponding point O' for the second impeller for the same ϕ and Ψ will give a head

$$H_2 = H_1 \left(\frac{r_2}{r_1}\right)^2$$

and a discharge

$$Q_2 = Q_1 \left(\frac{r_2}{r_1}\right)^3$$

by the rules of Eqs. (12.28). The efficiency at the new point again will be the same as that at the first and the entire H–Q curve for the larger machine may again be constructed point by point.

12.8 The Specific Speed

Let us now consider a somewhat different problem. Suppose that a certain impeller is given and that the operating characteristics are known so that a dimensionless ϕ–Ψ diagram can be constructed. It is now desired to find the operating point and impeller dimensions for a specified head H, discharge Q, and speed Ω. This type of problem occurs frequently in engineering because in addition to Q and H the speed also is often given. For example, many pumps are driven by electric motors, and it is often most desirable that the pump operate at the motor speed dependent on the electric current frequency. Another, and possibly more basic, limitation on the speed is imposed by the necessity of avoiding cavitation (partial vaporization) of the fluid being pumped. The effect of speed on cavitation will be discussed in some detail later, but it may suffice here to state that lower speeds tend to avoid the occurrence of cavitation. Finally, it may be mentioned that considerations regarding weight, size, and economy also often influence the selection of the speed.

We shall approach the present problem by first writing

$$\phi = \frac{Q}{Au} = \frac{Q}{\xi_2 \pi r_2^3 \Omega}, \tag{12.29}$$

where ξ_2 is a geometrical ratio defined as

$$\xi_2 \equiv \frac{A_2}{\pi r_2^2}.$$

A is the effective area normal to the meridional velocity v_m, and r_2, as usual, is the outer impeller radius. The total head coefficient is

$$\Psi = \frac{gH}{u_2^2} = \frac{gH}{r_2^2 \Omega^2}. \tag{12.30}$$

By eliminating the radius r_2 from Eqs. (12.29) and (12.30) and rearranging, we find that

$$\frac{\Omega Q^{1/2}}{(gH)^{3/4}} = \frac{(\pi \phi \xi_2)^{1/2}}{\Psi^{3/4}} \equiv N_s'. \tag{12.31}$$

The left-hand side of this equation is now completely determined, and this combination of terms has been called *specific speed*. We shall designate it by the symbol

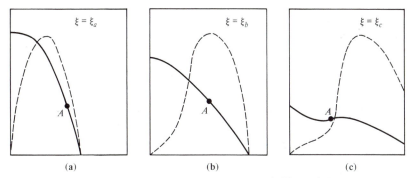

FIG. 12.30 Operating points as they would appear on a ϕ–Ψ graph (solid curves) of several pumps, for given values of Q, H, and Ω. The dashed curves indicate efficiencies. In general, the most suitable pump design is that for which the operating point occurs at a high value of efficiency.

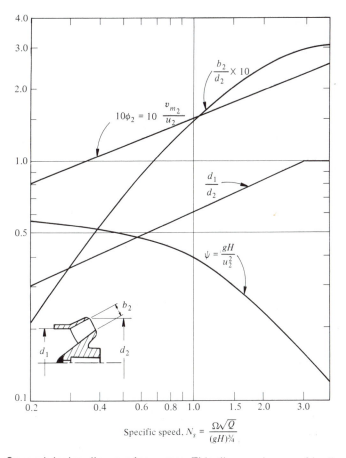

FIG. 12.31 General design diagram for pumps. This diagram is a combination of a large amount of empirical data. For a given specific speed this graph gives design information leading to a suitable pump. It is entirely possible that different designs are feasible which also would be suitable. (Diagram based partly on data from *Engineering Hydraulics*, H. Rouse, ed., John Wiley & Sons, Inc., New York, 1950.)

N_s'. We now have to match the specific speed as given by the operating conditions to an operating point on the ϕ–Ψ curve for the pump under consideration. Since ξ_2 is a geometrical ratio fixed for a given pump, the various points on the ϕ–Ψ curve represent the full range of possible specific speeds from zero to infinity. An operating condition at which the desired specific speed occurs can, therefore, always be found. However, as can be seen from the general shape of the ϕ–Ψ curve, a specified value of the specific speed will occur at one particular operating point only, say, at ϕ_a and Ψ_a. To this operating point there corresponds a definite value of efficiency. Unless the particular operating point coincides approximately with the point of highest efficiency for the pump design, the efficiency could well be quite low. For illustration assume that the pump design in question has a ϕ–Ψ curve and efficiencies as sketched in Fig. 12.30a, and that the operating point, as determined by the specified values of Ω, H, and Q, is the one indicated by point A. The efficiency of operation would then evidently be quite low. It would then be reasonable to examine the suitability of other pump designs, and the corresponding performance diagrams might appear as illustrated in Fig. 12.30b and c. Many different designs could, of course, be examined for the application in question, and the design selected would be that which gives the highest efficiency for the specified operating point or, in other words, for the given specific speed. Let us now call the specific speed, which corresponds to the maximum efficiency point of each pump, the *optimum* specific speed of the pump and let us designate this value by the special symbol N_s (without prime). In terms of this optimum specific speed N_s, the pump to be selected is the one whose optimum specific speed, N_s, matches the specific speed, N_s', required by the operating specifications.

The optimum specific speed, N_s, might be taken as an index characterizing the pump design and it is actually so used in the pump industry. One could imagine, of course, that several different types of pump could have the same optimum specific speed, N_s, and that all of those might have about the same high efficiency at that point. This possibility certainly exists. In the course of commercial pump development, however, a certain pump design has been evolved for each specific speed, and even designs by various manufacturers for the same specific speed are quite similar. The essential geometrical information for this series of pumps has been summarized in Fig. 12.31, where the various geometrical ratios are given as a function of the optimum specific speed, N_s, for each type of pump. The general appearance of impellers designed in this manner is as indicated in Fig. 12.32, and the optimum specific speed, N_s, may be regarded as a characterizing type parameter for the pump.

A few comments may be made to give some qualitative explanation for the changes in shape of the impellers as a function of their optimum specific speed. Let us start with an impeller designed for a low specific speed which at a given speed corresponds to a relatively high head and a low discharge rate. To obtain the high head, ample use will have to be made of the centrifugal effect, and therefore the ratio of outer to inner radius is expected to be large. The flow passages can well be small without causing excessive flow velocities because the discharge rate is small. As a consequence, the impeller will be similar to the one corresponding to $N_s \sim 0.25$ in Fig. 12.32. Let us assume now that the change in specific speed is brought about by increasing the rotative speed Ω, leaving Q and H constant. As the speed increases, the required change in angular momentum for a given head decreases. The necessary increase in angular momentum—which depends on the change in radius as well as on the change in tangential velocities—can then be brought about with a smaller discharge radius. The impeller diameter will therefore decrease, and the impeller

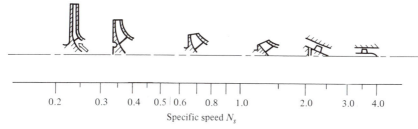

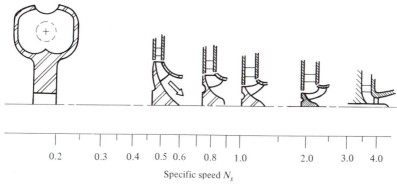

FIG. 12.32 Typical proportions for commercial pump designs at various values of the specific speed. The specific speed is given in dimensionless terms and is based on the rotative speed in rad/sec. (This chart is based on data from *Engineering Hydraulics*, H. Rouse, ed., John Wiley & Sons, Inc., New York, 1950.)

FIG. 12.33 Typical proportions for turbines at various values of the specific speed. The specific speed is given in dimensionless terms and is based on the rotative speed in rad/sec. Note the general similarity between pumps and turbines at given values of N_s. In both this figure and the one preceding, the size of the machine is adjusted to give the same head and flow rate when operated at a rotative speed proportional to the specific speed. (This chart is based on data from *Engineering Hydraulics*, H. Rouse, ed., John Wiley & Sons, Inc., New York, 1950.)

passages will become larger relative to the radius. The impeller then might appear as the second one in Fig. 12.32. As the speed is allowed to increase further, the ratio of passage width to diameter will increase further. The three-dimensional aspect of the flow will then have to be considered, and it has to be realized that it is no longer possible to effect a complete change of the flow from an axial to a radial direction within the impeller vanes. Since a purely radial outflow is not a fundamental requirement, the impeller discharge plane is allowed to be inclined, and the outflow will have an axial as well as radial and tangential velocity component. The third and fourth sketches in Fig. 12.32 illustrate such impellers, which are often called mixed flow impellers. As the speed increases still further, the radius change will become still less, the axial outflow component will increase further, and finally the impeller will evolve into a purely axial machine.

A word may be added about the efficiency of the series of pumps of different specific speeds. Each pump design, it will be recalled, represents the one that yields about the best efficiency obtainable at the given specific speed. For specific speeds between about 0.4 and 1.2 the values are of the order of 90 percent, provided that the Reynolds number ($u_2 r_2/\nu$) is at least 5×10^5. At lower specific speeds the passages within the impeller and diffusing passages of the volute tend to become narrow

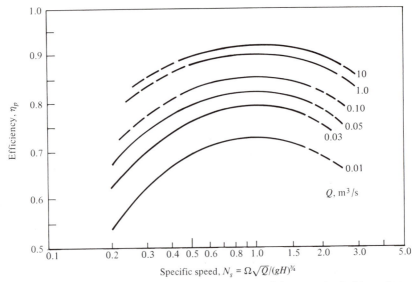

FIG. 12.34   Maximum efficiency of well-designed industrial pumps tested in water. (Adapted from Energy Research and Consultants Corp., H. H. Anderson, *Centrifugal Pumps*, Trade and Technical Press, Sutton, England, and others). The efficiency at the larger specific speeds is greatly dependent on specific test conditions, surface finish on the blades, running clearances, and the Reynolds number.

and long. The loss of total head due to friction then becomes quite pronounced within the impeller and diffuser, and the maximum efficiency of such pumps is lower than at the higher specific speeds. The maximum efficiency also tends to drop at high specific speeds. For such a machine the head produced becomes relatively smaller compared to, say, the axial velocity head. Friction losses of the flow past the surfaces of the surrounding casing and the vanes of the rotor as well as of the stator may then become an appreciable fraction of the total head rise. These trends are illustrated on the summary chart in Fig. 12.34, which shows the efficiency of some water pumps in current industrial use. The size and speed are implicit in the combination of specific speed and flow rate.

A comment should be repeated in regard to the use of the term "specific speed." The combination of terms

$$\frac{\Omega Q^{1/2}}{(gH)^{3/4}} = N'_s$$

was generally called *specific speed*, and the value corresponding to the maximum efficiency point of a pump was called the *optimum specific speed N_s*. Many authors do not make this distinction and, for example, automatically mean the *optimum specific speed N_s* when referring to *the* specific speed of a certain pump. Care has to be taken to ascertain the precise meaning wherever the term "specific speed" appears in the literature.

There is, furthermore, some confusion regarding the use of units. So far in computing the specific speed we have used consistent units for Q, H, and g and Ω in rad/sec. Many American pump designers omit g and give the discharge rate in gallons per minute and the pump speed in revolutions per minute. The values of specific speed as used in this book will have to be multiplied by 2732.8 to obtain the values of specific speed corresponding to this commonly used inconsistent set of units.

12.9 The Specific Radius

In the preceding section the selection of a pump for a given head, discharge, and speed was discussed. In certain applications the available space is limited, and then the radius r_2, instead of the speed Ω, should be regarded as given in addition to H and Q. Proceeding in a manner similar to that followed before, we now eliminate the speed from the defining equations for Ψ and ϕ and obtain

$$r_s' \equiv \frac{r_2(gH)^{1/4}}{\sqrt{Q}} = \frac{\Psi^{1/4}}{\sqrt{\pi\phi\xi_2}}, \tag{12.32}$$

where r_s' is called the *specific radius*. As before, there is a point on the performance curve of each pump which satisfies Eq. (12.32). Again, however, one would examine a set of pumps so that the operating point would coincide with a point of good efficiency. Furthermore, an *optimum* specific radius r_s could be defined as the specific radius at the point of maximum efficiency. A design chart could be prepared similar to that of Fig. 12.31. However, this chart itself may also be used directly by making use of the relationship

$$r_s N_s = \Psi^{-1/2}.$$

It is also interesting to note from Eqs. (12.31) and (12.32) that the pair of dimensionless quantities N_s' and r_s' can be used as an alternative set of parameters in place of Ψ and ϕ in characterizing the operation of a pump.

12.10 Hydraulic Turbines

All of the previous discussion has been centered around pumps, but a completely analogous development can be pursued for tubines. A specific speed could be defined in exactly the same way, for example. Since, however, the power output is generally of more immediate significance than the discharge rate, the turbine specific speed is defined in terms of shaft power output as

$$N_s' = \frac{\Omega\sqrt{P/\rho}}{(gH)^{5/4}}, \tag{12.33}$$

where P is the turbine power, H is the head imposed on the turbine, and ρ is the density of the medium. The quantity N_s would again designate the optimum specific speed of a given turbine.

It is of interest to note that in many instances it is possible to operate the same turbomachine either as a pump or a turbine by reversing the flow directions, the sense of rotation, and the pressure drop. If the magnitudes of the flow rate and speed of operation are the same for both types of operation, the specific speed computed by Eq. (12.31) for the pump operation will differ from that computed from the turbine operation, because the effects of friction decrease the head produced by the pump, whereas it increases that which has to be supplied to the turbine. When the same liquid is used, the specific speed for the turbine operation can be estimated from that of the pump operation by multiplying the latter by the product of $(\eta_p^{3/4}\eta_t^{5/4})$, where η_p and η_t are the pump and turbine efficiencies, respectively.

Hydraulic engineers accustomed to working with water only, generally neglect the density term in Eq. (12.33) as well as the gravitational constant.

The proportions of typical water turbine designs for various specific speeds are given in Fig. 12.33. The appearance of these machines may be explained qualitatively, as was done for the pumps. Thus the low-specific-speed machines are primarily radial, and the high-specific-speed machines are essentially axial, just as for the pump series. It may be noted that the shape of pumps and turbines are generally similar at the same specific speed. Design parameters representative of current hydraulic practice for liquid turbines may be found in Reference 1, and for other types of turbines References 4 and 7 may be consulted.

PROBLEMS

12.19 A pump discharges into a hydraulic circuit that has a head loss which varies as the square of the flow rate. Show that the flow coefficient ϕ is independent of pump speed.

12.20 Let point O of Fig. 12.29 represent the total head loss of a pipeline whose resistance varies as the square of the flow rate. For the larger impeller of this figure, what would be the flow rate for the same pipeline? Is this flow similar at this new operating point to the original one at point O?

12.21 Given the pump of Fig. 12.21, determine the diameter and speed of a similar pump to produce 0.3 m³/s at 80 m head at the best efficiency point.

12.22 The axial-flow pump with the performance curve shown is to produce a flow rate of 0.15 m³/s at 3 m head, but due to limitations of space, the impeller diameter cannot exceed 0.2 m. Determine the operating speed for this machine.

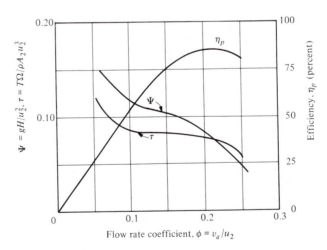

Problem 12.22

12.23 A water pump for a washing machine is to deliver 3 lbm/sec at a head of 20 ft. The pump will be belt driven and any required speed can easily be obtained by selecting a proper pulley size. However, to fit the arrangement, the pump size is limited and the impeller is to have a diameter of 4.0 in. For a reasonably efficient pump, find:
(a) The inlet (or eye) diameter d_1.

(b) The flow passage width b_2.

(c) The rotational speed in rpm.

12.24 A centrifugal pump of 9 in. impeller diameter has the performance curve shown. The diameter of the impeller is machined off to 8 in. without changing the breadth of the impeller or the rotative speed. It may be assumed that the discharge blade angle remains the same. Make a reasonable assumption about the flow leaving the impeller and calculate and plot the new performance curve.

   Note: The two pumps are no longer geometrically similar.

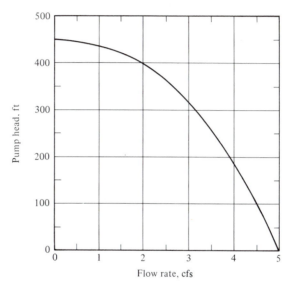

Problem 12.24

12.25 A drainage pump is to be driven by a 60-Hz induction motor. The flow rate is to be 0.1 m³/s at 6 m head. The motor can be supplied with a 4-, 6-, 8- or 10-pole winding. Determine the impeller diameter for each of these speeds in accordance with Fig. 12.31.

12.26 A circulating pump for a power plant is required to produce a flow of water of 2 m³/s at a total head increase of 10 m. The speed of rotation is determined by the 60-Hz power grid so that the pump can operate at 3600, 1800, 1200, and so on, rpm as determined by the number of poles of the electric motor drive.

   (a) Select a motor speed and thus determine the pump specific speed and head coefficient from the data in Fig. 12.31.

   (b) Calculate the impeller diameter in meters.

12.27 The liquid oxygen pump for the Titan I rocket engine has to deliver 5.8 ft³/sec at a total head rise of 1700 ft. The pump speed is set at 8255 rpm because of cavitation limitations.

   (a) Select the geometrical ratios b_2/d_2, d_1/d_2, as well as ϕ and Ψ at the operating point.

   (b) Determine the outside impeller diameter.

   (e) Make a sketch of the impeller *to scale*.

(d) Estimate the power (in hp) required to drive the pump. The density of liquid oxygen may be taken to be equal to that of water for a rough estimate.

12.28 A pump is to produce a pressure rise of 4.0 MPa at a flow rate of 0.6 m^3/s with a fluid of density 800 kg/m^3 with an impeller diameter of 0.75 m. Determine the specific speed for the pump and compute the corresponding rotative speed using the chart of Fig. 12.31.

12.29 The main features of the high-pressure oxidizer pump for the Space Shuttle main engine are given in Figs. 12.25 and 12.26. Assume further that the power input is 17.9 MW and the pump efficiency is 68 percent.

(a) Calculate the head coefficient.

(b) Find the specific speed (per side) of the pump.

(c) Calculate the specific radius.

REFERENCES

1. R. L. Daugherty and A. C. Ingersoll, *Fluid Mechanics*, 5th ed., McGraw-Hill, New York, 1954, Chaps. 15, 16, 17, 18.

2. D. G. Shepherd, *Principles of Turbomachinery*, Macmillan, New York, 1956.

3. A. J. Stepanoff, *Centrifugal and Axial Flow Pumps*, 2nd ed., Wiley, New York, 1957.

4. J. F. Lee, *Theory and Design of Steam and Gas Turbines*, McGraw-Hill, New York, 1954, Chap. 8.

5. G. T. Csnady, *Theory of Turbomachines*, McGraw-Hill, New York, 1964.

6. A. Betz, *Introduction to the Theory of Flow Machines*, Pergamon, London, 1966.

7. C. C. Warnick, *Hydropower Engineering*, Prentice-Hall, Elmsford, N.Y., 1984.

CHAPTER 13

Some Design Aspects of Turbomachines

13.1 Introduction

In the preceding chapter some overall characteristics of turbomachines were deduced from the angular momentum equation by assuming the flow to be guided perfectly by the vanes. Also, it was assumed for axial machines that the meridional streamlines were straight and did not change radius. By making these assumptions, it was possible to survey quickly various types of pumps and turbines and to gain a general knowledge of their characteristics. Neither of the assumptions mentioned above is satisfied in an actual machine with a definite number of vanes. The meridional streamlines in an axial-flow compressor, for example, can change radius, and the flow angle leaving a rotor or stator is usually not the same as the angle of the vane trailing edge. In fact, the flow in a turbomachine is essentially three-dimensional. To facilitate understanding and analysis of this complicated flow, we consider two simpler, two-dimensional flow problems which, taken together, pre-serve the essential features of the original problem. These two flows consist of the axisymmetric flow on stream surfaces of revolution, and the flow between the vanes of the rotor or stator on this surface of revolution. A sketch of these ideas is shown in Fig. 13.1. Figure 13.1b shows an oblique view of a particular streamline surface of revolution (say l–l of Fig. 13.1a) and the trace of the vanes on this surface. The array of vanes on the streamline surface of revolution as shown in Fig. 13.1b is sometimes called a *cascade*, or *lattice*. The problem then becomes one of determining the flow through the cascade and the location of the surface of revolution upon which the cascade flow takes place. These surfaces of revolution may be essentially axial, as for an axial-flow pump, or radial as for a centrifugal pump. For a "mixed-flow" machine they will be of a more general axisymmetric shape, as indicated in Fig. 13.1b. In the remainder of this chapter we shall discuss the foregoing two parts of the flow through a turbomachine, but shall limit ourselves to machines in which the flow is predominantly radial or axial and we shall not treat the machines with mixed flow in any detail.

Let us consider first the flow through the cylindrical cascade formed by the inter-section of the vanes on a cylindrical surface of essentially constant radius. The flow on such a cylinder is periodic, and it can be developed or unrolled onto a plane as was done in Section 12.3 when treating the impulse turbine. On this plane the flow pattern repeats itself indefinitely with a period equal to the circumferential spacing of the vanes on the original cylinder. Such a cascade is called a *linear cascade*. The flow through an element of an axial-flow machine of large hub ratio ($r_i/r_o \sim 1$) may be well approximated by such a cascade. Subsequently, we shall discuss *radial cas-cades*, which consist of an array of identical vanes in a radial plane, symmetrically

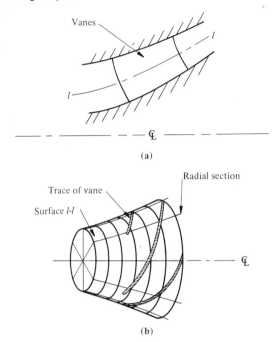

FIG. 13.1 Cross section (a), and oblique (b) view of general turbomachine showing stream surface of revolution.

located about a center. Such a cascade approximates the flow through the radial portions of a radial-flow turbomachine.

13.2 Linear Cascades

In Chapter 12 reference has been made to a somewhat imaginary cascade consisting of an infinite straight row of infinitesimally closely spaced vanes. Such a cascade may be regarded as the simplest kind of linear cascade, in which the flow is completely guided and the outlet flow angle must coincide with the direction of the tangent of the blades at the leaving edge. When vanes are relatively distant from each other, the leaving flow angle will deviate from the trailing-edge angle and furthermore may depend on the direction of the flow entering the cascade. An approximation to the flow through such a cascade may be obtained by neglecting friction and computing the potential flow past the row of vanes. This computation is always possible in principle, although usually difficult to carry out. When the vanes are spaced very far apart, the mutual influence of one blade upon another becomes negligible and the flow becomes equivalent to that over a single vane or airfoil. The many experimental as well as analytical results for single airfoils are then applicable, and it is only necessary to relate the parameters used in describing isolated airfoil behavior to the characteristic dimensions and angles that are of significance for the flow through a cascade.

We now consider some important characteristics of a linear cascade as illustrated in Fig. 13.2. The spacing between blades is given by the distance s, and the blade chord is designated by c. The ratio c/s is called the *solidity* and given the symbol σ. The angle between the chord and the direction normal to the cascade is called the stagger angle γ. The flow is assumed to be incompressible and purely two-

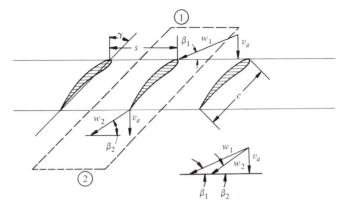

FIG. 13.2   Sketch of linear cascade.

dimensional, so that the velocity component v_a remains constant. The inlet and leaving velocities relative to the cascade are w_1 and w_2, respectively. For a stationary cascade the relative velocities will coincide with the absolute ones and in our previous notations we would designate these velocities by V and flow angles by α. The following discussion will, however, be applicable to either case. The velocities w_1, w_2, and so on, are measured sufficiently far upstream and downstream of the cascade, so that any local velocity perturbations due to the vanes have disappeared.

The force on any individual blade may now be determined with the aid of the momentum theorem. With the control surface as shown in Fig. 13.2 the components of the force exerted by the blade on the fluid are found to be

$$x\text{-direction: } F_x = (-w_2 \cos \beta_2 + w_1 \cos \beta_1)\rho s v_a, \qquad (13.1)$$

$$y \text{ direction: } F_y = -(p_2 - p_1)s, \qquad (13.2)$$

where all forces correspond to a unit depth of flow. The force exerted by the fluid on the blade is, of course, equal in magnitude and opposite in direction to that above. Let us further assume the flow to be *frictionless* and apply Bernoulli's equation between station (1) upstream and (2) downstream. Thus

$$p_2 - p_1 = \frac{\rho}{2} (w_1^2 - w_2^2) = \frac{\rho}{2} (w_1^2 \cos^2 \beta_1 - w_2^2 \cos^2 \beta_2)$$

and

$$F_y = -s \frac{\rho}{2} (w_1^2 \cos^2 \beta_1 - w_2^2 \cos^2 \beta_2).$$

So far we have dealt with the force of the blade on the fluid, but more often the force of the fluid on the blade is of more direct use. The components of this force are equal in magnitude and opposite in direction to F_x and F_y and we shall designate them by R_x and R_y, respectively. From these two force components the cotagent of the angle β_f between the force on the blade and the y-axis is

$$\cot \beta_f = \frac{-R_y}{R_x} = \frac{w_1 \cos \beta_1 + w_2 \cos \beta_2}{2v_a}. \qquad (13.3)$$

An inspection of the geometry of the velocity vectors w_1 and w_2 (see Fig. 13.3) now shows that the angle β_∞ formed by the vector mean w_∞ of w_1 and w_2 with the x-axis is equal to β_f. It follows that the force R is acting in a direction normal to

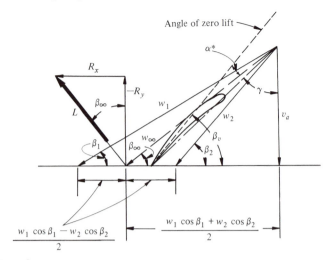

FIG. 13.3 Sketch illustrating vane and flow angles through a linear cascade.

the mean velocity w_∞. The relationship between the force and the velocity w_∞ for a blade in a frictionless cascade, therefore, is similar to that existing between the lift force and the velocity for an isolated airfoil in frictionless flow. In analogy to a single airfoil the force R acting on the blade may be called a lift force, which is usually designated by the symbol L. A lift coefficient C_L may be defined in terms of the velocity w_∞ as

$$C_L = \frac{2L}{\rho w_\infty^2 c}, \tag{13.4}$$

where c is the length of the chord. In general, this coefficient is expected to be a function of the shape of the individual blade, as well as of the stagger angle γ, the solidity σ, the direction of the approaching flow, and the Reynolds number. As in the case of a single airfoil, the dependence of lift coefficient on Reynolds number becomes slight when the Reynolds number based on chord and w_∞ exceeds about 500,000. Furthermore, as the solidity σ approaches zero, the effect of γ and σ become unimportant and the velocity vectors w_1 and w_2 tend to coincide with w_∞. The lift coefficient then approaches that of the single airfoil with the same shape.

In order to derive the relationships between the lift coefficient and the cascade geometry, let us write

$$L = \frac{-R_y}{\cos \beta_\infty}, \tag{13.5}$$

which when R_y is eliminated becomes

$$C_L = \frac{w_1^2 \cos^2 \beta_1 - w_2^2 \cos^2 \beta_2}{\sigma w_\infty^2 \cos \beta_\infty}. \tag{13.6}$$

Since

$$w_\infty \cos \beta_\infty = \frac{1}{2} (w_1 \cos \beta_1 + w_2 \cos \beta_2),$$

the equation above may also be written

$$C_L = \frac{2}{\sigma} \sin \beta_\infty (\cot \beta_1 - \cot \beta_2), \tag{13.7}$$

which represents a direct relationship between the flow angles and the lift coefficient. Equation (13.7) is applicable to both a moving or a stationary *frictionless* cascade. If the cascade in question is a moving one, we may further write (in the absence of friction)

$$\Psi = \phi(\cot \beta_1 - \cot \beta_2),$$

and consequently,

$$C_L = \frac{2}{\sigma} \frac{\Psi}{\phi} \sin \beta_\infty. \tag{13.8}$$

The relationships above are valid for any solidity σ, but they are mainly useful for the special case for which the solidity is sufficiently low, so that the lift coefficient is essentially that for a single airfoil. In any case the lift coefficient may be written as

$$C_L = m \sin \alpha^*,$$

where α^* is the angle of attack and m is a constant depending on the airfoil shape and cascade geometry. The angle α^* is equal to the difference between β_v, the direction of w_∞ which would lead to zero lift, and β_∞ the actual angle of approach of the velocity w_∞. (The symbol α^* is here being used for the angle of attack to avoid confusion with the absolute flow angles α_1 and α_2.)

Equating the expressions for C_L, we obtain

$$\Psi = \frac{2m\sigma \sin \beta_v}{4 + m\sigma \sin \beta_v} [1 - \phi(\cot \beta_v + \cot \alpha_1)]. \tag{13.9}$$

The first factor is sometimes abbreviated by setting

$$c_t \equiv \frac{2m\sigma \sin \beta_v}{4 + m\sigma \sin \beta_v}, \tag{13.10}$$

so that

$$\Psi = c_t[1 - \phi(\cot \beta_v + \cot \alpha_1)]. \tag{13.11}$$

This equation then gives the relationship between the total pressure coefficient and the flow coefficient of a frictionless cascade of arbitrary solidity. For a general cascade m and β_v must be found from experiment or estimated with the aid of a suitable theory. However, when the solidity is less than about 0.35, m and β_v are well approximated by the values for the isolated airfoil. For a flat plate, for example, $m = 2\pi$ and $\beta_v = (\pi/2 - \gamma)$, since the angle of zero lift coincides with the chord line for this simple shape. Equation (13.11) is quite similar in form to Eq. (12.18), which was written in terms of β_2, except that β_2 is now replaced by β_v and the coefficient c_t appears as a multiplier. This coefficient can also be written in the form

$$c_t = \frac{\cot \beta_1 - \cot \beta_2}{\cot \beta_1 - \cot \beta_v} \tag{13.12}$$

after some tedious manipulation. In this form the physical meaning of c_t becomes apparent: It represents the ratio of the change in the tangential velocity accomplished by the cascade to the change that would correspond to turning the inlet velocity vector into the direction β_v.

13.3 Linear Cascade with Drag

The developments of the foregoing section may be extended to include the effect of friction, in which case the drag force on the vane shapes will also have to be taken into account. In addition to the lift force, which is normal to the direction of the vector w_∞, we now have a drag force, which acts in a direction parallel to w_∞. Equations (13.1) and (13.2), which were derived from the momentum theorem, now become

$$-F_x = L \sin \beta_\infty + D \cos \beta_\infty = \rho s v_a(w_1 \cos \beta_1 - w_2 \cos \beta_2) \tag{13.13}$$

$$-F_y = L \cos \beta_\infty - D \sin \beta_\infty = (p_2 - p_1)s, \tag{13.14}$$

from which

$$p_2 - p_1 = \rho v_a \cot \beta_\infty(w_1 \cos \beta_1 - w_2 \cos \beta_2) \frac{1 - \varepsilon \tan \beta_\infty}{1 + \varepsilon \cot \beta_\infty}, \tag{13.15}$$

where

$$\varepsilon = \frac{D}{L}$$

is the ratio of drag to lift. The corresponding equations for a stator now may be obtained by replacing β by α. It should be noted that the velocity triangles and therefore the direction of the lift force have here been selected to correspond to a pressure rise through the cascade. This is the normal configuration for a compressor. For a turbine the lift force would be in the opposite direction.

Continuing with the example of a compressor cascade, we see that the pressure rise across the blade row, which would be achieved in the absence of friction, is obtained from Eq. (13.15) by letting $\varepsilon = 0$. We shall denote this rise by $(p'_2 - p_1)$. The ratio of the actual to ideal pressure rise may be called the "static pressure" efficiency of the blade row η_{st}, and therefore

$$\eta_{st} = \frac{p_2 - p_1}{p'_2 - p_1} = \frac{1 - \varepsilon \tan \beta_\infty}{1 + \varepsilon \cot \beta_\infty}. \tag{13.16}$$

The value of ε depends on the lift coefficient and Reynolds number. It may be as low as 0.01, but a typical value for fan or compressor cascade would be about 0.025. In actual impellers there will be additional losses due to boundary layers on the hub and casing, presence of a tip clearance, and so on, so that effective drag/lift ratios interpreted from overall efficiency measurements on a machine are somewhat higher than for isolated airfoils.

Similar calculations for η_{st} may be made for any number of rotor and stator rows and the overall efficiency of a machine may be computed in this manner. We content ourselves here with only one example, that of a repeating stage. By this is meant that the flow angle leaving the stator downstream of a rotor is equal to the absolute flow angle approaching the rotor. This is a common circumstance in multistage machines. The axial velocity v_a will also be assumed constant. Thus in the notation of Fig. 12.17, $\alpha_3 = \alpha_1$ and $V_3 = V_1$. Therefore, $p_3 - p_1$ is equal to the total pressure rise across the stage, since the velocities V_3 and V_1 are equal and cancel in the expression for the total pressure rise $p_{t_3} - p_{t_1}$. We obtain the static efficiency of the stator merely by replacing β_∞ by α_∞. Furthermore, letting ε_s be the drag/lift ratio of the stator, we obtain

$$(\eta_{st})_{stator} = \frac{p_3 - p_2}{p'_3 - p_2} = \frac{1 - \varepsilon_s \tan \alpha_\infty}{1 + \varepsilon_s \cot \alpha_\infty},$$

and as before for the rotor,

$$(\eta_{st})_{rotor} = \frac{p_2 - p_1}{p'_2 - p_1} = \frac{1 - \varepsilon_r \tan \beta_\infty}{1 + \varepsilon_r \cot \beta_\infty},$$

where ε_r now designates the drag/lift ratio of the rotor cascade. The overall stage efficiency then becomes

$$(\eta_{st})_{stage} = \frac{p_3 - p_1}{p'_3 - p_1} = \frac{(\eta_{st})_{stator}(p'_3 - p_2) + (\eta_{st})_{rotor}(p'_2 - p_1)}{p'_3 - p_1}.$$

If the drag/lift ratio for each of the rows is the same and equal to ε, this expression reduces to

$$\eta_{st} = \phi \frac{\cot \alpha_\infty - \varepsilon}{1 + \varepsilon \cot \alpha_\infty} + \frac{\cot \beta_\infty - \varepsilon}{1 + \varepsilon \cot \beta_\infty}. \tag{13.17}$$

For a given value of the flow coefficient ϕ and for small values of ε it can be shown that the maximum stage efficiency occurs when $\alpha_\infty = \beta_\infty$. That is, the velocity triangle must be symmetric and the reaction ratio equal to one-half (see Problems 12.9 and 12.10). Many axial-flow compressors are designed to take advantage of this fact by attempting to have the reaction ratio as nearly equal to one-half as possible for all radial stations. Finally, it can be shown that for a reaction ratio of one-half the flow coefficient should also be near one-half to achieve the best values of stage efficiency.

The efficiencies as defined above apply to compressor cascades. Equivalent definitions and analogous calculations may be made for turbine cascades. In addition, more refined calculations of blade row performance of this type can be carried out, but they are beyond the scope of this book. Excellent references for this type of work are available, as, for example, References 1 and 2.

13.4 Some Results for Linear Cascades

In the preceding sections some general relationships were established. These relationships still contain factors such as m, β_v, and ε, which depend on the cascade geometry and also on the effects of viscosity. In recent years much effort has been devoted toward understanding the flow through cascades. The theoretical analyses that have been carried out use as a first approximation the potential flow through the appropriate cascade. Even this calculation, however, is quite laborious, and the results of such theories are usually obtained by numerical procedures. In addition, numerous empirical rules based partly on experience and partly on theory have been advanced to obtain more rapid estimates of cascade behavior.

One of the few cases for which the potential solution can be carried out fairly readily is that for the flow through a cascade of flat plates. These results are summarized in Fig. 13.4, in which the coefficient c_t, Eq. (13.10), is given as a function of the solidity σ and the stagger angle γ ($\gamma = 90° - \beta_v$ for a flat plate). From this diagram the value of m may be determined for any β_v or σ from Eq. (13.10). It can be verified by calculation that $m \approx 2\pi$ for $\sigma < 0.35$, which means that the single airfoil assumption is applicable for $\sigma < 0.35$, as was stated earlier. The diagram indicates furthermore that for $\sigma > 2.0$ the flow leaves the cascade at approximately the angle of the flat plates themselves. This result means that the space between the plates now acts as a confining channel that compels the flow to move in the direction of the channel.

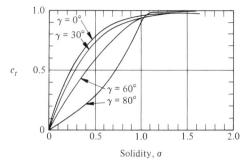

FIG. 13.4   Graph of coefficient c_t versus solidity for various stagger angles of a cascade of flat plates. [Information computed using the theory of E. König: "Potentialströmung durch Gitter," *Z. angew. Math. u. Mech.*, vol. **2**, 1922, p. 422. A more detailed graph can be found on p. 141 of Reference 4.]

For cascades of other types of vane shapes, the results are qualitatively similar in that single airfoil theory is applicable for $\sigma < 0.35$, and that for $\sigma > 2$ the leaving flow assumes the direction of the channel formed by adjacent blades. None of the potential flow calculations can, of course, predict any of the real fluid effects. On the basis of the potential flow field, however, estimates may be made of the boundary layer growth and its influence on lift and drag.

Aside from the analytical studies a considerable amount of experimental work has been performed, and from this some useful empirical rules have been devised. One of these, known as *Constant's rule* (see Reference 2), may be expressed in the form

$$\delta^* = 0.26 \frac{\theta}{\sqrt{\sigma}}. \tag{13.18}$$

In this equation δ^*, called the *deviation angle*, is the angle between the leaving flow and the tangent to the meanline of the blade shape, and θ is the camber angle of the blade. The angles are indicated in Fig. 13.5. Constant's rule [Eq. (13.18)] is limited to blade shapes whose mean profile line is a circular arc, and for which the camber angle and stagger angle are less than 60° and 65°, respectively.

Normally, the angle i between the incoming velocity and the tangent of the mean line will be between plus or minus 5°. Within the range the deviation angle is nearly independent of the entering angle if the solidity is greater than unity. This experimental finding considerably simplifies the design of blade systems to provide a definite amount of turning. The amount of turning that can be done by a given vane geometry is limited, however, by boundary layer growth and separation. If separation does not occur, the boundary layer growth and therefore energy loss of the blade system may be estimated with boundary layer theory. Usually, such calculations are too lengthy for design, and semiempirical criteria are used to determine the suitability of a particular design. One of these involves the so-called "diffusion" factor established from many tests carried out by NASA (Reference 3). This factor is defined as

$$k_d = \frac{w_1 - w_2}{w_1} + \frac{v_{\theta_2} - v_{\theta_1}}{2 w_1 \sigma} \tag{13.19}$$

and it is recommended that k_d should not exceed 0.6. The first of the terms in the diffusion factor relates to the recovery of the kinetic energy in the passage between the blades and the second term is nearly proportional to the lift coefficient. Taken

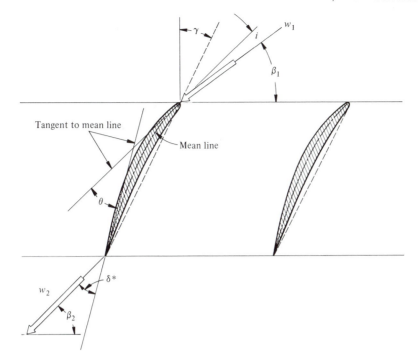

FIG. 13.5 Diagram showing definition of blade camber θ and deviation angle $\delta*$ for a linear cascade.

together the two terms represent the combined effect due to the reduction in the velocity of the stream and due to the pressure gradients imposed by the turning of the flow. These parameters are not the only ones to have been put forth, but are typical of those that have been found to be useful. Apart from these "rules of thumb" much experimental information on cascade flows is available and has been summarized in Reference 3.

13.5 Radial Cascades

A two-dimensional radial cascade consists of a set of blades or vanes placed symmetrically in a circular array as illustrated in Fig. 13.6. The cascade may either be stationary or rotating about its center. In the first instance the cascade would correspond to a set of guide vanes or nozzles and in the second it would represent a rotor or impeller. In principle, experiments and analyses can be carried out for radial cascades in a manner similar to those performed for linear cascades. Experimental results, however, are very scarce and calculations are more difficult to perform. We shall limit ourselves to outlining briefly some of the characteristics of a cascade of logarithmic spirals, which is amenable to some theoretical treatment.

The analysis of the rotor now differs from that of a stator and cannot be obtained by simple translation as was possible for the linear cascade. We shall first consider the stationary cascade as shown in Fig. 13.6. The equation of a logarithmic spiral is

$$\theta = A \log r,$$

from which it can be seen that the angle α_v between the normal to the radius and the tangent to the curve is a constant all along the curve. In this respect the spiral blade

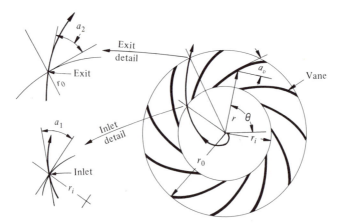

FIG. 13.6 Sketch showing radial cascade.

in the radial cascade is similar to the flat-plate blade of a linear cascade. In fact, it turns out that the potential flow through a cascade of flat plates can be transformed by methods of complex variables into that through a stationary radial cascade of logarithmic spiral blades. The solidity σ for the radial cascade becomes

$$\sigma = N \frac{\ln (r_o/r_i)}{2\pi \sin \alpha_v}, \tag{13.20}$$

where N is the number of blades, α_v the constant spiral angle, and r_o the outer radius and r_i the inner radius of the impeller. By letting r_o/r_i approach unity we can see that this definition of σ becomes identical to the one for a linear cascade. With this definition of σ the coefficient c_t, which is defined here as

$$c_t = \frac{\cot \alpha_1 - \cot \alpha_2}{\cot \alpha_1 - \cot \alpha_v},$$

is equal to the one for linear cascades of flat plates, where α_1 and α_2 now designate the flow directions in respect to a line normal to the radius. Consequently, the values of c_t for linear cascades, as given in Fig. 13.4, are applicable directly to the stationary radial cascade consisting of logarithmic spirals. This solution again, of course, neglects any effects of friction.

We now turn to the discussion of a rotating cascade. If friction is neglected, the absolute flow may still be considered irrotational, in spite of the fact that the cascade is rotating. The Kutta condition has to be fulfilled at the trailing edge, and the fluid velocity component normal to the blade surface must be the same as that of the solid surface itself in the same direction. This nonsteady potential flow problem has been solved and the total pressure rise through a rotor of this type may be written as

$$\Psi = \Psi_0 - c_t \phi(\cot \beta_v + \cot \alpha_1), \tag{13.21}$$

where β_v is the angle of the logarithmic spiral blade (corresponding to α_v for the stationary vane system) and α_1 is the entrance angle of the absolute flow entering the rotor. The function c_t is the one of Fig. 13.4, and Ψ_0 is another function obtained from the potential flow calculations. The function Ψ_0 is given in Fig. 13.7 in terms of the blade angle β_v and the number of vanes N. From Fig. 13.7 a few effects of the number of blades on performance can be estimated. For an infinite

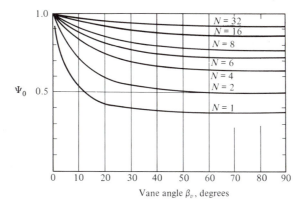

FIG. 13.7 Values of shutoff head coefficient Ψ_0 for various vane angles β_v and number of vanes for a radial cascade of logarithmic spiral blades. These results hold for values of the solidity [Eq. (13.20)] in excess of about 1.1. This coefficient decreases when the solidity becomes less than about 1.1 and for small values of the latter parameter becomes equal to c_t [Eq. (13.12)]. (Data computed from the theory of A. Busemann: "Das Förderhöhenverhältnis radialer Kreiselpumpen mit logarithmisch-spiraligen Schaufeln," *Z. angew. Math. u. Mech.*, vol. **8**, 1928, p. 372. Busemann's original graphs for Ψ_0 are given on p. 275 of Reference 4.)

number of vanes the coefficient Ψ_0 is equal to unity. If the vane angle β_v is small, for example, 6 or 8 vanes are sufficient to obtain values of Ψ_0 close to unity. For centrifugal gas compressors, because of the low density of the working substance, relatively high total head differentials are required to obtain a given pressure differential. This situation leads to relatively high rotor speeds and high values of Ψ at the operating point. A blade angle of 90° is therefore often selected for such compressors (1) because it lends itself to a design of high mechanical strength, and (2) because it reduces the effect of the negative term in the expression for Ψ. In order to maintain a high value of Ψ_0 for such a design, it is then necessary to select a larger number of vanes, as can be seen from Fig. 13.7. Air compressors with impeller blade angles of 90° generally have more blades than centrifugal pumps with small blade angles, and 10 to 20 blades are not uncommon.

In addition to the potential flow calculations discussed above, certain qualitative arguments may be given which provide a more direct physical picture of the flow in a rotating impeller. Ideally, the absolute flow, as was stated, is irrotational, but nonsteady. Relative to the impeller, which is rotating at a speed Ω, the flow will be steady. The vorticity, however, will now no longer be zero but will have a constant value of (-2Ω) throughout. This means that each fluid element will rotate about its center with an angular velocity Ω. The relative streamlines due to rotation alone for a simple cascade of straight radial blades at zero discharge are sketched in Fig. 13.8a. The streamlines form closed loops within the impeller since there is no net through flow and the relative circulation is clockwise (opposite to the direction of rotation). It is the existence of this relative rotation that is responsible for part of the departure from the infinite vane theory. Although the average radial velocity is zero, the average tangential component of the relative flow at the tip is not zero. In fact, it is opposite to the direction of rotation and a crude estimate of its magnitude may be obtained by the following argument: Let us imagine that the net effect of the relative circulation can be replaced by the circulation in a circle of diameter C, where C is somewhat arbitrarily selected as the distance between adjacent blades of the impeller at the outer rim. The flow within this circle is purely rotational, and we

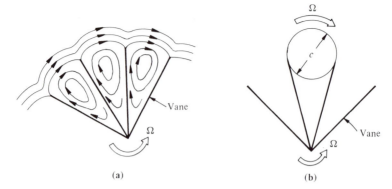

FIG. 13.8 Rotating cascade of straight radial vanes showing the rotational relative flow.

imagine it to be the same as the motion of a frictionless ball bearing mounted to the impeller at that point as shown in Fig. 13.8b. At the extremities of the diameter of this bearing, the relative velocity would be

$$\frac{\Omega C}{2} = \frac{\pi d_o}{N}\frac{\Omega}{2} = \frac{u_0 \pi}{N},\qquad(13.22)$$

where d_o is the outer diameter, u_0 the tip speed, and N the number of blades. If the number of blades is large, this should be approximately equal to the average tangential component of the relative velocity. Thus the absolute peripheral velocity is not u_0 but $u_0(1 - \pi/N)$, and the ideal frictionless head coefficient for an impeller of straight radial vanes is therefore less than unity. According to this simple approximation (due to Stodola, Reference 4), the head coefficient at zero flow for a vane angle of $90°$ would be $(1 - \pi/N)$. As the number of vanes becomes large, say 12 or more, the accuracy of Stodola's approximation becomes better, as may be seen by comparison with the theoretical results given in Fig. 13.7.

It should not be supposed that the flow pattern shown in Fig. 13.8 actually exists at zero flow rate in a real fluid. In the time required for a few revolutions, viscosity will completely obliterate the pattern shown and an irregular churning motion will take place. Near the point of best efficiency, however, that part of the total flow due to rotation alone will closely resemble that shown, and the approximation above is satisfactory.

13.6 Three-Dimensional Aspects in Axial-Flow Machines

In previous discussions on turbomachines we have considered two-dimensional flows in a plane. For example, in axial-flow machines the radial velocity components were neglected. In the present section we discuss a first step toward accounting for the three-dimensional nature of the flow. This is the problem referred to earlier, of determining the streamline surfaces of revolution. We limit ourselves to treating an essentially axial-flow machine.

The problem to be discussed is illustrated in Figs. 13.9 and 13.10. A flow with a given distribution of axial and tangential velocity approaches a blade system in an annular duct of constant cross section. The blade row may be either a rotor or a stator. The flow passes through the blade system and emerges downstream generally with an altered velocity and pressure distribution. We shall designate by (1) a station far upstream of the blade system and by (2) a station far downstream of this system. The two stations are assumed to be sufficiently far away from the blades so that any velocity variations in the tangential direction have disappeared. The flow at both stations is therefore assumed axisymmetric. It is furthermore assumed that at stations (1) and (2) the flow has attained its ultimate equilibrium state. Thus mathematically speaking, all derivatives with respect to the axial direction are zero. This assumption implies that there are no radial velocity components in existence at the two stations. Any radial velocity components, which are imparted by the blade system, are assumed to have disappeared at section (2). Although an equilibrium flow of this type can be expected to be established precisely only at an infinite distance from the blade system, one finds that for practical purposes it is established with sufficient accuracy at a distance of less than a few chords from the blades.

It should now be noted that whenever there is a change in the profile of the axial velocity component there must also take place a corresponding shift in th streamlines or stream surfaces. In Fig. 13.10 the axial velocity component near the hub decreases from station (1) to station (2), whereas it increases near the outer case. To satisfy the law of continuity, the flow rate between the hub and any particular streamline must remain constant. Fixing our attention on a streamline passing somewhere through the central portion of the blade, we see that the radial position of this streamline has to increase. We shall use the symbol r_1 to specify the radial position of any streamline at location (1) and r_2 to specify the position of the same

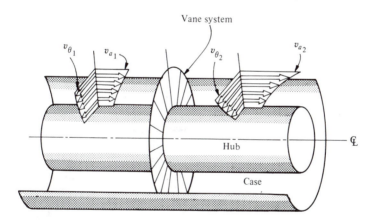

FIG. 13.9   General axisymmetric flow through axial-flow vane system.

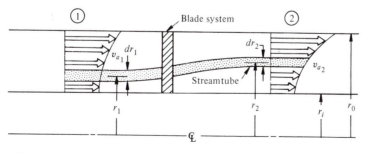

FIG. 13.10   Cross section of the flow through an axial compressor showing radial streamline shift through vane system.

streamline at location (2). Both r_1 and r_2 are variables. In general, r_1/r_2 will differ from unity, but the ratio always has to approach unity at $r = r_o$ and $r = r_i$, where r_o and r_i are the outer radius and the hub radius, respectively.

We now proceed to write the equations for an equilibrium flow that exists at either station (1) or (2). The Eulerian equation for the radial direction reduces to

$$\frac{v_\theta^2}{r} = \frac{1}{\rho} \frac{\partial p}{\partial r}. \tag{13.23}$$

(No subscript on r is needed here, as we are considering a single station only.) At the same station the Bernoulli constant, which is a measure of the energy may be written as

$$\frac{p}{\rho} + \frac{1}{2}(v_\theta^2 + v_a^2) = B(r). \tag{13.24}$$

In general, the Bernoulli constant at the station could differ from streamline to streamline and it could therefore be a function of the radius, as indicated in Eq. (13.24). In special cases which, however, are most important in turbomachine design, $B(r)$ is a constant. We consider this special case first. By differentiating Eq. (13.24) in respect to r and substituting into Eq. (13.23), we obtain in any case the relation

$$\frac{v_\theta^2}{r} + \frac{1}{2} \frac{\partial v_\theta^2}{\partial r} + \frac{1}{2} \frac{\partial v_a^2}{\partial r} + \frac{\partial B}{\partial r} = 0$$

or for B constant

$$\frac{v_\theta}{r} \frac{\partial(v_\theta r)}{\partial r} + \frac{1}{2} \frac{\partial v_a^2}{\partial r} = 0. \tag{13.25}$$

This differential equation has to be satisfied at stations such as (1) or (2) for a flow that has the same Bernoulli constant for all streamlines at this station.

To illustrate the use of Eq. (13.25), let us determine the tangential velocity component v_θ necessary to make the axial velocity component a constant. For $v_a = $ const, Eq. (13.25) shows that

$$\frac{\partial(v_\theta r)}{\partial r} = 0$$

and therefore

$$v_\theta = \frac{\text{const}}{r}. \tag{13.26}$$

This type of flow is known as "free vortex" flow, as the velocity component v_θ varies with the radius in the same way as the velocity in a free vortex. There is no other significance to be attached to this name.

As another example, let the tangential velocity component be given by

$$v_\theta = cr, \tag{13.27}$$

where c is a constant. This distribution is the same as that in a "forced vortex," and this type of flow is therefore often identified by this name. By solving Eq. (13.25) we then find that

$$v_a = \sqrt{\text{const} - 2c^2 r^2}. \tag{13.28}$$

The constant of integration has to be evaluated from the continuity condition, namely, that

$$Q = \int_{r_1}^{r_0} v_a \, 2\pi r \, dr, \tag{13.29}$$

where Q is the given flow rate.

Let us next consider the change in the Bernoulli constant when the flow passes through a rotor. Following a streamline from station (1) through the blade system to station (2), we find a gain in total pressure equal to

$$B_2 - B_1 = \left[\frac{p_2}{\rho} + \frac{1}{2}(v_{\theta_2}^2 + v_{a_2}^2)\right] - \left[\frac{p_1}{\rho} + \frac{1}{2}(v_{\theta_1}^2 + v_{a_1}^2)\right], \tag{13.30}$$

where all variables, including B_1 and B_2, still may be functions of the radius. We may now compare this expression with the Bernoulli equation as written in a rotating coordinate system Eq. (12.5),

$$\frac{p_2}{\rho} + \frac{1}{2}\left[(v_{\theta_2} - u_2)^2 + v_{a_2}^2\right] - \frac{u_2^2}{2} = \frac{p_1}{\rho} + \frac{1}{2}\left[(v_{\theta_1} - u_1)^2 + v_{a_1}^2\right] - \frac{u_1^2}{2}, \tag{13.31}$$

from which it is seen that

$$B_2 - B_1 = (v_{\theta_2} u_2 - v_{\theta_1} u_1) = \Psi u_0^2, \tag{13.32}$$

where u_1 and u_2 are the blade velocities at r_1 and r_2, u_0 is the velocity of the blade tips, and Ψ is the total head or total pressure rise coefficient based on tip speed we have used before.

13.7 Elementary Design Considerations

With the aid of the equations above we can now carry out some elementary design calculations for an axial-flow compressor or turbine. Strictly speaking, the equations are applicable only to the equilibrium flow far upstream and downstream of a blade system. As was pointed out earlier, however, they yield satisfactory results also for predicting the flow at relatively short distances from the blade edges. The point is even stretched by designers, and the foregoing equations are used to describe the flow between adjacent blade rows of a multistage axial machine. Experiments have shown that the results so obtained are still useful for engineering purposes.

(a) Free Vortex Design

Let us first consider the rotor of a stage and assume that the flow entering the rotor has a uniform Bernoulli constant ($B_1 = \text{const}$) and that the work addition or subtraction should be constant for all portions of the fluid, from which it follows that both Ψ and B_2 are constant. The flow rate and the dimensions of the compressor, as well as the tip speed, are also known. In terms of symbols this means that the quantities u_0, ϕ, Ψ, r_0, and r_i are given. In addition let it be required that the axial velocity component in front of the rotor is a constant. From Eq. (13.26) we see that this flow must be of the free vortex type. Since B_2 is constant, it follows from Eq. (13.32) that $r_2 v_{\theta_2}$ is constant as well. This implies that the flow at station (2) must also be of the free vortex type and therefore of constant axial velocity. As the axial velocity components are independent of the radius and the total flow rate is a constant, one may conclude that there will be no shift in the streamlines as they pass through the rotor, and a particular streamline that is located at a given radius at station (1) will be at this same radius at station (2).

The two tangential velocity components are then of the form

$$v_{\theta_1} = \frac{a_1}{r_1} \quad \text{and} \quad v_{\theta_2} = \frac{a_2}{r_2},$$

and from Eq. (13.32),

$$\Psi u_0^2 = \frac{a_2}{r_2} u_2 - \frac{a_1}{r_1} u_1 = \Omega(a_2 - a_1). \tag{13.33}$$

The equation above is to be regarded as a condition for the difference $(a_2 - a_1)$, where a_1 and a_2 are constants. A further condition could now be imposed arbitrarily on either a_1 or a_2 to fix their values, but in multistage machines this additional condition is determined by further considerations. Let us suppose that the rotor is considered to be part of a stage in a turbomachine consisting of many identical stages. The stator following the rotor may then be required to turn the flow in such a way that the velocity components at the entrance to the second rotor are equal to those at the entrance to the first rotor. Furthermore, we have seen earlier that because of efficiency considerations, it is advantageous to have a velocity diagram that leads to a reaction ratio of one-half. Since we are at liberty to select only one additional constant, this latter condition cannot be satisfied at all radii, but we can specify that a 50 percent reaction ratio be obtained at some specified radius r_m. Since, in our example, $v_{a_1} = v_{a_2}$, the velocity diagram at this radius r_m should then be symmetrical, as shown in Fig. 13.11. It then follows that

$$u_m - v_{\theta_{1m}} = v_{\theta_{2m}}$$

and

$$a_1 + a_2 = \Omega r_m^2. \tag{13.34}$$

This specification, together with the equation for $a_2 - a_1$ [Eq. (13.33)], now fully determines these constants; consequently, the velocity diagrams at all radii are now determined. Cascades to match the velocity diagrams at each radius are then usually selected by means of the procedures which apply to linear cascades. As a different cascade will be needed at each radius, the blades of the rotor will have a twisted shape with varying cross sections along the radius.

One special blade now remains to be considered. The rotor as conceived above requires an approach velocity as determined by v_{θ_1} and v_a. Generally, the fluid will approach a machine in a purely axial direction, and a special stator called a set of *entrance blades* is required to impart the desired turning to the flow. The flow enter-

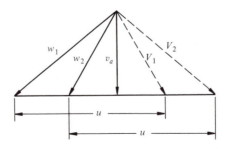

FIG. 13.11 Velocity diagram showing reaction ratio of one-half.

ing these blades is purely axial, and for the present machine the tangential component at the exit is to be $v_{\theta_1} = a_1/r$ and the axial component is to be constant. Proper cascade profiles may again be selected by the method of linear cascades. The type of design outlined is that for a "free vortex" machine, named again for the distribution of the tangential velocity component.

(b) Forced Vortex Design

Many other designs aside from the "free vortex" design are, of course, possible even with the stipulation that the Bernoulli constant be independent of radius and that the velocity profiles be repeating in subsequent stages. One additional example will be outlined here. Again the quantities Ψ, ϕ, u_0, r_o, and r_i are given. The tangential velocity component at the entrance to the rotor, however, is now specified to be

$$v_{\theta_1} = c_1 r_1. \tag{13.35}$$

Because of this distribution the resulting machine is often said to be of the "forced vortex" type, as mentioned earlier. From Eq. (13.28) it then follows that

$$v_{a_1} = \sqrt{c_2 - 2c_1^2 r_1^2},$$

where c_2 has to be evaluated from the total flow rate as described by Eq. (13.29). From the equation for the rotor work [Eq. (12.4)],

$$\Psi u_0^2 = v_{\theta_2} u_2 - v_{\theta_1} u_1$$
$$= \Omega(v_{\theta_2} r_2 - c_1 r_1^2)$$

and

$$v_{\theta_2} = \frac{\Psi u_0^2}{\Omega r_2} + \frac{c_1 r_1^2}{r_2}. \tag{13.36}$$

The radii r_1 and r_2 are the ones for a given streamline at stations (1) and (2), respectively. For a free vortex machine r_1 was equal to r_2 for all streamlines since the axial velocity components at the two stations were identical. For a forced vortex machine this equality can no longer be expected to exist. Because of the difference in axial velocity distribution the streamlines have to rearrange themselves to accommodate the flow, and r_1 is no longer equal to r_2 at all stations. Nevertheless, the two radii must be equal at the hub ($r_1 = r_2 = r_i$) and at the outer case ($r_1 = r_2 = r_o$). One may assume, therefore, as a first approximation that r_1/r_2 is approximately equal to unity for all streamlines. With this assumption

$$v_{\theta_2} = \frac{\Psi u_0^2}{\Omega r} + c_1 r, \tag{13.37}$$

and by applying the differential equation [Eq. (13.25)] to station (2), one finds that

$$v_{a_2} = \left[c_3 - 4c_1 \left(\frac{c_1 r^2}{2} + \Psi \frac{u_0^2}{\Omega} \ln r \right) \right]^{1/2}, \tag{13.38}$$

where the constant c_3 again has to be evaluated from the condition that

$$\int_{r_i}^{r_0} v_{a_2} 2\pi r \, dr = Q.$$

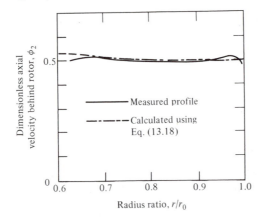

FIG. 13.12 Measured and calculated dimensionless axial velocity profile behind the rotor of a free-vortex axial-flow compressor. (Data from J. T. Bowen, R. H. Sabersky, W. D. Rannie, "Investigations of Axial Flow Compressors," *Trans. ASME*, vol. **73**, 1951, p. 1.)

The relationship between r_1 and r_2 may now also be established from the constancy of flow rate between the hub and any given streamline

$$\int_{r_i}^{r_1} v_{a_1} 2\pi r \, dr = \int_{r_i}^{r_2} v_{a_2} 2\pi r \, dr. \tag{13.39}$$

The relation derivable between r_1 and r_2 from this equation could then be inserted into Eq. (13.36) for a second approximation. But for most applications the accuracy of the first approximation is sufficient.

In Figs. 13.12 and 13.13 comparisons between calculated and measured velocity components are shown for axial-flow compressors designed for free vortex and for forced vortex patterns. The agreement is seen to be quite adequate over most of the range. There are definite deviations near the boundaries, of course, where the real fluid effects reduce the actual velocities to zero.

A word may be said here about the possible relative advantages of designs with different tangential velocity distributions such as the free vortex or forced vortex design. One may state in a very qualitative way that for a variety of hydrodynamic

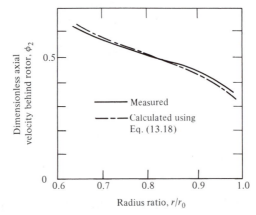

FIG. 13.13 Measured and calculated dimensionless axial velocity profile behind the rotor of a forced-vortex axial-flow compressor. (Data from J. T. Bowen, R. H. Sabersky, W. D. Rannie, "Investigations of Axial Flow Compressors," *Trans. ASME*, vol. **73**, 1951, p. 1.)

and mechanical reasons the forced vortex design is often desirable for multistage compressors of small hub ratio (r_i/r_o). Free vortex blading is most frequently encountered in high-specific-speed fans and pumps. It was also seen that the analytical work involved in designing a machine is the simplest for a free vortex distribution. This fact alone, however, should not be a determining factor in the selection, particularly since calculations usually also have to be performed for off-design flow rates. Under these conditions the free vortex blading does not lead to any more direct solutions.

(c) Flow for a Given Blade Row

Aside from designing blade systems for a desired flow pattern, the problem of predicting the flow pattern for a given blade system also occurs. As a simple example let us consider a stationary blade system which guides the flow such that the downstream angle is determined. If, in addition, the Bernoulli constant is assumed to be independent of the radius [Eq. (13.25)]; that is,

$$\frac{v_\theta}{r} \frac{\partial(r v_\theta)}{\partial r} + \frac{1}{2} \frac{\partial v_a^2}{\partial r} = 0$$

still holds. The axial and tangential velocities are related by the expression

$$\frac{v_\theta}{v_a} = \cot \alpha,$$

where $\cot \alpha$ is now a known function of r. When this substitution is made for v_θ, Eq. (13.25) becomes (as r is now the only variable, total derivatives will be used)

$$\frac{1}{v_a} \frac{d}{dr} (v_a) = -\frac{\cot^2 \alpha}{r} + \frac{d}{dr} (\log \sin \alpha), \tag{13.40}$$

which has the solution

$$v_a = A \sin \alpha e^{-\int (\cos^2 \alpha)/r \, dr}. \tag{13.41}$$

The constant A is determined from the continuity condition Eq. (13.29). The tangential velocity can now be computed from the known flow angle above to complete the solution.

The flow pattern is somewhat more difficult to predict if the fluid approaching the blade system has a nonconstant total energy, as $B_1(r)$ is then a function of radius, and if the blade system is a rotor which can change the energy of the fluid. Our procedure is the same as before, however. By differentiating the Bernoulli equation for rotating coordinates [Eq. (13.31)] with respect to the radius and eliminating the radial pressure gradient from Eq. (13.23), we obtain

$$v_{a2} \frac{dv_{a2}}{dr_2} = \frac{d}{dr_2} [B_1 + \Omega(r_2 v_{\theta_2} - r_1 v_{\theta_1})] = \frac{v_{\theta_2}}{r_2} \frac{d}{dr_2} (r_2 v_{\theta_2}), \tag{13.42}$$

which corresponds to Eq. (13.25) in the examples treated previously. The relative flow angle downstream of the rotor is now assumed to be known and with reference to the velocity triangle, Fig. 13.14, we see that

$$\cot \beta_2 = \frac{\Omega r - v_{\theta_2}}{v_{a2}}.$$

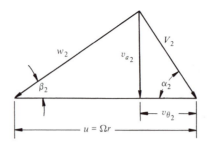

FIG. 13.14 Definition sketch of velocity components.

We can now eliminate v_{θ_2} in terms of the given angle β_2 and after some manipulation we are led to the equation

$$\frac{v_{a_2}^2}{\sin^2 \beta_2}\left(\frac{1}{v_{a_2}}\frac{dv_{a_2}}{dr_2} + \frac{\cos^2 \beta_2}{r_2} - \frac{d}{dr_2}\ln \sin \beta_2 - \frac{2\Omega \sin \beta_2 \cos \beta_2}{v_{a_2}}\right)$$

$$= \frac{d}{dr_2}[B(r_1) - \Omega r_1 v_{\theta_1}] \quad (13.43)$$

as the relation that corresponds to Eq. (13.40). This equation contains r_1 as a function of r_2 on the right-hand side, which makes it difficult to solve. As a first approximation r_1 may be put equal to r_2 on the right-hand side. Then after the solution of Eq. (13.43) is effected, values of r_2 in terms of r_1 can be found from the continuity equation as mentioned in Section 13.7(b). A more satisfactory first approximation, however, is obtained when the derivative with respect to r_2 on the right-hand side is transformed to one with respect to r_1 by utilizing the continuity relation for a stream tube, that is,

$$2\pi r_2 v_{a_2}\,dr_2 = 2\pi r_1 v_{a_1}\,dr_1. \quad (13.44)$$

The differential dr_2 in Eq. (13.43) can be eliminated by the relation above, and only now as the first approximation we put $r_1 = r_2 = r$ to obtain

$$\frac{dv_{a_2}}{dr} + v_{a_2}\left(\frac{\cos^2 \beta_2}{r} - \frac{d}{dr}\ln \sin \beta_2 - \frac{2\Omega \sin \beta_2 \cos \beta_2}{v_{a_2}}\right) = \frac{\sin^2 \beta_2}{v_{a_1}}\frac{d}{dr}[B_1(r) - \Omega r v_{\theta_1}].$$

$$(13.45)$$

The right-hand side is now a known function of radius, and v_{a_2} can now be determined by integration for any given distribution of β_2, $B_1(r)$, and $v_{\theta_1}(r)$.

The radial equilibrium equations described above enable one to make estimates of the radial streamline shifts through a rotor or stator row. As mentioned previously, these equations may be applied directly to a blade row by assuming that the flow leaving the blade row is given by these equations. In other words, it is assumed that the streamline shifts take place entirely within the space of the blade passages. The angles in Eq. (13.45), for example, are then identified with the leaving angles downstream of each cascade section for the radius in question. For many engineering applications this approximation gives satisfactory results.

13.8 Cavitation

A practical problem that arises when pumps or turbines are used with liquids is the occurrence of cavitation in the low-pressure regions of the flow. This phenomenon

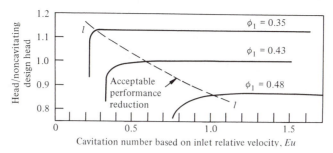

FIG. 13.15 The effect of cavitation on the performance of a centrifugal pump. The cavitation number is defined by Eq. (13.46) and the flow coefficient is $\phi_1 = v_{a_1}/u_1$. The dashed line *I-I* intersects the solid curves at the points of acceptable cavitation coefficients *k*. (Data computed from C. Blom, "Development of Hydraulic Design for the Grand Coulee Pumps," *Trans. ASME*, vol. **72**, 1950, p. 53.)

may be described as local boiling, resulting in the formation of bubbles and regions of vapor within the liquid as a result of the lowering of the local pressure to the vapor pressure of the liquid. There are two important consequences of cavitation: One is damage to the surrounding structure caused by pressure waves originating from the collapsing bubbles. The other is that the cavitating region through vigorous mixing with the main stream causes losses in total pressure and thereby a decrease in efficiency. A significant portion of the low-pressure zone may also become filled with vapor. The velocity in the remainder of the passage is thereby increased and the static pressure of the flow even further decreased. Eventually, the distortion of the flow pattern and energy loss caused by the cavitation lead to a significant and perhaps sudden decrease in the efficiency and head output of the machine. Figure 13.15 shows an example of the performance of a typical centrifugal pump for various flow rate coefficients and inlet pressures. It was shown in Chapter 5 that if the vapor pressure is important, a suitable similarity parameter is the Euler number, which is also called the *cavitation number* in hydrodynamics. This parameter is given by

$$\text{Eu} = \frac{p_1 - p_v}{\rho w_1^2/2}, \tag{13.46}$$

where p_1 is the inlet static pressure and w_1 is the inlet velocity relative to the impeller. A similar parameter can be defined for a turbine. Cavitation will commence when the minimum pressure within the fluid is equal to or slightly less than the vapor pressure of the liquid. Further reduction of the inlet pressure, or of Eu, causes the cavitating region to grow, but the pressure within it remains equal to the vapor pressure. A sketch of a centrifugal pump shown (Fig. 13.16) illustrates a possible location of the vapor region. When the effects of Reynolds number, surface tension, air diffusion, and heat transfer may be neglected, the inception of cavitation is determined by the minimum pressure on the vane surface, which depends on the geometry of the surface and the flow. Furthermore, when Eu is smaller than that for incipient cavitation, the extent of the cavitating region is again determined only by the flow geometry, provided that, as before, the real fluid effects mentioned above are negligible. Experiments using tap water as the working fluid confirm that the performance of geometrically similar machines is correctly scaled with the Euler number above.

Experiments using other fluids—especially petrochemicals—reveal that additional similarity parameters involving at least the thermal properties of the fluid are

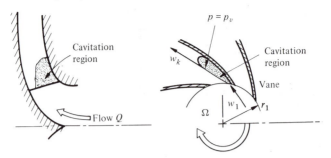

FIG. 13.16 Sketch showing location of cavitation in a centrifugal pump.

required to establish proper scaling. These are poorly understood as yet, but at least for water at room temperature the parameter Eu is sufficient for determining similar operating conditions. Acceptable values or typical values of Eu can be determined from experiments such as those illustrated in Fig. 13.15.

Let us suppose that an acceptable value of Eu has been found for a particular pump by experiment. We shall now develop some relationships between this special value of Eu, which we shall designate as k, and the operating conditions of a pump that are useful for engineering purposes. In our calculation we shall use a pump as an example, although equivalent relations may be developed for any turbomachine. Cavitation is assumed to be taking place, so that the minimum pressure is the vapor pressure of the fluid (at least as long as we deal with an essentially pure liquid). Let us begin our development by deriving the total pressure at the pump inlet as a function of a given flow rate Q, angular speed Ω (rad/sec), and impeller diameter d. We shall assume that the relative flow is frictionless and that any change in radius of the streamlines within the area of cavitation is negligible. If we let p_1 and V_1 be the pressure and velocity at the leading edge of the blades (see Fig. 13.16), the total pressure may be written

$$p_{t_1} = p_1 + \rho \frac{V_1^2}{2},$$

which is equal to the total pressure supplied to the pump if we neglect possible small losses in the inlet duct. We next write the Bernoulli equation relative to the moving blades between point 1 and the minimum pressure point and obtain

$$p_1 + \frac{\rho w_1^2}{2} = p_k + \frac{\rho w_k^2}{2}. \tag{13.47}$$

The subscript k designates conditions at the minimum pressure point. For cavitation p_k is equal to the vapor pressure p_v. If we substitute p_v for p_k and introduce the definition of the cavitation parameter k, Eq. (13.47) reduces to

$$w_k = w_1 \sqrt{1 + k}. \tag{13.48}$$

This relation lends a more physical interpretation to the parameter k in that $\sqrt{1 + k}$ is the factor by which w_k exceeds w_1. For a given vane shape this ratio may be predicted with fair accuracy from potential theory. If the blade had the characteristics of a circular cylinder in a nonseparated flow without circulation, for example, potential theory would give a value of $V_{max}/V_0 = w_k/w_1 = 2$, or from Eq. (13.48) a value of k equal to 3 at the point of cavitation inception. For the vane sections encountered in pumps, the range of k may be from about 0.25 to as low as 0.02.

However, the flow corresponding to these—and especially the lower value—may contain vapor-filled regions. In the latter case the entire vapor region will be at the minimum pressure. The foregoing equations and concepts, however, are still applicable and we may select any point on the surface of this region as the minimum pressure point for which Eqs. (13.47) and (13.48) are written.

We next rewrite the Bernoulli equation [Eq. (13.47)] using Eq. (13.48) and substituting

$$w_1^2 = V_1^2 + \Omega^2 r_1^2.$$

In so doing it is assumed that the vector V_1 is normal to the velocity of the blade leading edge Ωr_1. This is usually true, as pumps rarely have inlet vanes and the tangential velocity component is zero. It then follows that

$$p_{t_1} - p_v = \frac{\rho}{2} [V_1^2(1 + k) + k\Omega^2 r_1^2]. \tag{13.49}$$

This equation is useful in that it specifies the minimum total pressure p_{t_1} required for a given pump (with fixed r_1 and k) and given flow rate (which determines V_1). Let us now express Eq. (13.49) in dimensionless form and also eliminate the velocities V_1 and radius r_1 in terms of other quantities. Thus we let

$$V_1 = \frac{Q}{A_1} c_1,$$

where A_1 is the flow area normal to V_1. The coefficient c_1 is introduced since the velocity at the point in question is usually larger than the average velocity over the channel given by Q/A_1 due to curvature of the adjacent wall. An average inlet flow coefficient may be defined as

$$\phi_1 = \frac{Q}{A_1 u_1} = \frac{V_1}{c_1 u_1}.$$

The area A_1 may be expressed as a fraction of the duct area, that is,

$$A_1 = \pi r_1^2 \xi_1,$$

where ξ_1 is a geometrical characteristic of the inlet portion of the impeller. With these definitions V_1 and r_1 can be eliminated and we obtain after some rearrangement the expression

$$\frac{p_{t_1} - p_v}{\rho Q^{2/3} \Omega^{4/3}} = \frac{1}{2} \left(\frac{1}{\pi \phi_1 \xi_1} \right)^{2/3} [\phi_1^2 c_1^2(1 + k) + k].$$

The pressure difference is conveniently expressed as

$$g\Delta H_v \equiv \frac{p_{t_1} - p_v}{\rho},$$

and after an inversion we see that

$$\frac{\Omega Q^{1/2}}{(g\Delta H_v)^{3/4}} = 2^{3/4}(\pi \phi_1 \xi_1)^{1/2} \frac{1}{[(1 + k)c_1^2 \phi_1^2 + k]^{3/4}}. \tag{13.50}$$

The left-hand side now has the same form as the specific speed previously introduced, the only difference being that ΔH_v replaces the total head increase H. This quantity has been appropriately called the *suction specific speed* and has been given the symbol S:

$$\frac{\Omega Q^{1/2}}{(g\Delta H_v)^{3/4}} \equiv S. \tag{13.51}$$

As can be seen from Eq. (13.50), a definite value of S corresponds to each value of the flow coefficient ϕ_1 and k. The suction specific speed S, like the specific speed N_s, is a similarity parameter of the machine. Two geometrically similar machines, therefore, when operated at the same flow coefficient ϕ_1 will have the same cavitation performance (i.e., efficiency and total head coefficient) when the suction specific speeds are the same. Acceptable values of S can be determined by experiment such as led to the data of Fig. 13.15, or may be computed from these results with the use of Eq. (13.50). When experimental data for S are available, the appropriate size of a pump to discharge a given flow rate Q at speed Ω and at a known value of S is obtained from the foregoing definition of the flow rate coefficient ϕ_1,

$$r_1 = \left(\frac{Q}{\pi\Omega\phi_1\xi_1}\right)^{1/3},$$

where ϕ_1 is the flow rate coefficient corresponding to the value of S. As it is frequently desirable to have high values of S, S may be maximized in respect to ϕ_1. This can be accomplished by differentiating Eq. (13.50), although it must be kept in mind that the acceptable values of k as determined by experiment are a function of ϕ_1.

Pumps, since they operate with liquids, are designed with special attention to the cavitation problem. Conventional designs are intended to achieve a value of $S \approx 3$ [using a consistent set of units in Eq. (13.51)], but for special applications designs can be evolved which give satisfactory operation for S as high as about 15. These values correspond approximately to the values of $k = 0.25$ and 0.02, respectively. The latter design does not avoid cavitation but rather permits adequate operation even with cavitation being present in the inlet regions of the pump. Generally, small blade angles (and therefore small ϕ_1) correspond to high suction specific speeds. The quite high value of $S = 15$ is achieved only with pumps having long inlet blades with small vane angles ($\beta_{1_v} \approx 6°$). Conventional pumps tend to have inlet blade angles about the same as the exit angles, say 20 to 25°, and they operate with an inlet flow coefficient such that the inlet angle of attack is between 0 and 5°.

We have not yet mentioned values of the coefficient c_1. Axial inlets will have c_1 equal to unity, but in the curved portion of most centrifugal pumps c_1 may have different values depending on the radius of curvature of the meridional boundary. Well-designed pumps seem to have c_1 varying from about 1.2 to 1.4.

The foregoing remarks have been directed chiefly toward pumps. Any turbomachine when operated in a liquid can cavitate, and the problems that arise are similar to those described above for pumps. Hydraulic turbines have severe problems of cavitation, and the elevation of the turbine above the tailrace (the lower liquid level which determines the static pressure at discharge) is often determined by the amount of cavitation damage to the impeller that can be accepted over a certain period of time. Of course, the turbine could be set deep enough to prevent all occurrence of cavitation. The resulting cost of excavation and constructional difficulties could, however, offset the increased life and overhaul time of the impeller.

In closing, it should again be emphasized that in the foregoing considerations, effects such as those of heat transfer on evaporation rates have been neglected. The flow similarity expressed by the suction specific speed is consequently limited in this respect. One must, then, expect that the cavitation performance of a given pump may differ for different liquids even though it is operating at the same suction specific speed.

PROBLEMS

When not otherwise specified, the properties of the fluids are to be evaluated at 20°C and 1 atmosphere pressure.

13.1 A linear cascade moving at 30 m/s with a purely axial inlet velocity of 15 m/s is to produce 4.5 m total head rise at a lift coefficient of 0.6. The fluid is incompressible.
(a) Determine the required solidity of the cascade.
(b) Assuming that the cascade consists of flat-plate blades with a value of $m = 6$, determine the blade setting β_v.
(c) At what axial velocity will the total head be reduced to zero?
(d) What would be the total head rise in feet at zero axial velocity?

13.2 Show that for frictionless incompressible flow through a moving cascade with axial inlet the reaction ratio is $1 - \Psi/2$.

13.3 Derive Eq. (13.9) by eliminating β_2, β_1, and β_∞ in Eqs. (13.8) and in the expression for the lift coefficient $C_L = m \sin \alpha^*$.

13.4 An axial exhaust fan with proportions as illustrated in the figure is to be designed. The pressure in the plenum chamber is 0.5 in. of water (2.6 lb/ft²) below the outside atmosphere. The discharge from the fan passes into the atmosphere. The total discharge is to be 2000 ft³/min of standard air (assume a constant density of $\rho = 0.00238$ slug/ft³). The fan speed is 3600 rpm. Assume that there are no friction losses and that the swirl energy in the discharge is negligible. Determine the head and flow coefficients required at the tip of the fan, the solidity for a lift coefficient of 0.6, and the blade setting β_v if the value of m is equal to 2π.

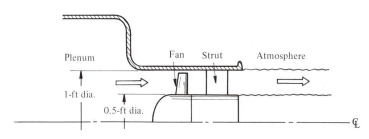

Problem 13.4

13.5 The tip section of an axial-flow pump has the following characteristics: The solidity is 0.4, the line passing through the nose and tail of the vane is inclined at 12° to the plane of the rotor disk, and the vane section is such that the angle for zero lift on the blade is 3.5° larger than this angle. The value of m in Eq. (13.9) may be taken as that for an isolated flat-plate airfoil. Plot the value of the frictionless head coefficient Ψ [from Eq. (13.11)] as a function of ϕ for values of $\alpha_1 = 70°$, 90°, and 110°. Compare the result for $\alpha_1 = 90°$ to the performance curve of the pump described in Problem 12.22. Assume the fluid to be incompressible and frictionless.

13.6 Fill in all of the steps leading to Eq. (13.17).

13.7 A multistage axial-flow *turbine* is to be designed in such a way that the velocity pattern for consecutive stages remains the same. In addition the pattern is to correspond to a 50 percent reaction ratio and the angles α_∞ and β_∞ are in each case equal to 45°. The flow can be considered purely two-dimensional, and the quantities computed for the radius at the center of the blade may be considered typical. The axial velocity component is 100 ft/sec and the total head coefficient per stage is to be $(-\Psi) = 0.04$. Treat the fluid as incompressible.

(a) Assuming frictionless flow, compute the velocity components v_{θ_1}, w_{θ_1}, and u.

(b) Draw to scale a complete velocity diagram labeling all vectors and angles.

(c) Assume "single airfoil" theory to be applicable and use the lift coefficient data from the accompanying graph. Select the lift coefficient such that $D/L = \epsilon$ is a minimum, but otherwise neglect the effects of drag in this portion of the problem. On this basis determine the angle of attack α^* and the solidity σ.

(d) Neglecting changes in w_∞ and β_∞, determine the percentage power reduction brought about by the drag.

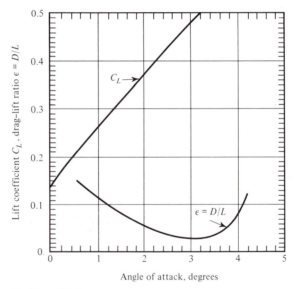

Problem 13.7

13.8 An axial compressor rotor row having a blade section with a value of $D/L = 0.025$ and operating at a lift coefficient of 0.7 is to produce the velocity diagram shown. Note that the absolute inlet velocity is purely axial.

(a) Calculate the ideal (without friction) total pressure rise coefficient $\Psi = \Delta p_t / \rho u^2$.

(b) Calculate the total pressure rise coefficient when the effect of friction is included.

(c) Determine the required solidity (including the effect of friction) and the angle of attack if $m = 2\pi$. The fluid is incompressible.

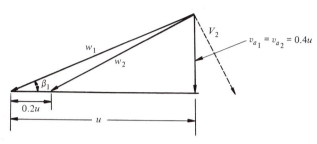

Problem 13.8

13.9 Show that for a rotor row the lift coefficient corresponding to Eq. (13.7) becomes

$$C_L = \frac{2}{\sigma} \sin \beta_\infty (\cot \beta_1 - \cot \beta_2) \frac{1}{1 + \varepsilon \cot \beta_\infty}$$

when friction is included.

13.10 Measurements of the total pressure loss through a stator row can be used to infer an effective drag coefficient. By use of Eqs. (13.13) and (13.14) show that this drag coefficient may be expressed as

$$C_D = \frac{D}{(\rho/2)V_\infty^2 c} = \frac{\sin \alpha_\infty}{\sigma} \frac{\Delta p_t}{(\rho/2)V_\infty^2},$$

where Δp_t is the total pressure loss through the blade row. An equivalent expression can be obtained for the flow through a rotor row by replacing α_∞ by β_∞ and V_∞ by w_∞.

Hint: Note that $\Delta p_t = p_1 + (\rho/2)V_1^2 - (p_2 + (\rho/2)V_2^2)$. Express the velocities in terms of the angles α_1 and α_2 and axial velocity v_a before using Eqs. (13.13) and (13.14).

13.11 With the use of Eq. (13.21) and Figs. 13.4 and 13.7, estimate the theoretical (frictionless) performance curve of the pump described in Fig. 12.21. For the relatively high solidity of this impeller, the discharge blade angle β_{2_v} should be used in Eq. (13.21).

13.12 A free vortex turbine is to produce the head coefficient $-\Psi_0 = a$ at the tip radius r_0 at a flow coefficient $\phi_0 = b$ also based on the tip radius. The reaction ratio is to be 50 percent at the radius $r_m < r_0$, and an appropriate set of entrance nozzles is provided to make this design possible.
(a) Sketch the dimensionless inlet and outlet velocity triangles at the 50 percent ratio radius r_m.
(b) Obtain expressions for the velocity ratios v_{θ_1}/u_0, v_{θ_2}/u_0 in terms of r/r_0, r_m/r_0, and a.

13.13 The turbine of Problem 13.12 has values of $a = 0.3$, $b = 0.4$, and a hub-to-tip ratio of 0.6. The working fluid is water, u_0 is 50 ft/sec, and $r_0 = 1.5$ ft.

(a) Calculate the horsepower produced neglecting friction.

(b) Draw to a suitable scale the inlet and outlet velocity diagrams across the rotor at the hub, tip, and at $r/r_o = 0.8$ given that the radius ratio for 50 percent reaction is 0.8.

13.14 An axial blower with a uniform axial inlet velocity of 50 ft/sec and having a tip speed of 100 ft/sec is to produce (ideally) 62 ft of head for all radii. The fluid is considered incompressible. Sketch accurately the relative and absolute velocity triangles for the entering and leaving velocity vectors for radius ratios of 1.0 (the tip), 0.80, and 0.6. Determine what rotor solidities are needed to keep the diffusion factor [Eq. (13.19)] less than 0.5 at each of these radii.

13.15 For the solidities of Problem 13.14, determine the blade setting γ and camber θ for blades of circular arc profile using Constant's rule [Eq. (13.18)]. For the purpose of definiteness make the inlet incidence angle zero for all radii.

13.16 An axial-flow fan receives a uniform *axial* velocity of 98 m/s. The rotor of the fan is 0.75 m in diameter and the hub diameter is 0.45 m in diameter. Assume that there are no losses. The inlet total temperature is 300 K, the density 1.0 kg/m^3, the fluid is air with $c_p = 1.0$ kJ/kgK, and the specific heat ratio is 1.4. The fan is to produce a total pressure rise for all radii of 6000 Pa when the rotor turns at 4000 rpm. The rotor is followed by a stator that turns the flow back to a purely axial direction. Note that this is a "free vortex" type of design.

(a) Calculate the total pressure ratio across the fan and the total temperature ratio. Because these are near unity, assume that the flow in what follows is incompressible (i.e., no change in density across the rotor or stator or change in axial velocity).

(b) Calculate the following at the tip and the hub and arrange your results in a table.

w_1 = inlet relative velocity, β_1 = inlet relative angle,
$v_{\theta 1}$ = inlet tangential velocity, β_2 = exit relative angle,
$v_{\theta 2}$ = exit tangential velocity, β_∞ = vector mean angle,
w_2 = exit relative velocity, α_2 = absolute flow exit angle

(c) Determine the solidity σ needed to keep the diffusion factor $k_d = 0.5$ for the rotor at the hub and tip.

(d) Calculate the lift coefficient needed for the rotor at hub and tip.

Note: It is very instructive to plot these velocity triangles to scale.

13.17 In Eq. (13.35) put $c_1 = \delta v_{avg}/r_o$, where r_o is the tip radius, v_{avg} is the average axial velocity in the duct, and δ is a number. Show that if δ is appreciably smaller than unity,

$$v_{a_1} \doteq v_{avg}\left[1 + \frac{\delta^2}{2}\left[1 + \left(\frac{r_i}{r_o}\right)^2 - 2\left(\frac{r}{r_o}\right)^2\right]\right],$$

where r_i is the inner radius and r_o the outer radius of the duct.

Hint: Use the binomial theorem and expand in terms of δ^2 to show first that

$$c_2 \doteq v_{avg}^2\left[1 + \delta^2\left(1 + \left(\frac{r_i}{r_o}\right)^2\right)\right].$$

13.18 A flow of constant total energy passes through a stationary set of inlet guide vanes which turns the flow to produce an axial velocity distribution

$$v_a = A\left(\frac{1 - r}{r_o}\right),$$

where r_o is a reference radius and A is a constant. Assuming that v_θ is zero at r equal to zero, determine the resulting distribution of v_θ.

13.19 The entrance blades of an axial-flow compressor turn a frictionless flow of constant total head into a constant angle α with respect to the axis. Show that

$$\frac{v_a}{v_{avg}} = \left(\frac{1 + \sin^2 \alpha}{2}\right)\frac{[1 - (r_i/r_o)^2](r_o/r)^{\cos^2 \alpha}}{1 - (r_i/r_o)^{1 + \sin^2 \alpha}},$$

where v_{avg} is the average velocity in the duct, r_i is the inner duct radius, and r_o is the outer duct radius.

13.20 The inlet velocity to an axial compressor is a purely axial velocity of 100 m/s. The tip radius is 1.0 m and the hub radius is 0.6 m. The flow is to considered *incompressible*, and the Bernoulli constant is the same for all streamlines. This axial flow is then deflected by a set of inlet guide vanes to a constant angle of $\alpha = 55°$ (i.e., $35°$ from the axial direction).
(a) Calculate and *plot* the axial velocity distribution from hub to tip.
(b) Calculate the radial shift of the streamline that has an initial radius of $r_1 = 0.82462$ m, the rms radius of the compressor.

13.21 Show from Eq. (13.42) that for the case $B_1 = $ const, $r_1 v_{\theta_1} = $ const, dv_{a_2}/dr_2 is positive or negative as v_{θ_2} is less or greater than $r_2 \Omega$ when dB_2/dr_2 is positive.

13.22 Consider a rotor composed of *helical* blades for which $\tan \beta = (r_o/r) \tan \beta_0$. Assuming that B_1 and $r_1 v_{\theta_1}$ are constants in Eq. (13.45), show that

$$\frac{v_a}{r_o \Omega} = \tan \beta_0 - \frac{(\tan \beta_0 - v_{avg}/r_o \Omega)[1 - (r_i/r_o)^2]}{\left[\left(\frac{r}{r_o}\right)^2 + \tan^2 \beta_0\right] \ln \dfrac{1 + \tan^2 \beta_0}{(r_i/r_o)^2 + \tan^2 \beta_0}}.$$

Note: The resulting differential equation for v_a is linear and may be solved as indicated in Dwight's Table of Integrals No. 891.1.

13.23 A small portable pumping plant is to be designed for firefighting. The pump is to be gas turbine driven. The speed of a gas tubine is usually high, and in order to reduce transmission problems it is desirable to operate the pump also at high speeds. In addition, high speed makes possible small and light-weight components. The pump has to deliver 0.5 ft³/sec of water at a pressure of 100 psig. The pump has to be capable of sucking water from a lake or other large reservoir without appreciable effects of cavitation. The inlet to the pump may be assumed to be at the same level as the lake. An acceptable suction specific speed of the pump may be taken to be 3. The vapor pressure of the water is given at 0.7 psia and the atmospheric pressure is equal to 12.0 psia at the altitude of the lake.
(a) Determine the maximum speed for this pump consistent with the suction specific speed.
(b) Using the speed determined in part (a) as the operating speed, determine the dimensionless operating point (Ψ and ϕ) and the outside diameter of the impeller which would be suitable for the above application.

13.24 A centrifugal pump of conventional design and having a specific speed $N_s = 0.45$ is to pump 0.85 m³/s at a total head of 152 m. The pump will take water with a vapor pressure of 3.4 kPa from a storage basin at sea level. For the pump speed consistent with the requirements above, calculate the elevation of the pump inlet relative to the water level based on an acceptable value of suction specific speed equal to 3.2.

13.25 Liquid oxygen with a vapor pressure of 1 atmosphere and weighing 80 lbf/ft³ is to be pumped from a pressurized tank. Experiments have shown that a pump can be made to operate well at values of $k = 0.05$ and $\phi_1 = 0.15$. The flow rate is to be 4.5 ft³/sec at a rotative speed of 8000 rpm. The inlet to the pump may be assumed purely axial with $\xi_1 = 0.85$ and $c_1 = 1.0$. Calculate the gage pressure in the tank at the pump centerline and the inlet diameter.

13.26 Assume that k is independent of ϕ_1 and show that the inlet radius that minimizes $p_{t_1} - p_v$ at constant flow rate Q is

$$r_1 = \left[\frac{2Q^2}{\pi^2 \Omega^2} \left(\frac{1+k}{k} \right) \right]^{1/6}$$

(provided that $\xi_1 = c_1 = 1$) and furthermore that the value of ϕ_1 so obtained is

$$\phi_1 = \sqrt{\frac{k}{2(1+k)}}.$$

REFERENCES

1. C. Keller, *The Theory and Performance of Axial Flow Fans*, McGraw-Hill, New York, 1931.
2. J. Horlock, *Axial Flow Compressors*, and also *Axial Flow Turbines*, Butterworth, London, 1958.
3. I. A. Johnsen and R. O. Bullock, eds., *Aerodynamic Design of Axial Flow Compressors*, Report R.M. E56B03a, Nat. Aero. and Space Adm., formerly NACA.
4. G. F. Wislicenus, *Fluid Mechanics of Turbomachinery*, McGraw-Hill, New York, 1947.
5. S. L. Dixon, *The Thermodynamics of Turbomachines*, Pergamon, London, 1966.
6. B. Eck, *Fans*, Pergamon, London, 1973.
7. J. L. Kerrebrock, *Aircraft Engines and Gas Turbines*, MIT Press, Cambridge, Mass., 1977.

Dimensions and Units

A-1.1 The International System of Units (SI) The SI unit system uses seven quantities of measure as primary, or *base*, units; all other units are based on, or derived from, the primary set. For most applications in fluid mechanics only three of the base units are needed: mass (kilogram), length (meter), and time (second). When dealing with compressible fluids, a fourth base unit, temperature (Kelvin), is required.

The quantities defined in the base units are specified by the General Conference on Weights and Measures (1960, 1967, 1968) as follows:

1. *Mass.* One kilogram (kg) is the mass of a special platinum–iridium bar kept under precise conditions at the International Bureau of Weights and Measures in France.
2. *Length.* One meter (m) is 1,650,763.73 times the wavelength of the orange-red radiation from a krypton-86 atom.
3. *Time.* One second (s) is 9,192,631,770 periods of the radiation from an excited cesium-133 atom.
4. *Temperature.* One Kelvin (K) is 1/273.16 of the temperature interval from absolute zero to the triple point of water (where solid, liquid, and vapor phase coexist). The Kelvin scale is an *absolute* scale; zero Kelvin is at absolute zero. The Celsius scale has the same interval as the Kelvin, but zero Celsius is taken as the freezing point of water at atmospheric pressure, which is 273.15 K ($K = {}^\circ C + 273.15$).

The most important of the derived units in the SI system is that of *force*. Newton's second law states that force equals mass times acceleration, so that in SI units of force must be the same as $kg \cdot m/s^2$. For convenience the unit of force is called the newton (N), in honor of Isaac Newton's many contributions to mechanics ($1 \text{ N} = 1 \text{ kg} \cdot 1 \text{ m/s}^2$).

The force per unit area, or *pressure*, is also a derived unit and is the same as N/m^2. This unit is called the pascal (Pa) in honor of Blaise Pascal ($1 \text{ Pa} = 1 \text{ N}/ 1 \text{ m}^2$). A useful reference value for pressure is that of the standard atmosphere, which is $1.01325 \cdot 10^5$ Pa.

The usual mechanical definition of *work* is that of a force acting through a distance. In SI the units of work are then the same as $N \cdot m$; the unit of work is called the joule (J) after James P. Joule ($1 \text{ J} = 1 \text{ N} \cdot 1 \text{ m}$). The first law of thermodynamics, which is discussed more completely in Appendix 4, relates work directly to heat and energy. Thus, in SI, the units of energy and heat are also the joule.

The rate at which work is done is defined as *power*, and in SI the units of power must be equal to J/s. This unit of power is called the watt (W) in honor of James Watt (1 W = 1 J/1 s).

Mention should be made of the acceleration produced by the earth's gravitational field. This acceleration, given the symbol g, varies somewhat from place to place on the earth's surface, but for most practical applications may be taken as $g = 9.80665$ (or 9.81) m/s^2 (this is the value at sea level at a latitude of 45°). Upon applying Newton's second law, it is seen that the force of attraction exerted on a unit mass (1 kg) is equal to

$$F = 1 \text{ (kg)} \cdot 9.81 \text{ (m/s}^2),$$

$$= 9.81 \text{ N}.$$

In other words, 1 kg "weighs" 9.81 N in the earth's gravitational field.

Finally, it should be noted that SI units are frequently given a prefix denoting a unit that is the base unit multiplied by various powers of 10. The most common use is that of the millimeter, written as mm, meaning a unit that is 0.001 times 1 m. In this book use is also made of the kilopascal, written as kPa, which is 1000 Pa, and the megapascal, written as MPa, which is 1,000,000 Pa. These prefixes (m, k, and M) can be attached to other base units to imply multiples of 10 in the same way.

A-1.2 The British, or Engineering Unit, System The British system has been the most common set of units in use in the English-speaking world for several hundred years. However, most countries are encouraging (through legislation) use of the SI system in order to promote international commerce. In the United States the British system has been modified slightly with regard to units of fluid volume: The modified system is sometimes referred to as the U.S. Customary System (USCS) and is still in frequent use for current engineering work.

The units used as base dimensions in the British system are length (foot), time (second), and force (pound-force). In addition, temperature (Rankine) is used with compressible flows. These units are defined to be:

1. *Length.* The foot (ft) is defined in terms of the meter as 1 foot = 0.3048 meter. The conversion constant is exact; no further decimal places are required.

2. *Time.* The second (sec) has the same meaning as the SI second.

3. *Force.* The pound-force (lbf) is specified in terms of the gravitational force on a standard mass, introducing the ambiguity of requiring a standard mass as well as a standard gravitational acceleration in this definition. Despite this ambiguity, it is nevertheless possible to develop a logical set of units using the pound-force as a base unit.

4. *Temperature.* The Rankine degree (°R) is taken as 1/180 of the temperature interval between the freezing and boiling point of water at 1 atmosphere. The Rankine scale is absolute: zero Rankine is at absolute zero. The Farenheit scale (the most common in everyday use) has the same interval but arbitrarily uses the freezing point of water, 459.67 °R, as zero (°R = °F + 460, approximately).

The concept of mass is defined on the basis of Newton's second law as being proportional to the force applied to a body divided by the acceleration produced. A logical unit of mass is then one that experience a unit acceleration (1 ft/sec^2) when acted upon by a unit force (1 lbf). This mass unit is then simply the ratio of the unit force to the unit acceleration, that is, lbf sec^2/ft. This unit has been given a special name and is called the *slug*. The density of a substance, for example, which is the mass per unit volume, is expressed as slugs/ft^3.

In the British system the standard acceleration due to the earth's gravitational field becomes $g = 32.17$ (or 32.2) ft/sec^2, so that the force required to hold up a slug becomes

$$F = 1 \text{ (slug)} \cdot 32.2 \text{ (ft/sec}^2),$$
$$= 1 \text{ (lbf sec}^2\text{/ft)} \cdot 32.2 \text{ (ft/sec}^2),$$
$$= 32.2 \text{ lbf},$$

or 1 slug "weighs" 32.2 lbf. A smaller unit of mass, the pound-mass (lbm) is frequently used, and is arbitrarily defined to be 1/32.2 of a slug. It can be readily seen that the pound-mass has the convenient feature that 1 lbm weighs 1 lbf in the standard gravitational field. The USCS unit system, in addition, adopts the pound-mass as a base unit, so that a conversion constant of 1/32.2 must be used in Newton's second law [F(lbf) = (lbm)/32.2 · acceleration (ft/sec^2)]. The use of the name "pound" for mass (lbm) and force (lbf) adds to an already sufficiently confusing situation; care must be taken to distinguish between mass and force when using the USCS units.

The concept of work, or energy, follows directly as with the SI system. The units of work are ft-lbf; no special name is attached to this unit. Unfortunately, the units of heat are different from those of work or energy. The unit of heat is called the *British Thermal Unit* (Btu), and historically was defined as the amount of heat needed to raise the temperature of 1 pound mass (lbm) of water by 1°F at atmospheric pressure. Since the pound mass is used in this definition, many thermodynamic quantities in the British system, therefore, involve this unit of mass. From the first law of thermodynamics it follows that the Btu must have the dimensions of work or energy, and measurements show that 1 Btu is equal to 778.16 ft-lbs, or approximately 778 ft-lbf.

Finally, we note that power in the British system of units is ft-lbf/sec, but the more common unit is *horsepower* (hp), where the unit value is arbitrarily taken as 1 hp = 550 ft-lbf/sec.

A-1.3 Some Useful Equivalents Using the specified units in either the SI or British system, tables of equivalent values can be derived for ease in future use. The reader is encouraged to do so; such a short table for conversion of British units to SI equivalents follows (see the References for further discussion of units and conversion factors):

Mass	1 slug = 14.59390 kg
	1 lbm = 0.45359 kg
Length	1 ft = 0.3048 m
Force	1 lbf = 4.44822 N
Pressure	1 lbf/ft^2 = 47.88026 Pa
	1 std. atm = 101.325 kPa
Volume	1 US gallon = 3.78541 · 10^{-3} m^3
	1 Imp. gallon = 4.54609 · 10^{-3} m^3
Work	1 ft-lbf = 1.35582 J
Heat	1 Btu = 1.05506 kJ
Power	1 hp = 0.7457 kW
Temperature	$1°F = 1°R = \frac{5}{9}°C = \frac{5}{9} K$

References

1. *Marks' Standard Handbook for Mechanical Engineers*, 8th ed., McGraw-Hill, New York, 1978.
2. American Society for Testing and Materials, ASTM E380-74, 1974.
3. E. A. Mechtly, "The International System of Units: Physical Constants and Conversion Factors," NASA, 1969.

Physical Properties of Various Fluids

TABLE 1-A Effect of Pressure on the Absolute Viscosities of Water and of a Typical Lubricating Oil (Similar to SAE 30). The values in the table represent the absolute viscosity at the pressure divided by the absolute viscosity at 1 atmosphere

Substance	Pressures in Atmospheres					
	100	*300*	*500*	*750*	*1000*	*2000*
Water at 30°C	1.0	1.01	1.02	1.04	1.05	1.13
Water at 10°C	1.0	0.98	0.97	0.96	0.95	0.965
Representative lubricating oil (similar to SAE 30) at 55°C	1.45	2.50	4.7	9.4	19	~150

[a] Data adapted from *International Critical Tables*, McGraw-Hill Book Company, New York, 1926 (courtesy of the National Academy of Sciences, National Research Council, Washington, D.C.); H. A. Everett, *High Pressure Viscosity as an Explanation of Apparent Oiliness*, Soc. Aut. Eng., Trans., vol. **41**, 5, p. 531, 1937; R. B. Dow, *The Effect of Temperature and Pressure on the Viscosity of Lubricating Oils*, Rheology Bulletin, Am. Inst. of Physics, 1937.

TABLE 2-A Surface Tension of Several Fluids at 1 Atmosphere Pressure[a]

Substance	σ (M/m)	
Water at:		
0°C	0.076	in contact with air
20°C	0.073	in contact with air
100°C	0.059	in contact with air
Mercury at:		
20°C	0.466	in contact with air
20°C	0.375	in contact with water
Carbon tetrachloride at:		
20°C	0.268	in contact with air
20°C	0.449	in contact with water

[a] Abstracted from Am. Soc. Civ. Eng., Manual 25, 1942.

TABLE 3-A   Approximate Physical Constants of Some Common Fluids[a]

Substance	Temperature (°C)	Density (kg/m³)	Viscosity (Pa · s)	Bulk Modulus (MPa)
Ethyl alcohol	20	788	11.9×10^{-4}	
Benzene	20	878	6.5×10^{-4}	
Carbon tetrachloride	20	1,595	9.6×10^{-4}	1,385
Glycerine	20	1,260	1.49	
Mercury	20	13,545	15.7×10^{-4}	26,889
Oil, linseed	20	933	141.0×10^{-4}	1,606
Oil, SAE 30	38	919	670.0×10^{-4}	1,723
Water	20	998	11.3×10^{-4}	2,206
Air (1 atm.)	20	1.206	0.18×10^{-4}	

[a] Abstracted from Am. Soc. Civ. Eng., Manual 25, 1942.

TABLE 4-A   U.S. Standard Atmosphere[a]

Geometric altitude m	Temperature deg K	Pressure kPa	Density kg/m³	Viscosity (Dynamic) Pa · s · (10^5)
−1,000	294.6	111.3	1.135	1.821
−500	291.4	107.5	1.285	1.805
0	288.2	101.3	1.225	1.789
500	284.9	95.5	1.167	1.774
1,000	281.7	89.9	1.112	1.758
1,500	278.4	84.5	1.058	1.742
2,000	275.2	79.5	1.001	1.726
2,500	271.9	74.7	0.957	1.709
3,000	268.7	70.1	0.909	1.694
3,500	265.4	65.8	0.863	1.678
4,000	262.2	61.7	0.819	1.661
4,500	258.9	57.8	0.777	1.645
5,000	255.7	54.0	0.736	1.628
10,000	223.3	26.5	0.414	1.416
11,000	216.7	22.7	0.365	1.422
11,100	216.7	22.4	0.359	1.422
15,000	216.7	12.1	0.195	1.422
20,000	216.7	5.3	0.073	1.422
25,000	221.6	2.6	0.039	1.448
30,000	226.5	1.2	0.018	1.448
35,000	236.5	0.6	0.008	1.529

[a] "U.S. Standard Atmosphere, 1976," N.O.A.A., U. S. Government Printing Office, Washington, D.C.

TABLE 5-A Some Properties of Gases That May Be Considered Perfect for Many Applications

Gas	Chemical Symbol	M	R	$c_p^\star$	$c_v^\star$	γ
Helium	He	4.0	2.077	5.225	3.147	1.66
Argon	A	39.9	0.208	0.519	0.311	1.67
Air	—	29.0	0.287	1.004	0.717	1.40
Oxygen	O_2	32.0	0.260	0.910	0.650	1.40
Nitrogen	N_2	28.0	0.297	1.039	0.742	1.40
Hydrogen	H_2	2.0	4.124	14.180	10.064	1.40
Carbon monoxide	CO	28.0	0.297	1.039	0.742	1.41

M = molecular weight.
R = gas constant, kJ/kg K.
c_p = Specific heat at constant pressure, kJ/kg K.
c_v = Specific heat at constant volume, kJ/kg K.
$\gamma = c_p/c_v$.
* At room temperature.

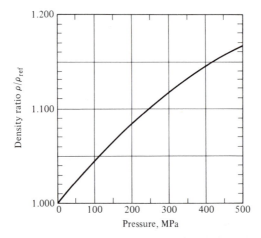

FIG. 1-A   Density of water at 0°C versus pressure. The density is given in terms of the density at 1 atmosphere pressure. (Data from *International Critical Tables*, McGraw-Hill Book Company, 1929; Courtesy of the National Academy of Sciences, National Research Council, Washington, D.C.)

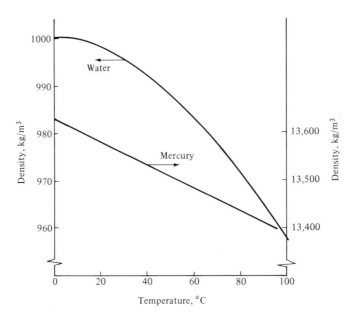

FIG. 2-A   Density of water and liquid mercury versus temperature at a pressure of 1 atmosphere. (Data on water from *International Critical Tables*, McGraw-Hill Book Company, 1929; Courtesy of the National Academy of Sciences, National Research Council, Washington, D.C.; on mercury from *Smithsonian Physical Tables*, W. E. Frosythe, ed., 9th ed., 1954, Smithsonian Institution, Washington, D.C.)

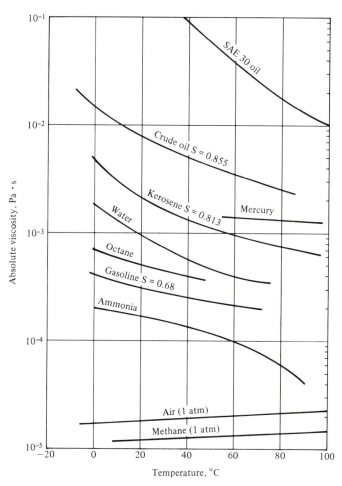

FIG. 3-A   Absolute viscosity of various fluids. Here *S* refers to the density of the substance relative to that of water at 60°F. (Prepared from data in R. L. Daugherty and A. C. Ingersoll, *Fluid Mechanics*, McGraw-Hill Book Company, New York, 1954.)

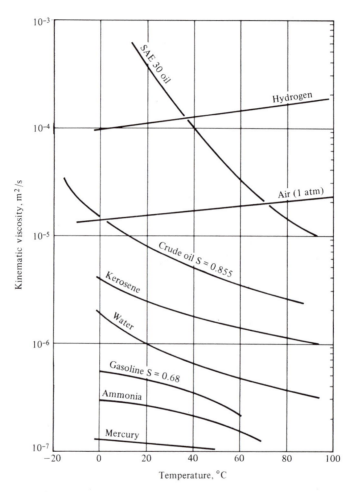

FIG. 4-A Kinematic viscosity of various fluids. *S* has the same meaning as in Fig. 3-A. Prepared from data in R. L. Daugherty and A. C. Ingersoll, *Fluid Mechanics*, McGraw-Hill Book Company, New York, 1954.

Summary of the Properties
of Vectors

A-3.1 Included herein is a brief summary of the principal operations of vector algebra and vector calculus used in this book. It is by no means intended to be exhaustive and is included only to serve as a handy reference for the more frequently used properties of vectors for the convenience of the reader. For proofs and other related questions, reference should be made to one or more of the works listed at the end of this appendix.

A-3.2 Vectors A vector may be expressed in terms of components parallel to the axes of the coordinate system. Thus a vector $\mathbf{A}$ in a Cartesian system can be written

$$\mathbf{A} = \mathbf{A}_x + \mathbf{A}_y + \mathbf{A}_z,$$

where $\mathbf{A}_x$, $\mathbf{A}_y$, and so on, are vectors in the direction of the x-, y-, z-axes. These component vectors are usually written as

$$\mathbf{A}_x = A_x \mathbf{i},$$
$$\mathbf{A}_y = A_y \mathbf{j},$$
$$\mathbf{A}_z = A_z \mathbf{k},$$

where $\mathbf{i}$, $\mathbf{j}$, and $\mathbf{k}$ are unit vectors or base vectors of unit magnitude parallel to the x-, y-, z-directions, respectively, as indicated in the diagram, and A_x, A_y, and A_z are the magnitudes of the components of the vector $\mathbf{A}$ in the x-, y-, z-directions, respectively.

In a cylindrical coordinate system $\mathbf{A}$ may be expressed as

$$\mathbf{A} = A_r \mathbf{i}_r + A_\theta \mathbf{i}_\theta + A_z \mathbf{i}_z,$$

where $\mathbf{i}_r$, $\mathbf{i}_\theta$, and $\mathbf{i}_z$ are base vectors in the radial, tangential, and axial directions. Note that the directions of the base vectors $\mathbf{i}_r$, $\mathbf{i}_\theta$ change with position.

A-3.3   Vector Algebra   We list here some of the more common and useful vector operations. For convenience some of these operations are expanded in Cartesian form. $\mathbf{A}$, $\mathbf{B}$, and so on, are vectors; A and B are scalar quantities. We have first the scalar product of two vectors

$$\mathbf{A} \cdot \mathbf{B} = AB \cos \theta, \tag{A-3.1}$$

where θ is the angle between the vectors $\mathbf{A}$ and $\mathbf{B}$. The scalar or "dot" product of two vectors is a scalar quantity having only magnitude but no direction. The vector product of two vectors is

$$\mathbf{A} \times \mathbf{B} = \mathbf{i}_n AB \sin \theta \tag{A-3.2}$$

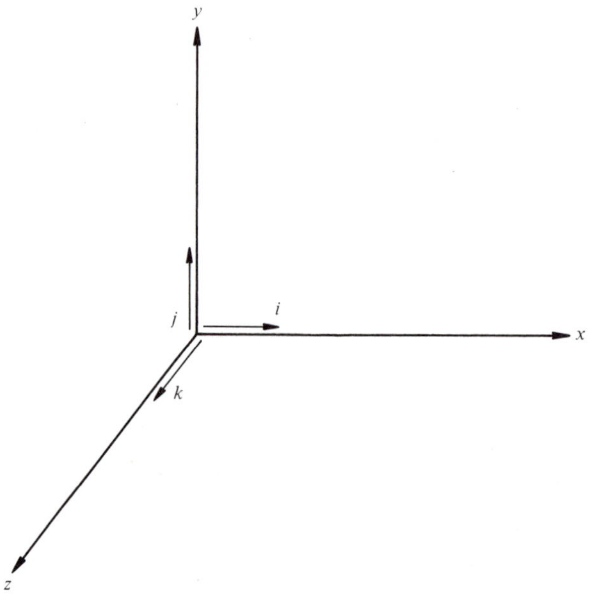

FIG. 5-A Unit vectors for the Cartesian coordinate (x, y, z) system.

where $\mathbf{i}_n$ is a unit vector perpendicular to the plane of $\mathbf{A}$ and $\mathbf{B}$ and in a direction given by the right-hand rule in which θ is the angle measured from $\mathbf{A}$ to $\mathbf{B}$. From these definitions it follows that

$$\mathbf{i} \cdot \mathbf{i} = \mathbf{j} \cdot \mathbf{j} = \mathbf{k} \cdot \mathbf{k} = 1,$$
$$\mathbf{i} \cdot \mathbf{k} = \mathbf{i} \cdot \mathbf{j} = \mathbf{j} \cdot \mathbf{k} = 0,$$
$$\mathbf{i} \times \mathbf{j} = \mathbf{k} = -\mathbf{j} \times \mathbf{i}, \qquad \text{(A-3.3)}$$
$$\mathbf{j} \times \mathbf{k} = \mathbf{i} = -\mathbf{k} \times \mathbf{j},$$
$$\mathbf{k} \times \mathbf{i} = \mathbf{j} = -\mathbf{i} \times \mathbf{k}.$$

The following results are then easily established:

$$\mathbf{A} \cdot \mathbf{B} = A_x B_x + A_y B_y + A_z B_z,$$
$$\mathbf{A} \times \mathbf{B} = \mathbf{i}(A_y B_z - A_z B_y) + \mathbf{j}(B_x A_z - B_z A_x) + \mathbf{k}(A_x B_y - A_y B_x), \quad \text{(A-3.4)}$$

and also

$$\mathbf{A} \cdot (\mathbf{B} \times \mathbf{C}) = \mathbf{B} \cdot (\mathbf{C} \times \mathbf{A}) = \mathbf{C} \cdot (\mathbf{A} \times \mathbf{B}), \qquad \text{(A-3.5)}$$
$$\mathbf{A} \times (\mathbf{B} \times \mathbf{C}) = -(\mathbf{B} \times \mathbf{C}) \times \mathbf{A} = \mathbf{B}(\mathbf{A} \cdot \mathbf{C}) - \mathbf{C}(\mathbf{A} \cdot \mathbf{B}). \qquad \text{(A-3.6)}$$

A-3.4 Vector Calculus The notation grad φ, where φ is a scalar function of position indicates that the *gradient* operation is to be performed on φ. In Cartesian coordinates the gradient may be expressed

$$\text{grad} \equiv \mathbf{i}\frac{\partial}{\partial x} + \mathbf{j}\frac{\partial}{\partial y} + \mathbf{k}\frac{\partial}{\partial z}. \qquad \text{(A-3.7)}$$

The symbols

$$\nabla\varphi \equiv \text{grad } \varphi \qquad \text{(A-3.8)}$$

are used interchangeably, where the "del" symbol stands for the operator

$$\nabla \equiv \mathbf{i}\,\frac{\partial}{\partial x} + \mathbf{j}\,\frac{\partial}{\partial y} + \mathbf{k}\,\frac{\partial}{\partial z}.$$

The *divergence* of a vector is given by

$$\text{div } \mathbf{A} \equiv \nabla \cdot \mathbf{A} = \frac{\partial}{\partial x}\,A_x + \frac{\partial}{\partial y}\,A_y + \frac{\partial}{\partial z}\,A_z. \tag{A-3.9}$$

The curl of a vector may be expressed in component form

$$\text{curl } \mathbf{A} \equiv \nabla \times \mathbf{A} = \mathbf{i}\!\left(\frac{\partial A_z}{\partial y} - \frac{\partial A_y}{\partial z}\right) + \mathbf{j}\!\left(\frac{\partial A_x}{\partial z} - \frac{\partial A_z}{\partial x}\right) + \mathbf{k}\!\left(\frac{\partial A_y}{\partial x} - \frac{\partial A_x}{\partial y}\right), \tag{A-3.10}$$

which may be remembered by the determinant form

$$\text{curl } \mathbf{A} = \begin{vmatrix} \mathbf{i} & \mathbf{j} & \mathbf{k} \\ \dfrac{\partial}{\partial x} & \dfrac{\partial}{\partial y} & \dfrac{\partial}{\partial z} \\ A_x & A_y & A_z \end{vmatrix}. \tag{A-3.11}$$

Several useful vector identities involving these operations are summarized below: **A**, **B**, and so on, are vectors: φ and ψ are scalars.

$$\text{div } (\mathbf{A} + \mathbf{B}) \equiv \nabla \cdot (\mathbf{A} + \mathbf{B}) = \nabla \cdot \mathbf{A} + \nabla \cdot \mathbf{B}, \tag{A-3.12a}$$

$$\text{grad } (\varphi\psi) \equiv \nabla(\varphi\psi) = \varphi\nabla\psi + \psi\nabla\varphi, \tag{b}$$

$$\text{div } (\varphi\mathbf{A}) \equiv \nabla \cdot (\varphi\mathbf{A}) = \varphi(\nabla \cdot \mathbf{A}) + \mathbf{A} \cdot (\nabla\varphi), \tag{c}$$

$$\text{curl } (\varphi\mathbf{A}) \equiv \nabla \times (\varphi\mathbf{A}) = \varphi(\nabla \times \mathbf{A}) + (\nabla\varphi) \times \mathbf{A}, \tag{d}$$

$$\text{grad } (\mathbf{A} \cdot \mathbf{B}) \equiv \nabla(\mathbf{A} \cdot \mathbf{B}) = (\mathbf{A} \cdot \nabla)\mathbf{B} + (\mathbf{B} \cdot \nabla)\mathbf{A} + \mathbf{A} \times (\nabla \times \mathbf{B}) + \mathbf{B} \times (\nabla \times \mathbf{A}). \tag{e}$$

Note that

$$(\mathbf{A} \cdot \nabla)\mathbf{B} = A_x\,\frac{\partial \mathbf{B}}{\partial x} + A_y\,\frac{\partial \mathbf{B}}{\partial y} + A_z\,\frac{\partial \mathbf{B}}{\partial z}$$

and

$$\text{div grad} = \nabla \cdot \nabla \equiv \nabla^2 \equiv \frac{\partial}{\partial x^2} + \frac{\partial}{\partial y^2} + \frac{\partial}{\partial z^2},$$

so that

$$\nabla \cdot \nabla\varphi = \nabla^2\varphi,$$

$$\nabla^2\mathbf{A} = \frac{\partial^2 \mathbf{A}}{\partial x^2} + \frac{\partial^2 \mathbf{A}}{\partial y^2} + \frac{\partial^2 \mathbf{A}}{\partial z^2},$$

$$\nabla^2\mathbf{A} = \nabla(\nabla \cdot \mathbf{A}) - \nabla \times (\nabla \times \mathbf{A}),$$

$$\text{curl grad } \varphi \equiv \nabla \times \nabla\varphi = 0,$$

$$\text{div curl } \mathbf{A} \equiv \nabla \cdot (\nabla \times \mathbf{A}) = 0.$$

For general reference the gradient, divergence, and curl operations are also given in cylindrical coordinates r, θ, and z:

$$\text{grad } \varphi = \nabla\varphi = \mathbf{i}_r \frac{\partial\varphi}{\partial r} + \mathbf{i}_\theta \frac{\partial\varphi}{r\,\partial\theta} + \mathbf{i}_z \frac{\partial\varphi}{\partial z}, \tag{A-3.13}$$

$$\text{div } \mathbf{A} = \nabla\cdot\mathbf{A} = \frac{1}{r}\frac{\partial}{\partial r}(rA_r) \frac{1}{r}\frac{\partial A_\theta}{\partial\theta} + \frac{\partial A_z}{\partial z}. \tag{A-3.14}$$

A_r, A_θ, and A_z are the radial, tangential, and axial components of $\mathbf{A}$.

$$\text{curl } \mathbf{A} = \nabla\times\mathbf{A} = \mathbf{i}_r\left(\frac{1}{r}\frac{\partial A_z}{\partial\theta} - \frac{\partial A_\theta}{\partial z}\right) + \mathbf{i}_\theta\left(\frac{\partial A_r}{\partial z} - \frac{\partial A_z}{\partial r}\right) + \mathbf{i}_z\left(\frac{1}{r}\frac{\partial rA_\theta}{\partial r} - \frac{\partial A_r}{r\,\partial\theta}\right). \tag{A-3.15}$$

A-3.5 Surface, Volume, and Line Integrals In the following integrals $\mathbf{n}$ represents an *outward* pointing unit vector from the surface considered. $\mathbf{A}$, $\mathbf{B}$, and so on, are vectors and φ is a scalar function. $\mathscr{V}$ is the volume considered and R is the surface area of the volume:

$$\int_\mathscr{V} \nabla\varphi\,d\mathscr{V} = \int_S \varphi\mathbf{n}\,dS, \tag{A-3.16}$$

$$\int_\mathscr{V} \nabla\cdot\mathbf{A}\,d\mathscr{V} = \int_S \mathbf{A}\cdot\mathbf{n}\,dS, \tag{A-3.17}$$

$$\int_\mathscr{V} \nabla\times\mathbf{A}\,d\mathscr{V} = \int_S \mathbf{n}\times\mathbf{A}\,dS, \tag{A-3.18}$$

$$\int_S \mathbf{n}\cdot(\nabla\times\mathbf{A})\,dS = \int_C \mathbf{A}\cdot d\mathbf{r}. \tag{A-3.19}$$

Equation (A-3.17) is usually called Gauss' integral theorem and Eq. (A-3.19) is known as Stokes' theorem. The quantity $d\mathbf{r}$ in the latter equation is the vector line element of the contour C of integration.

REFERENCES

1. H. B. Phillips, *Vector Analysis*, 3rd ed., Wiley, New York, 1934.
2. C. E. Weatherburn, *Advanced Vector Analysis*, Bell, London, 1928.
3. J. H. Wayland, *Differential Equations Applied in Science and Engineering*, Van Nostrand, New York, 1957.

Summary of Thermodynamic Relations

A-4.1 This short summary is intended to give the reader more information on the thermodynamic results used in dealing with compressible flows. The discussion is carried only far enough to meet this aim, and further information, if required, can be obtained from the references listed at the end of this appendix.

The interplay of thermodynamic effects is governed by four laws that are thought to be universally valid. As in the study of mechanics, where Newton's laws predict the motion of bodies, thermodynamic laws are used to predict the thermal behavior of matter. In should be emphasized that the laws of thermodynamics, like Newton's laws, are empirical laws and cannot be derived or proven in any mathematical sense. The four laws govern the following situations: The zeroth law relates to the concept of thermal equilibrium between bodies and establishes temperature as a common characteristic of these bodies; the first law relates heat and work as energy transfers during a change in a body and establishes internal energy as a property of matter; the second law places a limit on the direction in which natural changes of a body can proceed and establishes entropy as a further property of matter; and the third law establishes the absolute quantitative values of properties of a substance. Of these laws, the first and second are most important for engineering problems, since they deal with energy conversion and the direction in which it occurs naturally.

A-4.2 Terminology The laws of thermodynamics apply to a fixed mass of substance, which is termed a thermodynamic *system*. The *surroundings* are everything outside the system. When a system and its surroundings have been in contact for a long time and no spontaneous change in the system alone is observed, the system and surroundings are said to be in thermal *equilibrium*. Substances in thermal equilibrium can be described by their *properties*, for example, temperature or pressure. The collection of system properties describes the *state* of the system. For most cases of interest in engineering, not all system properties are relevant to energy conversion, for example, color or odor. It has been found by experience that the state of a substance is sufficiently well specified by any *two* independent properties, all other properties being thus determined. The relationship between properties for a specific substance is called the *equation of state* of that substance. Some properties, such as temperature or pressure, do not depend on the amount of substance in the system and are called *intensive* properties. *Extensive* properties are proportional to the mass of the system, for example, internal energy, volume, and entropy. Any extensive property can be made intensive by dividing by the mass. The intensive properties formed in this way are called specific properties and are noted by the lowercase form of the symbol referring to the extensive property. For example, volume (extensive) $= V$, specific volume (intensive) $= v$.

If a substance undergoes a transformation from one state to another, the successive incremental changes are called the transformation process. A useful idealization is that of a *reversible* process, where the process can return a substance to its initial state with no effect on the surroundings. In reality the effects of friction and heat flow cause all processes to be *irreversible*. For example, when work is done against friction by moving a block over a rough surface, work is not regained if the direction of motion is reversed. In the same way heat flow as a result of temperature difference between the system and surroundings cannot be reversed, since this would require a jump in surrounding temperatures. Thus a reversible process is viewed as a succession of equilibrium states, the system being in thermal equilibrium with the surroundings at all times. Heat flow must occur without a temperature difference and there are no frictional forces. Although these may seem to be very restrictive conditions, calculations made using assumptions of reversibility are useful indications of trends and relationships.

When a process, whether reversible or irreversible, causes a system to finally return to its initial state, it is called a *cyclic* process. Although there may be no change in the system itself, an irreversible cyclic process must result in a change of the surroundings. It should be noted that the change in a property during a cyclic process is necessarily zero, since the system returns to its initial state. For example, if we view the pressure p as changing in increments dp along the process path, the total change in p is $p_2 - p_1 = \int_1^2 dp$; that is, p must be a *perfect differential*. For a cyclic process $\oint dp = 0$, where the symbol $\oint$ is taken to mean an integration around the cycle, returning to the initial state. Thus we conclude that a differential increment of any property is a perfect differential, or $\oint d(\text{property}) = 0$. Conversely, if the cyclic integral of a quantity is zero, that quantity must be a property of the system.

A-4.3 Work and Heat Work and heat represent different forms of energy exchange between the system and its surroundings. With regard to work, it is useful to have an expression for the work done on a system during an incremental reversible process. We imagine a system to be contained in a closed cylinder with a piston at one end as shown in Fig. 6-A. The piston is slowly withdrawn a distance

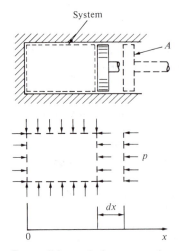

FIG. 6-A Reversible work done on a closed system.

dx. Since the process is assumed reversible, there are no frictional forces present and the normal stress on the system boundary is simply the negative of the pressure. In accordance with mechanics, we take work done on a system as positive and equal to the dot product of the force and displacement vectors. Thus, in this case, we can write

$$dW = \mathbf{F} \cdot \mathbf{dx} = -pA\ dx.$$

However, the increment in the volume of the system dV can be written as $dV = A\ dx$, so that for a reversible, incremental process

$$dW = -p\ dV.$$

To avoid confusion of the volume symbol V for velocity, we note that $V = m/\rho$, so that

$$dw = -pd\left(\frac{1}{\rho}\right) \tag{A-4.1}$$

where w is the work per unit mass.

The total work done during a process between two equilibrium states is

$$w_{1\to 2} = -\int_1^2 pd\left(\frac{1}{\rho}\right)$$

and this work must depend on the particular *path* of the process; that is, on the specific relationship of p and ρ during the process. Thus work is not a property of the system, and dw is not a perfect differential. The net work done on a system during a cyclic process becomes

$$w_{\text{net}} = \oint dW = -\oint pd\left(\frac{1}{\rho}\right).$$

Heat is also an exchange of energy between the system and surroundings, heat addition to a system being taken as positive. For a reversible incremental process the increment in heat added to a system dQ can also be represented in terms of the system properties. However, at this point we simply note that when two substances at different temperatures are brought into contact and allowed to come to equilibrium, energy is transferred in the form of heat. As in the case of work, heat is not a property of the system, that is, it is not possessed by a system and the increment dQ is not a perfect differential.

Historically, heat was thought to be a physical entity contained within a body. This gave rise to the concept of *specific* heat as an increment in heat per unit mass per unit temperature rise

$$C = \frac{dQ}{m\ dT}.$$

Since heat can be added to a system in many ways (e.g., with a gas, holding either pressure or volume constant), several different specific heats can be formed. Rather than continue the discussion at this point, a more precise definition of specific heats will be given later.

A-4.4 First Law As stated earlier, the first law relates heat and work as energy transfer during a process. It is also simply stated as the principle of conservation of energy. The law was first suggested by Joule after a series of experiments were con-

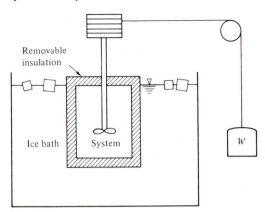

FIG. 7-A Apparatus for verifying the first law of thermodynamics.

ducted to determine how heat and work were related. Joule's apparatus was equivalent to that shown in Fig. 7-A. A fixed amount of water is placed in a vessel fitted with a paddle wheel. The vessel itself is immersed in an ice-water bath, which stays at a constant temperature by the melting of the ice. The system is taken as the water in the vessel and is initially in equilibrium with the water-ice bath. The vessel is then insulated so that no heat transfer can take place, and an amount of work is done on the system through the paddle-wheel arrangement. The vessel is then put back into the bath and the insulation removed so that heat transfer can take place and bring the system back to equilibrium. The process described is evidently a *cyclic* process, and Joule determined that the amount of heat removed from the system was directly proportional to the work done by the paddle.

Stated mathematically,

$$\oint dW = -\oint dQ.$$

If we rewrite this expression as

$$\oint (dQ + dW) = 0, \tag{A-4.2}$$

the first law states that the total heat and work added to a system during a cyclic process must add up to zero. Care should always be exercised in relating units of heat and work in the British or engineering system of units (1 Btu = 778 ft lbf).

Although dW and dQ are not perfect differentials, we see that their sum must represent the perfect differential of a property of the system. If we write

$$dQ + dW = dE,$$

where E is a property, we see from physical grounds that this property must be the energy of the system. In general, a system can have several forms of energy; for example, kinetic, potential, electrical, and so on, and the symbol E stands for the sum of all these. Beside the forms of energy just described, we see, for example, in Joules experiment, where there need be no change in kinetic or potential energy that another form of energy, the *internal energy U*, must also be included. If the system is completely isolated, $dQ + dW = dE = 0$ or, in other words the energy of an isolated system remains constant.

For an incremental process where all other forms of energy except internal energy remain constant,

$$dQ + dW = dU. \tag{A-4.3}$$

The change in internal energy between two equilibrium states is independent of the process, since U is a property of the system

$$U_2 - U_1 = \int_1^2 dU = \int_1^2 (dQ + dW).$$

For the special case of a reversible incremental process

$$dU = dQ - p d\left(\frac{m}{\rho}\right). \tag{A-4.4}$$

When dealing with flowing substances, it is convenient to define another property called enthalpy H as

$$H = U + p\left(\frac{m}{\rho}\right),$$

or, in terms of specific quantities

$$h = u + \frac{p}{\rho}. \tag{A-4.5}$$

Thus the first law in the form of Eq. (A-4.4) can also be written as

$$dh = dq + \frac{dp}{\rho}.$$

A-4.5 Second Law Returning to Joule's experiment, we note that if we calculate the ratio of increment in heat dQ to the temperature of the surroundings T_{sur}, the cyclic integral of this ratio is a negative quantity, since heat flows out of the system

$$\oint \frac{dQ}{T_{sur}} < 0. \tag{A-4.6}$$

Although this appears as a result of the present experiment, many other experiments with different apparatus and under different conditions yield exactly the same result, so that Eq. (A-4.6) is apparently correct for all situations.

In order to relate Eq. (A-4.6) to more familiar ideas, we examine two statements that have been successfully tested in many ways.

Statement 1: "Perpetual motion is impossible," that is, it is impossible in a continuously operating cyclic process to convert energy from a single source into work without a resulting change in the surroundings.

Statement 2: "Heat flows from hot to cold," that is, it is impossible to continuously operate a device in a cyclic process and transfer heat from a cold to a hotter body without a resulting change in the surroundings.

Examine the first statement and suppose that we have a system undergoing a cyclic process, receiving energy from a single source, and producing work as indicated in Fig. 8-Aa. The ratio Q/T_{sur} is evidently positive in contravention with Eq. (A-4.6). Thus, Eq. (A-4.6) does not permit such operation, in agreement with statement 1. We conclude that Eq. (A-4.6) and statement 1 are equivalent.

Upon examining statement 2, imagine a cyclic process of a system that receives an amount of heat Q_H at a high temperature T_H and rejects an amount of heat Q_L to

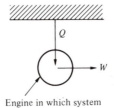

Q

W

Engine in which system
undergoes cyclic process

(a)

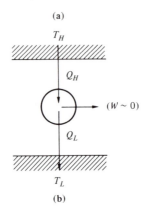

T_H

Q_H

$(W \sim 0)$

Q_L

T_L

(b)

FIG. 8-A   (a) A perpetual motion machine; (b) heat flow from hot to cold.

a lower temperature T_L, as shown in Fig. 8-A(b). If Q_H and Q_L have different magnitudes, the cycle produces positive work (negative work input), as can be easily verified by applying the first law to the process. Suppose, however, that the cycle is so inefficient that very little work is produced, and the magnitudes of Q_H and Q_L are nearly the same. In the limit as the work tends to zero, the ratio of dQ/T_{sur} over the cycle becomes

$$\oint \frac{dQ}{T_{sur}} = \frac{Q_H}{T_H} - \frac{Q_L}{T_L} \to Q\left(\frac{1}{T_H} - \frac{1}{T_L}\right) < 0.$$

The net result of this cycle is simply a flow of heat from one body to the other, without any other change in the surroundings. Thus Eq. (A-4.6) states that $T_H > T_L$, or the heat flow can only be from the hotter to the cooler body. Again we conclude that Eq. (A-4.6) and statement 2 are equivalent.

We may summarize as follows: Either statement 1 or 2 can be taken as statements of the second law, which is expressed mathematically as Eq. (A-4.6). Note the parallel with the first law, which is stated as a principle of conservation of energy, but is expressed mathematically by Eq. (A-4.2).

In the limit as all processes approach being reversible processes, the cyclic integral Eq. (A-4.6) very nearly equals zero. We suppose that in the limit of reversible processes

$$\oint \left(\frac{dQ}{T}\right)_{rev} = 0, \tag{A-4.7}$$

where the temperature of the surroundings T_{sur} and the system temperature T are now equal. The integrand in Eq. (A-4.7) must be the perfect differential of a property of the system. This property is called *entropy* and the increment in entropy of a

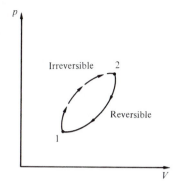

FIG. 9-A Reversible and irreversible process paths.

system dS is thus defined to be

$$dS = \left(\frac{dQ}{T}\right)_{\text{rev}}.$$ (A-4.8)

We would like to calculate the entropy change during an irreversible process. In order to do this, consider a cyclic process having two legs, one reversible and one irreversible as indicated in Fig. 9-A. The entropy change between points 1 and 2 can be computed along the reversible path as

$$S_2 - S_1 = \int_1^2 \left(\frac{dQ}{T}\right)_{\text{rev}}.$$

On the other hand, the cyclic integral for a cycle beginning at 1, proceeding along the irreversible path to 2, then back to 1 along the reversible path, can be written as

$$\oint \frac{dQ}{T_{\text{sur}}} = \int_1^2 \frac{dQ}{T_{\text{sur}}} + \int_2^1 \left(\frac{dQ}{T}\right)_{\text{rev}} < 0.$$

The second integral can be written in terms of the entropy difference, so that we obtain

$$\int_1^2 \frac{dQ}{T_{\text{sur}}} + S_1 - S_2 < 0$$

or

$$S_2 - S_1 > \int_1^2 \frac{dQ}{T_{\text{sur}}}.$$

Thus, for an incremental process,

$$dS > \frac{dQ}{T_{\text{sur}}}.$$

In particular, if the system is isolated the heat flow $dQ = 0$, so that

$$dS > 0.$$

Thus the entropy of an isolated system must always increase, or in the limit of reversible processes, the entropy remains constant.

A-4.6 Properties of a Substance The laws of thermodynamics are independent of the substance under consideration. However, every substance in an equilibrium state will have its properties related to each other in some manner, called the equation of state for that substance. The equation of state thus poses a further restriction on the variation of properties during incremental processes.

It has been determined experimentally that properties of interest in most energy conversion problems are specified by fixing any other two properties. For example, most gases such as air, helium, carbon dioxide, and so on, behave as so-called *perfect gases* and satisfy the relationship

$$p = \rho R T$$

over a wide range of pressure and temperatures. The constant R has a different value for each gas, and values of R for a number of common gases are listed in Appendix 2 (for air, $R = 287$ J/kgK $= 53.3$ ft.-lbf/lbm°R).

It has also been established from experiments that perfect gases have very special thermal behavior; that is, their internal energy and enthalpy are dependent only on temperature. This behavior was brought out in a second experiment by Joule, who used an apparatus as shown in Fig. 10-A. One side of the apparatus was filled with the gas in question, the other evacuated. The gas was allowed to come to equilibrium with the surrounding water. The valve was then opened and left until the whole system again came to equilibrium with the bath. From his measurements Joule deduced that no heat transfer to the gas occurred during the process, and of course no work was done by the gas on the surroundings. Thus, from the first law, Eq. (A-4.4), $dU = 0$, or the internal energy U remains constant. Assuming that the internal energy U is determined by two other properties, say T and volume V, we may write

$$dU(T, V) = \left(\frac{\partial U}{\partial T}\right)_V dT + \left(\frac{\partial U}{\partial V}\right)_T dV = 0.$$

However, since $dT = 0$ in this experiment, we conclude that $(\partial U/\partial V)_T = 0$, and so the internal energy of a perfect gas is a function only of temperature. Further, from Eq. (A-4.4), we have in general for a constant volume process $dQ = dU$. Thus, for a perfect gas,

$$\frac{dQ}{dT} = \left(\frac{\partial U}{\partial T}\right)_V.$$

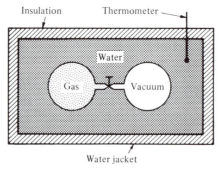

FIG. 10-A Experimental apparatus to measure the effect of pressure on the internal energy of gases.

In accordance with the historical ideas of specific heat, we define $C_V = (\partial U/\partial T)_V$ to be the specific heat at constant volume. Finally, we obtain for a perfect gas

$$du = c_v \, dT$$

or

$$u = \int_r c_v \, dT + u_r,$$

where r denotes some reference state. For many gases, c_v itself does not depend strongly on temperature, so that

$$u = c_v(T - T_r) + u_r.$$

In a similar way we note that the enthalpy h of a perfect gas is a function only of temperature, since both u and p/ρ are functions of temperature. Thus we may write

$$h = u + \frac{p}{\rho} = (c_v + R)T + u_r - c_v T_r$$

or

$$h = c_p(T - T_r) + h_r,$$

where $c_p = c_v + R$ is another constant. Note that from the first law expressed in terms of enthalpy, the enthalpy change at constant pressure equals the heat added, so that c_p is the usual specific heat at constant pressure.

A-4.7 Some Special Results We first wish to establish relationships useful in determining the speed of sound of a gas. From the rules of differential calculus, if z is some function of x and y, $z = z(x, y)$, we can write

$$dz = \left(\frac{\partial z}{\partial x}\right)_y dx + \left(\frac{dz}{\partial y}\right)_x dy$$

or

$$\left(\frac{\partial x}{\partial y}\right)_z \left(\frac{\partial z}{\partial x}\right)_y \left(\frac{\partial y}{\partial z}\right)_x = -1.$$

Using this rule in succession, we have

$$\left(\frac{\partial p}{\partial \rho}\right)_s = -\frac{(\partial s/\partial \rho)_p}{(\partial s/\partial p)_\rho} = -\frac{(\partial s/\partial T)_p (\partial T/\partial \rho)_p}{(\partial s/\partial T)_\rho (\partial T/\partial p)_\rho} = \frac{(\partial s/\partial T)_p}{(\partial s/\partial T)_\rho} \left(\frac{\partial p}{\partial \rho}\right)_T. \quad \text{(A-4.10a)}$$

However, if we assume for a general substance $U = U(T, 1/\rho)$, we can write Eq. (A-4.4) combined with Eq. (A-4.8) as

$$T \, ds = c_v \, dT + \left[\left(\frac{\partial u}{\partial(1/\rho)}\right)_T + p\right] d(1/\rho). \quad \text{(A-4.10b)}$$

It may be noted that while the above equation was derived by restricting the incremental process to be reversible, the result involves only properties of the system, and so must be independent of the process. In particular, the changes in properties during an irreversible process also follow Eq. (A-4.10b).

As a general result,

$$T\left(\frac{\partial s}{\partial T}\right)_\rho = c_v. \quad \text{(A-4.10c)}$$

In the same way, from the form of the first law using enthalpy,

$$T\left(\frac{\partial s}{\partial T}\right)_p = c_p.$$
(A-4.10d)

Finally, combining (A-4.10a), b, c, and d we obtain

$$\left(\frac{\partial p}{\partial \rho}\right)_s = \frac{c_p}{c_v}\left(\frac{\partial p}{\partial \rho}\right)_T = \gamma\left(\frac{\partial p}{\partial \rho}\right)_T$$
(A-4.11)

where

$$\gamma = \frac{c_p}{c_v}.$$

Next we wish to calculate some properties of a p..fect gas. For such a gas the internal energy depends only on temperature $u = u(T)$, so that from Eq. (A-4.10b)

$$ds = c_v \frac{dT}{T} + \frac{p}{T} d\left(\frac{1}{\rho}\right)$$

or

$$ds = c_v \frac{dT}{T} + R \frac{d(1/\rho)}{(1/\rho)}.$$

Integrating, we obtain

$$s_2 - s_1 = c_v \ln \frac{T_2}{T_1} + R \ln \frac{\rho_1}{\rho_2}.$$
(A-4.12)

Note that Eq. (A-4.12) represents the true entropy change between two equilibrium states of a perfect gas, no matter what process is used in going between the two states.

For adiabatic reversible processes, $s_2 - s_1 = 0$, so that

$$\frac{T_2}{T_1} = \left(\frac{\rho_1}{\rho_2}\right)^{-R/c_v}$$

but

$$-\frac{R}{c_v} = -\frac{c_p - c_v}{c_v} = -(\gamma - 1),$$

so that

$$\frac{T_2}{T_1} = \left(\frac{\rho_1}{\rho_2}\right)^{-(\gamma-1)}.$$
(A-4.13a)

Using the perfect gas law, we can also write Eq. (A-4.13a) as

$$\frac{T_2}{T_1} = \left(\frac{p_2}{p_1}\right)^{(\gamma-1)/\gamma}.$$
(A-4.13b)

REFERENCES

1. R. H. Sabersky, *Engineering Thermodynamics*, McGraw-Hill, New York, 1957.
2. W. C. Reynolds, *Thermodynamics*, McGraw-Hill, New York, 1965.
3. J. F. Lee, and F. W. Sears, *Thermodynamics*, 2nd ed., Addison-Wesley, Reading, Mass., 1962.

5

Gas Dynamic Tables

TABLE 6-A Property variations across a normal shock wave in air ($\gamma = 1.4$)

M_1	M_2	p_2/p_1	T_2/T_1	ρ_2/ρ_1	p_{02}/p_{01}
1.00	1.000	1.000	1.000	1.000	1.000
1.05	0.953	1.120	1.033	1.084	1.000
1.10	0.912	1.245	1.065	1.169	0.999
1.15	0.875	1.376	1.097	1.256	0.997
1.20	0.842	1.513	1.128	1.342	0.993
1.25	0.813	1.656	1.159	1.429	0.987
1.30	0.786	1.805	1.191	1.516	0.979
1.35	0.762	1.960	1.223	1.603	0.970
1.40	0.740	2.120	1.255	1.690	0.958
1.45	0.720	2.286	1.287	1.776	0.945
1.50	0.701	2.458	1.320	1.862	0.930
1.55	0.684	2.636	1.354	1.947	0.913
1.60	0.668	2.820	1.388	2.032	0.895
1.65	0.654	3.010	1.423	2.115	0.876
1.70	0.641	3.205	1.458	2.198	0.856
1.75	0.628	3.406	1.495	2.279	0.835
1.80	0.617	3.613	1.532	2.359	0.813
1.85	0.606	3.826	1.569	2.438	0.790
1.90	0.596	4.045	1.608	2.516	0.767
1.95	0.586	4.270	1.647	2.592	0.744
2.00	0.577	4.500	1.687	2.667	0.721
2.05	0.569	4.736	1.729	2.740	0.698
2.10	0.561	4.978	1.770	2.812	0.674
2.15	0.554	5.226	1.813	2.882	0.651
2.20	0.547	5.480	1.857	2.951	0.628
2.25	0.541	5.740	1.901	3.019	0.606
2.30	0.534	6.005	1.947	3.085	0.583
2.35	0.529	6.276	1.993	3.149	0.561
2.40	0.523	6.553	2.040	3.212	0.540
2.45	0.518	6.836	2.088	3.273	0.519

TABLE 6-A   (*cont.*)

M_1	M_2	p_2/p_1	T_2/T_1	ρ_2/ρ_1	p_{02}/p_{01}
2.50	0.513	7.125	2.137	3.333	0.499
2.55	0.508	7.420	2.187	3.392	0.479
2.60	0.504	7.720	2.238	3.449	0.460
2.65	0.500	8.026	2.290	3.505	0.442
2.70	0.496	8.338	2.343	3.559	0.424
2.75	0.492	8.656	2.397	3.612	0.406
2.80	0.488	8.980	2.451	3.664	0.389
2.85	0.485	9.310	2.507	3.714	0.373
2.90	0.481	9.645	2.563	3.763	0.358
2.95	0.478	9.986	2.621	3.811	0.343
3.00	0.475	10.333	2.679	3.857	0.328
3.50	0.451	14.125	3.315	4.261	0.213
4.00	0.435	18.500	4.047	4.571	0.139
4.50	0.424	23.458	4.875	4.812	0.092
5.00	0.415	29.000	5.800	5.000	0.062
5.50	0.409	35.125	6.822	5.149	0.042
6.00	0.404	41.833	7.941	5.268	0.030
6.50	0.400	49.125	9.156	5.365	0.021
7.00	0.397	57.000	10.469	5.444	0.015
7.50	0.395	65.458	11.879	5.510	0.011
8.00	0.393	74.500	13.387	5.565	0.008
8.50	0.391	84.125	14.991	5.612	0.006
9.00	0.390	94.333	16.693	5.651	0.005
9.50	0.389	105.125	18.492	5.685	0.004
10.00	0.388	116.500	20.387	5.714	0.003
∞	0.378	∞	∞	6.000	0.000

TABLE 7-A   Property variations for isentropic flow of air ($\gamma = 1.4$)

Subsonic

M_1	A/A^*	p/p_0	ρ/ρ_0	T/T_0	V/V^*
0.00	∞	1.000	1.000	1.000	0.000
0.05	11.591	0.998	0.999	1.000	0.055
0.10	5.822	0.993	0.995	0.998	0.109
0.15	3.910	0.984	0.989	0.996	0.164
0.20	2.964	0.972	0.980	0.992	0.218
0.25	2.403	0.957	0.969	0.988	0.272
0.30	2.035	0.939	0.956	0.982	0.326
0.35	1.778	0.919	0.941	0.976	0.379
0.40	1.590	0.896	0.924	0.969	0.431
0.45	1.449	0.870	0.906	0.961	0.483
0 50	1.340	0.843	0.885	0.952	0.535
0.55	1.255	0.814	0.863	0.943	0.585
0.60	1.188	0.784	0.840	0.933	0.635
0.65	1.136	0.753	0.816	0.922	0.684
0.70	1.094	0.721	0.792	0.911	0.732
0.75	1.062	0.689	0.766	0.899	0.779
0.80	1.038	0.656	0.740	0.887	0.825
0·85	1.021	0.624	0.714	0.874	0.870
0.90	1.009	0.591	0.687	0.861	0.915
0.95	1.002	0.559	0.660	0.847	0.958

TABLE 7-A (cont.)

Supersonic

M_1	$A/A^\star$	p/p_0	ρ/ρ_0	T/T_0	$V/V^\star$	v(deg.)
1.00	1.000	0.528	0.634	0.833	1.000	
1.05	1.002	0.498	0.608	0.819	1.041	0.487
1.10	1.008	0.468	0.582	0.805	1.081	1.336
1.15	1.017	0.440	0.556	0.791	1.120	2.381
1.20	1.030	0.412	0.531	0.776	1.158	3.558
1.25	1.047	0.386	0.507	0.762	1.195	4.830
1.30	1.066	0.361	0.483	0.747	1.231	6.170
1.35	1.089	0.337	0.460	0.733	1.266	7.561
1.40	1.115	0.314	0.437	0.718	1.300	8.987
1.45	1.144	0.293	0.416	0.704	1.333	10.438
1.50	1.176	0.272	0.395	0.690	1.365	11.905
1.55	1.212	0.253	0.375	0.675	1.395	13.381
1.60	1.250	0.235	0.356	0.661	1.425	14.860
1.65	1.292	0.218	0.337	0.647	1.454	16.338
1.70	1.338	0.203	0.320	0.634	1.482	17.810
1.75	1.386	0 188	0.303	0.620	1.510	19.273
1.80	1.439	0.174	0.287	0.607	1.536	20.725
1.85	1.495	0.161	0.272	0.594	1.561	22.163
1.90	1.555	0.149	0.257	0.581	1.586	23.586
1.95	1.619	0.138	0.243	0.568	1.610	24.992
2.00	1.687	0.128	0.230	0.556	1.633	26.380
2.05	1.760	0.118	0.218	0.543	1.655	27.748
2.10	1.837	0.109	0.206	0.531	1.677	29.097
2.15	1.919	0.101	0.195	0.520	1.698	30.425
2.20	2.005	0.094	0.184	0.508	1.718	31.732
2.25	2.096	0.086	0.174	0.497	1.737	33.018
2.30	2.193	0.080	0.165	0.486	1.756	34.283
2.35	2.295	0.074	0.156	0.475	1.775	35.525
2.40	2.403	0.068	0.147	0.465	1.792	36.747
2.45	2.517	0.063	0.139	0.454	1.809	37.946
2.50	2.637	0.059	0.132	0.444	1.826	39.124
2.55	2.763	0.054	0.125	0.435	1.842	40.280
2.60	2.896	0.050	0.118	0.425	1.857	41.415
2.65	3.036	0.046	0.112	0.416	1.872	42.528
2.70	3.183	0.043	0.106	0.407	1.887	43.621
2.75	3.338	0.040	0.100	0.398	1.901	44.694
2.80	3.500	0.037	0.095	0.389	1.914	45.746
2.85	3.671	0.034	0.090	0.381	1.927	46.778
2.90	3.850	0.032	0.085	0.373	1.940	47.790
2.95	4.038	0.029	0.080	0.365	1.952	48.783
3.00	4.235	0.027	0.076	0.357	1.964	49.757
3.50	6.790	0.013	0.045	0.290	2.064	58.530
4.00	10.719	0.007	0.028	0.238	2.138	65.785
4.50	16.562	0.003	0.017	0.198	2.194	71.832
5.00	25.000	0.002	0.011	0.167	2.236	76.920
5.50	36.869	0.001	0.008	0.142	2.269	81.245
6.00	53.180	0.001	0.005	0.122	2.295	84.955
6.50	75.134	0.000	0.004	0.106	2.316	88.168
7.00	104.143	0.000	0.003	0.093	2.333	90.973
7.50	141.842	0.000	0.002	0.082	2.347	93,440
8.00	190.110	0.000	0.001	0.072	2.359	95.625
8.50	251.086	0.000	0.001	0.065	2.369	97.572

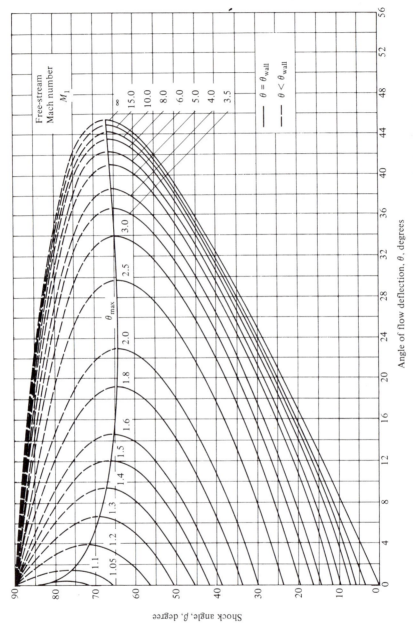

FIG. 11-A The relation between wave angle β and deflection angle θ for the flow of air ($\gamma = 1.4$) through an oblique shock at various free-stream Mach numbers M_1. (From *NACA TN 1373*.)

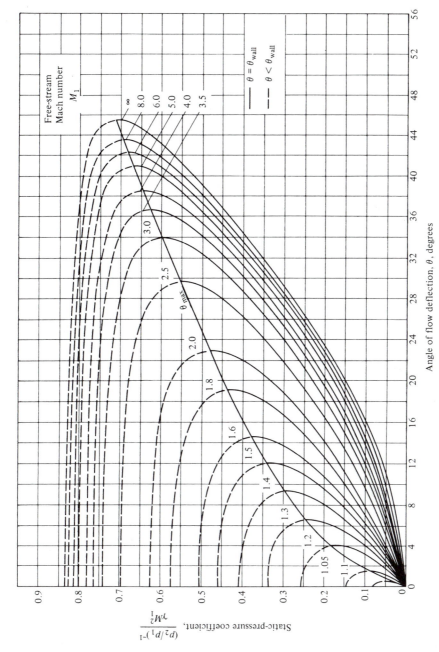

FIG. 12-A A dimensionless pressure rise across an oblique shock in air ($\gamma = 1.4$) for various flow deflection angles θ and free-stream Mach numbers M_1. (From *NACA TN 1373*.)

Answers to Selected Problems

Chapter 1

1.3 At $10°C$, $v = 1.4 \times 10^{-6}$ m^2/s,
at $65°C$. $v = 4.4 \times 10^{-6}$ m^2/s.

1.5 0.722.

1.6 3.78 hp.

1.9 12 mm.

1.10 11.9 mm; 29.8 mm; 24 percent; 60 percent.

1.13 3.04 kPa.

1.15 $y = h \exp\left(-\sqrt{\dfrac{\rho g}{\sigma}}\, x\right)$, $h = \dfrac{\cot \theta}{\sqrt{\rho g/\sigma}}$.

Chapter 2

2.2 $\dfrac{\partial \rho}{\partial t} + \dfrac{\rho}{r^2}\dfrac{\partial}{\partial r}(r^2 v_r)$

$\qquad + \dfrac{\rho}{r^2 \sin \theta}\dfrac{\partial}{\partial \theta}(v_\theta \sin \theta)$

$\qquad + \dfrac{\rho}{r \sin \theta}\dfrac{\partial v_\phi}{\partial \phi} = 0.$

2.4 (a) $\dfrac{V d^2}{d_o^2 - d^2}$. (b) $\dfrac{V d_o^2}{d_o^2 - d^2}$.

2.5 (a) Yes. (b) Yes.

2.6 $v = -2axy$.

2.7 (a) $V_1 = \dfrac{4Q}{\pi A^2}$, $V_2 = \dfrac{4Q}{\pi B^2}$,

$\qquad V_3 = \dfrac{4Q}{\pi C^2}$.

(b) $V_1 = 3.18$ m/s, $V_2 = 12.73$ m/s,
$V_3 = 1.21$ m/s.

(c) $V_1 = 3.18$ m/s, $V_2 = 21.22$ m/s,
$V_3 = 1.01$ m/s.

2.8 (a) $v_r = \dfrac{c}{r}$. (b) $v_\theta = c$.

2.9 $\rho = \rho_1 u_1/u$.

2.10 No, correct value $= \dot{w}\,\dfrac{(a^2 - b^2)}{a^2}$.

2.17 (a) $v = e^{-x} \sinh y$.
(b) $\psi = e^{-x} \sinh y + y$.

2.20 $a_x = 0$; $a_y = 50$ ft/sec^2.

2.23 (a) $\mathbf{a} = 4\mathbf{r}$.
(b) $p = -2\rho r^2 + \text{const.}$

2.24 $p = -\rho g y + \text{const.}$

2.25 $\dfrac{\partial p}{\partial r} = \dfrac{\rho v_\theta^2}{r}$.

2.26 (a) $p = +\rho \Omega^2\, \dfrac{r^2}{2} + \text{const.}$

(b) $p = \dfrac{-\rho c^2}{2r^2} + \text{const.}$

2.31 $p = 100 e^{-y/8443}$.

2.32 139.9 kPa.

2.33 $p_{max} = 120.2$ kPa;
$p_{min} = 96.5$ kPa; slope $= 17.0°$.

2.34 3.77 rev/sec.

2.35 $h = La/g$.

2.36 $p = p_s\left[1 - \dfrac{(\gamma - 1)}{\gamma}\left(\dfrac{\rho_s g y}{p_s}\right)\right]^{(\gamma - 1)/\gamma}$

2.37 $\Delta p = 24.69$ kPa.

2.38 $x/y = 1/26.2$.

2.39 $\rho = \rho_L[1 - \alpha(1 - \rho_A/\rho_L)]$;
$\alpha = h/H$.

2.40 94 kPa.

2.41 176.8 MPa.

2.42 $\rho g \pi d^3/24$.

2.43 $F = \sqrt{W^2 + W\rho g A h} - W$.

2.45 $79.3°$.

2.48 $h_c = \sqrt{3}$.

2.50 $a/b = (\frac{112}{9})^{1/2}$.

2.51 $b = \left[\dfrac{\rho h^2}{3(\rho_d - \rho)}\right]^{1/2}$.

2.53 201.3 kN; 1.89 m.

2.54 (a) $L = \dfrac{\pi D^3}{6}\left(\dfrac{l}{D}\right)$.

(b) $L_{\max} = \pi D \sigma_{\max}$.

2.56 Center of buoyancy is 7.72 ft below top of barge; metacenter is 0.35 ft below top.

2.57 Stable for $B/h > \sqrt{2}$.

2.58 $u = \dfrac{\rho g \sin \theta}{2\mu}(2hy - y^2)$;

$\tau_w = \rho g h \sin \theta$;

$u_{avg} = \dfrac{\rho g h^2 \sin \theta}{3\mu}$.

2.59 (b) $\sin \theta_{cr} = \tau_y/\rho g H$.

(d) $Q = \dfrac{V}{3}(H + 2h)$.

2.60 (a) $\mu r \dfrac{\partial}{\partial r}\left(\dfrac{v_\theta}{r}\right)$.

(b) $\mu \dfrac{1}{r}\dfrac{\partial v_r}{\partial \theta}$;

$-p - \dfrac{2}{3}\mu\Theta + 2\mu\dfrac{\partial v_r}{\partial r}$.

2.62 (a) $\tau =$ const.

(b) $u = 0$ at $y = 0$, $u = U_0$ at $y = h$, u is continuous at $y = h/2$.

(c) For $y < h/2$,

$u = \dfrac{2U_0\mu_2}{h(\mu_1 + \mu_2)}y$.

For $y > h/2$,

$u = U_0 - \dfrac{2U_0\mu_1}{h(\mu_1 + \mu_2)}(h - y)$.

(d) $\dfrac{2U_0\mu_1\mu_2}{h(\mu_1 + \mu_2)}$.

2.63 $v_\theta = \dfrac{a^2\Omega}{a^2 - b^2}\left(r - \dfrac{b^2}{r}\right)$;

on inner cylinder $\tau = \dfrac{2\mu\Omega b^2}{a^2 - b^2}$;

on outer cylinder $\tau = \dfrac{2\mu\Omega a^2}{a^2 - b^2}$.

2.64 (a) A.

(b) A.

(c) $\sigma_{rr} = \sigma_{\theta\theta} = \sigma_{zz} = -p$,
$\tau_{r\theta} = -2\mu B$.

2.65 (b) $p_{th} = -\rho\dfrac{A^2(x^2 + y^2)}{2}$.

(c) $\sigma_{xx} = -p_0 + 2\mu A$.

2.70 $\tau_0 = \sqrt{\rho\mu K}\, U' e^{Kt}$.

2.71 (a) $v_r = \dot{R}\left(\dfrac{R}{r}\right)^2$.

(b) $p_b = p + \dfrac{2}{R}(\sigma + 2\mu V)$.

Chapter 3

3.1 294.4 kPa.

3.2 (a) $v = \sqrt{\dfrac{2(p_1 - p_a)}{\rho}}$

$\times \tanh\left[\dfrac{t}{2l}\sqrt{\dfrac{2(p_1 - p_a)}{\rho}}\right]$.

(b) 9.3 sec.

3.3 $A = A_0\sqrt{\dfrac{h}{h + x}}$.

3.4 0.05 ft^3/sec.

3.5 $\Delta p = \dfrac{\rho v^2}{2}$.

3.6 $t = \dfrac{2x_2(x_2 - x_1)}{x_1\sqrt{2gy}}\tanh^{-1}\eta$.

3.7 $V = \sqrt{2gy\left(\dfrac{\rho_1}{\rho_2} - 1\right)}$.

3.8 $p = \rho\left(\dfrac{V}{b}\right)^2(L^2 - x^2)$.

3.9 (a) $v_r = \dfrac{B}{r}$, $v_\theta = -\dfrac{A}{r}$.

(c) $a_r = -\dfrac{A^2 + B^2}{r}$, $a_\theta = 0$.

(d) $p = p_\infty - \dfrac{\rho(A^2 + B^2)}{2r^2}$.

3.10 $\dfrac{V^2}{2} + \dfrac{p}{\rho_0}\left[\dfrac{(p/p_0)^{-a}}{1 - a}\right] - \dfrac{c}{x} =$ const.

3.11 $\ddot{x} + \dfrac{2g(\rho_a - \rho_b)}{(\rho_a l_a - \rho_b l_b)}x = 0$.

3.12 $R\ddot{R} + \dfrac{3}{2}\dot{R}^2 = \dfrac{p_1 - p_\infty}{\rho}$.

3.13 3.69 ft^3/sec.

3.14 $p_1 - p_2$
$$= -\rho(L - x_0 \sin \omega t)x_0 \omega^2 \sin \omega t.$$

3.15 (c) $p_R = \rho\left(\dfrac{V_0 R_0}{2h}\right)^2$
$$\times \left[\ln\left(\dfrac{R_0}{R}\right) + 1 - \left(\dfrac{R_0}{R}\right)^2\right].$$

3.16 $Q = \dfrac{\pi D^2/4}{\sqrt{\left(\dfrac{D^2}{D^2 - d^2}\right)^2 - 1}}\sqrt{2(s-1)gh}.$

3.17 (a) 10.5 kPa. (b) -3.59 kPa.

3.18 $V_1 = 0.808$ m/s; $V_2 = 2V_1$.

3.19 $p_1 - p_2 = 1.545\,\dfrac{\rho V_1^2}{2}$;

$b_1 = 0.314b; b_2 = 0.186b.$

3.23 2.646 m/s.

3.24 (a) $\dfrac{dB}{dr} = 2\left[\dfrac{a^2\Omega}{a^2 - b^2}\right]^2\left(r - \dfrac{b^2}{r}\right)$

(b) $\dfrac{p - p(a)}{\rho} = \dfrac{1}{2}\left[\dfrac{a^2\Omega}{a^2 - b^2}\right]^2\left[(r^2 - a^2)\right.$
$$\left. + b^4\left(\dfrac{1}{a^2} - \dfrac{1}{r^2}\right) - 4b^4 \ln\left(\dfrac{r}{a}\right)\right].$$

3.25 (a) 0.083 ft.
(b) 97.3 ft/sec.
(c) 0.065 ft.

3.26 (a) 7.36 m.
(b) 2.16 kW.
(c) $h_1 = 2.54$ m, $h_2 = 9.90$ m.

3.27 (a) $V_j = \dfrac{2gh}{\sqrt{\dfrac{1}{c_v^2} + \dfrac{k}{9}}}.$

(b) 15.1 m/s.

3.28 (a) $V_j = \left(\dfrac{2p_0}{\rho\left[\dfrac{1}{c_v^2} + k\left(\dfrac{A_n}{A_p}\right)^2\right]}\right)^{1/2}.$

(b) $P = \dfrac{\rho A_n}{2}\, V_j^3.$

(c) $\dfrac{A_p}{c_v\sqrt{2k}}.$

3.29 (a) 52.2 m/s.
(b) 2.230 kW.
(c) 671 kPa.

3.30 89 percent.

3.31 3.52 miles.

3.32 $p = p_a = \dfrac{(c_c^2 - 1)(p_1 - p_a)}{\left(\dfrac{1}{c_v^2} - 1 + c_c^2\right)};$

164.9 kPa.

3.33 (a) $p = p_a - 945h.$
(b) 0.0655 ft^3/sec.

3.35 $\dfrac{A_1}{c_d A_2}\sqrt{\dfrac{2}{g}}(y_0^{1/2} - y_1^{1/2}).$

3.36 $\tfrac{2}{3}bc_d[\sqrt{2g(h+a)^3} - \sqrt{2gh^3}].$

3.37 (a) 31.0 m.
(b) 6.08 kW.

3.38 (a) $h - (1 + k)\dfrac{V^2}{2g}.$

(b) $\dfrac{\rho g A}{495}\left[hv - (1 + k)\dfrac{V^3}{2g}\right].$

(c) $V^2 = \dfrac{2gh}{3(1 + k)}.$

(d) $\tfrac{2}{3}.$

3.39 (a) $t =$
$$\dfrac{A}{c_d A_0}\sqrt{\dfrac{2}{g}}\left[\sqrt{\dfrac{W}{\rho g A} + y_0} - \sqrt{\dfrac{W}{\rho g A}}\right].$$
(b) 2.06×10^{-3} m^2.

3.40 $V_j = \left[\left(\dfrac{2F}{\rho A_p}\right)\left(\dfrac{1}{(1/c_v^2) - (A_j/A_p)^2}\right)\right]^{1/2}.$

3.42 (a) 0.03 m^3/s.
(b) 12.58 kW.

3.43 (a) 0.436 psi.
(b) 610 hp.

3.44 (a) $3.06\,\dfrac{dV_2}{dt} + 21\,\dfrac{V_2^2}{2g}$
$$+ 1000\,\dfrac{V_2}{A_{pipe}} = 30.$$
(b) 9.80 m.
(c) 4.09 m.

3.45 (a) 44.3 m^3/s.
(b) 144.8 kW.

3.46 (a) $t = \dfrac{2l}{\sqrt{2gMk}}\tan^{-1}\dfrac{V_0}{\sqrt{2gM/k}}.$
(b) 4.15 sec.

3.47 $Q = -\dfrac{B}{2a} + \left[\left(\dfrac{B}{2a}\right)^2 + \dfrac{A}{a}\right]^{1/2}$, where

$a = \dfrac{\rho}{2A_1^2}\left[k + \left(\dfrac{A_1}{C_v A_2}\right)^2\right].$

3.48 (a) 9.72×10^{-3} m^3/s. (b) 5.1 kW.

3.49 (a) Total head $= \dfrac{V_1^2}{2g} + (h + c)$.

(b) $\dfrac{V^2}{2g} + y = \dfrac{V_1^2}{2g} + (h + c)$.

(c) $Q = \dfrac{c_c b}{3g} \{[V_1^2 + 2g(h + c)]^{3/2}$
$\qquad\qquad - [V_1^2 + 2gc]^{3/2}\}$.

3.50 (a) $V_j = \left[\dfrac{2gh}{1 + kr^2}\right]^{1/2}$.

(b) Power $= \sqrt{2}\,\rho r A_p \left[\dfrac{gh}{1 + kr^2}\right]^{3/2}$.

(c) $r_{max} = 1/\sqrt{2k}$.

(d) Max power $= \left(\dfrac{2}{3}\right)^{3/2} \dfrac{\rho A_p}{\sqrt{k}} (gh)^{3/2}$.

3.53 $Q = \left(\dfrac{8}{15}\right) c_d \sqrt{2gh^5} \tan\left(\dfrac{\alpha}{2}\right)$.

Chapter 4

4.1 $2\rho g h c_v^2\, c_c\, A$.

4.2 2.1×10^6 N.

4.3 1990 lbf.

4.4 $\frac{1}{2}$.

4.5 $h_1 = h_2/2$.

4.6 (a) No.
(b) Yes.
(c) Yes.

4.7 (b) 7360 N.
(c) 5160 N.

4.8 $R_x = a(p_1 - p_2)$.
$R_y = \rho a V_1^2 \cos^2 \beta_1 (\tan \beta_1 - \tan \beta_2)$.

4.9 (a) $F = -\rho a A l + \rho(V_0 - at)^2 A$.
(b) 404 N.

4.10 (a) $p + \dfrac{F}{A_r} - \dfrac{\rho V^2}{2} + p_a$.

(b) $F = \dfrac{\rho V^2 (A_r A_c)(A_r + A_c)}{2(A_c - A_r)^2}$.

4.12 (a) 8.4 psig.
(b) $F_x = 2.8$ lbf; $F_y = -9.1$ lbf.

4.13 (a) $V = c_v \sqrt{2gh}$;
$Q = c_v A \sqrt{2gh}$.
(b) $V_1 = \sqrt{2g(c_v^2 h - h_1)}$.
(c) $W = 2\rho g A (c_v^2 h)$.

4.14 (a) $V_2 =$
$\sqrt{\dfrac{2(p_1 - p_a)}{\rho}} \Bigg/ \left[1 + k - \left(\dfrac{d_2}{d_1}\right)^4\right]$;
$F = A_2\Bigg[\rho V_2^2\left(1 - \dfrac{A_2}{A_1}\right)$
$\qquad\qquad + (p_a - p_1)\left(\dfrac{d_1}{d_2}\right)^2\Bigg]$.

(b) $F = -4016$ N.

4.15 (a) $\dot{m} = \dfrac{\pi \rho V_b d_5^2}{8}$
$\qquad + \sqrt{\left(\dfrac{\pi \rho V_b d_5^2}{8}\right)^2 + \dfrac{\pi \rho D}{4}\, d_5^2}$.

(b) $h = \dfrac{V_5^2}{2g}\left[\dfrac{1}{c_b^2} + (k_{1-2} + k_{2-3})\left(\dfrac{d_5}{d_1}\right)^4\right]$
$\qquad - \dfrac{V_b^2}{2g}$.

(c) 27.25 kW.

4.16 $V = V_0 \Bigg/ \sqrt{1 + \dfrac{\gamma V_0 t}{m}}$.

4.17 $\dfrac{\rho g(h - l)^3}{2(h + l)}$.

4.18 $\sqrt{h^2 + \dfrac{2V^2 h}{g}}$.

4.19 $D = D_0\left(1 - \dfrac{x}{L}\right)^{1/2}$.

4.20 $D = (p_0 - p_v)h + \rho V_0^2 h$
$\qquad - \rho V_0 h \sqrt{V_0^2 + \dfrac{2(p_0 - p_v)}{\rho}}$.

4.21 $x = \dfrac{M}{4\rho A}(1 - e^{-4\rho V_0 At/M})$;
$x_{max} = \dfrac{M}{4\rho A}$.

4.23 $a = \dfrac{\dot{m} V_e}{m_0 - \dot{m} t}$;
$V = V_e \ln\left(\dfrac{m_0}{m_0 - \dot{m} t}\right)$.

4.24 (a) 139.26 kPa.
(b) $R_x = -147.1$ kN, $R_y = 17.33$ kN.
(c) $R_x = -151.34$ kN, $R_y = 15.58$ kN.

4.25 $m_b \ddot{y} + \dot{m}_b \dot{y} + ky = ky_0 + \dot{m}_b V_b$.

4.27 (a) 900 N. (b) 2.7 kW.

4.28 (a) 346 ft/sec. (b) 1452 ft/sec.

4.29 (a) 14 m/s. (b) $p_2 = 424$ kPa.

4.30 (a) 76.72 m/s.

(b) $R_x = A_j(p_a + \rho V_j^2) - \dfrac{A_p}{\sqrt{2}}(p_p + \rho V_p^2),$

$R_y = \dfrac{A_p}{\sqrt{2}}(p_p + \rho V_p^2).$

(c) $R_x = -2648$ kN, $R_y = 3121$ kN.

4.31 (a) $V_1 = 141.4$ ft/sec, $V_2 = 191.74$ ft/sec, $V_3 = 101.5$ ft sec.

(b) $R_y = \rho V_0 L[u - V_0(\cot \beta_2 + \cot \alpha_1)].$

(c) $R_y = 14.8$ lbf.

4.32 $D = \dfrac{\rho V_0^2 a}{3}.$

4.34 $T = 2\rho A h V^2$ c.c.w.

4.35 (a) $\rho Q(v_{\theta_2} R_2 - v_{\theta_1} R_1).$

(b) $r = R_2 e^{[(\theta - \text{const})/\cot \alpha]}.$

(c) No in (a) provided that shear stress on surfaces is neglected; generally yes in (b) depending on boundary conditions.

4.36 $T_{0-z} = \rho Q R^2 \Omega \sin^2 \alpha.$

4.37 (a) $F = \rho h V^2 \sin \alpha;$

$a = \dfrac{h}{2}(1 - \cos \alpha);$

$b = \dfrac{h}{2}(1 + \cos \alpha).$

(b) $l = -\dfrac{b^2 - a^2}{2h \sin \alpha}.$

4.39 (a) $\rho Q(v_{\theta_2} R_2 - v_{\theta_1} R_1) + \rho Q \Omega(R_2^2 - R_1^2).$

(b) $M = \dfrac{\Omega}{g}\left(\dfrac{Q \cot \alpha}{2\pi} - v_{\theta_1} R_1\right)$

$+ \dfrac{\Omega^2}{g}(R_2^2 - R_1^2).$

4.40 $\Omega = V_e r \ln\left(\dfrac{m + m_0}{m + m_0 - \dot{m}t}\right).$

Chapter 5

5.4 (a) 0.54 m/s; 0.54 cm.

(b) 0.72 hr.

5.5 (a) 1882 rpm.

(b) $H_m/H_p = 0.0138;$
$P_m/P_p = 0.406.$

5.6 (b) 54.0 ft/sec.; 4.3×10^5 ft³/sec.

5.7 (a) 1000 rpm.

(b) $10^{-7}.$

5.8 (a) 379 ft/sec.

(b) 2000 lbf.

(c) 1.956 psia.

5.9 11,000 N; 3.75 m/s.

5.10 (a) 348 rpm.

(b) 1.25 meters.

(c) 0.38 m³/s.

5.11 12.8 kPa; 0.00318.

5.12 0.0358 ft³/sec.

5.13 (a) 6.12 m/s, 184 rpm.

(b) 25,920 Nm, 54 kN.

5.14 (b) 7.14 m/s.

5.15 (b) 42.7 ft³/day.

5.16 (b) 4 m/s.

(c) 8×10^5 N/m².

5.20 $\dfrac{t}{t_\infty} = f\left(\dfrac{\mu V}{\sigma}, \dfrac{\rho g t_\infty^2}{\mu V}\right).$

5.21 $\theta = f\left(\dfrac{\mu V}{\sigma}, \dfrac{\mu V}{T}\right).$

5.22 (a) 0.0126 m³/s.

(b) 0.26 kW.

5.23 936 m; 4858 kW.

5.24 3.5×10^{-4} m³/s.

5.25 (a) 35.7 ft.

(b) 398 hp.

5.26 (a) $V = \sqrt{\left(\dfrac{4Q}{\pi d^2}\right)^2\left(f\dfrac{l}{d} + 1\right) + 2gh}$

(b) 3.34 m/s.

5.27 (a) 787 psig.

(b) 4270 hp.

5.28 Yes; 1.15 m.

5.29 4.12 m/s.

5.30 (a) 1.43 m³/s.

(b) Suggest $f = 0.029.$

5.31 7.85 ft/sec.; 43.7 psig.

5.32 From 390 m reservoir to junction $Q = 0.0179$ m³/s, from 360 m to junction $Q = 0.0055$ m³/s, from junction to 300 m reservoir, $Q = 0.0234$ m³/s; $h = 356$ m.

5.38 3.37 kN.

5.39 2.33 m/s.

Chapter 6

6.1 No.

6.3 (a) $\psi = \dfrac{y^2 - x^2}{2}.$

(b) $\psi = 3x^2 y - y^3.$

(c) $\psi = -\dfrac{y}{x^2 + y^2}$.

(d) $\psi = -\dfrac{2xy}{(x^2 + y^2)^2}$.

6.5 $\psi = Ar^{\pi/\alpha} \sin \dfrac{\pi\theta}{\alpha}$.

6.6 $\psi = \dfrac{\Gamma}{2\pi} \ln r + \dfrac{q}{2\pi} \theta$.

6.7 (a) 57.7 m/s.
(b) 1.0 s.

6.8 $-\rho V_0 \Gamma - \dfrac{\rho \Gamma^2}{4\pi h}$; $-\rho V_0 \Gamma$.

6.9 (b) $u = \dfrac{q}{\pi} \left[\dfrac{x - h}{(x - h)^2 + h^2} \right.$

$\left. + \dfrac{x + h}{(x + h)^2 + h^2} \right]$.

(c) $p = p_0 - \dfrac{\rho u^2}{2}$.

6.10 $p = p_0 + \dfrac{\rho}{2} \left(\dfrac{V_0 q}{\pi x} - \dfrac{q^2}{4\pi^2 x^2} \right)$.

6.11 (a) q/V_0; $q/2V_0$; $q/2V_0 \pi$.

6.12 $p_0 q/V_0$.

6.13 $\left(-0.0796 \dfrac{q}{V_0}, 4.55 \dfrac{q}{V_0} \right)$;

$4.55 \dfrac{q}{V_0}$.

6.14 (b) $4b^2 \pi A$.

6.15 (a) $\dfrac{qh}{V_0 h + q}$.

(b) $\dfrac{qh}{V_0 h + q} \left(p_0 - \dfrac{\rho V_0 q}{2h} \right)$;

qp_0/V_0.

6.16 $\sqrt{\dfrac{4Q}{\pi V_0}}$; $\sqrt{\dfrac{Q}{4\pi V_0}}$.

6.17 $l \sim a$; $b \sim q/V_0$;

$p \sim p_0 - \dfrac{\rho V_0^2}{2} \cdot \dfrac{4b}{\pi l}$.

6.18 Stagnation points at

$\theta = -\sin^{-1} \dfrac{\Gamma}{4\pi r_0 V_0}$.

6.19 (b) $v_{\theta max} = \frac{3}{2} V_0$;

$\theta = -\sin^{-1} \frac{2}{3}$.

For cylinder $v_{\theta max} = 2V_0$;

$\theta = -\sin^{-1} \frac{1}{2}$.

6.20 (a) $\alpha = 30°$.
(b) $-1.732 \rho V_0^2$.

6.21 54.6°.

6.26 (a) $\alpha = \tan^{-1} \left(\dfrac{\Gamma}{q} \right)$.

(b) $p - p_0 = \dfrac{\rho[\Gamma^2 + q^2]}{2(2\pi r)^2}$.

6.28 (a) $\varphi = -\dfrac{Q}{4\pi r}$.

(b) $V = \left(\dfrac{R}{b} \right) \left[\dfrac{Q}{2\pi(R^2 + b^2)} \right]$.

(d) $R_{min} = b$.

6.32 $\dfrac{\pi}{2} \rho V_0^2 r_0^2$.

6.34 (a) $u = \dfrac{\Gamma}{2\pi} \left[\dfrac{(y - h_0)}{x^2 + (y - h_0)^2} \right.$

$\left. - \dfrac{(y + h_0)}{x^2 + (y + h_0)^2} \right]$,

$v = \dfrac{\Gamma x}{2\pi} \left[\dfrac{1}{x^2 + (y + h_0)^2} \right.$

$\left. - \dfrac{1}{x^2 + (y - h_0)^2} \right]$.

(b) Speed $= \dfrac{\Gamma}{4\pi h_0}$.

6.36 (a) As $r \to \infty$, $\varphi \to Ar^2 \cos 2\theta$;

$v_\theta = 0$ on $\theta = \pm \dfrac{\pi}{2}$;

$v_r = 0$ on $r = 1$.

(b) All a_n's $= 0$ except for $a_2 = a_{-2} = A$; all b_n's $= 0$;

$v_{\theta max} = 4A$.

6.37 (a) On inner cylinder $v_r = V_0 \cos \theta$;
on outer cylinder $v_r = 0$.

(b) $\varphi = \dfrac{r_0^2 V_0 \cos \theta}{r_0^2 - R_0^2} \left(r + \dfrac{R_0^2}{r} \right)$.

6.38 p_{max} at $\dfrac{x}{\lambda} = \dfrac{1}{4}, \dfrac{5}{4}, \dfrac{9}{4} \cdots$;

p_{min} at $\dfrac{x}{\lambda} = \dfrac{3}{4}, \dfrac{7}{4}, \dfrac{11}{4}, \cdots$.

6.40 Az; z^2; $z^{1/2}$; $Az^{\pi/\alpha}$.

6.42 $V_0 \left(z + \dfrac{r_0^2}{z} \right) + i \dfrac{\Gamma}{2\pi} \log z$.

6.44 $V_0 \left(z + \dfrac{r_0^2}{z} \right)$.

6.45 $V_0 \left[e^{-i\alpha} z + r_0^2 \dfrac{e^{i\alpha}}{z} \right]$.

6.48 (a) Yes.

(b) $r_s = \left[\dfrac{q}{8\pi u_0}\right]^{1/4}$, $\theta_s = \dfrac{\pi}{4}$.

6.53 (a) $\dfrac{V_0 r_0}{2}$.

(b) 2.28°.

(c) $l = 9.2 r_0$.

(d) 0.024.

6.55 $c_p = 1 - \left(0.995 + 0.0995 \sqrt{\dfrac{c/2 - \xi}{c/2 + \xi}}\right)^2$

on upper surface;

$c_p = 1 - \left(0.995 - 0.0995 \sqrt{\dfrac{c/2 - \xi}{c/2 + \xi}}\right)^2$

on lower surface.

6.56 $y = -\dfrac{k}{cV_0}\dfrac{x^2}{2} + \text{const}$;

$C_L = \dfrac{k\pi}{2V_0}$.

Chapter 7

7.2 (a) $\dfrac{du}{dr} = -\dfrac{\rho g r}{4\mu}$.

(b) $u = \dfrac{\rho g}{8\mu}(r_0^2 - r^2)$.

(c) $\mu = \dfrac{\pi \rho g r_0^4}{16Q}$.

7.3 $u = \dfrac{\Delta p}{4\mu}\left[b^2 - r^2 + \dfrac{b^2 - a^2}{\ln(b/a)}\ln\left(\dfrac{r}{b}\right)\right]$.

7.4 (a) $u = V_0 \dfrac{\ln\left(\dfrac{r}{r_i}\right)}{\ln\left(\dfrac{r_0}{r_i}\right)}$.

(b) $F = \dfrac{2\pi V_0}{\ln(r_0/r_i)}$.

7.5 3.7×10^{-5} m²/s.

7.6 $Q = \dfrac{-\Delta p}{l}\dfrac{\pi r_0^4}{8\mu}\left(1 + \dfrac{a}{r_0}\right)$.

7.9 10.24 W.

7.10 (a) $nA_n^2 = \text{const}$, $V_n = \dfrac{\text{const}}{\sqrt{n}}$.

7.11 $\Delta p = \dfrac{12 l \mu}{h^3}\left(\dfrac{\Omega R h}{2} - Q\right)$;

$\text{torque} = lR\left(\dfrac{\Delta p h}{2l} + \dfrac{\Omega R \mu}{h}\right)$;

$\text{input} = \dfrac{(\text{torque})\,\Omega}{550}$;

$\text{efficiency} = 6\,\dfrac{Q}{\Omega R h}\dfrac{\left(1 - \dfrac{2Q}{\Omega R h}\right)}{\left(4 - 6\dfrac{Q}{\Omega R h}\right)}$.

7.12 (b) $y = 0, u = -V_0$; $y = h$,

$\dfrac{du}{dy} = 0$.

(c) $u = \dfrac{\rho g y}{\mu}\left(h - \dfrac{y}{2}\right) - V_0$.

(d) $Q = \dfrac{\rho g h^3}{3\mu} - V_0 h$.

7.13 (a) $q = -\dfrac{p_m h_1^3}{12\mu l} + \dfrac{V_0 h_1}{2}$.

(b) $q = \dfrac{p_m h_2^3}{12\mu l} + \dfrac{V_0 h_2}{2}$.

(c) $p_m = \dfrac{6\mu l V_0(h_1 - h_2)}{h_1^3 + h_2^3}$.

(d) $L = p_m l$.

(e) $L = \dfrac{2\mu l^2 V_0}{3h_2^2}$.

7.14 (a) $3\mu V_0 \left(\dfrac{l}{h}\right)^2$.

(b) $\dfrac{4h}{3l}$.

(c) $F = 6130$ lbf;

$\dfrac{D}{F} = 2.67 \times 10^{-3}$.

7.17 (b) $Q = \dfrac{\Delta p}{12\mu l}\displaystyle\int_0^B h^3\,dz$.

(c) $\dfrac{Q_\epsilon}{Q_0} = 1 + \dfrac{3\epsilon^2}{2}$.

7.18 (a) $p = \dfrac{3\mu V_0(r_0^2 - r^2)}{h^3}$.

(b) $F = \dfrac{3\pi\mu V_0 r_0^4}{2h^3}$.

(c) $t = \dfrac{3\pi\mu r_0^4}{4Fh^2}$.

7.20 $f = \dfrac{8}{[4.75 - 2.5\log_e(2\epsilon_s/d)]^2}$.

7.21 (a) 8.6×10^{-6} m; 4.3×10^{-5} m.

(b) 1.34×10^{-4} m;

8×10^{-4} m.

7.22 (a) 0.17 N/m².

(b) 0.03.

7.23 2.07×10^5.

7.28 $\bar{u} = \sqrt{\dfrac{\tau_0}{k\rho}} \tanh \dfrac{y}{\mu} \sqrt{\tau_0 k\rho}.$

Chapter 8

8.1 (a) $v = \text{const} = v_w$.

    (c) $v_w = \dfrac{\text{const}}{\sqrt{t}}$.

8.4 $n = 0,\ m = 1,\ A = \left(\dfrac{2V_0}{Rv}\right)^{1/2}$,

    $B = \left(\dfrac{2V_0\, v}{R}\right)^{1/2}$.

8.5 (a) $n = \dfrac{m}{2}$.

    (b) $n = \frac{1}{3},\ f''' + \frac{1}{3}ff'' + \frac{1}{3}f'^2 = 0$.

8.6 Shear:

    (a) $\dfrac{\rho U_0^2}{2} \dfrac{(0.658)}{\sqrt{\mathrm{Re}_x}}$.

    (b) $\dfrac{\rho U_0^2}{2} \dfrac{(0.73)}{\sqrt{\mathrm{Re}_x}}$.

    (c) $\dfrac{\rho U_0^2}{2} \dfrac{(0.578)}{\sqrt{\mathrm{Re}_x}}$.

    Boundary layer δ:

    (a) $\dfrac{4.79x}{\sqrt{\mathrm{Re}_x}}$.

    (b) $\dfrac{5.48x}{\sqrt{\mathrm{Re}_x}}$.

    (c) $\dfrac{3.46x}{\sqrt{\mathrm{Re}_x}}$.

    Drag:

    (a) $\dfrac{\rho U_0^2 l}{2} \dfrac{(1.31)}{\sqrt{\mathrm{Re}_l}}$.

    (b) $\dfrac{\rho U_0^2 l}{2} \dfrac{(1.46)}{\sqrt{\mathrm{Re}_l}}$.

    (c) $\dfrac{\rho U_0^2 l}{2} \dfrac{(1.16)}{\sqrt{\mathrm{Re}_l}}$.

    drag $= 0.063$ N; $\delta = 3.1 \times 10^{-3}$ m.

8.9 (a) $\dot{m} = \dfrac{\rho g \delta^3}{3v}$.

    (b) $\delta = \left(\dfrac{4k\Delta Tv}{\lambda \rho g} x\right)^{1/4}$.

8.13 737 N; 0.069 m; 70.6 N.

8.14 1.74 ft.

Chapter 9

9.1 341.4 m/s.

9.2 1439 m/s.

9.3 78 mph.

9.5 $\Delta V = 9.8 \times 10^{-5}$ m/s, $\Delta T = 3.4 \times 10^{-5}$ K.

9.6 343.4 m/s; 343.7 m/s.

9.7 $2\Delta p$.

9.12 $T = T_w - A\left(\dfrac{u^2}{2} + \dfrac{q_w}{\tau_w} u\right).$

9.13 433 K, 1320 K.

9.15 (a) 789 kPa. (b) 512.6 m/s.
    (c) 220.7 kg/s.

9.16 81.4 lbm/sec ft^2.

9.17 1.75 kg/s.

9.20 683.9 m/s; 0.937.

9.23 0.53.

9.24 $+1.01$ percent; $+5.26$ percent; -4.76 percent.

9.30 $\dfrac{p_s}{p_A} = \left[\dfrac{(\gamma + 1)^{\gamma + 1}(M_A^2/2)^\gamma}{2\gamma M_A^2 - \gamma + 1}\right]^{1/\gamma - 1}.$

9.32 (b) 6.0.

9.33 (a) 750 m/s.
    (b) $M_1 = 1.89,\ M_2 = 0.60$.

9.35 2950 ft/sec, 2098 ft/sec.

9.36 (a) 1029 m/s, 266.8 m/s.
    (b) 979.6 kPa.
    (c) 3321 kPa.

9.40 (a) 560 K, 528.4 m/s, 782.4 m/s.

9.43 $\dfrac{A_t}{A_e} = \left(\dfrac{2}{\gamma + 1}\right)^{\gamma + 1/2(1 - \gamma)} \left(\dfrac{2}{\gamma - 1}\right)^{1/2}$

    $\times \left(\dfrac{p_2}{p_0}\right)^{1/\gamma} \left[1 - \left(\dfrac{p_2}{p_0}\right)^{\gamma - 1/\gamma}\right]^{1/2}$

9.44 6.71.

9.45 0.456 kg/s.

9.46 (a) 25.5 psia, 13.5 psia, 3.27 psia, 0.577.
    (b) 18.4 psia.

9.49 Before shock, 79.5 K, 2.56 kPa, $M = 5.36$; after shock, 186 m/s, 518.3 K, 85.5 kPa, entropy rise $= 0.85$ kJ/kgK.

9.50 4.01 ft.

9.52 $\dfrac{A_e}{A_t} = \dfrac{1}{M_e}\left(\dfrac{2 + (\gamma - 1)M_e^2}{\gamma + 1}\right)^{(\gamma + 1)/[2(\gamma - 1)]}$

9.54 (a) 0.54 kg/s.
    (b) 0.57 kg/s, 528 kPa.

9.55 2.27×10^{-3} lbm/sec,
1370 ft/sec, $M = 1.45$.

9.58 4.17.

9.62 (a) 484 m/s, 0.32 kg/s.
(b) Unchanged. (c) Decreases.

9.72 45.6 ft adiabatic,
42.0 ft isothermal.

Chapter 10

10.1 228.7 m/s.

10.2 (a) $M = 2.24$, $\mu = 26.6°$.
(b) 0.5 mile before observer.

10.4 3.38, 0.125.

10.5 $C_L = \dfrac{4\alpha}{\sqrt{M^2 - 1}}$, $C_D = \dfrac{4\alpha^2 + 2\beta^2}{\sqrt{M^2 - 1}}$,

3.23×10^5 N.

10.6 $(c/h)_{max} = 5.65$.

10.7 (a) $C_L = \dfrac{4\alpha}{\sqrt{M^2 - 1}}$,

$C_D = \dfrac{4\alpha^2 + 6(t/c)^2}{\sqrt{M^2 - 1}}$.

10.8 (a) $C_D = \dfrac{4(t/c)^2}{\sqrt{M^2 - 1}}$.

10.9 $10.8°$.

10.10 0.03.

10.11 $C_D = \dfrac{4(\alpha^2 + 2\alpha\epsilon e + \epsilon^2 e)}{\sqrt{(M^2 - 1)^{1/2}}}$.

10.12 $M_0 = 2.23$.

10.15 (a) $M_2 = 1.63$, $p_2 = 120.4$ kPa,
2.31 J/kgK, 1.295.

10.17 $\psi(r, \theta) = r\left[\cos\left(\dfrac{\gamma - 1}{\gamma + 1}\right)^{1/2} \theta \right]^{\gamma + 1/\gamma - 1}$

$= \text{const.}$

10.18 142.5 kPa normal shock only, 199.2 kPa oblique plus normal shock.

10.19 $17.2°$.

10.20 $(h/c)_{min} = 0.52$.

10.21 69.8 kPa. $5.4°$.

10.22 (a) Approx. 6100 m.
(b) 1.47, $14.9°$ from horizontal.

10.23 (a) $u = \left(\omega\delta \cos \dfrac{\omega x}{a} \right.$

$\left. - \omega \cot \dfrac{\omega L}{a} \sin \dfrac{\omega x}{a} \right) \cos \omega t$,

$p_1 = \rho_0 \omega \sin \omega t \left(a\delta \sin \dfrac{\omega x}{a} \right.$

$\left. + a \cot \dfrac{\omega L}{a} \cos \dfrac{\omega x}{a} \right).$

10.26 (b) $p = p_0 \left(1 + \dfrac{2\gamma M_0^2 \alpha}{\sqrt{M_0^2 - 1}} \right)$.

10.28 $\varphi_1 = - \dfrac{V_0 y \cos (2\pi x/\lambda)}{\sqrt{1 - M_0^2}}$

$\times \dfrac{\cosh \dfrac{2\pi(y - h)}{\lambda} \sqrt{1 - M_0^2}}{\sinh \dfrac{2\pi h}{\lambda} \sqrt{1 - M_0^2}}$.

on $y = 0$, $p_1 =$

$-\rho_0 V_0^2 y_0 \left(\dfrac{2\pi}{\lambda} \right) \sin \dfrac{2\pi x}{\lambda}$

$\times \dfrac{\cosh \dfrac{2\pi h}{\lambda} \sqrt{1 - M_0^2}}{\sqrt{1 - M_0^2} \sinh \dfrac{2\pi h}{\lambda} \sqrt{1 - M_0^2}}$,

on $y = h$, $p_1 =$

$-\dfrac{\rho_0 V_0^2 y_0 \left(\dfrac{2\pi}{\lambda} \right) \sin \dfrac{2\pi x}{\lambda}}{\sqrt{1 - M_0^2} \sinh \dfrac{2\pi h}{\lambda} \sqrt{1 - M_0^2}}$.

Chapter 11

11.1 43.8 ft/sec; 700 ft^3/sec; 5.46.

11.4 17.2 ft/sec; 17.2 ft^3/sec.

11.5 $y = \dfrac{V^2}{2g} + h - \dfrac{3(Vhg)^{2/3}}{2g}$;

1.10 m.

11.9 No.

11.10 51.3 ft; 1.88 ft; 3.0 ft.

11.15 (a) $V = \left\{ \left(\dfrac{g}{2d} \right) \left[\left(h + \dfrac{3}{2} d \right)^2 \right. \right.$

$\left. \left. - \left(\dfrac{d}{2} \right)^2 \right] \right\}^{1/2}$.

(b) $V \to (gd)^{1/2}$.

(c) $h_f = \dfrac{d}{4} \left(\dfrac{h}{d} \right)^3 + \cdots$.

11.18 0.0817 m^3/s.

11.19 (a) 1.01 m.
(b) It rises.

11.20 0.6 m^3/s/m.

11.21 2.14×10^{-4}.

11.22 8.55 m/s.

11.23 (a) $\eta = a \sin \dfrac{2\pi(x - ct)}{\lambda}$.

11.24 (a) Changing part $= \dfrac{\eta}{\cosh \dfrac{2\pi h}{\lambda}}$.

11.25 (a) 282.7 s.
(b) 0.074 m/s.

11.26 (a) 6.84 m/s.
(b) 0.22 m/s.
(c) To the left.

11.27 $\dfrac{V^2}{2g} = 2.5 \times 10^{-3}$ m; yes.

11.29 $\lambda = 0.0171$ m; $c = 0.23$ m/s.

Chapter 12

12.3 $u/V_1 = 0.52$; $V_2/V_1 = 0.265$.

12.4 $\eta = c_v^2 \eta_b$.

12.6 (a) $\dfrac{u}{V_1} = \dfrac{1}{6}$.
(b) 0.754.

12.7 Diameter of jet $= 0.638$ ft.

12.11 $\dfrac{\Psi}{2}$.

12.12 (a) Tip; $\phi = 0.182$, $\Psi = 0.0675$; hub $\phi = 0.274$, $\Psi = 0.152$; midway $\phi = 0.219$, $\Psi = 0.0974$; $\alpha_1 = 90°$, $\beta_2 = 13.6°$.
(b) 0.71 hp.
(c) 1.625 hp; 0.715 ft water.

12.13 (a) $\left(\dfrac{v_a}{u}\right)^2 \dfrac{1}{\sin^2 \beta^2} + 2 \cot \alpha_1 \dfrac{v_a}{u}$
$$- \left(1 + \dfrac{2gz}{u^2}\right) = 0.$$
(b) $-H =$
$$\dfrac{u^2}{g}\left[\dfrac{v_a}{u}(\cot \alpha_1 + \cot \beta_2) - 1\right].$$

12.15 $\Psi_2 = \left(\dfrac{r_1}{r_2}\right)^2$
$$- \phi_2\left(\dfrac{b_2}{b_1}\cot \beta_1 + \cot \alpha_1\right).$$

12.16 $\phi_d = 0.155$.

12.17 (a) $d_2 = 0.35$ m; $b_2 = 0.044$ m.
(b) 204.8 kW.

12.21 0.36 m; 2127 rpm.

12.22 2316 rpm.

12.25 0.21; 0.25; 0.29; 0.32 m.

12.27 (a) $b_2/d_2 = 0.083$; $d_1/d_2 = 0.48$; $\phi = 0.12$; $\Psi = 0.48$.
(b) 0.78 ft.
(d) 1240 hp.

12.28 $N'_s = 0.32$; 2345 rpm.

Chapter 13

13.1 (a) $\sigma = 0.152$.
(b) $\beta_v = 32.8°$.
(c) $v_a = 19.33$ m/s.
(d) 20.9 m.

13.4 $\Psi_2 = 0.0762$; $\phi_2 = 0.30$; $\sigma = 0.252$; $\beta_v = 23.3°$.

13.7 (a) $v_{\theta_1} = 104$ ft/sec; $w_{\theta_1} = 96$ ft/sec; $u = 200$ ft/sec.
(c) $\alpha^* = 3°$; $\sigma = 0.235$.
(d) 3 percent.

13.8 (a) 0.2.
(b) 0.189.
(c) 0.552; $\alpha^* = 6.38°$.

13.11 $\Psi = 0.82 - 2.3\phi$.

13.12 (b) $\dfrac{v_{\theta_1}}{u_0} = \dfrac{r_m^2}{2rr_0} + \dfrac{ar_0}{2r}$;
$$\dfrac{v_{\theta_2}}{u_0} = \dfrac{r_m^2}{2rr_0} - \dfrac{ar_0}{2r}.$$

13.13 (a) 234 hp.

13.14 Tip, $\sigma = 0.26$; $\dfrac{r}{r_0} = 0.8$,
$$\sigma = 0.7; \dfrac{r}{r_0} = 0.6,\ \sigma = 2.2.$$

13.15 Tip, $\theta = 11.3°$, $\gamma = 57.9°$;
$$\dfrac{r}{r_0} = 0.8,\ \theta = 18.1°,\ \gamma = 49°;$$
$$\dfrac{r}{r_0} = 0.6,\ \theta = 52.2°,\ \gamma = 21.9°.$$

13.18 $v_\theta = A\sqrt{\dfrac{2}{3}}\left[\dfrac{r}{r_0}\left(1 - \dfrac{3r}{4r_0}\right)\right]^{1/2}$.

13.23 (a) 6330 rpm.
(b) $\Psi = 0.47$; $\phi = 0.118$; $d_2 = 0.385$ ft.

13.24 4.0 ft below surface.

13.25 26.9 psi; 0.475 ft.

Index

F

G

Vortex (*continued*)
 free, 199
 sheet, 223
 spiral, 201
 strength of, 198
Vortex lines, 86
Vorticity, 60

W

Wake, 185
Wake, law of, 293
Watt, J., 488
Watt, unit of power, 489
Wave angle, of oblique shocks, 367
Wave drag, 178
 in compressible flow, 358

Wave equation, in compressible flow, 377, 378
Wavelength, 406
Wave velocity, 408
Waves
 acoustic, 377
 deep water, 410
 progressive, 410
 shallow water, 410
 standing, 410
 surface, 406
 with surface tension, 411
 tidal, 410
 traveling, 410
Weber number, 160
Weir, 112, 113
 broad-crested, 399
 undershot, 84
Wetting angle, 12
Work, 302, 502